T0202746

# Statistical Mechanics *of* Membranes *and* Surfaces

*Second Edition*

# Statistical Mechanics *of* Membranes *and* Surfaces

*Second Edition*

*Edited by*

## D. Nelson
Harvard University, USA

## T. Piran
Hebrew University, Israel

## S. Weinberg
University of Texas at Austin, USA

 **World Scientific**

NEW JERSEY · LONDON · SINGAPORE · BEIJING · SHANGHAI · HONG KONG · TAIPEI · CHENNAI

*Published by*

World Scientific Publishing Co. Pte. Ltd.

5 Toh Tuck Link, Singapore 596224

*USA office:* 27 Warren Street, Suite 401-402, Hackensack, NJ 07601

*UK office:* 57 Shelton Street, Covent Garden, London WC2H 9HE

**British Library Cataloguing-in-Publication Data**
A catalogue record for this book is available from the British Library.

First published 2004
Reprinted 2008

**STATISTICAL MECHANICS OF MEMBRANES AND SURFACES**
**2nd Edition**

ISBN-13 978-981-238-760-8
ISBN-10 981-238-760-9

ISBN-13 978-981-238-772-1 (pbk)
ISBN-10 981-238-772-2 (pbk)

Printed in Singapore

# PREFACE TO THE FIRST EDITION

The Fifth Jerusalem Winter School on the Statistical Mechanics of Membranes and Surfaces was held from December 28, 1987 to January 6, 1988. The School focused on the theory of the highly convoluted surface fluctuations which appear in such diverse condensed matter systems as microemulsions, wetting and growth interfaces, bulk lyotropic liquid crystals, chalcogenide glasses and sheet polymers. The delicate interplay between geometry and statistical mechanics in these systems can be described using tools from the fields of polymer physics, differential geometry, and critical phenomena. Our theoretical understanding of these problems can be tested by a wide variety of laboratory experiments, which probe fluctuations ranging from relatively benign capillary waves at interfaces, to wild undulations in biological membranes. The School was fortunate to have many lecturers who were outstanding teachers as well as distinguished scientists: Jürg Fröhlich spoke on the roughening transition, as well as on his extensive and pioneering work on random surfaces. Michael Fisher lectured on the wetting transition and on interfacial wandering. Stanislas Leibler discussed fluctuations in liquid membranes, lyotropic smectics, and other lipid systems. David Andelman spoke about Langmuir–Blodgett films and the physics of microemulsions. Yacov Kantor gave a very thorough review of the theory of polymerized surfaces; I followed with a discussion of the crumpling transition. Francois David gave a beautiful series of talks on differential geometry, and its application to liquid and hexatic membranes. Bertrand Duplantier described important recent work on epsilon expansions for polymerized membranes. Virtually all the lecturers contributed manuscripts to this volume, which can serve as a useful introduction for theorists and experimentalists who wish to learn more about this rapidly developing field. I would in conclusion like to thank Steven Weinberg, who made this School possible, and Tsvi Piran, who helped make the School a reality.

*David Nelson*
Cambridge, Massachusetts
December, 1988

# PREFACE TO THE SECOND EDITION

I was very pleased when K. K. Phua at World Scientific Publishing suggested issuing a second edition of "Statistical Mechanics of Membranes and Surfaces", a book I edited (with Tvsi Piran and Steve Weinberg) and contributed to in 1988. Over the intervening 15 years, I received many compliments on the excellent work described by Michael E. Fisher, Stanislas Leibler, David Andelman, Yacov Kantor, François David and Bertrand Duplantier in this account of a Jerusalem Winter School which took place in 1988. My students still consult this book frequently, and the sales during the past decade and a half suggest that there is still considerable interest in the relevant theory and experimental systems.

To capture important additional developments in the statistical mechanics of membranes and surfaces, I was fortunate to persuade four excellent researchers to write three new chapters for the second edition. Leo Radzihovsky contributed a chapter on the fascinating effects which arise when anisotropy and heterogeneity are incorporated in polymerized membranes. Mark Bowick surveyed the physics of fixed connectivity membranes in general, including very recent theory and experiments probing crystalline ground states on curved surfaces. This second edition concludes with an authoritative survey of triangulated surface models of fluctuating membranes (including studies of liquid and hexatic phases) by Gerhard Gompper and Dan Kroll.

I am particularly grateful to Betrand Duplantier, who provided a very extensive update for his chapter on self-avoiding crumpling manifolds. Thanks are also due to Michael E. Fisher who kindly provided some additional references for Chapter 3. Although several previous authors took the opportunity of a second edition to revise or correct their contributions, most of the older chapters should not be viewed as comprehensive updates. Rather, they are "snapshots" of progress in a field which was just beginning to emerge and confront real experiments. Nevertheless, I feel there is a timeless quality about all the early chapters which makes them as relevant today as when the first edition was published.

*David R. Nelson*
Cambridge, Massachusetts
November, 2003

# CONTENTS

## Chapter 3   Equilibrium Statistical Mechanics of Fluctuating Films and Membranes     49

*Stanislas Leibler*

## Chapter 8  Statistical Mechanics of Self-Avoiding Crumpled Manifolds — Part I                                              211

*Bertrand Duplantier*

# CHAPTER 1

# THE STATISTICAL MECHANICS OF
# MEMBRANES AND INTERFACES

David R. Nelson

*Department of Physics*
*Harvard University*
*Cambridge, Massachusetts 02138*

An enterprise of considerable current interest in theoretical physics is the study of interfaces and membranes. In condensed matter physics, an "interface" usually means a boundary between two phases, whose fluctuations can be studied by methods adapted from equilibrium critical phenomena. The statistical mechanics is typically controlled by a surface tension, which insures that such surfaces are relatively flat. Recently, however, there has been increasing interest in membrane-like surfaces. "Membranes" are composed of molecules different from the medium in which they are imbedded, and they need not separate two distinct phases. Because their microscopic surface tension is small or vanishes altogether, membranes exhibit wild fluctuations. New ideas and new mathematical tools are required to understand them.

In this Chapter, we first sketch the physics of "flat" interfaces, and then discuss some issues which arise in the description of crumpled membranes. Although related problems arise in field theory models of elementary particles,[1] most of the models discussed here have explicit experimental realizations in condensed matter physics. Much of the vitality of this subject arises because of a delicate interplay between theory and experiment: theoretical predictions can often be checked by inexpensive but revealing laboratory experiments in a matter of months. Most of the topics sketched in this Chapter are discussed in more detail elsewhere in this book.

## 1. Flat Surfaces

### 1.1. *The Roughening Transition*

Interesting problems in statistical mechanics arise even for surfaces constrained by surface tension to be fairly flat. A particularly well-studied example is the roughening transition of crystalline interfaces.[2] As shown in Fig. 1, we imagine a crystal in equilibrium, with, say, its own vapor. The position of the interface is described by a height function $h(x^1, x^2)$. Such a description implicitly ignores "overhangs" (which cannot be described by a single-valued $h(x^1, x^2)$), islands of crystal in the

*D. R. Nelson*

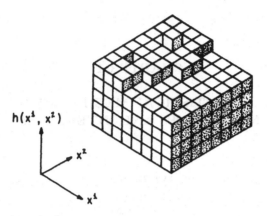

Fig. 1.  Height function $h(x^1, x^2)$ used to describe the configuration of a crystal–vapor interface.

vapor phase, and islands of vapor in the crystal. These complications are believed to be irrelevant variables in the long wavelength limit.[2] Microscopically, the interface height is quantized in units of the spacing between the Bragg planes normal to the $h$-axis.

At high temperatures, this discreteness is washed out by thermal fluctuations, and we can describe the free energy of the interface by a surface tension $\sigma$. It is a useful pedagogic exercise to describe this free energy using differential geometry, which, although inessential here, is often the language of choice for crumpled membranes. For an arbitrary parameterization of the surface $\vec{r}(\zeta^1, \zeta^2)$, the free energy is the surface tension times the surface area,

$$F = \sigma \int \sqrt{g} d^2\zeta \qquad (1.1)$$

where $g$ is the determinant of the metric tensor, $g = \det g_{ij}$,

$$g_{ij} = \frac{\partial \vec{r}}{\partial \zeta^i} \cdot \frac{\partial \vec{r}}{\partial \zeta^i} \qquad (1.2)$$

The formula for the surface area in terms of the metric tensor is derived in many textbooks.[3] For the particular parameterization embodied in Fig. 1, *i.e.*,

$$\vec{r}(x^1, x^2) = (x^1, x^2, h(x^1, x^2)), \qquad (1.3)$$

we have

$$g_{ij} = \begin{pmatrix} 1 + \left(\frac{\partial h}{\partial x^1}\right)^2, & \left(\frac{\partial h}{\partial x^1}\right)\left(\frac{\partial h}{\partial x^2}\right) \\ \left(\frac{\partial h}{\partial x^1}\right)\left(\frac{\partial h}{\partial x^2}\right), & 1 + \left(\frac{\partial h}{\partial x^2}\right)^2 \end{pmatrix}. \qquad (1.4)$$

With this coordinate system (called the Monge representation in differential geometry), Eq. (1.2) assumes the familiar form

$$F = \sigma \int d^2x \sqrt{1 + |\vec{\nabla} h|^2}. \qquad (1.5)$$

At temperatures sufficiently high so that (1.1) is an appropriate description, we can expand the square root in (1.5),

$$F \approx \text{const.} + \frac{1}{2}\sigma \int d^2x |\vec{\nabla}h|^2 \qquad (1.6)$$

and calculate, for example the height-height correlation function

$$\langle (h(\vec{y}) - h(\vec{0}))^2 \rangle = \frac{\int \mathcal{D}h(\vec{x}) |h(\vec{y}) - h(\vec{0})|^2 e^{-F/k_B T}}{\int \mathcal{D}h(\vec{x}) e^{-F/k_B T}}. \qquad (1.7)$$

The effects of higher order gradients in Eq. (1.6) can be absorbed into a renormalized surface tension. The Gaussian functional integral is easily carried out in Fourier space, with the result

$$\langle (h(\vec{y}) - h(\vec{0}))\rangle^2 = \frac{2k_B T}{\sigma} \int \frac{d^2q}{(2\pi)^2} \frac{1}{q^2} (1 - e^{i\vec{q}\cdot\vec{y}})$$

$$\approx \frac{k_B T}{\pi\sigma} \ln(y/a), \quad \text{as } y \to \infty, \qquad (1.8)$$

where $a$ is a microscopic length. The large $y$ behavior is the signature of a high temperature rough phase.

At low temperatures, on the other hand, one might expect a "smooth" interface, i.e., one that has become localized at an integral multiple of $a$, the spacing between Bragg planes. To see how quantization of the interface height affects the prediction (1.8), we add a periodic perturbation to Eq. (1.6) which tends to localize the interface at $h = 0, \pm a, \pm 2a, \ldots$, and consider the free energy

$$F = \text{const.} + \frac{1}{2} \int d^2x [\sigma |\vec{\nabla}h|^2 + 2y(1 - \cos(2\pi h/a))]. \qquad (1.9)$$

This sine-Gordon model can be solved directly by renormalization group methods, or by first mapping the problem via a duality transformation onto an XY-model or the two-dimensional Coulomb gas.[2,4] There is a finite temperature roughening transition which is in the universality class of the Kosterlitz–Thouless vortex unbinding transitions. At sufficiently high temperatures ($T > T_R \simeq \pi\sigma a^2/k_B$), the periodicity is irrelevant and the interface behaves according to Eq. (1.8). For $T < T_R$, however, the interface localizes in one of the minima of the periodic potential and the effective free energy at long wavelengths can be approximated by expanding the cosine

$$F \approx \text{const.} + \frac{1}{2} \int d^2x \left[\sigma |\vec{\nabla}h|^2 + \left(\frac{4\pi^2 y}{a^2}\right) h^2\right]. \qquad (1.10)$$

It is easily shown from Eq. (1.10) that height–height correlation function (1.7) now tends to constant,

$$\langle (h(\vec{y}) - h(\vec{0}))^2 \rangle \approx \text{const.}, \quad \text{as } y \to \infty, \qquad (1.11)$$

in contrast to Eq. (1.8).

The analogy with vortex unbinding transitions leads to many detailed predictions about the roughening transition.[2] This analogy is only approximate, however, so it is important to have rigorous proofs of phase transitions in this and related models.[5] Although Eq. (1.9) is a plausible model of roughening, a more faithful representation of the microscopic physics is the solid-on-solid model, where interface heights $\{h_i\}$ sit on a lattice of sites $\{i\}$ and are themselves quantized at all temperatures, $h_i = 0, \pm a, \pm 2a, \ldots, \forall i$. The Hamiltonian is

$$H = J \sum_{\langle ij \rangle} |h_i - h_j|, \qquad (1.12)$$

where the sum is over nearest neighbor lattice sites and $J > 0$ is a microscopic surface energy. Equation (1.12) measures directly the increase in interfacial area associated with discrete steps in the interface. We call (1.12) a "Hamiltonian" because it is a microscopic energy, in contrast to "free energies" like (1.9), which are supposed to be coarse-grained descriptions, embodying both energy and entropy. Rigorous proofs of phase transitions in, *e.g.*, the solid-on-solid model are particularly valuable, because its connection with the more easily solved Eq. (1.9) are not yet clearly established.

## 1.2. Wetting Transitions

Interesting transitions in interfacial surfaces also occur in wetting layers.[6] Consider, in particular, the approach to a liquid-gas phase boundary in the presence of a wall which microscopically prefers to be wet by the liquid, as opposed to the gas. The interfacial profile is shown in Fig. 2a.

Because the wall prefers the denser liquid, there is a thin layer of liquid present, even though the chemical potential of the bulk gas is slightly slower than the bulk liquid. This wetting layer extends a distance $l(T, p)$ into the gas phase, terminating at a liquid-gas interface whose width is comparable to the correlation length $\xi(T, p)$.

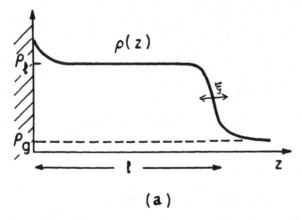

**(a)**

Fig. 2a.   Density profile near a wall in the bulk gas phase close to liquid-gas coexistence. The density starts at a large value $\rho_l$ appropriate to the nearby liquid phase and drops to a smaller value $\rho_g$ appropriate to the gas at distance $\ell$ from the wall.

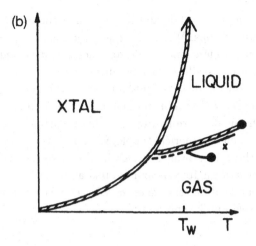

**Fig. 2b.** Pressure temperature phase diagram with regions of first order and continuous wetting transitions along the liquid-gas coexistence curve indicated by dashed and solid lines, respectively. A first order "prewetting" transition terminating in a critical point extends into the gas phase. The density profile in Fig. 2a corresponds to the situation near a wall at the point $x$.

Two distinct behaviors are possible as the liquid-gas coexistence curve is approached from the gas phase (see Fig. 2b). Far from the critical point, $\ell(p, T)$ usually remains finite at the liquid-gas coexistence curve (*i.e.*, along the dashed line in Fig. 2b). Closer to the critical point, however, $\ell(p, T)$ diverges (logarithmically, in simple model calculations) as the coexistence curve is approached (along the solid line in Fig. 2a). This divergence may be preceded by a first order "prewetting" transition in the bulk liquid signaled by an upward jump in the liquid density at the wall. The point at which $\ell(p, T)$ diverges to infinity along the coexistence curve, at $T = T_w$, locates a wetting transition, which has been the subject of considerable theoretical interest recently. This transition can be first order, or it occurs via a rather exotic second order transition.[7] More information about wetting is contained in the chapters of M.E. Fisher and S. Leibler.

## 2. Crumpled Membranes

### 2.1. *Experimental Realizations*

Membranes can be regarded as two-dimensional generalizations of linear polymer chains, for which there is a vigorous theoretical and experimental literature.[8,9] Flexible membranes should exhibit even more richness and complexity, for two basic reasons. The first is that important geometric concepts like intrinsic curvature, orientability and genus, which have no direct analogue in linear polymers, appear naturally in discussions of membranes: Our understanding of the interplay between these concepts and the statistical mechanics of membranes is still in its infancy. The second reason is that surfaces can exist in a variety of different phases. The possibility of a two-dimensional shear modulus in planar membranes shows that we must distinguish between solids and liquids when these objects are allowed to crumple into three dimensions. We shall argue later that hexatic membranes, with extended six-fold bond orientational order, are another important possibility. There are no such sharp distinctions for linear polymer chains.

Figure 3 shows two examples of *liquid* membranes. Figure 3a is an erythrocyte or red blood cell. The cell wall is a membrane, composed of a bilayer of amphiphillic molecules, each with one or more hydrophobic hydrocarbon tails and a polar head group. The membrane has a spherical topology, as do artificial vesicles formed from bilayers. Although these membranes could, in principle, crystalize upon cooling, they exhibit an almost negligible shear modulus at biologically relevant temperatures. The small shear modulus that is observed for erythrocytes may be due to an additional protein skeleton like spectrin.[10]

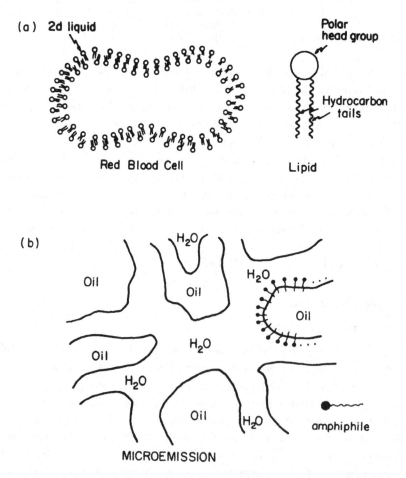

Fig. 3.  Examples of liquid-like membranes: (a) red blood cell and (b) microemulsion.

Figure 3b illustrates the topology of a microemulsion, which is a transparent solution in which oil (*e.g.*, dodecane) and water mix in essentially all proportions.[11] This remarkable mixing is only possible because of the addition of significant amounts of an amphiphile like SDS (sodium dodecyl sulfate), which sits at the interface between oil and water and reduces the surface tension almost to zero. The size of the oil-rich and water-rich regions, which are constantly shifting as the interface fluctuates, is of order 100 Angstroms. Usually, a cosurfactant like pentanol is necessary to stabilize the microemulsion.

For more about liquid membranes, see the chapters of S. Leibler and F. David, and the collections of papers in Refs. 12 and 13.

Although careful experimental investigations are only just beginning, there are also many examples of *solid* membranes. One can, for example, explore the properties of flexible sheet polymers, the "tethered surfaces" described in the lectures of Kantor. Tethered surfaces can be synthesized by polymerizing Langmuir–Blodgett films or amphiphillic bilayers.[14] Although lipid monolayers polymerized at an air-water interface would be initially flat, they could be inserted into a neutral solvent like alcohol and their fluctuations made visible by attaching a fluorescent dye. There are fascinating accounts of cross-linked methyl-methacrylate polymer assembled on and then extracted from the surface of sodium montmorillonite clays.[15]

Two less familiar examples of solid membranes are illustrated in Fig. 4. Figure 4a shows a model of large sheet molecule believed to be an ingredient of glassy $B_2O_3$.[16] Similar structures, also in crumpled form, may exist in chalgogenide glasses such as $As_2S_3$. Although it may be difficult to obtain dilute solutions in a good solvent, we might hope to produce a dense melt of such surfaces, in analogy with polymer melts or models of amorphous selenium.[17]

Figure 4b illustrates an idea for synthesizing a large number surfaces of two-dimensional polyacrylamide gel, which I have pursued in collaboration with R.B. Meyer at Brandeis University. We first form a lyotropic smectic liquid crystal of amphiphillic bilayers, similar to those discussed above. The bilayers are separated by water, and if necessary can be pushed further apart by the addition of oil or water.[18,19] If the lipids have multiple double bonds, one could of course polymerize the bilayers as discussed above. An attractive alternative for producing flexible surfaces is to introduce polyacrylamide gel into the watery interstices between the bilayers. Meyer and I have succeeded in stabilizing a smectic phase in which each ≈20 Angstrom thick water-rich region contains about 15 weight percent acrylamide and bis-acrylamide monomers. By shining ultraviolet light on this mixture, it may be possible to produce many slabs of 2d cross-linked polyacrylamide gel. The lipid bilayers, which are used simply as spacers in this experiment, would then be washed away.

A third class of membrane surfaces is possible if we replace fixed covalent cross links like those in Fig. 4a by weaker van der Waals forces. Van der Waals interactions will tend to crystalize the lipid bilayers discussed above at sufficiently low temperatures. Although these surfaces will have a nonzero shear modulus when

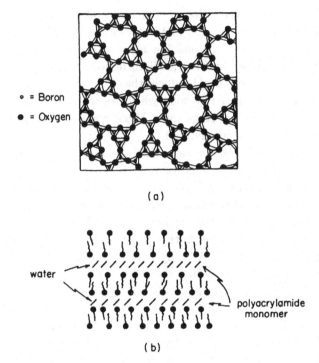

Fig. 4.   Examples of solid-like membranes: (a) planar section of boron-oxide which, when crumpled describes a glass and (b) lyotropic smectic phase with polymerizable polyacrylamide/monomer in the watery interstices.

confined to a plane, they are unstable to the formation of free dislocations when allowed to buckle into the third dimension.[20] Dislocations necessitate broken bonds, and thus would require prohibitively large energies in covalently bonded systems. The presence of a finite concentration of unbound dislocations at any temperature means that unpolymerized lipid bilayers will in fact be hexatic liquids with residual bond-orientational order at low temperatures.[20,21] The properties of hexatic membranes are intermediate between liquid and solid surfaces, and will be discussed in a later Chapter by myself and in the lectures of F. David.

## 2.2. Plaquette Surfaces

One route toward understanding crumpled membranes is to generalize various results from polymer physics. There are both lattice[9] and continuum[8] formulations of polymer statistical mechanics, and it turns out that the natural generalizations lead to two distinct classes of membranes. We first review the lattice generalization.[5]

As illustrated in Fig. 5a, we can catalogue polymer configurations on a lattice by first counting the number of self-avoiding walks starting at the origin and terminating at position $R$. The function $\vec{r}(s)$ gives the position of the walk after the $s$th step. If $\mathcal{N}_N(\vec{R})$ is the number of walks of length $N$ starting at the origin and

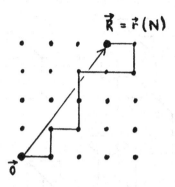

Fig. 5a. Polymer configuration extending from the origin to $\vec{R}$ on a square lattice.

terminating at $\vec{R}$, the *total* number of walks of length $N$ is given by

$$\mathcal{N}_N^{\text{tot}} = \sum_{\vec{R}} \mathcal{N}_N(\vec{R}). \tag{1.13}$$

A typical polymer size is given by the radius of gyration $R_G$,

$$R_G = \left[ \frac{1}{N^2} \sum_{s=1}^{N} \sum_{s'=1}^{N} \langle |\vec{r}(s) - \vec{r}(s')|^2 \rangle \right]^{1/2}, \tag{1.14}$$

where the average is over all polymer configurations. Polymer critical exponents are defined by the asymptotic large $N$ behavior of $R_G$ and $\mathcal{N}_N^{\text{tot}}$,

$$R_G \sim N^\nu \tag{1.15}$$

$$\mathcal{N}_N^{\text{tot}} \sim (\bar{z})^N N^{\gamma - 1}. \tag{1.16}$$

Here, $\bar{z}$ is a nonuniversal effective "coordination number", reduced from the actual coordination number by self-avoiding constraints. The radius of gyration exponent $\nu$ is increased by self-avoidance from the random walk result $\nu = 1/2$ to the universal result $\nu \approx 0.59 \approx 3/5$ in three dimensions. The exponent $\gamma \approx 1.18$ is also universal for polymers with free ends, although it changes for ring polymers.[9] The effect of self-avoidance on the exponents vanishes for $d > d_c = 4$, which is the upper critical dimension for linear polymers.

Figure 5b shows a similar counting problem for a surface consisting of contiguous plaquettes on cubic lattice. The ensemble of surfaces with $N$ plaquettes is now subdivided into varying numbers of surfaces $\mathcal{N}_N(\Gamma)$ with a fixed boundary contour $\Gamma$. A particular surface can be specified by a function $\vec{r}(P)$ which locates the center of an occupied plaquette $P$. In analogy with linear polymers on a lattice, we can

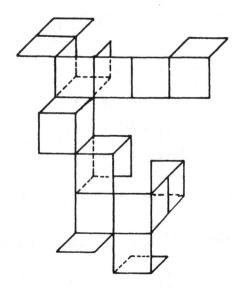

Fig. 5b.   Configuration of a plaquette surface on a cubic lattice.

ask for the total number of surfaces with $N$ plaquettes,

$$\mathcal{N}_N^{\text{tot}} = \sum_\Gamma \mathcal{N}_N(\Gamma) \qquad (1.17)$$

and the radius of gyration

$$R_G = \left[ \frac{1}{N^2} \sum_{P,P'} \langle |\vec{r}(P) - \vec{r}(P')|^2 \rangle \right]^{1/2} \qquad (1.18)$$

where the sums are over occupied plaquettes. Critical exponents are defined by the asymptotic large $N$ behaviors,

$$R_G \sim N^\nu \qquad (1.19)$$
$$\mathcal{N}_N^{\text{tot}} \sim \mu^N N^{-\theta} \qquad (1.20)$$

where $\mu$ is the nonuniversal parameter analogous to $\bar{z}$, and $\nu$ and $\theta$ are expected to be universal critical exponents.

The most important theoretical results concerning plaquette surfaces are reviewed by Fröhlich in Ref. 5. The dominant configurations of plaquette surfaces are thin, branched objects, like the bark of a tree. These configurations appear because they are favored entropically, and because there is no energy penalty for long, thin cylindrical tubules of surface. It is safe to neglect self-avoidance above $d_c = 8$, where the critical exponents assume the mean field values appropriate to noninteracting branched polymers. For $d < d_c$, it is believed that the exponents are those of self-avoiding branched polymers. This conjecture is supported by careful

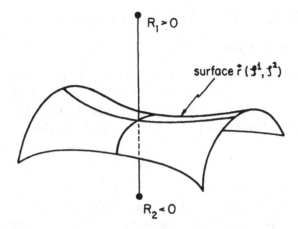

Fig. 6.  Principal radii of curvature associated with one point on the surface $\vec{r}(\zeta^1, \zeta^2)$.

numerical work in $d = 3$ by Glaus,[22] who finds

$$\nu \simeq 0.504 \qquad \theta \simeq 1.48. \tag{1.21}$$

Although the underlying theory is very beautiful, it is hard to think of direct experimental realizations of plaquette surfaces in condensed matter physics. Because they resemble branched polymers, plaquette surfaces are rather wild, unruly objects compared to the experimentally realizable membranes discussed above. Some sort of bending energy is needed to control the tubular fluctuations. Bending energy appears naturally in continuum models of liquid membranes. As shown in Fig. 6, we associate with every point on a surface $\vec{r}(\zeta^1, \zeta^2)$ the two signed principal radii of curvature $R_1(\zeta^1, \zeta^2)$ and $R_2(\zeta^1, \zeta^2)$. The parametrization-independent bending energy is then[23]

$$F_b = \frac{1}{2} \int \sqrt{g} d^2\zeta \left[ \kappa \left( \frac{1}{R_1} + \frac{1}{R_2} \right)^2 + \bar{\kappa} \frac{1}{R_1 R_2} \right]. \tag{1.22}$$

The parameter $\kappa$ (which multiplies four times the mean curvature squared) is called the bending rigidity, while $\bar{\kappa}$ (which multiplies the Gaussian curvature) is called the Gaussian rigidity. In simple models, $\bar{\kappa}$ is usually negative, which is necessary to stabilize saddle distortions of an initially flat membrane.

## 2.3. *Perturbation Theory for Tethered Surfaces*

Membrane generalizations of the *continuum* theory of polymer chains are conveniently presented in the language of differential geometry. The partition function of the resulting tethered surfaces are a special case of a more general partition function which first arose in the study of bosonic strings, namely[24]

$$Z = \int \mathcal{D}g_{0,ab} \int \mathcal{D}\vec{r}(\zeta^1, \zeta^2) e^{-\frac{1}{2}K \int d^2\zeta \sqrt{g_0} g_0^{ab} \partial_a \vec{r} \cdot \partial_b \vec{r}}. \qquad (1.23)$$

The "action" is composed of surface gradients $\partial_a \vec{r}$ contracted with a metric tensor $g_0^{ab}$. The integrations are over all possible metrics $g_{0,ab}$, as well as over all possible surface configurations $r(\zeta^1, \zeta^2)$. Although the underlying metric and the surface are independent variables, Polyakov[24] has shown that a relation analogous to Eq. (1.2), i.e.,

$$g_{0,ab} = \frac{\partial \vec{r}}{\partial \zeta^a} \cdot \frac{\partial \vec{r}}{\partial \zeta^b} \qquad (1.24)$$

is recovered in the low temperature, strong coupling ($K \to \infty$) limit.

A microscopic physical interpretation of Eq. (1.23) is illustrated in Fig. 7. For a fixed metric $g_{0,ab}$, the surface is represented by a fixed triangulation of particles, connected by harmonic springs. The action in Eq. (1.23) is the continuum limit of the energy associated with these springs. The particle positions can be arranged to approximate *any* particular simply-connected surface with free boundaries $\vec{r}(\zeta^1, \zeta^2)$; there is, however, a significant energetic cost associated with large deviations from the surfaces preferred by the underlying connectivity or "background metric". To carry out the functional integral (1.23) on a computer, one would first integrate

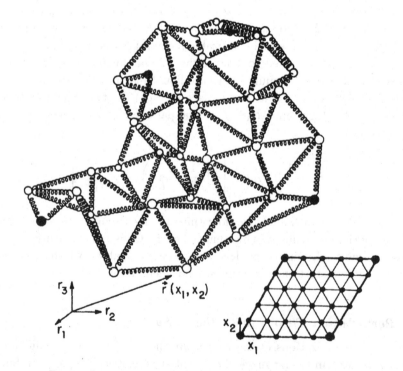

Fig. 7.   Lattice of Gaussian springs with a fixed connectivity.

over all particle positions for a fixed triangulation, and then sum over different triangulations.

Tethered surfaces, discussed more completely in the lectures of Kantor, are an example of the string partition function (1.23), specialized to a single "flat" triangulation, where every particle is connected to exactly six nearest neighbors. The "background metric" is thus

$$g_{0,ab} = \delta_{ab} \tag{1.25}$$

and the partition function is

$$\mathcal{Z}_0 = \int \mathcal{D}\vec{r}(x_1, x_2) e^{-F_0} \tag{1.26}$$

where

$$F_0 = \frac{1}{2} K \int d^2x \left[ \left( \frac{\partial \vec{r}}{\partial x_1} \right)^2 + \left( \frac{\partial \vec{r}}{\partial x_2} \right)^2 \right]. \tag{1.27}$$

As it stands, we now have a model for "phantom" polymerized membranes, without self-avoiding interactions between distant particles. To obtain a model for a real self-avoiding membrane, we replace the free energy $F_0$ by

$$F = \frac{1}{2} K \int d^2x \left( \frac{\partial \vec{r}}{\partial \mathbf{x}} \right)^2 + \frac{1}{2} v \int d^2y \int d^2y' \delta[\vec{r}(\mathbf{y}) - \vec{r}(\mathbf{y}')]. \tag{1.28}$$

The second term assigns a positive energetic penalty $v$ whenever two elements of the surface occupy the same position in the three-dimensional embedding space.

To make analytic progress with the statistical mechanics associated with (1.28), it is useful to generalize (1.28), and consider *manifolds* $\vec{r}(\mathbf{x})$ with a $D$-dimensional flat internal space embedded in a $d$-dimensional external space.[25-27] The associated free energy is

$$F = \frac{1}{2} K \int d^D x \left( \frac{\partial \vec{r}}{\partial \mathbf{x}} \right)^2 + \frac{1}{2} v \int d^D y \int d^D y' \delta^d[\vec{r}(\mathbf{y}) - \vec{r}(\mathbf{y}')], \tag{1.29a}$$

or

$$F = \frac{1}{2} \int d^D x \left( \frac{\partial \vec{R}}{\partial \mathbf{x}} \right)^2 + \frac{1}{2} v K^{d/2} \int d^D y \int d^D y' \delta^d[\vec{R}(\mathbf{y}) - \vec{R}(\mathbf{y}')], \tag{1.29b}$$

where we have introduced the $d$-dimensional rescaled variable,

$$\vec{R}(x^1, x^2) = \sqrt{K} \vec{r}(x^1, x^2). \tag{1.30}$$

When $v = 0$, we have a free field theory, and it is easy to show that the mean squared distance between points with internal coordinates $\mathbf{x}_A$ and $\mathbf{x}_B$ is

$$\langle |\vec{r}(\mathbf{x}_A) - \vec{r}(\mathbf{x}_B)|^2 \rangle \underset{x_{AB} \to \infty}{\simeq} \frac{2dS_D}{(2-D)K}[|\mathbf{x}_{AB}|^{2-D} - a^{2-D}] \qquad (1.31)$$

where $S_D = 2\pi^{D/2}/\Gamma(D/2)$ is the surface area of a $D$-dimensional sphere, $\mathbf{x}_{AB} = \mathbf{x}_A - \mathbf{x}_B$ and $a$ is a microscopic cutoff. If we take $\mathbf{x}_A$ and $\mathbf{x}_B$ to be close to opposite sides of the manifold (in the internal space), Eq. (1.31) becomes a measure of the squared radius of gyration. When $D = 1$, we are dealing with a linear polymer chain and we see that the size $R_G$ increases as the square root of the linear dimension $L \sim |\mathbf{x}_{AB}|$, i.e., $R_G \sim L^{1/2}$. The same argument, however, shows that the characteristic membrane size $R_G$ increases only as the square root of the logarithm of the linear dimension $L$ for $D = 2$,

$$R_G \sim \frac{1}{K} \ln^{1/2}(L/a). \qquad (1.32)$$

To see how self-avoiding corrections affect Eq. (1.31), we can carry out perturbation theory in the excluded volume parameter. Each term can be represented as in Fig. 8, where the dotted lines represent self-avoiding interactions between different pieces of the manifold. Dimensional analysis using the rescaled free energy Eq. (1.29b) shows that this perturbation theory becomes singular in the limit of large internal linear dimension $L$: The correction to Eq. (1.31) must take the form

$$\langle |\vec{r}(\mathbf{x}_A) - \vec{r}(\mathbf{x}_B)|^2 \rangle$$
$$\simeq \frac{2dS_D}{(2-D)K}|\mathbf{x}_{AB}|^{2-D}[1 + \text{const.} \times vK^{d/2}L^{2D-(2-D)\frac{d}{2}} + \cdots]. \qquad (1.33)$$

Whenever

$$2D > (2-D)\frac{d}{2} \qquad (1.34)$$

the corrections to the free field result (1.31) diverge as $L \to \infty$, signaling a breakdown of perturbation theory. If this inequality is reversed, however, we expect self-avoidance to be asymptotically irrelevant in large systems. This is the case for

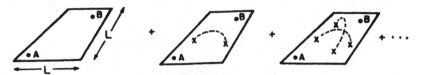

Fig. 8.   Graphical representation of the perturbative calculation of the mean square distance in the embedding space between the points $\vec{r}(\mathbf{x}_A)$ and $\vec{r}(\mathbf{x}_B)$.

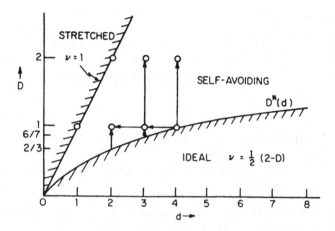

Fig. 9. Different regimes in the $(d, D)$-plane for self-avoiding tethered surfaces.

polymers $(D = 1)$ when $d > 4$. Note, however, that the perturbative correction in (1.33) is always large for membranes, *i.e.*, for $D = 2$.[28]

Figure 9 shows the critical curve

$$D^*(d) = \frac{2d}{4 + d} \qquad (1.35)$$

which separates ideal from self-avoiding behavior in the $(d, D)$-plane. Also shown is the line $D = d$, along which the manifold becomes fully stretched due to self-avoidance. The critical line $D^*(d)$, of course, passes through the point $(d^* = 4, \ D^* = 1)$, which is the basis for epsilon expansions of polymers,[8] but in fact *any point* on this line is an equally good expansion candidate. We could, for example, stay in three dimensions $(d = 3)$, and change the manifold dimensionality $D$. Self-avoidance dominates for solid elastic cubes $(D = 3)$, but is less important for elastic surfaces $(D = 2)$. It produces relatively small corrections to the Gaussian result for linear manifolds $(D = 1)$, and becomes formally negligible when $D < D^* = 6/7$! This idea forms the basis for a $6/7 + \epsilon$ expansion for tethered surfaces,[25-27] as will be discussed in more detail in the chapter of Duplantier. The result is that the radius of gyration scales with the linear dimension according to

$$R_G \sim L^\nu \qquad (1.36)$$

where

$$\nu = \frac{2 - D}{2} + 0.469 \left( D - \frac{6}{7} \right) + \mathcal{O} \left( D - \frac{6}{7} \right)^2. \qquad (1.37)$$

This novel epsilon expansion gives excellent results for linear polymers in three dimensions $(\epsilon = 1/7, \nu = 0.567)$, but is not very accurate for polymerized membranes $(\epsilon = 8/7, \nu = 0.536)$. More precise numerical methods for extracting the exponent $\nu$ is are described in the lectures of Kantor.

## Acknowledgments

The work described in Sec. II.C was carried out in collaboration with M. Kardar and Y. Kantor, and received support from the National Science Foundation, through Grant DMR85-14638 and the Harvard Materials Research Laboratory.

## References

1. M.B. Green, J.H. Schwarz, and E. Witten, *Superstring Theory 1 and 2* (Cambridge University Press, Cambridge, 1987).
2. J. Weeks, in *Ordering in Strongly Fluctuating Condensed Matter Systems*, ed. T. Riste (Plenum, New York, 1980).
3. For an introduction useful to physicists, see B.A. Dubrovin, A.T. Fomenko, and S.P. Novikov, *Modern Geometry — Methods and Applications* (Springer, New York, 1984) and the lectures of F. David at these Proceedings.
4. J.V. Jose, L.P. Kadanoff, S. Kirkpatrick, and D.R. Nelson, *Phys. Rev.* **B16**, 1217 (1977).
5. J. Fröhlich, in *Applications of Field Theory to Statistical Mechanics*, edited by L. Garido, Lecture Notes in Physics, Vol. 216 (Springer-Verlag, Berlin, 1985).
6. P.G. de Gennes, *Rev. Mod. Phys.* **57**, 827 (1985).
7. E. Brézin, B.I. Halperin, and S. Leibler, *Phys. Rev. Lett.* **50**, 1387 (1983); R. Lipowsky, D.M. Kroll and R.K.P. Zia, *Phys. Rev.* **B27**, 4499 (1983).
8. Y. Oono, *Adv. Chem. Phys.* **61**, 301 (1985).
9. P.G. de Gennes, *Scaling Concepts in Polymer Physics* (Cornell University Press, Ithaca, New York, 1979).
10. E. Evans and R. Skalak, *Mechanics and Thermodynamics of Biomembranes* (CRC Press, Boca Raton, 1980).
11. P.G. de Gennes and C. Taupin, *J. Phys. Chem.* **86**, 2294 (1982).
12. *Physics of Complex and Supermolecular Fluids*, eds. S.A. Safran and N.A. Clark (John Wiley and Sons, New York, 1987).
13. *Physics of Amphiphillic Layers*, eds. J. Meunier, D. Langevin, and N. Boccara (Springer-Verlag, Berlin, 1987).
14. J.H. Fendler and P. Tundo, *Acc. Chem. Res.* **17**, 3 (1984).
15. A. Blumstein, R. Blumstein, and T.H. Vanderspurt, *J. Colloid Interface Sci.* **31**, 236 (1969).
16. M.J. Aziz, E. Nygren, J.F. Hays, and D. Turnbull, *J. Appl. Phys.* **57**, 2233 (1985).
17. R. Zallen, *The Physics of Amorphous Solids* (Wiley, New York, 1983).
18. J. Larche, J. Appell, G. Porte, P. Bassereau, and J. Marignan, *Phys. Rev. Lett.* **56**, 1700 (1986).
19. C.R. Safinya, D. Roux, G.S. Smith, S.K. Sinha, P. Dimon, and N.A. Clark, *Phys. Rev. Lett.* **57**, 2718 (1986).
20. D.R. Nelson and L. Peliti, *J. Physique* **48**, 1085 (1987); S. Seung and D.R. Nelson, *Phys. Rev.* **A38**, 1055 (1988).
21. D.R. Nelson and B.I. Halperin, *Phys. Rev.* **B19**, 2457 (1979); D.R. Nelson, *Phys. Rev.* **B27**, 2902 (1983).
22. U. Glaus, *Journal of Statistical Physics* **50**, 1141 (1988).
23. W. Helfrich, *Z. Naturforsch.* **28c**, 693 (1973).
24. A.M. Polyakov, *Phys. Lett.* **103B**, 207 (1981).
25. M. Kardar and D.R. Nelson, *Phys. Rev. Lett.* **58**, 1289 (1987); and *Phys. Rev.* **A38**, 966 (1988).

26. J.A. Aronovitz and T.C. Lubensky, *Europhys. Lett.* **4**, 395 (1987).

27. D. Duplantier, *Phys. Rev. Lett.* **58**, 2733 (1987).

28. There are logarithmic corrections to the result of naive dimensional analysis in this case. See the Appendix of Y. Kantor, M. Kardar, and D.R. Nelson, *Phys. Rev.* **A35**, 3056 (1987).

# CHAPTER 2

# INTERFACES: FLUCTUATIONS, INTERACTIONS AND RELATED TRANSITIONS

Michael E. Fisher

*Institute for Physical Science and Technology*
*University of Maryland*
*College Park, Maryland 20742, USA*

A brief, informal review of the theory of interface fluctuations and their effects is presented. Topics touched upon include mean field theory for interfaces, interface models, complete wetting transitions, fluctuations and the wandering exponent, random media, effective interface–interface interactions, the shape of vicinal crystal faces, critical wetting transitions and renormalization group theory for interfaces.

## Introduction

A natural starting point in building a theory of membranes, surfaces and their statistical mechanical behavior is the consideration of *asymptotically flat interfaces*. An equilibrium *interface* separates two coexisting but distinct phases, say $\alpha$ and $\beta$, e.g. gas and liquid, fluid and crystal, etc. An interface is formed essentially, of the same molecules that constitute the bulk phases; it has limited internal structure. By contrast, a *membrane* is normally made of molecules *not* characteristic of the bulk phases, furthermore, it has crucially important internal structure, entailing rigidity, ordering of various sort, etc. Asymptotically flat interfaces (or membranes) may, and generally will fluctuate significantly away from a smooth planar configuration but on macroscopic scales they may still be associated with a definite plane. For that reason their theory is appreciably simpler than that of fully convoluted surfaces. Here we present an informal review of selected aspects of the theory of asymptotically flat interfaces. The reader is cautioned that references are given only to a small fraction of the articles that might be cited, including, however, a few more systematic and thorough reviews which should provide an entry to the literature.[1-7]

The first object of study is a single interface which is (a) *free*; but the effects of (b) *external fields* is of central interest: these may be (i) *slowly varying* e.g. gravitational, van der Waalsian decaying with a power law, etc., (ii) *periodic*, e.g. as imposed by a crystalline substrate, leading to the study of roughening phenomena, (iii) *rapidly varying* representing, typically, the presence of attractive and repulsive walls giving rise to wetting and deepening transitions. Another aspect (c) is the

behavior of a single interface in a *random medium* which can greatly modify the
long-wavelength fluctuations.

The interplay of *two nearby interfaces*, separating sequential phases $\alpha, \beta$ and $\gamma$,
is of major importance: note that a rigid wall may often be viewed as a limiting
case of an equilibrium interface. The dominant feature is the appearance of effective, *fluctuation-induced interactions* which drive various wetting and delocalization
transitions, etc. Three *distinct* interacting interfaces provides an interesting theoretical problem which can be solved in $d = 2$ dimensions.[8] Finally, systems of *many
neighboring interfaces* underly the behavior of commensurate-incommensurate transitions, determine the shapes of vicinal crystal faces, etc.

A range of theoretical techniques may be brought to bear on these problems.
Phenomenological analyses play a valuable foundational role.[1-3] For special microscopic models, like the lattice gas, exact results are available[4] and various simplified
semiphenomenological models can also be analyzed precisely.[5,6] Recently, renormalization group methods have also been developed and exploited for interfaces and
membranes,[9-11,53,55] as we will discuss.[57]

## 1. Interface Models, Mean Field Theory, and Wetting

### 1.1. *Levels of Theory*

There is a natural hierarchy of theoretical approaches. Most basic are (a) *microscopic theories* founded on a Hamiltonian, $\mathcal{H}_{micro}[s(\vec{r})]$, for the bulk degrees of
freedom, $s(\vec{r})$, e.g. molecular coordinates, Ising spin variables, etc. where $\vec{r}$ is a
$d$-dimensional coordinate. It proves convenient to write $\vec{r} = (\vec{y}, z)$ where $\vec{y}$ is a
$d' = (d-1)$-dimensional vector parallel to the interface plane while $z$ is the normal or
perpendicular coordinate. Next (b) come *density functional theories* based on some
order-parameter field or density, say $m(\vec{r})$, and a postulated Landau–Ginzburg–
Wilson effective Hamiltonian, $\mathcal{H}_{LGW}[m(\vec{r})]$ or free-energy functional $\mathcal{F}[m(\vec{r})]$; the
nature and form of the mean interface *profile*, $\bar{m}(z)$, as $m(\vec{r})$ varies between the
coexisting bulk values $m_\beta$ and $m_\alpha$, is a principal object of study. Lastly, it is much
easier and usually more instructive to work with (c) *interface models* in which a
function $l(\vec{y})$ represents the local departure of the interface from a reference plane
$(z = l = 0)$ and one constructs an effective *interface Hamiltonian*, $\mathcal{H}_{int}[l(\vec{y})]$, or
corresponding *interfacial free-energy functional*, $\mathcal{F}_{int}$. The appropriate form of this
functional will be discussed and put to use in the subsequent developments where
fluctuations are the main focus.

### 1.2. *Mean Field Theory for Order Parameters*

In mean field theory one supposes that the true free energy, $F(T, H, \ldots)$, can be
found by minimizing an appropriate order-parameter functional so that

$$F(T, H, \ldots) = \min_{m(\vec{r})} \mathcal{F}[m(\vec{r}); T, h, \ldots], \qquad (1.1)$$

where $T$ is the temperature and $h$ represents an external ordering field like the chemical potential difference between two species, a magnetic field, etc. Ideally, $\mathcal{F}$ would be deduced from some $\mathcal{H}_{micro}$; but, in practice, one is forced to rely on general phenomenological principles. Thus, in the absence of external fields causing departure from two-phase equilibrium the one must have

(A) *translational invariance,* $\mathcal{F}_0[m(\vec{r})] = \mathcal{F}_0[m(\vec{r} + \vec{c})]$ and
(B) *stationarity of the optimal profile,* $\check{m}(z)$, i.e. $\delta\mathcal{F} = \mathcal{F}[\check{m} + \delta m] - \mathcal{F}[\check{m}] = O[\delta m^2]$.

In addition, one normally postulates

(C) *locality* so that a gradient expansion is valid in the form

$$\mathcal{F}_0 = \int d^d r \left[ \Phi(m; T) + \frac{1}{2} K(m)(\nabla m)^2 + \cdots \right], \qquad (1.2)$$

where $\Phi(m)$ has two equally deep (but, in general differently curved) minima at $m = m_\alpha$ and $m_\beta$ with $\Phi = 0$. Then on minimizing $\mathcal{F}_0$ to find $\check{m}(z)$ (see, e.g. Ref. 1) the equilibrium *interfacial free-energy* (per unit area) or *surface tension*, which is of primary interest, is found to be

$$\Sigma(T) = \mathcal{F}_0[\check{m}]/A = \int_0^\infty K[\check{m}(z)] \left( \frac{d\check{m}}{dz} \right)^2 dz, \qquad (1.3)$$

where $A$ is the (projected) interfacial area and higher-order gradient terms have been neglected.

## 1.3. *Derivation of Interface Models*

To derive an effective Hamiltonian $\mathcal{H}_{int}[l(\vec{y})]$ for an interface, as a preliminary to studying the nature of interface fluctuations, one would like to start with a more basic theory. Ideally one should constrain (by some appropriate means!) the interface away from its natural, flat position into the configuration $l(\vec{y})$ and then, using $\mathcal{H}_{micro}[s(\vec{r})]$ or $\mathcal{F}[m(\vec{r})]$, take a partial trace over the bulk variables, $s(\vec{r})$, or order parameter, $m(\vec{r})$ and thence deduce the correct $\mathcal{H}_{int}$. Finally, a trace [or functional integral $\int \mathcal{D}l(\vec{y})$] over the configurations $l(\vec{y})$ would lead to the correct $\Sigma(T)$. Of course, this is an over-ambitious program in realistic situations. One may, however, at least ask for a derivation that would be consistent at the mean-field level, i.e. so that mean field theory for $\mathcal{F}[m]$, in the form (1.1), is equivalent to mean field theory based on minimizing $\mathcal{H}_{int}[l]$. This program can be carried through fairly completely, although, as will be indicated below, a fully satisfactory and transparent treatment in the presence of rapidly varying fields does not seem to be in the literature.[58]

To start, one may consider a free interface and make the variational ansatz[2,12,13]

$$m = \tilde{m}(\vec{r}) = \tilde{m}(z - l(\vec{y})),   \tag{1.4}$$

to describe a "constrained equilibrium" interface. Then using (1.1)–(1.3) and the principles (A) and (B) one finds

$$\Delta \mathcal{H}_{int} = \mathcal{F}_0[\tilde{m}] - \mathcal{F}_0[\check{m}] = \int d^{d'}y \frac{1}{2}\Sigma[\nabla l(\vec{y})]^2,   \tag{1.5}$$

correct to leading order in the gradients. This form agrees with the intuitive picture of a nonplanar interface embodied in the assertion that the interfacial Hamiltonian should just be $\int \Sigma dS$, where $dS = [1 + (\nabla l)^2]^{1/2} d^{d'}y$. Note, however, that in an *anisotropic* medium, in which the tension $\Sigma$ depends on the orientation of the interface, an expansion shows that one must replace $\Sigma$ in (1.5) by the *interfacial stiffness*

$$\tilde{\Sigma} = \Sigma(0) + \Sigma''(0)   \tag{1.6}$$

where, for simplicity, we suppose $\Sigma = \Sigma(\nabla l)$ and the primes denote differentiation (see e.g. Ref. 14).

## 1.4. *External Forces*

External forces which vary slowly on the scales of the correlation lengths, $\xi_\alpha$ and $\xi_\beta$ for the bulk phases may be incorporated fairly readily. Note that $\xi_\alpha$ and $\xi_\beta$ control, *within mean field theory*, the exponential decay of the equilibrium profile $\check{m}(z)$ to the values $m_\alpha$ and $m_\beta$, respectively. One must also suppose that the external fields do not significantly modify the bulk phases themselves.

It is convenient to distinguish, first, the bulk ordering field $h$ which vanishes at coexistence and is positive in, say, phase $\beta$. Secondly, consider a gravitational field, which couples to $\int zm(z)dz$. Lastly, suppose a rigid wall of specified composition is located at $\ell = z = 0$. Molecules in phases $\alpha$ and $\beta$ will interact with those in the wall via long-range van der Waals forces with a pair-potential decaying as $\varphi(r) \sim 1/r^{d+\sigma}$ with $\sigma = 3$ (or, for retarded potentials, $\sigma = 4$). In general, a net force on the interface, proportional to $\Delta m = (m_\beta - m_\alpha)$ will result.

In total one obtains a contribution to the interface Hamiltonian due to the external forces of

$$\mathcal{H}_E[\ell(\vec{y})] = \int d(\vec{y}) V[\ell(\vec{y})],   \tag{1.7a}$$

with

$$V(\ell) = \bar{h}\ell + \tfrac{1}{2}\bar{g}\ell^2 + V_W(\ell)   \tag{1.7b}$$

where $\bar{h} = -h\Delta m, \bar{g} = g\Delta m$ and, for large $\ell$,

$$V_W(\ell) \approx V_0/\ell^{\sigma-1},   \tag{1.8}$$

where $V_0$ is a constant.

## 1.5. *The Complete Wetting Transition*

Now suppose that the long-range wall forces are such that phase $\beta$ is attracted more strongly than phase $\alpha$. If $\beta$ is a one-component liquid while $\alpha$ is its vapor, this is just the typical situation. Then phase $\beta$ will normally form a layer bounded by the wall at $z = 0$ and by the interface at $z = \ell(\vec{y})$. Correspondingly, the amplitude $V_0$ in (1.8) will be *positive* implying a *repulsion* of the $\beta|\alpha$ interface away from the wall. In the absence of ordering and gravitational fields ($\bar{h} = \bar{g} = 0$) the equilibrium interface will thus sit *infinitely far* from the wall, which is then said to be *wet* (by $\beta$). If, however, the ordering field, $h$, is imposed in such a sense as to favor the bulk phase $\alpha$ ($\bar{h} > 0$) the wetting layer becomes of only finite mean thickness, $\ell_W(\bar{h})$. When $\bar{h}$ approaches zero the thickness diverges: this is the *complete wetting transition*.[15] A power law,

$$\ell_W(\bar{h}) \sim 1/\bar{h}^\psi, \qquad (1.9)$$

may be anticipated but what is the value of $\psi$?

This question can be answered within mean field theory simply by using the wall potential (1.8) and minimizing the total potential, $V(\ell)$, in (1.7) with $\bar{g} = 0$. One finds immediately

$$\psi = 1/\sigma. \qquad (1.10)$$

This result corresponds to a one-third law for the divergence of the wetting layer thickness when ordinary van der Waals forces are acting. In fact, various experiments on liquid films absorbed from the vapor onto good substrates confirm this exponent value. One interesting verification[16] has been made in the context of *triple point wetting*.[17] One approaches the triple point of vapor–liquid–crystal coexistence at $(p_t, T_t)$ along the vapor-solid phase boundary below $T_t$. A substrate, a gold fiber in the experiments of Krim, Dash and Suzanne,[16] is wet by the liquid but *not* by the crystalline solid. Accordingly, the temperature displacement $(T_t - T)$ can be regarded as proportional to the ordering field, $h$ (a chemical potential difference, here) which measures the deviation from the liquid–vapor phase boundary (say, extended metastably below $T_t$). Thus one expects[17] and observes[16] a liquid layer thickness, $\ell_W$, diverging as $l/(T_t - T)^{1/3}$.

Why should a crude mean-field argument in which all effects of fluctuations are suppressed give, apparently, the correct exponent value? The answer must lie in the nature of the fluctuations and their interplay with long-range forces as, indeed, will be seen below.

## 1.6. *Wall Effects and the Interface Hamiltonian*

A situation of particular theoretical interest which is also experimentally relevant in the absence, or cancellation, of long-range power-law forces is a *hard wall* which simply excludes the phases $\alpha$ and $\beta$ from the region, say, $z < 0$. Within an order-parameter theory, the bulk behavior can be described by a free energy functional $\mathcal{F}_0^>$, which has the same form as in (1.2) but with the integration restricted to $z \geq 0$.

However, the representation of the wall itself, even under the assumption that it has *only short-range interactions* with the bulk phases, requires thought. The correct phenomenological approach (see e.g. Ref. 4) is to add to $\mathcal{F}_0^>$ a specific wall term

$$\mathcal{F}_1 \equiv \mathcal{F}_1[m_1 = m(z = 0)] = \int d^{d'} y \, \Phi_1[m_1(y)], \tag{1.11}$$

in which the integration runs over the wall plane at $z = 0$. Then, in leading order, neglecting gradients at the surface, etc. One must, at least, take

$$\Phi_1(m_1) = -h_1 m_1 - \tfrac{1}{2} g_1 m_1^2. \tag{1.12}$$

Here $h_1$ represents a *surface field* (favoring phase $\beta$) and $g_1$ represents a *surface enhancement* of the ordering (or phase separation tendencies of the bulk medium).

Mean field theory for this situation can be worked out explicitly (see e.g. Ref. 18). A surprisingly complex phase diagram in the space $(T, h, h_1, g_1)$ is revealed[18] with (i) complete wetting transitions, (ii) *critical wetting transitions*, at which $\ell_W(T)$ diverges to $\infty$ as $T = T_{cW}$ is approached along the coexistence curve, $h = 0$; (iii) *tricritical wetting points* at which the system goes over to (iv) a *first-order wetting transition* on the coexistence curve, while (v) *critical prewetting transitions* appear for $h \neq 0$. In addition, various surface multicritical points (*ordinary*, *special* and *extraordinary*) exist[18] associated with the bulk critical point at $T = T_c$.

Our specific question here is "What is the appropriate *interface Hamiltonian* that could provide a description of short-range critical wetting that would allow for the incorporation of fluctuations and be amenable to renormalization group treatment?"

The answer that has been accepted[13,19] is to add to the effective potential $V(\ell)$ in (1.7) a short-range wall contribution of the form

$$W(\ell) = w_1 e^{-\ell/\xi_\beta} + w_2 e^{-2\ell/\xi_\beta} + w_3 e^{-3\ell/\xi_\beta} + \cdots, \tag{1.13}$$

in which $w_1, w_2, \ldots$ vary smoothly with $T, h_1$ and $g_1$; for small values, the ordering field $h$ enters only through the term $\bar{h}\ell$ in (1.7). (Here we neglect gravitational forces: $\bar{g} = 0$.) The leading exponential decay of $W(\ell)$ can be understood intuitively by noting that the tail of the free mean-field interface profile, $m(z)$, in the wetting layer is truncated by the presence of the wall; but the profile, decays as $\exp(-|z - \ell|/\xi_\beta)$. The second term, likewise, is reasonable when one recognizes contributions to the free-energy functional varying as $m^2, (\nabla m)^2$, etc. For the renormalization group theory of the critical wetting transition in $d = 3$ dimensions discussed below, however, it turns out to be crucial that no other comparable terms appear. Furthermore, the intuitive argument overlooks the significant distortions of the profile actually induced by the wall itself.[58]

Clearly, a full derivation of $W(\ell)$ should provide the actual forms of $w_1, w_2, \ldots$. However, Refs. 13 and 19 reach somewhat different conclusions. The arguments of Brézin *et al.*[19] where $h = 0$ is presupposed, are appealing but inadequate. The nature of the constraint actually implied is unclear and a rescaling of the expressions for $w_1$ and $w_2$ by an arbitrary factor can be seen to leave intact the acceptability criterion employed. Lipowsky *et al.*[13] present a fuller discussion but it is predicated on the ansatz (1.4); it is not obvious that this constitutes a fully acceptable constraint since the free mean-field profile, $\bar{m}(z)$, will be distorted by the wall, as mentioned, but also as a result of deflections of the interface from its optimal position.

Despite these apparently open conceptual questions,[58] the validity of the general form (1.13) is very plausible. Furthermore, the critical, tricritical and first-order wetting transitions of the full mean field theory,[18] can be reproduced within interfacial mean field theory by treating $w_1$ and $w_2$ as thermodynamic fields which may change sign.[13,19] Let us illustrate some aspects of this.

## 1.7. *Wetting Transitions with Short-range Forces:*
##      *Mean Field Theory*

For simplicity suppose $w_2 > 0$ so that $w_3$ and higher order terms in (1.13) can be neglected. (This prevents us from treating *tricritical* wetting.) Now mean field theory for the interfacial Hamiltonian amounts merely to minimizing

$$V(\ell) = W(\ell) + \bar{h}\ell. \tag{1.14}$$

Consider, first, the case $w_1 > 0$: when the wetting layer thickness, $\ell_W$, is large, $w_2$ can be neglected and $V(\ell)$ has a minimum at

$$\ell_W(h) \approx \xi_\beta \ln(w_1/\xi_\beta \bar{h}). \tag{1.15}$$

When $h \to 0$ this corresponds to a complete wetting transition with $\psi = 0$ in (1.9) representing a *logarithmic* divergence of $\ell_W$. This result is in precise accord with mean field theory for the full order-parameter functional.

Suppose, next, that $h = 0$ and $w_2 > 0$ but $w_1$ is negative: $W(\ell)$, and hence $V(\ell)$, then has a minimum at finite $\ell_W$ so the substrate is only *partially wet* even at coexistence. If, as one expects, $w_1$ decreases in magnitude and changes sign as $T$ rises towards $T = T_{cW}$, the layer thickness diverges according to

$$\ell_W \approx \xi_\beta \ln(2w_2/|w_1|) \sim \xi_\beta \ln(T_{cW} - T)^{-1}. \tag{1.16}$$

This evidently represents a *critical wetting transition* on the phase boundary (i.e., at coexistence) of the sort originally proposed by Cahn.[20] Above $T_{cW}$ the substrate is completely wet at $h = 0$.

As a final illustration, suppose a long-range force, as in (1.8), is added to $W(\ell)$ (still in the regime $w_1 < 0$ and $h = 0$). Since the power law with $V_0 > 0$ dominates at long distances, $V(\ell)$ is always repulsive for $\ell \to \infty$. Nevertheless, if $w_1$ is sufficiently negative the interface can be bound near the origin with $V(\ell_W) < 0$; however, as $T$ increases and $w_1$ approaches zero, the minimum rises. At some temperature $T_W$ one has $V_{min}(\ell < \infty) \to 0$; then the wetting layer thickness jumps discontinuously to infinity. One concludes that van der Waals forces always convert a critical wetting transition into a first-order wetting transition. In real fluids the transitions observed appear to be of this sort.[20b]

The challenge now is to understand how far, if at all, these conclusions of mean field theory remain valid when the fluctuations of the interface are taken into account.

## 2. Fluctuations and Steric Repulsions

In this section the effects of fluctuations on an asymptotically flat interface will be discussed. The central concepts are the *wandering exponent*, $\zeta$, and the induced steric repulsive forces between an interface and a wall and between interfaces.

### 2.1. The Wandering Exponent

Consider an interface of finite dimensions, $L \times L \times \cdots = L^{d'} (L \equiv L_{\parallel})$ held (or bound) by, say, a gravitational field with potential $\Delta V(\ell) = \frac{1}{2}\bar{g}\ell^2$. More generally consider a potential well of curvature

$$\bar{g} = (d^2V/d\ell^2)_0 = \bar{g}(T, h, h_1, \ldots) \tag{2.1}$$

at the minimum, approximated by a parabola. Owing to fluctuations, the interface will acquire a (root mean square) thickness or transverse extension, $L_\perp$. One can also identify, in standard ways (see below), correlation lengths, $\xi_{\parallel}$ and $\xi_\perp$, for fluctuations parallel and perpendicular to the interface plane. Then, as $\bar{g} \to 0$ the fluctuations increase and as $L$ and $\xi_{\parallel}$ diverge one finds, rather generally,

$$\frac{\xi_\perp}{b} \approx \left(\frac{\xi_{\parallel}}{a}\right)^{\zeta} \quad \text{and} \quad \frac{L_\perp}{b} \approx \left(\frac{L_{\parallel}}{a}\right)^{\zeta}. \tag{2.2}$$

where $a$ and $b$ are two fixed or slowly varying scale lengths. Here $\zeta(d)$ is the wandering exponent which plays a basic role in determining the effects of the fluctuations.

For *thermal fluctuations* this result has its origins simply in the Boltzmann weight

$$\mathcal{P}[\ell(\bar{y})] = e^{-\beta\Delta\mathcal{H}}/\text{Tr}_\ell[e^{-\beta\Delta\mathcal{H}}], \tag{2.3}$$

with, as discussed in Sec. 1,

$$\Delta\mathcal{H}[\ell(\bar{y})] = \int d^{d'}y \left[\frac{1}{2}\tilde{\Sigma}(\nabla\ell)^2 + \frac{1}{2}\bar{g}\ell^2\right], \tag{2.4}$$

while the trace operation corresponds to a functional integral over the interface configurations, $\ell(y)$. A short wavelength cutoff on momenta, namely,

$$|\vec{q}| \leq \pi/a, \tag{2.5}$$

is always understood, $a$ being a lattice spacing or other appropriate microscopic length in the interfacial plane.

The Gaussian integrals entailed in (2.3) and (2.4) are straightforward when $\ell(\bar{y})$ is decomposed into Fourier components, which are usually called "capillary waves" (although the connection to real, dynamical, capillary waves is *not* very direct). The equilibrium interface pair correlation function is found to be

$$G(\vec{y}) \equiv \langle \ell(\vec{0})\ell(\vec{y}) \rangle = \frac{k_B T}{\tilde{\Sigma}} \frac{1}{L^{d'}} \sum_{\vec{q}} \frac{e^{i\vec{q}\cdot\vec{y}}}{q^2 + \bar{g}/\tilde{\Sigma}}. \qquad (2.6)$$

Consider, first, $\bar{g} = 0$ for a free finite interface and evaluate $L_\perp^2 \equiv \langle \ell^2(\vec{0}) \rangle = G(\vec{0})$. One readily reproduces the second part of (2.2) finding, explicitly, a wandering exponent[7,14]

$$\zeta = \zeta_0(d) \equiv \tfrac{1}{2}(3-d) \quad \text{for } d \le 3, \qquad (2.7)$$

where the subscript zero indicates thermal fluctuations. On the borderline dimension, $d = 3$, the value $\zeta_0 = 0$ corresponds to

$$L_\perp \approx (k_B T/2\pi\tilde{\Sigma})^{1/2}[\ln(L/a)]^{1/2}. \qquad (2.8)$$

Above the borderline, i.e. for $d > 3$, one has $\zeta_0 \equiv 0$ but

$$L_\perp \approx c_d b_0 \quad \text{with } b_0^2 = k_B T/\tilde{\Sigma}a^{d-3}, \qquad (2.9)$$

where $c_d$ is a constant; the interface is then said to be *smooth*; conversely, for $d \le 3$ where $L_\perp$ diverges as $L_\parallel \to \infty$ the interface is *rough*.

The power-law fluctuations of rough interfaces as embodied in the capillary wave description, lead to subtle effects in the thermodynamics of *finite*, $L < \infty$ interfaces, which are of potential significance in both experiments on fluids and in simulations.[21]

## 2.2. *Interfaces in Two-dimensions: Random Walks*

For $d = 2$ the result (2.7) yields $\zeta = \tfrac{1}{2}$. or $L_\perp^2 \sim L_\parallel$. This can be verified by a different route which also serves to check that the neglect of higher order terms, $\ell^3, \ell^4$, etc. in (2.4), does not matter asymptotically. Specifically, one may regard the ($d' = 1$)-dimensional $y$ axis as time-like and consider a random walk along the $z$ axis.[6] The walk trajectory, $z(y)$, represents the interface. The mean-square displacement $(\Delta Z_n)^2 \equiv \langle (z_n - z_0)^2 \rangle$ after $n = y/a = L/a$ steps of the walk corresponds simply to $L_\perp^2$. The standard diffusive properties of random walks dictate $(\Delta z_n)^2 = nb^2$, where $b^2$ is the diffusivity. Evidently we have $\zeta = \tfrac{1}{2}$ and, by the central limit theorem, this is independent of the details of the single-step distribution. The random-walk picture of interfaces in $d = 2$ dimensions proves very fruitful[5–8] and we will allude to it again.

## 2.3. *Correlations and Correlation Lengths*

If, now, we retain $\bar{g} > 0$ but let $L \to \infty$ the fluctuations remain bounded and one finds

$$G(\vec{y}) \approx D \frac{k_B T}{\tilde{\Sigma} \xi_{\parallel}^{d-3}} \frac{e^{-y/\xi_{\parallel}}}{y^{(d-1)/2}}, \tag{2.10}$$

where $D$ is a constant while the longitudinal correlation length is given by

$$\xi_{\parallel} = a_{cap}/\sqrt{2} = (\tilde{\Sigma}/\bar{g})^{1/2}. \tag{2.11}$$

Here $a_{cap}$ is the standard *capillary length*,[1] which for water at $0°C$ is about $4\,\text{mm}$.

The fluctuations "smear" the interface yielding a mean profile which can be expressed in terms of the probability density, $P(z)$, of finding the interface between $z$ and $z + dz$. One obtains

$$P(z) = \exp\left[-\frac{1}{2}(z/\xi_{\perp})^2 / \sqrt{2\pi \xi_{\perp}^2}\right] \tag{2.12}$$

which also serves to define $\xi_{\perp}$ explicitly. The calculations confirm the first part of (2.2), with the same thermal wandering exponent (2.7). On the borderline $d = 3$ the result $\xi_{\perp} \sim b_0 [\ln(\xi_{\parallel}/a)]^{1/2}$ complements (2.8).

Evidently $\xi_{\parallel}$ and $\xi_{\perp}$ diverge as $\bar{g} \to 0$ (for $d < 3$). Thus an infinite free interface is a system at *criticality*. Since local sections of the interface can drift infinitely far from the mean interfacial plane, $G(y)$ becomes ill-defined; the difference correlation function,

$$\Delta G(\vec{y}) = \frac{1}{2}\langle [\ell(\vec{y}) - \ell(\vec{0})]^2 \rangle = \lim_{\bar{g} \to 0} [G(\vec{0}) - G(\vec{y})], \tag{2.13}$$

then serves to characterize the critical fluctuations. All is still determined by the wandering exponent since

$$\Delta G(\vec{y}) \sim b_0^2 (y/a)^{2\zeta}, \tag{2.14}$$

with $\zeta = \zeta_0(d)$; at $d = 3$ one finds again logarithmic behavior, namely,

$$\Delta G(\vec{y}) \approx (k_B T/2\pi\tilde{\Sigma}) \ln(y/a). \tag{2.15}$$

## 2.4. *The Stiffening or Roughening Transition*

The low index crystal faces of a real $(d = 3)$-dimensional crystal, in equilibrium with its vapor (or melt) are found to be *smooth or stiff*, corresponding to $\tilde{\Sigma} = \infty$ at low enough temperatures. At first sight this fact contradicts the logarithmically rough behavior predicted by $\zeta_0(3) = 0$ with (2.8) and (2.15). The resolution of the paradox lies in recognizing that the underlying bulk crystal effectively imposes a *periodic potential* on the interface which is, thus, no longer truly free. Accordingly, let us add to the interface Hamiltonian a term

$$V_{periodic}[\ell(\vec{y})] \propto X_p[\ell] = \cos(p\ell), \tag{2.16}$$

where $p = 2\pi/a_{\perp}$ corresponds to the interlayer periodicity of the crystal normal to the face in question. Does such a perturbation change the wandering behavior?

To answer this question we appeal to general renormalization group and scaling theory for critical phenomena (see e.g. Ref. 22). A perturbation with a local density $X(y)$ is *relevant* at a critical point, and hence changes the critical behavior, if and only if the correlation function

$$G_{XX}(\vec{y}) \sim 1/|\vec{y}|^{2w_x} \quad \text{(in } d' \text{ dimensions)} \qquad (2.17)$$

evaluated at the critical point in question satisfies the criterion

$$\omega_X < d'. \qquad (2.18)$$

If $\omega_X > d'$ the perturbation is *irrelevant*; at $\omega_X = d'$ it is *marginal*.

The Gaussian character of the free interface allows the explicit calculation of $G_{XX}$ with the result

$$G_{XX}(\vec{y}) \approx \frac{1}{2} \exp[-p^2 \Delta G(\vec{y})]. \qquad (2.19)$$

For $d < 3$ one thus has $G_{XX} \sim \exp(-By^{d-3})$ or $\omega_X = \infty$ so that the periodicity is strongly irrelevant: the interface is always rough. For $d = 2$ this is, indeed, confirmed by the random walk picture.

However, when $d = 3$ the logarithmic behavior (2.15) yields a power-law decay as in (2.17). The exponent is

$$\eta(T) \equiv 2\omega_X = p^2 k_B T / 2\pi \tilde{\Sigma}(T). \qquad (2.20)$$

At low temperatures $\eta(T)$ will drop below $2d' = 4$ and the rough interface becomes unstable to the perturbations which then stiffen it. Conversely, as the smooth, stiff, low-temperature interface is heated it undergoes a roughening transition at a temperature $T_R$ which satisfies

$$k_B T_R = 8\pi p^{-2} \tilde{\Sigma}(T_R+) \quad \text{or} \quad \pi b_0^2(T_R+) = 2a_\perp^2. \qquad (2.21)$$

To elucidate the nature of the roughening transition itself one must, of course, treat $V_{periodic}$ beyond the level of an infinitesimal perturbation. A transition of Kosterlitz–Thouless character is predicted: see e.g. Refs. 23 and 24.

## 2.5. Random Media

Impurities and localized imperfections are ubiquitous in real materials; in crystalline systems they are frequently immobile or frozen in random locations. In such cases one has a *quenched random medium*. When considering phase coexistence and interfaces in such media, two sorts of randomness must be distinguished: first, if the impurities favor one phase over the other, one speaks of *random fields*, picturing, say, a distribution of random local magnetic fields, $\delta h_i$, on sites $i$ of an Ising model, with mean $[\delta h_i]_{av} = 0$; near coexistence, different regions of the medium will prefer different bulk phases. On the other hand, if the impurities do *not* distinguish between the phases, which circumstance must normally reflect some underlying symmetry, one has a *random bond* medium; random exchange couplings, $J_{ij}$, in an Ising model encapsulate this situation. Random bonds which are weaker than average tend to pin interfaces even though they are indifferent to the local bulk phases. Interesting exemplars of random field and random bond systems arise in the study of phase transitions in near-monolayer adsorbates on crystalline substrates: see e.g. Ref. 7, especially Fig. 10.

It is intuitively clear that impurities in a random medium may have a profound effect on the behavior of an interface. It appears, however, that the principal changes can be summarized in new values for the wandering exponent.[7] For *random fields* the thermal expression (2.7) is replaced by

$$\zeta(d) = \begin{cases} \dfrac{1}{3}(5-d), & \text{for } 2 \le d \le 5, \\ 0, & \text{for } d \ge 5. \end{cases} \qquad (2.22)$$

This result can be obtained heuristically by balancing the reduction of local bulk free energy gained by taking advantage of random fluctuations, against the cost incurred by the excursions of the interface around the deformed domains.[7,25,26] Such an "Imry-Ma" argument can be trusted here because long-wavelength fluctuations dominate the behavior. However, it may also be checked by renormalization group arguments including $\epsilon$-expansion calculations.[25,26]

Note that (2.22) yields $\zeta = 1$ for $d = 2$, but a single interface for which $L_\perp$ has the same magnitude as $L_\parallel$ must, essentially, fill space. What this really means is that the supposedly coexisting domains are no longer spatially separated on macroscopic length scales. Consequently, the interface is destroyed as, indeed, is the bulk phase transition itself. This inference is consistent with our knowledge of the random-field Ising model for which $d = 2$ is the lower borderline dimension for phase separation and criticality at $T > 0$.[27,28]

The influence of random bonds on interfaces is more subtle;[7,29] the longest length scales alone do not suffice to determine the behavior. Renormalization-group analysis[29] for *random bonds* indicates

$$\zeta(d) = \begin{cases} 0, & \text{for } d \ge 5, \\ c_1(5-d) + O[(5-d)^2], & \text{for } d \le 5, \end{cases} \qquad (2.23)$$

where $c_1 \simeq 0.21$. For $d = 2$ this has been supplemented by Huse, Henley and D.S. Fisher[30] who transformed the diffusion-type equation for the interface in the random walk picture into Burgers equation with additive noise.[7,30] That leads to the exact result

$$\zeta(2) = \frac{2}{3} \quad (\textit{random bonds}). \qquad (2.24)$$

Interpolation suggests $\zeta(3) \simeq 0.4$.

The application of these results for $\zeta(d)$ to wetting transitions, etc. in random media will be discussed, below. (See also Ref. 7 for a more detailed and complete account.)

## 2.6. *Fluctuations at Complete Wetting under Long-range Forces*

Let us return to the description of complete wetting transitions in the presence of intermolecular forces decaying as $1/r^{d+\sigma}$. The mean-field theory in Sec. 1.5 was based on the interface potential

$$V(\ell) \approx V_0/\ell^{\sigma-1} + \bar{h}\ell, \qquad (2.25)$$

from which the wetting layer thickness was predicted to diverge with an exponent $\psi = I/\sigma$ when $\bar{h} \to 0$: see Eq. (1.10). The correlation lengths should also diverge and so we may write

$$\xi_\parallel \sim 1/\bar{h}^{\nu_\parallel} \quad \text{and} \quad \xi_\perp \sim 1/\bar{h}^{\nu_\tau}. \qquad (2.26)$$

To obtain the exponents $\nu_\parallel$ and $\nu_\perp$, let us approximate $V(\ell)$ by a parabola near its minimum and thus use (2.1), (2.11) and (2.2): one finds

$$\nu_\parallel = \frac{1}{2}(1 + \sigma^{-1}) \quad \text{and} \quad \nu_\perp = \zeta\nu_\parallel = \frac{1}{2}\zeta(1 + \sigma^{-1}). \qquad (2.27)$$

For van der Waals forces in $d = 3$ dimensions this yields $\nu_\parallel = \frac{2}{3}$.

What is the validity of this purely harmonic, mean-field calculation which, in particular neglects the presence of a hard wall at $z = \ell = 0$? Following Lipowsky,[31] we may recall Ginzburg's criterion for the validity of mean-field theory at a normal critical point and argue as follows: If the typical transverse displacements of the fluctuating interface, which are measured by $\xi_\perp$, are small compared with the average distance of the interface from the wall, which is, $\ell_W$, the interface should not "feel" the wall. In other words, if one has $\xi_\perp \ll \ell_W$ when $\bar{h} \to 0$, then mean field theory should be valid. This translates into the condition

$$\nu_\perp < \psi \quad \text{or} \quad \zeta < \zeta^* \equiv 2/(\sigma + 1) \qquad (2.28)$$

which specifies the *mean-field regime* for complete wetting.

For standard van der Waals forces we have a $\sigma = 6 - d$. Then, on using the previous results for $\zeta$, we conclude that the mean-field treatment is valid in $d = 3$ dimensions for both thermally fluctuating interfaces or for a random-bond medium. However, any interfaces in $d = 2$ dimensions and interfaces in a three-dimensional random-bond medium lie outside the mean-field regime. A fuller justification of these conclusions calls for a more detailed understanding of the effects of a hard wall on a fluctuating interface: to that topic we now turn.

## 2.7. *Constrained Interfaces and the Wall-interface Potential*

Consider a fluctuating interface constrained to lie between two parallel planes, *alias*, repulsive hard walls, at spacing $2\ell$. For simplicity suppose the two phases separated by the interface are symmetrically related so the average interfacial plane sits centrally. Owing to the constraint, there will be an increase, $\Delta\Sigma(\ell; T, \ldots)$, in the free energy per unit area of the interface (relative to a free interface which is realized when $\ell = \infty$). One can then interpret $W(\ell; T, \ldots) = \frac{1}{2}\Delta\Sigma(\ell; T, \ldots)$ as the potential of an *effective repulsive force* between each wall and the interface.

To estimate the free energy increase and thence $W(\ell)$ for a medium that might be random, note that there is a loss of entropy owing to "collisions" of the interface with the walls and, in a random medium, a loss of energy-reducing excursions of the interface into randomly favored regions cut off by the walls.[7,14] Now a scale area, $A_{coll}(\ell) \equiv L_{coll}^{d'}$, is set by the density of collisions with the walls; but $L_{coll}$ may be estimated by noting that the interface must wander a transverse distance $L_\perp \sim L_{coll}^\zeta$ of typical magnitude $\ell$ between collisions; thus we have

$$L_{coll} \approx a(\ell/b)^{1/\zeta}. \qquad (2.29)$$

From the interface Hamiltonian (2.4) the free energy increase in a collision area may be estimated as

$$\Delta F_{coll} \approx \frac{1}{2}A_{coll}\tilde{\Sigma}(\nabla\ell)^2_{coll} \approx \frac{1}{2}A_{coll}\tilde{\Sigma}(2\ell/L_{coll})^2, \qquad (2.30)$$

since interface displacements of size $2\ell$ on scales longer than $L_{coll}$ are truncated. Finally, for the potential of the fluctuation-induced force we obtain

$$W(\ell) = \frac{1}{2}\Delta\Sigma = \frac{\Delta F_{coll}}{2A_{coll}} \approx \tilde{\Sigma}\left(\frac{\ell}{L_{coll}}\right)^2 \approx \frac{W_0}{\ell^\tau}, \qquad (2.31)$$

where, by (2.29), the exponent of the power-law decay is

$$\tau = 2(1-\zeta)/\zeta, \qquad (2.32)$$

while $W_0 \propto \tilde{\Sigma}b^{2/\zeta}/a^2$.

For a *thermally fluctuating interface* in a pure medium one has $b \equiv b_0$ so that from (2.9) and (2.7) one finds $W_0 \propto k_BTb_0^\tau/a^{d'}$. When $d = 2$ the wall-interface potential then reduces to the inverse square law

$$W(\ell) \approx \frac{1}{2}\frac{k_BT}{a}\left(\frac{b_0}{\ell}\right)^2. \qquad (2.33)$$

For a thermal interface in $d = 3$ dimensions the result, $\zeta_0 = O(\log)$ leads to a gaussian contribution, $W_0\ell^2\exp(-2\ell^2/b^2)$ to $W(\ell)$;[14] however, at long distances this is always dominated by bulk-correlation-induced terms decaying as $\exp(-\ell/\xi_\beta)$ which reflect the mean-field result (1.13). (Such terms may even be oscillatory when the bulk phases have an ordered structure: see, e.g. Ref. 32.)

## 2.8. Membranes

It is worth commenting that if one models a rigid but tensionless *membrane*[33] by replacing the $\frac{1}{2}\tilde{\Sigma}(\nabla\ell)^2$ term in the interface Hamiltonian (2.4) by $\frac{1}{2}\kappa(\nabla^2\ell)^2$, the parallel arguments go through. First one finds a thermal wandering exponent

$$\zeta = \begin{cases} \frac{1}{2}(5-d), & \text{for } 3 \le d \le 5, \\ 0, & \text{for } d \ge 5, \end{cases} \tag{2.34}$$

which results from $q^4$ replacing $q^2$ in the expression (2.6). Then, with a corresponding change in (2.30), one obtains the decay exponent

$$\tau = 2(2-\zeta)/\zeta, \tag{2.35}$$

for the sterically-induced repulsive potential, $W(\ell)$. For $d = 3$ dimensions an inverse square law, $\tau = 2$, is predicted. Consequences of this force, first proposed for membranes by Helfrich, are discussed in the lectures by Leibler.[33]

## 2.9. *Checks of the Wall-interface Potential*

The sketchy, heuristic derivation of the wall-interface decay law (2.32), which we have presented may not inspire great confidence in its validity! Nevertheless, a variety of more detailed and analytically complete checks confirm the results. For the case of a pure thermal interface in $d = 2$ dimensions, the random walk picture proves most efficacious. Thus the problem of a random walk on the half-line, $z > 0$, corresponding to a hard wall at the origin excluding the region $z \le 0$, is readily solved by the method of images.[6] The most probable position of the walker, $z_0 \equiv \ell_0$, increases as $bn^{1/2}$ while the corresponding free energy per step varies as $\frac{1}{2}\ln(1+n^{-1})$. This can be translated directly into the repulsive potential $W(\ell)$ and confirms (2.33) precisely. Full details are presented in Refs. 6 and 7 where other checks are also described.

Generally, the form of (2.33) is always recaptured but the numerical factor becomes slightly modified in different circumstances. Thus for the case of two confining walls considered in Sec. 2.7, a correction factor of $\pi^2/8 \simeq 1.234$ is found in an exact analytical treatment.[6,14]

More surprising, perhaps, are the checks that can be made for *random media* in $d = 2$ dimensions following notable exact calculations by Kardar and Nelson[34,35] using the method of replicas to treat critical wetting and the commensurate — incommensurate transition. By accepting the expression (2.32) with $\zeta = \frac{2}{3}$ for random bonds, so that $\tau = 1$, one can interpret the exact results found[34,35] as simply arising from a balance of effective forces.[7] We close this chapter with two applications illustrating the power of the approach.[31]

## 2.10. *Complete Wetting Revisited*

To analyze complete wetting in the presence of only a hard wall we may now merely replace the long-range term, $V_0/\ell^{\sigma-1}$, in (2.25) by the effective wall-interface potential $W_0/\ell^\tau$ and minimize the total effective potential. Of course, this amounts just to replacing $\sigma$ by $\tau + 1$. Thus the wetting layer thickness, $\ell_W$, diverges with an exponent

$$\psi = 1/(\tau + 1) = \zeta/(2 - \zeta), \qquad (2.36)$$

while, for the correlation lengths one obtains

$$\nu_{\parallel} = \frac{1}{2} + \frac{1}{2}(\tau + 1)^{-1} = 1/(2 - \zeta) \quad \text{and} \quad \nu_{\perp} = \zeta \nu_{\parallel}. \qquad (2.37)$$

Note that the last result implies $\nu_{\perp} = \psi$ which confirms the underlying physical picture, namely, that it is the repeated collisions of the interface with the wall due to the fluctuations that keep the layer thickness, $\ell_W$, of same magnitude as $\xi_{\perp}$.

For $d = 2$ one finds $\psi = \nu_{\perp} = \frac{1}{3}$ which result may be cross-checked in the random walk picture by solving the corresponding Schrödinger-like diffusion equation[6,7] in an external potential $U(\ell) = \bar{h}\ell$ with the aid of Airy functions.

For complete wetting in a *random bond* medium, the result (2.24) for $\zeta$ together with (2.36) and (2.37) yield the predictions[31] $\psi = \nu_{\perp} = \frac{1}{2}$ and $\nu_{\parallel} = \frac{3}{4}$. These values cannot, as yet, be checked by experiment! However, following the lead of Huse and Henley[36] for a free interface in a random medium, effective Monte Carlo calculations can be performed.[37] It proves possible to explore three or more decades in $\bar{h}$: the data[37] indicate $\psi = \nu_{\perp} = 0.500 \pm 0.008$ in excellent agreement with the predictions. One can also test the underlying relation $\xi_{\perp} \approx \xi_{\parallel}^{\zeta}$ which yields[37] $\zeta = 0.65 \pm 0.02$ again confirming the theoretical expectation $\zeta = \frac{2}{3}$.

Finally, suppose the direct long-range wall interactions are reintroduced. To elucidate the interplay of the direct interactions and the hard wall we may add the term $V_0/\ell^{\sigma-1}$ to the total effective potential. It is clear, on minimization, that when $\ell \to \infty$ the long-range term dominates if $\sigma - 1 < \tau$ but is otherwise irrelevant. In the latter case the fluctuations determine all the wetting exponents. But the inequality $\sigma - 1 < \tau$ is not new: indeed, it is precisely equivalent to the condition $\zeta < \zeta^*$ in (2.28), which delineates the mean-field regime for complete wetting! This serves to confirm the Ginzburg-inspired interpretation. Various regimes for *critical* wetting can be analyzed along similar lines.[7]

## 2.11. Many Walls and the Shape of a Vicinal Crystal Face

To close this section we pose and solve a problem entailing the *shape* of a real interface separating a three-dimensional crystal from its vapor or liquid melt. On a crystal of finite total volume, $V_{tot}$, an interface below its particular roughening transition, say $T_{R1}$, forms a smooth, flat, *facet* of finite area. At low enough temperatures adjacent crystalline faces may also be smooth and flat: however, if such a *vicinal face* is *above* its own roughening temperature it will, on macroscopic scales, appear *rounded*.[38,39] How does the rough, rounded vicinal face join onto the smooth flat facet?

To answer this question, orient the crystal so that the flat facet lies in the $(x, y)$ plane with the positive $z$ axis pointing into the crystal. We may suppose that the vicinal face is cylindrical with the edge abutting the flat facet parallel to the $y$ axis and located at $x = x_0, z = 0$. the profile $z(x)$ for fixed $y$ and $x \geq x_0$, then describes the desired shape of the vicinal face. We may anticipate a power law,

$$z(x) \approx C\Delta x^{\Lambda} \quad \text{for } \Delta x = x - x_0 \geq 0, \tag{2.38}$$

but then need to find the exponent $\Lambda$.

To proceed one must recognize that a rough crystal face close to a flat facet can be regarded as composed of a series of roughly parallel atomic-sized *steps* separating flat *terraces* of structure matching the facet in question: see e.g., Refs. 23 and 39. Now successive terraces of increasing height, $a, 2a, \ldots$, above the $(x, y)$ plane can be regarded as distinct $(d = 2)$-dimensional phases; the lines of steps separating them then correspond to $(d' = 1)$-dimensional interfaces. Thus we have a two-dimensional system of dimensions, say, $L_x \times L_y$, consisting of many asymptotically parallel but thermally fluctuating interfaces. If the mean spacing of the steps/interfaces is $\bar{\ell}$, the slope of the corresponding vicinal surface is just

$$q \equiv (dz/dx) = a/\bar{\ell}. \tag{2.39}$$

Since $q$ varies microscopically along the profile we need to know how the tension, $\Sigma$, of the vicinal face depends on $q$. Now this can be answered by invoking the wall-interface potential, $W(\ell)$, since, up to a factor, say $c_0$, it should apply *equally* as an effective *interface–interface repulsive potential.*[6,7] Thus the total free energy for a system of $N = L_x/\ell$ steps/interfaces of length $L_y$ can be estimated by minimizing

$$\mathcal{F}\ell \approx (L_x/\ell)L_y[-|\sigma| + c_0 W_0/\ell^{\tau}], \tag{2.40}$$

where $\sigma(T) < 0$ is the effective *line tension* of an isolated step/interface. One obtains $\ell \approx [(\tau + 1)c_0 W_0/|\sigma|]^{1/\tau}$ and thence finds

$$\Delta\Sigma(q) \equiv \Sigma(q) - \Sigma(0) \approx \sigma_0 q + Bq^{\tau+1}, \tag{2.41}$$

where $B$ is a constant and $\sigma_0$ is the true step tension.

As a last step the overall shape of the crystal surface must be determined by minimizing the total surface free energy: this contains a vicinal surface contribution

$$\Delta\mathcal{F}[z(x)] = L_y \int \Delta\Sigma(q)ds = L_y \int \Delta\Sigma(q)(1 + q^2)^{1/2}dx, \tag{2.42}$$

where $ds$ is an element of arc length along the profile $z(x)$. The minimization is subject to the constraint of constant total volume which, for our present purposes, may be expressed as

$$V_{tot} = V_R - L_y \int z(x)dx = \text{const}, \tag{2.43}$$

where $V_R$ is a constant reference volume. If one neglects the factor $(1 + q^2)^{1/2}$ in (2.42), which is easily justified *post facto,* the variational equation determining

$z(x)$ is

$$\Delta\Sigma''(dz/dx)(d^2z/dx^2) = \lambda, \tag{2.44}$$

where $\Delta\Sigma'' = d^2\Delta\Sigma/dq^2$ and $\lambda$ is a Lagrange multiplier. On using our result (2.41) this finally yields the profile (2.38) with exponent

$$\Lambda = 1 + (1/\tau) = (2 - \zeta)/2(1 - \zeta). \tag{2.45}$$

In summary, the shape of a vicinal crystal interface as it approaches a flat facet is determined by the wandering exponent, $\zeta$, for the one-dimensional lines of atomic steps forming the macroscopically rounded surface. For thermal fluctuations we have $\zeta = \frac{1}{2}$ and hence predict $\Lambda = \frac{3}{2}$.[38,39] By contrast, the corresponding mean-field prediction is $\Lambda = 2$. Experimental tests are possible! Carmi, Lipson, and Polturak[40] have observed crystals of helium-four in equilibrium with their superfluid melt. In the asymptotic regime near a dominant facet, $z$ varied over about 1 mm while $x$ changed by 8 to 10 mm. A careful analysis, which entails an unbiased determination of $x_0$, yields[40] $\Lambda = 1.55 \pm 0.06$. The agreement with theory is most gratifying!

As a postscript we mention that the problem of many fluctuating parallel interfaces is also relevant to the description of commensurate-incommensurate transitions.[6,7] The exact solution of Pokrovsky and Talpov[41] for $d = 2$ dimensions, which utilized a representation in terms of fermions moving on a $(d' = 1)$-dimensional line, serves to confirm our use of $W(\ell) \sim 1/\ell^\tau$ with $\tau$ given by (2.32), for the interface–interface potential in the many-interface situation. From a quite different starting perspective, Kasteleyn's exact solution[42] of the hard dimer problem on a brick (or honeycomb) lattice provides an independent verification. Further details are given in Refs. 6 and 7.

One must notice, however, that in bulk three-dimensional systems arrays of statistically parallel interfaces, as typically characterize commensurate, spatially modulated phases, also fluctuate but much more weakly so that only exponentially decaying interactions are found — as in (1.13). But, as mentioned after (2.33), these interactions can readily gain oscillatory factors. Furthermore, there arise three-body (and higher order) potentials, $W_3(l, l')$, etc., where $l$ and $l'$ are the spacings between successive interfaces, that may be calculated for simple lattice models: these may, indeed, play a significant role in determining cascades of ordered phases: see the general theory developed in Ref. 32.

## 3. Critical Wetting and Renormalization Groups for Interfaces

In a *critical wetting situation*, as explained within mean field theory in Sec. 1.7, an interface is bound to a wall by direct interactions even when the ordering field $h$ vanishes, i.e., at coexistence. The interface then unbinds as $T$ rises towards $T = T_{cW}$ via a divergence of the layer thickness, $\ell_W(T)$. Of course, one anticipates that the mean-field prediction (1.16) for short-range forces will be modified by fluctuations. It is clear, however, that the long-range effective wall-interface interaction, $W(\ell)$, found in the last section, does not suffice for treating this problem since it presupposes only isolated repulsive collisions with the wall rather than entropically favored escape from sojourns in a potential well with a minimum close to the wall. The treatment of such problems of the *strong fluctuation regime*[7,11] is the aim of this section.

### 3.1. *Critical Wetting in d = 2 Dimensions*

A most important result for $d = 2$ dimensions is Abraham's[43] exact solution of critical wetting in the full, nearest-neighbor, planar Ising model. Both interfacial and bulk fluctuations are taken into account and many features can be precisely elucidated (although an external bulk field cannot be handled analytically). Abraham's work serves to check a variety of heuristic calculations based only on an effective Hamiltonian for the interface. The random walk approach (Sec. 2.2) leads to a picture of the bound interface as a "necklace" of "beads", representing bubbles of phase $\beta$ bulging out from the wall, linked by "string" corresponding to segments of interface bound tightly to the wall. As explained in Ref. 6, all Abraham's qualitative and universal results can be derived from that starting point which then provides a basis for treating more elaborate and analytically less tractable problems.[6,44]

Another approach[23,45–47] is based on the recursion relation for $Q_n(\ell)$, the partition function for an interface or random walk of length $y = na$ which starts at the origin $z = 0$ and wanders in an external potential, $U(\ell) = V_E(\ell)/k_BT$, up to $z = \ell$. In an appropriate continuum limit this reduces to the Schrödinger-like diffusion equation alluded to previously.[6] After the initial transients have decayed the equation assumes the pure Schrödinger form

$$\frac{d^2Q}{dz^2} + \frac{2}{b^2}[E - U(z)]Q = 0. \tag{3.1}$$

The ground state "wavefunction", $Q_0(z)$, describes the equilibrium distribution of the interface displacement, $\ell \cong z$. [Compare with (2.12).] The ground state energy, $E_0(b; U)$, determines the fluctuation-induced wall contribution to the interfacial tension via[6,44–47]

$$\Delta\Sigma = k_BT E_0. \tag{3.2}$$

To analyze critical wetting it suffices to represent the short-range wall couplings by a square-well potential, $U(z)$, of depth, say, $\epsilon = \epsilon(T)$ and fixed width $c$. Standard quantum mechanics then yields a bound state, corresponding to a bound interface, for $\epsilon < \epsilon_c = \pi^2 b^2/8c^2$. As $(\epsilon_c - \epsilon) \sim (T_{cW} - T) \to 0$ one finds

$$\Delta\Sigma(T) \sim (T_{cW} - T)^{2-\alpha}, \quad \text{with } \alpha = 0(\text{discon.}). \tag{3.3}$$

At the same time the wetting layer thickness diverges as

$$\ell_W(T) \sim (T_{cW} - T)^{-\psi} \quad \text{with } \psi = \nu_\perp = 1. \tag{3.4}$$

Correspondingly one finds $\nu_\parallel = \nu_\perp/\zeta = 2$ and thence sees that hyperscaling is valid since $d'\nu_\parallel = 2 - \alpha$.

For $\epsilon > \epsilon_c$ or $T > T_{cW}$ there is no bound state. This corresponds to delocalization of the interface and complete wetting of the wall. The energy sticks at $E_0 = 0$ so $\Delta\Sigma = 0$. The specific heat thus undergoes a simple jump discontinuity at $T_{cW}$. As mentioned, all these predictions are confirmed by Abraham's (earlier!) exact Ising model results. However, one can also use (3.1) to study the effects of direct wall forces decaying less rapidly with $z$ or $\ell$. A variety of interesting modifications of the critical wetting transition occur especially in the marginal case[48,8] where $U(\ell)$ decays as $1/\ell^2$.

## 3.2. *Renormalization Groups for Critical Wetting*

Renormalization group theory currently provides the only analytically based route to progress in the strong fluctuation regime for $d \neq 2$. The aim of the remainder of this section is to outline the approach initiated by Brézin, Halperin and Leibler[9] for $d = 3$ as systematized and extended by D.S. Fisher and Huse,[10] and developed in approximate numerical form by Lipowsky and the author[11] for general $d$.

The starting point is the interface Hamiltonian

$$\mathcal{H}[\ell(\vec{y})] = \int d^{d'}y \left[ \frac{1}{2}\bar{\Sigma}(\nabla\ell)^2 + V(\ell) \right], \tag{3.5}$$

where, following the discussion in Sec. 1.6, the potential to be studied is

$$V(\ell) = -w_A e^{-\kappa\ell}\Theta(\ell) + w_R e^{-2\kappa\ell}\Theta(\ell) + w_0\Theta(-\ell), \tag{3.6}$$

where $\Theta(\ell) = 1$ for $\ell > 0$ but vanishes otherwise, while the inverse correlation length of the bulk $\beta$ phase is

$$\kappa \equiv 1/\xi_\beta. \tag{3.7}$$

The transition is controlled by the initially positive attractive amplitude $w_A$ which we may suppose decreases smoothly as $T$ increases, while the repulsive amplitude $w_R$ remains positive. The wall medium occupying $z, \ell < 0$ is represented by $w_0 > 0$. More generally one might envisage a $W_0(\ell)$ rising rapidly to large values for $\ell < 0$; ideally, one might also wish to take $w_0 \to \infty$.

The form (3.5) may be constrasted with the usual LGW effective Hamiltonian for bulk critical phenomena in which $\ell$ is replaced by the order parameter $\phi$, while $V(\phi)$ takes the form $r\phi^2 + u\phi^4 + \cdots$. Under renormalization, based on integrating out the highest momentum components of $\phi(y)$, the parameters $r, u, \ldots$ change. More generally we may contemplate a full *functional renormalization* transformation, as originally described by Wilson:[49] then one obtains a renormalized potential, $V'(\ell)$, associated with the usual rescaling of the spatial coordinates

$$\vec{y} \Rightarrow \vec{y}' = \vec{y}/b \tag{3.8}$$

with $b \geq 1$. (Of course, in principle one must also generate further nonlocal terms like $(\nabla^2\ell)^2$, $V_2(\ell)(\nabla\ell)^2$, etc;[58b,c,59] but see further below.) In addition to the spatial scaling one must rescale the field variable, in this case $\ell$, via

$$\ell \Rightarrow \ell' = \ell/b^\zeta. \tag{3.9}$$

As usual,[22(b)] $\zeta$ must be adjusted to yield an appropriate nontrivial fixed point. Since critical wetting occurs at $\ell \to \infty$ with $V \to 0$, i.e. in the limit the interface becomes free, one finds that

$$\zeta = \tfrac{1}{2}(3 - d), \tag{3.10}$$

applies for all $d$. Again it is useful to go to the continuum limit by writing

$$b = e^\tau \approx 1 + \tau, \tag{3.11}$$

where $\tau$ is now the continuous renormalization group flow parameter.

## 3.3. The Linearized Functional Renormalization Group

On following the procedures indicated and integrating out the highest momentum components of $\ell(\bar{y})$, one finds the flow equation for the renormalized potential, $V_\tau(\ell)$, to be[10]

$$\frac{\partial V_\tau}{\partial \tau}(\ell) = (d - 1)V + \zeta\ell\frac{\partial V}{\partial \ell} + \frac{1}{\sigma}\frac{\partial^2 V}{\partial \ell^2}, \tag{3.12}$$

up to corrections of quadratic order in the potential $V$. The basic scale factor, $\sigma$, and an associated numerical parameter, $\omega$, which turns out to be crucial to the critical behavior for $d = 3$, are defined via

$$\frac{1}{\sigma} \equiv \omega\xi_\beta^2 \equiv \frac{1}{2}a_\perp^2 = c_d b_0^2(T) = c_d\frac{k_B T}{\tilde{\Sigma}a^{d-3}}, \tag{3.13}$$

where $a$ enters through the lattice cutoff, $|q| < \pi/a$, while $c_d$ is a numerical constant with $c_3 = 1/4\pi$. Thus one has

$$\omega = k_B T/4\pi\tilde{\Sigma}\xi_\beta^2 \quad \text{for } d = 3, \tag{3.14}$$

which is *independent* of the cutoff.

The flow equation (3.13) can be integrated exactly to yield[10]

$$V_\tau(\ell) = \frac{e^{(d-1)\tau}}{\sqrt{(2\pi)\delta_\tau}} \int d\bar{\ell} \exp\left[\frac{(e^{\zeta\tau\ell} - \bar{\ell})^2}{2\delta_\tau^2}\right] V_0(\bar{\ell}), \tag{3.15}$$

in which

$$\delta_\tau^2 = \begin{cases} a_\perp^2\tau, & \text{for } d = 3, \\ a_\perp^2\frac{e^{(3-d)\tau} - 1}{3 - d}, & \text{for } d \neq 3. \end{cases} \tag{3.16}$$

One sees that under renormalization the original potential, $V_0(\ell)$ given by (3.6), is both translated to larger values of $\ell$ and smeared over a growing region (for $d \leq 3$). This conclusion of the linearized equation should be valid at least in the regime where $V_\tau \to 0$.

One can use (3.15) in the way originally prescribed by Wilson[10,22(b),49]: i.e., one integrates up to a matching point $\tau = \tau^\dagger(w_A)$ at which

$$(\partial^2 V_{\tau\dagger}/\partial \ell^2)_{min} = \tilde{\Sigma}/\xi_\beta^2, \tag{3.17}$$

so that the renormalized correlation length, $\xi_{\parallel\tau}$, matches the noncritical bulk correlation length $\xi_\beta$. Then the original longitudinal correlation length of the interface is given by

$$\xi_\parallel(T) \approx e^{\tau^\dagger}\xi_\beta. \tag{3.18}$$

### 3.4. Critical Wetting in $d = 3$ Dimensions

Following Ref. 10 let us now exploit the linearity of (3.12) and (3.15) to follow, for $d = 3$, the renormalization of each of the three terms in (3.6), on the natural scale set by $x = \ell/\xi_\beta$. By using a steepest descents estimation for $\tau \to \infty$, the wall term is found always to have a gaussian tail of the form $(w_0/X)\exp(-x^2/4w\tau)$ where $X = x/2w\tau$. On the other hand, the second, repulsive term has a gaussian tail, with a prefactor $w_R/(2 - X)$, only for $x < 4w\tau$; the original exponential tail is found for $x > 4w\tau$ with prefactor $w_R e^{4w\tau}$; a similar changeover occurs in the attractive term but when $x \approx 2w\tau$. Note how the parameter $w$, defined in (3.14), enters in a crucial manner. As a consequence of these changes in asymptotic behavior *three distinct regimes* of critical wetting emerge:

*Regime I* is defined by $0 \le w < \frac{1}{2}$. The potential $V_\tau$ retains the original competition between two exponential tails and one finds a correlation length exponent

$$\nu_\parallel = 1/(1 - w), \tag{3.19}$$

which evidently varies continuously with $w$!

*Regime II* is specified by $\frac{1}{2} < w < 2$. The renormalized potential in the region of its minimum is now a sum of two gaussian pieces and only one exponential term. The correlation length diverges as

$$\xi_\parallel \sim 1/[t(\ln t^{-1})^{(w/8)^{1/2}}]^{\nu_\parallel}, \tag{3.20}$$

with $t \sim w_A - w_A^c$ while the exponent becomes

$$\nu_\parallel = 1/[2 + w - (8w)^{1/2}], \tag{3.21}$$

which actually joins smoothly on to the form (3.19). In both regimes I and II the critical point occurs at the trivial mean-field value $w_A^c = 0$ and the wetting layer thickness, $\ell_w$, diverges in mean-field fashion as $\ln t^{-1}$. Finally,

*Regime III* results for $w > 2$. All terms have Gaussian tails near the minimum of the total renormalized potential. The critical point occurs at the nontrivial value, $w_A^c \sim (w - 2)$, and one finds $\nu_\parallel = \infty$ corresponding to the strong divergence

$$\xi_\parallel \sim \exp\{(ct)^{-1}[\ln(ct)^{-1} + \ln \, \ln(ct)^{-1} + \cdots]\}, \tag{3.22}$$

where $c$ is a constant. Correspondingly, the wetting layer now diverges as $\ln(ct)^{-1}/t$ which is much stronger than mean-field theory indicates.

These three regimes, uncovered by analysis of the exact linearized functional renormalization group,[10] (3.12) were first proposed by Brézin, Halperin and Leibler[9] on the basis of somewhat *ad hoc* arguments. The systematic treatment of Fisher and Huse[10] also allows an analysis of the crossovers between the different regimes occurring at $\omega = \frac{1}{2}$ and $\omega = 2$. (See Ref. 10 for details but note Ref. 58.)

### 3.5. Test of the $d = 3$ Critical Wetting Predictions

The striking prediction[9,10] of a continuous variation of $\nu_{\parallel}$ with the parameter $\omega$ across three distinct regimes invites careful testing. Unfortunately, in wetting of walls by real fluids the effects of long-range van der Waals forces are hard to escape. Interfaces between magnetic or other symmetrically related domains in a crystalline solid might serve to eliminate long-range interactions but no appropriate experimental systems seem to have been identified as yet.

On the other hand, Binder, Landau and Kroll[50] have simulated a wall and an interface in a three-dimensional Ising model at two fixed temperatures below $T_c$. In order to observe a transition they varied the reduced surface field, $h_1 \equiv H_1/k_B T$ (see Sec. 1.6).

According to the mean-field phase diagram,[18] this should carry one through a point, $h_{1c}(T)$, of critical wetting. For the transitions in question, Binder *et al.* estimated $\omega > \frac{1}{2}$ so that the predictions of Regime II should be applicable. Since it was not possible to observe the correlation length, $\xi_{\parallel}$, the local surface magnetization, $m_1(h)$, was studied as the bulk field $h \equiv H/k_B T$ was reduced to 0. The behavior is described by a surface critical wetting exponent, $\beta_1$, which, in Regime II should, neglecting nonlinearities, take the value[51] $(2\omega)^{1/2} - \frac{1}{2}\omega > \frac{3}{4}$.

By contrast, the simulations[50] found $\beta_1 \simeq 0.5$. This actually agrees with the mean-field value[18] $\beta_1 = \frac{1}{2}$; indeed, no deviations from the mean-field predictions could be seen! An obvious concern is the necessary finite size of the simulations: as $h \to 0$ one predicts $\ell_w \approx \xi_\beta \ln |h|^{-1}$ but the excursions of the simulated interface are restricted by the finite extent of the lattice normal to the wall. On the other hand, Brézin and Halpin-Healy[52] developed a Ginzburg criterion and argued that the true critical region was not accessed in the simulations; however, they did not compute a crossover scaling function which is really needed to establish the point. By contrast, later simulations of a *pure interface model* by Gompper and Kroll[53] did exhibit strongly nonclassical exponents depending on $\omega$.

Lastly, of course, one must recognize that the predictions described in the last section rest on a linearization of the exact functional renormalization group. Fisher and Huse[10] address this issue and argue that the nonlinearities are almost certainly irrelevant in Regime I, so that nonclassical effects should be observable. However, this aspect of the theory requires further analysis and, in particular, a truly "hard" wall, with $w_0 \to \infty$ in (3.6) can, clearly, not be treated within a linearization in $V(\ell)$.

### 3.6. *Approximate Nonlinear Renormalization Group*

The remarks just made motivate the development of a renormalization group scheme in which nonlinear effects are explicitly incorporated: some form of approximation, however, must be accepted. A good model is provided by Wilson's original approximate renormalization group transformation for ordinary bulk critical points, which proved to be exact to first order in the $\epsilon = (4 - d)$ expansion.[49]

On following Wilson's arguments using "wave packets in phase space" but with a special eye to interface problems, Lipowsky was led to the transformation[11]

$$V'(\ell) = -\tilde{v}b^{d-1}\ln\left\{\int_{-\infty}^{-\infty}\frac{d\ell'}{\sqrt{(2\pi)\tilde{a}}}\exp\left(-\frac{\ell'^2}{2\tilde{a}^2} - K[\ell,\ell';b;V]\right)\right\} \qquad (3.23)$$

where, as before, $b > 1$ is the spatial rescaling factor and

$$K[\ell,\ell';b;V] = \frac{1}{2\tilde{v}}[V(b^\varsigma\ell - \ell') + V(b^\varsigma\ell + \ell')], \qquad (3.24)$$

in which $\varsigma$ is still given by (3.10), while

$$\tilde{v}(b) = k_BT\int^>d^{d'}p/(2\pi)^{d'} \qquad (3.25)$$

measures the volume of the momentum shell specified by $q_\Lambda \geq |\vec{p}| > q_\Lambda/b$ where $q_\Lambda = \pi/a$ is the standard cutoff. The normalization of the integral in (3.23) preserves the condition $V(\ell) \to 0$ as $\ell \to \infty$, which is clearly necessary in treating interfacial wetting problems. Lastly, the scale length $\tilde{a}$, which was treated as arbitrary by Wilson, is defined here by

$$\tilde{a}^2(b) = (k_BT/\tilde{\Sigma})\int^>d^{d'}p/p^2(2\pi)^{d'} = \delta_{\tau_0}^2, \qquad (3.26)$$

where $b = \exp(\tau_0)$ and $\delta_\tau$ was given in (3.16). This assignment of $\tilde{a}$ ensures that the transformation (3.23) is *exact* to *linear order* in $V$ for *all* $b$ and $d$.

A weakness of Wilson's original approximate transformation is that the critical point decay exponent,[22] $\eta$, is forced to vanish. For interface delocalization problems, however, that is of no concern since $\eta = 0$ corresponds to the correct wandering exponent, $\varsigma$, for the free, critical interface and hence should characterize the fixed points and flows of interest.

One may take the continuum limit of (3.23) as before. This leads to

$$\frac{\partial V}{\partial \tau} = (d-1)V + \varsigma\ell\frac{\partial V}{\partial \ell} + \frac{1}{2}v_0\ln\left[1 + \frac{2}{\sigma v_0}\frac{\partial^2 V}{\partial \ell^2}\right], \qquad (3.27)$$

in which $v_0 = \lim_{b\to 1}\tilde{v}(b)d'/(1 - b^{-d'})$. It is obvious that this equation reduces to the exact form (3.12) on linearization. Essentially this same equation has been derived for bulk critical phenomena by Hasenfratz and Hasenfratz.[54] It has been explored somewhat by them in that context but has not, so far, been much studied for interface problems.

The balance of this chapter will be devoted to a report on a numerical study of the discrete approximation $(b > 1)$ for interfaces.[11] Since the transformation is nonlinear there is no problem in allowing $V$ to approach $+\infty$. Accordingly, the starting potential employed is (3.6) with $w_0 = \infty$. We remark first, however, that one can obtain[11] certain analytic bounds on the behavior of the transformation (3.23) that serve to show that the effects of the nonlinearities are comparatively weak.

## 3.7. Numerical Studies

To understand the critical wetting behavior predicted by the approximate nonlinear renormalization group transformation (3.23)–(3.24) one varies the parameters of the initial potential, specifically $w_A$ and $\omega$ at, say, fixed $w_R$ [see (3.6) and (3.14)], and studies the behavior of the renormalized potential, $V^{(N)}(\ell)$, after $N$ iterations as $N$ becomes large. For fixed $d < 3$ one discovers a critical locus $w_A^c(\omega)$ in the $(w_A, \omega)$ plane and *three* types of asymptotic behaviour.[11] (i) For $w_A > w_A^c$ the initial minimum of $V^{(N)}(\ell)$ becomes deeper and deeper: this clearly corresponds to the bound, partially wet phase in which $\ell_W$ is finite and small. (ii) For $w_A = w_A^c$ the potential eventually approaches a definite *fixed point potential*, $V_c^*(\ell)$, with an attractive well, which is seen to govern the critical wetting transition at which $\ell_W \to \infty$. Lastly, (iii) for $w_A < w_A^c$ the original attractive well shrinks and disappears and $V^{(N)}(\ell)$ approaches a purely repulsive fixed point potential, $V_0^*(\ell) > 0$, which controls the completely unbound phase with $\ell_W = \infty$. (Graphical plots of $V_c^*(\ell)$ and $V_0^*(\ell)$ are presented in Refs. 11.)

One can examine eigenperturbations about the critical fixed point which are identified by their behavior

$$[V^*(\ell) + \theta E_j(\ell)]^{(N)} = V^*(\ell) + \theta b^{N\lambda_j} E_j(\ell) + O(\theta^2). \qquad (3.28)$$

There is only one relevant, i.e. positive, eigenvalue, $\lambda_1$, which as usual, yields the correlation length exponent via $\nu_\| = 1/\lambda_1$. In addition, there is always a redundant[22(b)] eigenvalue $\lambda_2 = -\zeta$ which becomes marginal as $d \to 3$.

For $d = 2$ the numerical analysis for $b = 2$ and $b = 4$ yields

$$\nu_\| = 2.04 \pm 0.05 \qquad (3.29)$$

This compares very favorably with the exact value $\nu_\| = 2$ found in Sec. 3.1. Despite the approximations inherent in (3.23) the transformation seems to be a surprisingly accurate representation of the corresponding exact renormalization groups.

## 3.8. Approach to $d = 3$: Anomalous Bifurcation

In the usual analysis of bulk critical phenomena, in which one takes $V(\phi) = r\phi^2 + u\phi^4$, the behavior above the borderline dimension, $d^> = 4$, is controlled by a simple Gaussian $(u = 0)$ fixed point; classical exponent values such as $\nu = \frac{1}{2}$ prevail. When $d$ falls below $d^>$ a bifurcation occurs: a new, nontrivial fixed point, with $r^*$ and $u^*$

of order $\epsilon = d^> - d$, splits off smoothly from the gaussian fixed point. By the same token, the critical exponents depart continuously from their classical values: thus, for an Ising-like system, one has $\nu = \frac{1}{2} + \frac{1}{12}\epsilon + O(\epsilon^2)$.

By contrast, numerical study of the approximate renormalization group for interfaces with $d$ near $d^> = 3$ reveals quite different behavior. Above $d^>$ simple mean-field behavior still applies with, e.g. $\nu_{||} = 1$, but below $d^>$ matters change. In fact, the numerical data suggest[11]

$$\nu_{||} \approx \tfrac{1}{2}[\ln(3/\epsilon) + c_1\epsilon]^{1/2}/\epsilon^{1/2} \quad \text{with } \epsilon = 3 - d, \qquad (3.30)$$

where $c_1 \simeq 3.65$. In other words, the exponent $\nu_{||}(d)$ diverges to $\infty$ as $d \to 3-$. This striking behavior is quite unexpected and, as yet, not understood analytically. It is, however, consistent with the Brézin *et al.* and Fisher-Huse predictions of Sec. 3.4 that $\nu_{||}$ can take *any* value in the range $(1, \infty)$ when $d = 3$, depending on the parameter $\omega$ (which is irrelevant for $d \neq 3$).

The observed behavior of the fixed point potentials, $V_c^*(\ell)$ and $V_0^*(\ell)$, is also unexpected. When $d \to 3$, the location of the minimum in $V_c^*(\ell)$, moves outwards as $1/\epsilon$ while its depth decreases like $\epsilon^3/[\ln(B/\epsilon)]^{1/2}$. At the same time $V_c^*(\ell)$ and $V_0^*(\ell)$, the critical and unbound fixed-point potentials, become increasingly close together. In the limit $d = 3$ they appear to merge into a single potential, $V^\dagger(\ell)$, which, although fixed in form, *drifts* steadily along the $\ell$ axis under renormalization! This behavior does not obviously correspond to that found by Fisher and Huse; however, the relation of the observed, nonlinear drifting fixed point to the linear theory in $d = 3$ dimensions has not yet been elucidated analytically. Unfortunately, for $d$ close to $d^> = 3$ the numerical analysis also becomes hard to control. One does see[11] definite evidence of the boundary between Regimes II and III at $\omega = 2$ but even the prediction that $w_A^c(\omega)$ sticks at 0 for $\omega < 2$ and departs linearly for $\omega > 2$ has not been convincingly checked. In the classic refrain, "More work remains to be done"![57]

### 3.9. *Concluding Remarks*

Although the behavior of interfaces undergoing critical wetting in the presence of only short-range wall forces in $d = 3$ dimensions is not yet fully resolved, the approximate functional renormalization group certainly has more to teach us. In particular, the influence of long-range $1/\ell^{\sigma-1}$ forces on a critical wetting fixed point can be studied analytically[11]; one confirms the existence of the mean-field regime and finds a *weak fluctuation regime*[7,11] for $\sigma - 1 < \tau(d)$, where $\tau(d)$ is the effective interface-wall repulsion exponent computed in Sec. 2.7: see also Sec. 2.10. Furthermore, the transformation can be adapted to the treatment of *membranes* with instructive results.[55,53] One can, additionally, contemplate the exploration of tricritical wetting[18,57b] and, in the future, hope to obtain further insight into critical wetting in three dimensions both analytically and numerically.[60]

In summary, in the last decade, 1980–89, much has been learned about the fluctuations of asymptotically flat interfaces under the influence of various sorts of external forces, walls, and other interfaces, even though not all the important questions have been answered.[61] But when one goes beyond simple, flat interfaces to membranes with structure, to crumpled and convoluted surfaces, etc., one enters a new exciting domain: other authors in this volume address those questions and report on insights freshly gained and novel ideas in the making.

## Acknowledgments

I am grateful to Professor David R. Nelson and Professor Tsvi Piran for the invitation to speak at the Jerusalem School and for their consideration and patience. Collaborative work with Dr. Reinhard Lipowsky has played a crucial role in many of the research results and ideas described in these lectures. Stimulating interactions with Professor Daniel S. Fisher, Dr. Matthew P.A. Fisher, Martin P. Gelfand, Dr. Timothy J. Halpin-Healy, Professor Christopher L. Henley, Dr. David A. Huse and Dr. Stanislas Leibler on various topics have much appreciated. Last but not least, I acknowledge gratefully the support of the National Science Foundation through the Condensed Matter Theory Program.[56]

## References

1. J. S. Rowlinson and B. Widom, *Molecular Theory of Capillarity* (Clarendon Press, Oxford, 1982).
2. D. Jasnow, *Repts. Prog. Phys.* **47**, 1059 (1984); and in *Phase Transitions and Critical Phenomena*, Vol. 10, Eds. C. Domb and J. L. Lebowitz (Academic Press, New York, 1986), p. 270.
3. D. Sullivan and M. M. Telo de Gama, in *Fluid Interfacial Phenomena*, Ed. C. A. Croxton (Wiley, New York, 1985), p. 45.
4. K. Binder, in *Phase Transitions and Critical Phenomena*, Vol. 8, Eds. C. Domb and J. L. Lebowitz (Academic Press, New York, 1983), p. 1.
5. D. B. Abraham, in *Phase Transitions and Critical Phenomena*, Vol. 10, *loc. cit.*, p. 1.
6. M. E. Fisher, *J. Stat. Phys.* **34**, 667 (1984); and in *Fundamental Problems in Statistical Mechanics VI*, Ed. E. G. D. Cohen (North Holland Publ. Co., Amsterdam, 1985), p. 1.
7. M. E. Fisher, *J. Chem. Soc. Faraday Trans.* 2, **82**, 1569 (1986) (Faraday Symposium 20).
8. M. E. Fisher and M. P. Gelfand, *J. Stat. Phys.* **53**, 175 (1988); *ibid.* **55**, 472 (1989) [E].
9. E. Brézin, B. I. Halperin and S. Leibler, *Phys. Rev. Lett.* **50**, 1387 (1983).
10. D. S. Fisher and D. A. Huse, *Phys. Rev. B* **32**, 247 (1985).
11. R. Lipowsky and M. E. Fisher, (a) *Phys. Rev. Lett.* **57**, 2411 (1986); (b) *Phys. Rev. B* **36**, 2126 (1987).
12. H. W. Diehl, D. M. Kroll and H. Wagner, *Z. Phys. B* **36**, 329 (1980).
13. R. Lipowsky, D. M. Kroll and R. K. P. Zia, *Phys. Rev. B* **27**, 4499 (1983); R. Lipowsky, *Z. Phys. B* **55**, 345 (1984).
14. M. E. Fisher and D. S. Fisher, *Phys. Rev. B* **25**, 3192 (1982).
15. R. Lipowsky, *Phys. Rev. B* **32**, 1731 (1985).
16. J. Krim, J. G. Dash and J. Suzanne, *Phys. Rev. Lett.* **52**, 640 (1984).
17. R. Pandit and M. E. Fisher, *Phys. Rev. Lett.* **51**, 1772 (1983).
18. H. Nakanishi and M. E. Fisher, *Phys. Rev. Lett.* **49**, 1565 (1982).
19. E. Brézin, B. I. Halperin and S. Leibler, *J. Physique* **44**, 755 (1983).

20. (a) J. W. Cahn, *J. Chem. Phys.* **66**, 3667 (1977); (b) M. R. Moldover and J. W. Cahn, *Science* **207**, 1073 (1980).

21. M. P. Gelfand and M. E. Fisher, (a) *Int. J. Thermophys.* **9**, 713 (1988); (b) *Physica A* **166**, 1 (1990): note that the factors $\frac{1}{2}$ in (3.3.6) and (3.3.10) should be deleted and in Eq. (D.26) $L_2$ should read $L_1$.

22. M. E. Fisher, in (a) *Proc. Nobel Symp.* **24**, *Collective Properties of Physical Systems*, Eds. B. Lundqvist and S. Lundqvist (Academic Press, New York, 1974), p. 16 and; (b) *Critical Phenomena*, Ed. F. J. W. Hahne (Springer Verlag, Berlin, 1983), p. 1.

23. J. D. Weeks, in *Ordering in Strongly Fluctuating Condensed Matter Systems*, Ed. T. Riste (Plenum Press, New York, 1980), p. 293.

24. D. R. Nelson, in *Phase Transitions and Critical Phenomena*, Vol. 7, *loc. cit.* (1983), p. 1.

25. G. Grinstein and S.-K. Ma, (a) *Phys. Rev. Lett.* **49**, 685 (1982); (b) *Phys. Rev. B* **28**, 2588 (1983).

26. J. Villain, *J. Physique* **43**, L551 (1982).

27. J. Z. Imbrie, *Phys. Rev. Lett.* **53**, 1747 (1984).

28. J. Bricmont and A. Kupiainen, *Phys. Rev. Lett.* **59**, 1829 (1987).

29. D. S. Fisher, *Phys. Rev. Lett.* **56**, 1964 (1986).

30. D. A. Huse, C. L. Henley and D. S. Fisher, *Phys. Rev. Lett.* **55**, 2924 (1985).

31. R. Lipowsky and M. E. Fisher, *Phys. Rev. Lett.* **56**, 472 (1986).

32. A. M. Szpilka and M. E. Fisher, *Phys. Rev. Lett.* **57**, 1044 (1986); *Phys. Rev. B* **36**, 644, 5343, 5363 (1987).

33. S. Leibler in this volume: chap. 3.

34. M. Kardar and D. R. Nelson, *Phys. Rev. Lett.* **55**, 1157 (1985).

35. M. Kardar, *Phys. Rev. Lett.* **55**, 2235 (1985).

36. D. A. Huse and C. L. Henley, *Phys. Rev. Lett.* **54**, 2708 (1985).

37. M. Huang, M. E. Fisher and R. Lipowsky, *Phys. Rev. B* **39**, 2632 (1989).

38. C. Jayaprakash, W. F. Saam and S. Teitel, *Phys. Rev. Lett.* **50**, 2017 (1983).

39. C. Rottman and M. Wortis, *Phys. Rev. B* **29**, 328 (1984); *Phys. Repts.* **103**, 59 (1984).

40. Y. Carmi, S. G. Lipson and E. Polturak, *Phys. Rev. B* **36**, 1894 (1987).

41. V. L. Pokrovsky and A. L. Talapov, *Zh. Eksp. Teor. Fiz.* **78**, 269 (1980) [*Sov. Phys. JETP* **51**, 134 (1980)].

42. P. W. Kasteleyn, *J. Math. Phys.* **4**, 287 (1964).

43. D. B. Abraham, *Phys. Rev. Lett.* **44**, 1165 (1980).

44. D. A. Huse and M. E. Fisher, *Phys. Rev. Lett.* **49**, 793 (1982); *Phys. Rev. B* **29**, 239 (1984).

45. S. T. Chui and J. D. Weeks, *Phys. Rev. B* **23**, 2438 (1981).

46. T. W. Burkhardt, *J. Phys. A* **14**, L63 (1981).

47. J. M. J. van Leeuwen and H. J. Hilhorst, *Physica* **107A**, 319 (1981).

48. R. Lipowsky and T. M. Nieuwenhuizen, *J. Phys.* **A21**, L89 (1988).

49. K. G. Wilson, *Phys. Rev. B* **4**, 3184 (1971); K. G. Wilson and M. E. Fisher, *Phys. Rev. Lett.* **28**, 240 (1972).

50. K. Binder, D. P. Landau and D. M. Kroll, *Phys. Rev. Lett.* **56**, 2272 (1986).

51. R. Lipowsky, *Ferroelectrics* **73**, 69 (1987): see App. B.

52. T. J. Halpin-Healy and E. Brézin, *Phys. Rev. Lett.* **58**, 1220 (1987).

53. G. Gompper and D. M. Kroll, *Phys. Rev. B* **37**, 3821 (1988); *Europhys. Lett.* **5**, 49 (1988).

54. A. Hasenfratz and P. Hasenfratz, *Nucl. Phys. B* **270** [FS16], 687 (1986).

55. R. Lipowsky and S. Leibler, *Phys. Rev. Lett.* **56**, 2541 (1986).

56. Under Grant No. DMR 87-01223/96299, CHE 03-01101 and earlier grants.

57. For some subsequent developments not specifically discussed here, see: (a) R. Lipowsky, *Europhys. Lett.* **7**, 255 (1988) and (b) F. David and S. Leibler, *Phys. Rev. B* **41**, 12 926 (1990).

58. Subsequent to this review, the issue of deriving an effective interface Hamiltonian in a more satisfactory way was investigated further: see (a) M. E. Fisher and A. J. Jin, *Phys. Rev. B* **44**, 1430 (1991). Although the *effective potential* $W(l)$, between the wall and the interface, gains terms more general than proposed in (1.13), these do not play such an important role in the renormalization group analysis. However, further examination reveals that the *interfacial stiffness*, introduced in (1.6), gains a dependence on $l(\bar{y})$, the interfacial profile, that does prove significant: see (b) M. E. Fisher and A. J. Jin, *Phys. Rev. Lett.* **69**, 792 (1992); (c) A. J. Jin and M. E. Fisher, *Phys. Rev. B* **47**, 7365 (1993); (d) M. E. Fisher, A. J. Jin and A. O. Parry, *Ber. Bunsenges. Phys. Chem.* **98**, 357 (1994).

59. A. J. Jin and M. E. Fisher, *Phys. Rev. B* **48**, 2642 (1993).

60. The issue of critical wetting in $d=3$ dimensions has proved difficult to resolve. Some subsequent theoretical developments are reported in Refs. 57, 58 and 59 and in (a) K. Binder and D. P. Landau, *Phys. Rev. B* **37**, 1745 (1988); (b) G. Gompper *et al.*, *Phys. Rev. B* **42**, 961 (1990); (c) A. O. Parry *et al.*, *Phys. Rev. B* **43**, 11535 (1991); (d) M. E. Fisher and H. Wen, *Phys. Rev. Lett.* **68**, 3654 (1992); (e) K. Binder *et al.*, *Phys. Rev. Lett.* **68**, 3655 (1992). The situation as viewed in the light of Monte Carlo simulations of Ising models has been reviewed a decade later by K. Binder, D. P. Landau and M. Müller, *J. Stat. Phys.* **110**, 1411 (2003) who summarize the theoretical questions as yet unanswered.

61. In addition to the reviews 1–7 and other specific references cited above, a detailed review including some further developments may be mentioned: see *The Behavior of Interfaces in Ordered and Disordered Systems* by G. Forgacs, R. Lipowsky and Th. M. Nieuwenhuizen in *Phase Transitions and Critical Phenomena*, Vol. 14, Eds. C. Domb and J. L. Lebowitz (Academic Press, New York, 1991), p. 135.

## CHAPTER 3

## EQUILIBRIUM STATISTICAL MECHANICS
## OF FLUCTUATING FILMS AND MEMBRANES

Stanislas Leibler
*Service de Physique Théorique*
*Institut de Recherche Fondamentale*
*C.E.A.-CEN Saclay*
*91191 Gif-sur-Yvette Cedex, France*

## Introduction

The lectures presented here do not by any means attempt to be an exhaustive summary of the physics of amphiphilic systems. They are rather intended to form an introduction to a restricted class of problems which have recently been studied. In fact, we shall limit the discussion to a very particular choice of systems and phenomena:

(a) we shall consider only model systems, such as pure lipid membranes;
(b) we shall discuss only the equilibrium properties of those systems;
(c) we shall concentrate mainly on phase transitions, critical phenomena and other "universal" or generic features.

Even after those drastic restrictions many subjects will still not be discussed. Some of them appear in other lectures of this school (see e.g. the contributions to these Proceedings by D.R. Nelson or F. David), others, unfortunately, will have to be passed over.

It is sometimes tempting to think that the physical phenomena described here are of some importance for understanding of biological systems such as cell membranes, and the phenomena observed in them (e.g. exo- and endo-cytosis, cell fusion, etc.). In our opinion one should be very cautious when trying to apply the results obtained here to real biological systems, and that is for exactly the same reasons as enumerated above:

(a) real systems are complex, contain many types of molecules with highly non-trivial properties;
(b) they are in general out of equilibrium;

(c) their important features are specific rather than generic and obviously non-universal.

We should like therefore to think of the phenomena described below as a part of *physics done with biological material*, rather than of biophysics.

Cell membranes and the phenomena in which they are involved, however, can be a rich source of inspiration for physicists. This is why we start our lectures in Sec. 1 with a short description of *biological membranes*. The history of the discovery of the membrane structure will allow us to introduce the basic facts about amphiphilic molecules, and bilayer membranes as well as the orders of magnitude of related physical quantities. After a brief description of membrane proteins and their role in certain biological phenomena we shall turn towards model amphiphilic systems, and show that pure membranes can also present a surprisingly rich behavior.

Section 2 deals with the *elastic properties* of fluid membranes. We shall first introduce the elastic model due to Helfrich and others. Then, we shall describe one of the principal achievements of this model, namely the explanation it provides of the non-trivial *shapes of vesicles* and red blood cells. Finally, we shall present a summary of recent numerical studies of fluctuating two-dimensional vesicles.

The role of thermal fluctuations on the behavior of (fluid) membranes will be considered in Sec. 3. We shall summarize there recent calculations of the renormalization of elastic constants and the theoretical consequences of the results. After having discussed a single fluctuating sheet we shall present more speculative description of the behavior of *an ensemble of fluctuating films* (membranes). In particular, we shall show how the topological properties of such ensembles can be included in a simple model. A generalization of such a model which incorporated the intermembrane interactions could be useful in describing some physical systems such as *microemulsions* or *lyotropic liquid crystals*.

Interactions between membranes are considered in Sec. 4. After having described several experimental techniques which made it possible to measure these interactions, we shall examine a simple problem of the swelling of lamellar systems. In particular, we shall discuss the competition between *molecular forces* and fluctuations-induced *entropic repulsion*, which can lead to new critical phenomena: *the unbiding transitions*. We shall describe the theory of these transitions and connect them to similar phenomena in interface physics, e.g. wetting transitions.

Finally, Sec. 5 is concerned with what we call *"fat"* membranes, i.e. membranes whose internal degrees of freedom cannot be neglected, and which simply *cannot* be described as thin, fluctuating surfaces as assumed before. We present three simple examples of theories which deal with such systems: a model *curvature instabilities* in membranes built from a mixture of amphiphilic molecules, a Landau theory of *lamellar phases* of lyotropic liquid crystals, and a geometrical calculation of "frustration" effects in *cubic phases* (which can be viewed as 3d of membranes).

## 1. Cell Membranes as an Inspiration for Physics

### 1.1. *A History of the Discovery of Membrane Structure: A Few Basic Facts about Membranes*

Figure 1.1 shows an electron microscope photograph of human erythrocytes or red blood cells (RBC). These cells, of an average diameter close to $8\,\mu m$, are in fact kinds of small "balloons" made of the *plasma membrane*.[1] Since the plasma membrane is the only membrane of these cells it is relatively easy to study: its structure and physical properties are much better known than those of any other biological membranes. In the following sections we shall see for instance that a simple elastic model for the plasma membrane explains the shapes shown on Fig. 1.1. Qualitatively however, before describing the structure of the RBC membrane we wish to summarize here the history of its discovery.[2] In our opinion, this history is not only interesting in itself, but it also provides an opportunity to introduce basic facts (and orders of magnitudes!) about all membranes:

(1) 1890s: Ch. E. Overton, working on cells of plant root hairs, discovered that cells are enveloped in a *selectively permeable* layer. He found a strong correlation between the *permeability* with respect to different molecules and the *solubility* of these molecules in lipids (e.g. lipophilic molecules penetrate into the cell easily).

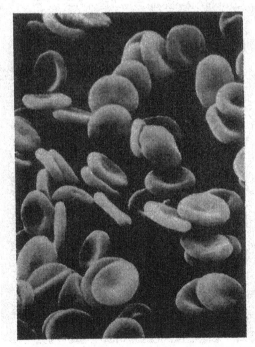

Fig. 1.1. An electron microscope photograph of human red blood cells. [Chailley, B., Thesis (1976), Université Paris VI.] The cells with this typical shape are called discocytes. Note also a very well defined size of order of few microns.

He even suggested that cholesterol and lecithins are among the principal components of the envelope!

In fact, we now know that phospholipids (and cholesterol) are the main components of membranes. Transport through membranes is not in general a purely "passive" (i.e. diffusive) phenomenon, it can involve the intramembrane channels, complex biochemical processes etc.[1] If, however, it occurs through diffusion, then small, non-polar molecules, which are lipid-soluble, have indeed a very high permeability coefficient. In artificial membranes it is of order $10^{-5}$–$10^{-3}$ cm/s compared to $10^{-14}$ cm/s of charged, or polar molecules or ions. The factor $10^{-9}$ between these numbers explains the observations of Overton.[1]

(2) ~1905: I. Langmuir dissolves the phospholipid membranes in benzene and spreads them on a water surface. (One can easily "mimic" these experiments in one's own kitchen introducing a tiny amount of liquid soap into a container full of water, on the surface of which some pepper has previously been sprinkled: one can then observe how rapidly the soap molecules spread on this surface repelting the pepper!) By studying this lipid monolayer (after evaporation of the benzene) Langmuir discovered its main physical properties:

(i) the *amphiphilic nature of the molecules*: lipid molecules, made of a polar head and hydrocarbon chains (see Fig. 1.2) remain on the air-water interface, so that the heads are in contact with water and the chains with air (see below for the description of the hydrophobic effect).

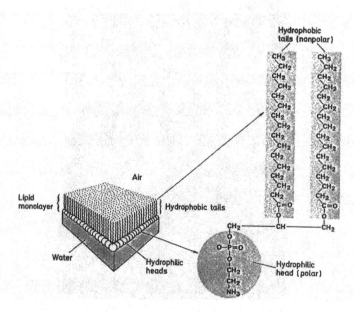

Fig. 1.2. A schematic view of amphiphilic molecules forming a monolayer on a water/air interface. The molecules shown here are phospholipids (DPPC) with two hydrophobic chains. [Redrawn from Ref. 2.]

(ii) the molecules exert a *lateral pressure* which can easily be measured by microbalance techniques (the so-called Langmuir trough). The surface tension of the air/water interface is lowered by an amount equal to this pressure.

(iii) the typical *area occupied by a lipid molecule* is $50\,\text{Å}^2$, and it can be changed by compressing (or decompressing) the monolayer. Using Langmuir techniques, one can study *phase transitions taking place within the layer*.[3] Under high lateral pressures, for instance, the *fluid layer* is transformed into a solid-like phase where the lateral diffusion coefficient, $D$ (typically $D \sim 10^{-8}\,\text{cm}^2/\text{s}^{(1)}$) is strongly increased.[3]

(3) 1925: E. Gorter and F. Grendel dissolve red blood cells in acetone and spread them on a water surface. They compare the area of the monolayer with the area of dried RBCs (observed under microscope). Although they made two important mistakes, their conclusion is correct (thanks to the compensation of errors): the cell membrane consists indeed of *lipid bilayer*. In the bilayer *the hydrophobic chains* are isolated from water by the layers of *hydrophilic heads*. We now know that bilayer structures are often observed in lipid/water mixtures. They constitute one of the main structures of amphiphilic *polymorphism*: the mixtures of amphiphiles and water can form a large variety of thermodynamically stable phases.[4] Figure 1.3 shows an example of a rich phase diagram of the monoolein/water system.[5] In addition to a *lamellar phase* ($L_\alpha$) built of parallel bilayers, one encounters here: an *inverted micellar phase* ($L_2$) were water forms small dispersed droplets, an *inverted hexagonal phase* ($H_{II}$) where water forms a crystal of parallel tubes and *cubic phases* (G and D) in wich bilayers form gyroid- and diamond-like crystals of cubic

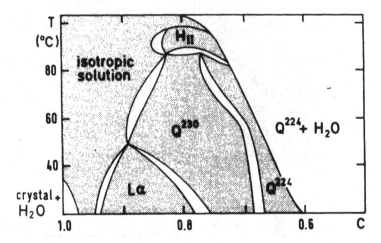

Fig. 1.3. A schematic phase diagram of monoolein/water mixtures. [Ref. 71 (reprinted with permission of Springer-Verlag).] The rich thermodynamic behavior of this simple binary mixture is due to the amphiphilic nature of the monoolein molecules. $L_\alpha$ phase is the liquid lamellar phase, $H_{II}$ is the inverse hexagonal phase, while Q-phases have different cubic symmetries (see Sec. 4). All these phases are ordered at the length scales of hundreds of Ångströms.

symmetries (see Sec. 5), etc. In these lectures we will mainly concentrate our attention on the systems where amphiphiles forms bilayers,[6] i.e. on lamellar (and — to a lesser extent — cubic) phases;

(4) 1935, 1954: H. Dawson and J. Danielli developed a "sandwich" model of biological membranes in order to explain the observed differences in *permeability* and *resistivity* between artificial (pure lipidic) membranes and the ones found in cells. They evoked the presence of *proteins* and proposed that the lipid bilayers are coated on both sides with protein layers. Although such a structure seemed to be observed universally in electron microscope pictures (Robertson's experiments in 1950s) it was soon realized that the proteins do not cover the bilayers (completely). It was, however, established that the membranes have a hydrophobic core of a thickness of order 30–35 Å, and two hydrophilic layers each ~20 Å thick. The *asymmetricity* of the inner and outer layer was also established;

(5) 1972: thanks to the rapid developments in microscopy and molecular biology. S.J. Singer and G. Nicolson proposed the (final) *fluid mosaic model* of biological membranes.[7,8] This model describes a membrane as a fluid, lipidic bilayer (made of lipids and cholesterol) in which other macromolecules are incorporated. The *protein/lipid ratio* can vary drastically in different cell membranes (from <1/4 to >4). Both layers are strongly asymmetric and can preserve their own identity due to the slow *"flip-flop" rate* of lipid exchange. Proteins diffuse more or less freely in the membrane, their diffusion constant is usually much smaller than that of lipids

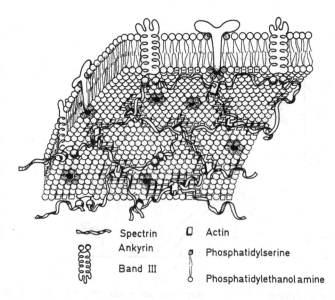

| | Spectrin | ▯ Actin |
| --- | --- | --- |
| | Ankyrin | Phosphatidylserine |
| | Band III | Phosphatidylethanolamine |

Fig. 1.4. A schematic picture of the plasma membrane of human red blood cells. [Ref. 40.] Note the network of proteins attached to the bilayer. This is an example of a cytoskeleton. For human RBCs the cytoskeleton is two-dimensional and does not extend inside the cells. A red blood cell is a simple model system to study the elastic properties of biomembranes.

$(D \sim 10^{-10} \, \text{cm}^2/\text{s})$. They can form pores through which small molecules or ions penetrate (or be "pumped") into the cell.

Figure 1.4 shows schematically the structure of the RBC plasma membrane, it is now established.[1,2] We can see that in addition to lipids, cholesterol and intramembrane proteins, this membrane also consits of glycoproteins and of a *spectrin network*, which is attached to the inner layer through (Band III) proteins. This protein network is important in determining the elastic and mechanic properties of the membrane.[9] For instance the shear modulus of $10^{-2} \, \text{erg/cm}^2$ is largely due to this network.

### 1.2. *Some Physical Properties of Membranes and Amphiphilic Films*

We shall now briefly discuss some physical properties of phospholipid bilayers which underlie the phenomena described in the next sections:

**(1) The amphiphilic nature of the constituents.** Each lipid (or surfactant) molecule has two well-defined parts: a *hydrophobic* and a *hydrophilic* one. The hydrophobic effect is mainly due to entropy:[10] the nonpolar parts of amphiphiles modify the structure of the surrounding water. This effect also induces a strong attraction between nonpolar molecules or surfaces. In contrast, polar molecules (or polar parts of amphiphiles), soluble in water due to their electrostatic interactions,[10] strongly repel each other. If one introduces the amphiphiles onto the water/air interface their hydrophilic parts will be dissolved in water, while the hydrophobic parts will stay outside the solvent. The free energy of $N$ amphiphiles can then be written as a function of the area per molecule, $\Sigma$:[11]

$$F = N\varphi(\Sigma) = N[\varphi_{\text{phob}}(\Sigma) + \varphi_{\text{phil}}(\Sigma) + \varphi_{\text{int}}(\Sigma)], \qquad (1.1)$$

where $\varphi_{\text{phob}}$ and $\varphi_{\text{phil}}$ are effective attractive and repulsive parts, respectively, mediated by water, while $\varphi_{\text{int}}(\Sigma)$ is direct, generically repulsive interaction between the amphiphiles. Therefore $\varphi(\Sigma)$ has a minimum at some value $\Sigma = \Sigma^*$. If the amplitude membrane, or film, does not exchange its molecules with a reservoir, *and can* freely adjust its total area, $A$, then in equilibrium where $\frac{\partial F}{\partial \Sigma}\big|_{\Sigma_{\text{eq}}} = 0$:

$$\Sigma_{\text{eq}} = \Sigma^*, \qquad (1.2)$$

which means that the surface tension of the membrane $\gamma(\Sigma)$ vanishes:[11]

$$\gamma(\Sigma_{\text{eq}}) = \frac{\partial F}{\partial A}\bigg|_{\Sigma_{\text{eq}}} = \frac{\partial \varphi}{\partial \Sigma}\bigg|_{\Sigma^*} = 0. \qquad (1.3)$$

This simple thermodynamical argument is usually evoked to explain why the tension of the amphiphilic films is very small. In the next section, we shall see that the fluctuations of such films role membranes can governed by their *curvature energy* and not the surface tension;

(ii) **Fluidity of the membranes.** Most biological membranes are found in the fluid phase, where the molecules can diffuse freely and their hydrocarbon chains are disordered.[6] When the temperature is lowered below some value $T_m$ (which can cary from one lipid to another, e.g. ($T_m(DPPC) = 41°C$, $T_m(DOPC) = -22°C$) the chains become ordered, i.e. they are in all-"trans" configuration and the "cis" excitations are rare. The value $T_m$ depends on several parameters such as:[6]

(a) the length of the hydrocarbon chains, $n : T_m$ increases with increasing $n$;
(b) the degree of saturation of the chains; completely saturated chains are the easiest to order;
(c) the concentration of impurities, e.g. of cholesterol. There is a lot of experimental evidence for the influence of cholesterol on the membrane fluidity: its presence increases the viscosity in the fluid phase, while on the other hand it decreases the value of $T_m$. (Note: it seems that many organisms, from bacteria to some cold-blooded animals, regulate the fluidity of their cell membranes by modifying *all* these parameters.[1,2])

In Sec. 5 we shall discuss the melting transition in lamellar stocks of interacting membranes.

(iii) **The possibility of topology variations.** Fluid membranes can vary their topology; for instance two membranes can fuse with one another.[12] Figure 1.5 shows three elementary events happening in cells which consist in changing the topology of cell membranes: endocytosis, exocytosis and bugging.[1] Detailed understanding of the biochemistry and the physics of such events is an open issue and an important one: the study of the fusion of *artificial* membranes and vesicles could help in clarifying this issue.[12] It is important, however, to stress that in biological systems the curving of membranes is generally due to changes in the protein structure.[1] For instance, endocytosis is often accompanied by a polymerization of the membrane proteins, e.g. clathrins,[1,2] and the importance of the lipids in this process is not obvious. Still, the phenomena of membrane fusion, pore formation, endo- or exocytosis etc., do have counterparts in pure amphiphilic systems. Phase transitions in many physico-chemical systems are connected with the variation of the topology of films and membranes. We shall study this aspect of amphiphilic polymorphism in Sec. 3.

From now on we shall consider only pure amphiphilic systems, without proteins, cholesterol, and other "complications" which would make a quantitative, physical approach impossible.

## 2. The Elastic Properties of Fluid Membranes and the Shapes of Vesicles

### 2.1. *Curvature Energy and a Simple Elastic Model*

Consider a piece of pure *fluid* membrane freely fluctuating in a solvent. We shall suppose that the thickness $d$ of the membrane is small compared to the length

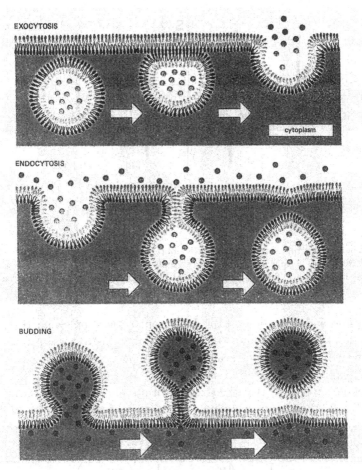

Fig. 1.5. A schematic picture of the elementary processes which change the topology of membranes: endo- and exo-cytosis and budding. [Redrawn from Ref. 1.]

scales describing its shape and its undulations. It was noticed a long time ago by Canham,[13] Helfrich[14] and others, that the statistical behavior of such a "thin" *membrane* can be studied within a continuum model based on the following Hamiltonian:

$$\mathcal{H}_{el} = \int d^2\sigma \left[ \frac{\kappa}{2} (H - H_0)^2 + \bar{\kappa} K \right], \qquad (2.1)$$

where $H$ and $K$ are respectively the mean and the Gaussian curvature at the point $\vec{r}(\sigma)$ of the membrane ($\sigma = (\sigma^1, \sigma^2)$ being local coordinates of the surface). If $R_1$ and $R_2$ are the principal radii of curvature at $\vec{r}(\sigma)$, then $H = R_1^{-1} + R_2^{-1}$ and $K = (R_1 \cdot R_2)^{-1}$. $H_0$ is the spontaneous curvature connected to the asymmetricity of the membrane. The elastic constants $\kappa$ and $\bar{\kappa}$ are the bending (rigidity) coefficient and the Gaussian bending coefficient, respectively.

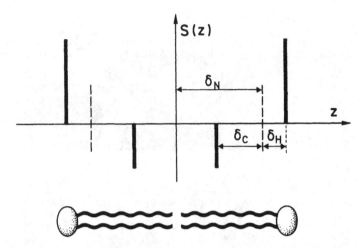

Fig. 2.1.  The simplest microscopic model for membrane elasticity, developed in Ref. 17, supposes that the stresses S(z) within a membrane are localized in a discrete set of regions. The broken lines on this schematic figure correspond to the position of the neutral surfaces described in the text.

Derivations and discussions of this Hamiltonian based on symmetry and invariance considerations will be found in other contributions to these proceedings.[15,16] Here, we would like to describe briefly a simple microscopic model which, although oversimplified, can give an important insight into the physics of Eq. (2.1).

The model supposes[17] that the stresses within a lipid layer are localized in two regions (Fig. 2.1): one corresponding to the heads of the molecules, the other to their chains. In a harmonic approximation the elastic free energy per molecule can then be written as

$$\varphi = \frac{1}{2} k_H \left( \frac{A_H}{A_H^0} - 1 \right)^2 + \frac{1}{2} k_C \left( \frac{A_C}{A_C^0} - 1 \right)^2, \qquad (2.2)$$

where $A_H$ and $A_C$ are actual area per molecule at the level of heads and chains respectively, $A_H^0$ and $A_C^0$ are "preferential" areas and $k_H$, $k_C$ are elastic constants.

If we now call $A$ the areas over the so-called neutral surface (i.e. the surface on which the sum of the moments of stresses vanishes), and $\delta_H$ and $\delta_C$ the distances of the heads and the chains from this neutral surface ($\delta = \delta_H + \delta_C$), then

$$\begin{aligned} A_H &= A\left[1 + \delta_H \cdot H + \delta_H^2 K\right] \\ A_C &= A\left[1 - \delta_C \cdot H + \delta_C^2 K\right] \end{aligned} \qquad (2.3)$$

(Note that these formulae can be viewed as a simple *geometrical interpretation of the mean and Gaussian curvatures, H and K.*) For a flat membrane the free energy (per unit area) (2.2) can be rewritten therefore as:

$$f \equiv \varphi / A_0 = cst + \frac{1}{2} K_s \left( \frac{A}{A_0} - 1 \right)^2, \qquad (2.4)$$

which is just a *stretching energy* with the stretching constant $K_s = \frac{k_H}{A_H^0} + \frac{k_C}{A_C^0}$, and $A_0 = K_s/\left(\frac{k_H}{A_H^{0^2}} + \frac{k_C}{A_C^{0^2}}\right) \equiv K_s/(\beta_H + \beta_C)$. However, if the membrane is bent one obtains an extra contribution due to the fact that the areas per head and per chains are not the same! By introducing Eq. (2.3), and the equalities $\delta = \delta_H + \delta_C$ and $\delta_H\beta_H = \delta_C\beta_C$ (which comes from the equilibration of the momenta on the neutral surface) into the Eq. (2.2) *one obtains the elastic free energy* Eq. (2.1), with

$$\kappa \equiv \kappa_m = \delta^2 K_s \frac{\beta_H\beta_C}{(\beta_H+\beta_C)^2} = \delta_H\delta_C K_s \qquad (2.5a)$$

$$\bar{\kappa} \equiv \bar{\kappa}_m = \left(A_H^0 - A_C^0\right)\frac{\beta_H\beta_C}{(\beta_H+\beta_C)^2}(\beta_H - \beta_C) \qquad (2.5b)$$

$$H_0 = H_0^m = \frac{\bar{\kappa}}{\kappa}\frac{\beta_H+\beta_C}{\delta(\beta_C-\beta_H)} = \frac{\bar{\kappa}}{\kappa}\frac{1}{\delta_H-\delta_C} \qquad (2.5c)$$

(a subscript $m$ standing for monolayers).

Although some features of Eq. (2.5), e.g. the exact relation between $\kappa_m$, $\bar{\kappa}_m$ and $H_0^m$ of Eq. (2.5c), are obviously model-dependent, these three equations can provide an insight into elastic properties of amphiphilic monolayers. For instance, notice that $\kappa_m \propto \delta^2$, and not $\delta^3$ as for *solid plates* (shells). We expect that for more realistic models, in which the distribution of stresses is not restricted to two points, $\kappa_m \propto \delta^x$, with $2 < x < 3$.[18] Equation (2.5a) also tells us that $\kappa_m$ is positive and gives us a crude estimation of this quantity: for typical values $K_s \simeq 50\,\mathrm{erg/cm^2}$ and $\delta_H \sim \delta_C \sim 10\,\text{Å}$ we obtain $\kappa_m \sim 5.10^{-13}\,\mathrm{erg}$. In contrast with $\kappa$, the sign of $\bar{\kappa}$ can vary. This is a very important theoretical prediction: we shall see in the next section how varying the sign of $\bar{\kappa}$ of bilayers can influence the phase behavior of amphiphilic systems. The formula (2.5) also suggest that one can easily alter $\bar{\kappa}_m$ and $H_0^m$ of monolayers: since they are both proportional to $(A_H^0 - A_C^0)$ one can simply modify the interaction between the heads, or their lateral dimensions (e.g. through methylation[19]).

The connection between the elastic properties of two monolayers and those of bilayer can be treated in similar simplified fashion.[17] The detailed relation depends on whether the monolayers are connected (no slipping on one another) or unconnected, which in turn can depend on the contrains imposed on the system (e.g. no exchange of molecules between the monolayers i.e. no "flip-flops", etc.). For unconnected monolayers one obtains:[17]

$$\kappa = 2\kappa_m; \qquad \bar{\kappa} = 2(\bar{\kappa}_m - 2\kappa_m H_0^m \delta_N); \qquad H_0 = 0, \qquad (2.6)$$

where $2\delta_N$ is the distance between the neutral surfaces of the monolayers: $2\delta_N < d$.

This simplest of elastic models could in principe be generalized for a more realistic stress distribution across the bilayers, $s(z)$. An example of such a generalization is provided by the following relations:[20]

$$\kappa H_0 = - \int z s(z) dz \qquad (2.7a)$$

$$\bar{\kappa} = \int z^2 s(z) dz \qquad (2.7b)$$

derived under several simplifying assumptions (such as $\gamma = \int s(z)dz = 0$, and $R_z^{-1} = H_0$ with $R_2^{-1} = 0$). One should also mention some interesting attempts at developing a microscopic model for the elastic properties of co-polymers (which can also have amphiphilic properties).[21] These and other[22] microscopic models should make for a more precise understanding of the physics of Eq. (2.1) in the near future.

## 2.2. Shapes and Fluctuations of Vesicles

One of the first successful results of the theories based on Eq. (2.1) was the explanation of some-trivial shapes of red blood cells.[13,14,23] However, as it was already mentioned in Sec. 1 the membrane of a red blood cell consists not only of a phospholipid bilayer (which includes proteins, glycophorins etc.) but also of a cytoskeleton attached to this bilayer. One cannot therefore apply without restrictions the model of *fluid membranes* to the membranes of red blood cells, which have a non-zero shear modulus.[9] More sophisticated models, which try to take into account the full elastic properties of erythrocyte membranes, have recently been developed.[24] Here, we shall restrict our discussion to Eq. (2.1) only. This equation can describe the energy of pure, fluid *vesicles*, i.e. closed membranes, for instance those shown on Fig. 2.2. As we can see, the artificial vesicles can have a size ($\sim 1-10\,\mu$) and a shape similar to those of red blood cells. Moreover, by changing physical parameters one can induce the shape transformation, as is shown on Fig. 2.2.

The calculation of the shapes of closed membranes can be formulated as the minimization problem. If the membrane is supposed to be *impermeable* and *incompressible* then the variational problem to solve is

$$\delta \mathcal{H}_{el} = 0 \qquad \text{with } A = \text{const and } V = \text{const}, \qquad (2.8)$$

where $A$ is the total area of the vesicle, and $V$ is its volume. This problem was (partially) solved by several authors,[23,25] who assumed the rotational symmetry (around an axis) of the shapes. Figure 2.3 shows the results of these calculations in a schematic way: the energy of different shapes is plotted here as a function of $x \sim V/A^{3/2}$ (Fig. 2.3a),[23] and of $y \sim H_0 A^{1/2}$ (Fig. 2.3b)[25] (with $y$ and $x$ values fixed respectively in the first and second plot). Note, that the discocyte-like shapes are in fact of minimal energy for the value of $x$ corresponding to the real erythrocytes. Moreover, the variations of the vesicle shapes from a "discocyte", through a "stomatocyte", to a small sphere enclosed in a bigger one shown on Fig. 2.2 correspond exactly to the variations in $y$ of Fig. 2.3b! It is therefore natural to speculate that variations of temperature — which did induce these transformations in the

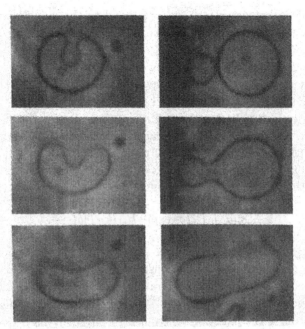

Fig. 2.2. Video microscopy photographs of shape transformations induced in an artificial vesicle (made of a DMPC bilayer). Different shapes correspond to different temperatures, all being superior but close to $T_m$ — the melting point of the bilayer. The size of the vesicle is bigger than $10\mu$. [Courtesy of H.-P. Duwe, H. Engelhardt and E. Sackmann; see Ref. 26.]

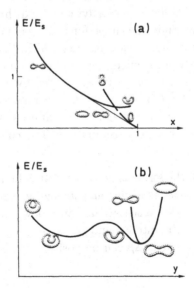

Fig. 2.3. The results of the minimization of the bending energy Eq. (2.8). are shown here schematically. The energy E of a vesicle is plotted as a function of $x \sim V/A^{3/2}$ (a) and $y \sim H_0 A^{1/2}$ (b), where V and A are the volume and the area of the vesicle. [The plots are based on the results of Refs. 23 and 25.]

experiments of Sackmann *et al.*[26] — correspond to the modification of the effective spontaneous curvature of the membrane (e.g. through the modification of the head-head or chain-chain interactions).

The problem defined by Eq. (2.8) (or similar ones, e.g. for permeable or compressible membranes) is in fact a purely *mechanical* problem. What really differentiates vesicles from ballons or closed shells is the fact that the *elastic constant* $\kappa$ *of bilayer membranes is of order* $k_B T$. Therefore, the thermal fluctuations can influence the shape and other mechanical properties of the vesicles. We shall now consider the thermodynamic behavior of these objects.

The equilibrium statistical mechanics of fluctuating membranes is based on the evaluation of the following statistical sum:

$$ Z = \int D\vec{r}(\sigma) \exp[-\mathcal{H}_{\text{el}}/k_B T]\delta(\cdots), \qquad (2.9) $$

where the integration is to be done over all configurations of random surfaces and $\delta(\cdots)$ takes into account all possible constraints imposed on the membrane. In the case of vesicles the considered configurations are restricted to *self-avoiding* ones and to those with the *sphere topology*. The constraints are, for instance, those of Eq. (2.8). Of course the partition function $Z$ is extremely hard to evaluate. Some difficulties connected with this question are discussed in the lecture by F. David.[15] Here we shall simplify the problem by considering *two-dimensional vesicles*[27] and thus avoiding the difficulty of integrating over random surfaces.

Even for this simple, two-dimensional problem (in which a vesicle is simply a polymer-like nonintersecting loops) no analytic methods had been developed which would make it possible to study the shape fluctuations and many other thermodynamic properties. With this aim in view R. Singh, M.E. Fisher and the author have recently carried out a study of the microscopically based model[28] using *Monte-Carlo methods*. In this model the membrane is represented by a closed string of $N$ hard, nonoverlapping spheres of radius $a$ ($N \leq 100$), connected by "tethers" of fixed, maximum length $b$. The model is therefore a two-dimensional analog of the models studied by Y. Kantor and his collaborators.[29,30] The constraints imposed on the vesicle are the following:

(a) the total length of the string is not fixed, but is bounded from above by $Nb$. One can thus consider vesicles as having an entropy-induced lateral compressibility;

(b) the total area of the vesicles can vary; however a pressure differential $\Delta p = p_{int} - p_{ext}$ between the interior and the exterior, conjugate to the enclosed area $A$, is supposed to be fixed. $\Delta p$ can be positive *or* negative. The total energy of a vesicle configuration is

$$ E = \frac{\kappa}{a} \sum_{i=1}^{N} (1 - \cos\theta_i) + \Delta p \cdot A, \qquad (2.10) $$

where $\theta_i$ are the angles between neighboring sphere-to-sphere vectors.

Here are the main results of this simulation:

(i) for $\Delta p = \kappa = 0$ the vesicles flaccid and behave as closed self-avoiding walks; their radius of gyration $R_G$ and the inclosed area behave as

$$\langle R_G^2 \rangle \sim N^{2\gamma}; \qquad \langle A \rangle \sim N^{2\gamma_A} \tag{2.11}$$

with $\gamma_A/\gamma = 1.007 \pm 0.013$ in accordance with a natural expectation of a "non-fractal" area determined by the mean linear size (e.g. vesicles do not collapse for $N \to \infty$);

(ii) for $\Delta p \neq 0$ the shapes change significantly. These changes can be described by the scaling functions:

$$\langle R_G^2 \rangle \sim N^{2\gamma} X(\Delta p \, N^{\varphi\gamma}); \qquad \langle A \rangle \sim N^{2\gamma} Y(\Delta p \, N^{\varphi\gamma}), \tag{2.12}$$

where the cossover exponent $\varphi = 2.13 \pm 0.17$. For $\Delta p < 0$ the functions $X$ and $Y$ describe *how vesicles shrink*; in the asymptotic regime

$$X(x) \sim \frac{1}{|x|^\sigma} \quad \text{and} \quad Y(x) \sim \frac{1}{|x|^\tau} \tag{2.13}$$

with $\sigma = 0.13 \pm 0.05$, $\tau = 0.25 \pm 0.04$. The vesicles look then like seaweeds with many branches: the values of exponents $\sigma$ and $\tau$ suggest that they behave as *branched polymers*;[31]

(iii) in the region $\Delta p \leq 0$ the bending rigidity $\kappa \neq 0$ renormalizes only the effective distance between spheres. However, in the deflated regime of large negative $\Delta p$ the rigidity drastically changes the behavior of vesicles. The loops display a variety of shapes (cytotypes) analogous to those observed in 3d experiments.

Figure 2.4 illustrates this phenomenon: by superimposing the instantaneous positions of the centers of the spheres ("snapshots") taken during a certain period of time we obtain information about the average shape *and* the thermal fluctuations around it. These strong fluctuations were observed in red blood cells as early in the 19th century[32] and are called *flickering*.[33] One can follow the changes of forms and fluctuations with variations of the reduced parameters

$$\bar{p} = \Delta p a^2 / k_B T \quad \text{and} \quad l_\kappa / a \equiv \kappa / (k_B T a). \tag{2.14}$$

In addition to the *rigidity (persistence) length* $l_\kappa$ the relevant length scale are:

$$l_p = |k_B T / \Delta p| \quad \text{and} \quad L = Na. \tag{2.15}$$

The cytotype regime is then characterized by $l_\kappa / L \leq 1$ and $p^*/L \leq 1$, where the typical minimal radius of curvature of a cytotype is

$$p^* = (l_\kappa l_p)^{1/3} \sim (\kappa / \Delta p)^{1/3}. \tag{2.16}$$

(Note that in two dimensions the effect of the spontaneous curvature $H_0$ is trivial: $\oint H_0 dl = \pm 2\pi$ and therefore the thermodynamic properties of $H_0$ are exactly the

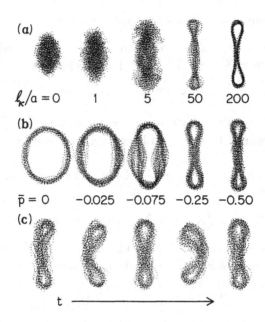

Fig. 2.4.   Vesicle shapes of two-dimensional vesicles obtained in a Monte-Carlo simulation. [Ref. 28.] (a) For $\bar{p} = -1.25$ as $\ell_\kappa$ increases; (b) for $\ell_\kappa/a = 50$ and increasing $|\bar{p}|$; (c) as a function of time, t, for $\bar{p} = -0.075$ and $\ell_\kappa/a = 10$. (Reprinted by permission of *Phys. Rev. Lett.* **59** (1987) 1989.)

same whether $H_0$ is present or absent up to a global shift of energy scale, this is not the case for real 3d vesicles as we have discussed before.)

(iv) Under the simple Monte-Carlo dynamics, the vesicles display appreciable harmonic flickering. More interestingly however, there are regions of parameters $l_K/a$ and $\bar{p}$ where one observes *nonlinear flickering*, that is thermally driven transformations between different cytotypes under fixed external conditions. An example of this phenomenon is shown on Fig. 2.4(c). It should be possible to observe nonlinear flickering in experimental systems by suitably fine-turning the control variables.

The question of generalizing these and similar results to three-dimensional vesicles certainly remains open. The difficulty of numerical approaches lies in the long times the shapes require to equilibrate when using a single-move Monte–Carlo dynamics. Also, a simulation of fluid membranes involves a constant changing of neighbors[39] which is indeed very time-consuming.

## 2.3. *Measuring of Elastic Constants*

The study of the thermal fluctuations of red blood cells[33] and artificial vesicles[35-38] makes it possible[34] to estimate the actual values of the rigidity constant $\kappa$. For instance, for *a quasi-spherical* (but still fluctuating) *vesicle* described by its shape function

$$\vec{r}(\Omega) = r_0(1 + \mathcal{U}(\Omega))\hat{r} = r_0\left[1 + \sum_{l,m}\mathcal{U}_{l,m}Y_{lm}(\Omega)\right]\hat{r},$$

where $Y_{lm}$ are spherical harmonic functions of the solid angle $\Omega$, one can in principle measure, through various video-microscopy techniques,[40] both $\langle|\mathcal{U}_{lm}|^2\rangle$ and $\langle\mathcal{U}_{lm}(t)\mathcal{U}_{lm}(0)\rangle$. One can then estimate these quantities within the framework of simple models based on the equipartition of energy among different spherical modes,[36–39,41] and the hydrodynamic theory of relaxation of these modes.[39] The bending constant $\kappa$ enters both the amplitudes of flickering modes $\langle|\mathcal{U}_{lm}|^2\rangle$[(39)]

$$\langle|\mathcal{U}_{lm}|^2\rangle = \frac{k_B T}{\kappa}\frac{1}{l(l+1) - 4w - 2w^2 - \lambda}, \tag{2.17}$$

(where $w = r_0 H_0$, and $\lambda$ is the Lagrange multiplier which ensures that the area remains (on average) constant) and the relaxation times $\tau_{lm}$ of these modes; $\langle\mathcal{U}_{lm}(t)\mathcal{U}_{lm}(0)\rangle = \langle|\mathcal{U}_{lm}|^2\rangle e^{-\tau/\tau_{lm}}$:

$$\tau_{lm} = \frac{\eta\tau_0^3}{\kappa}\frac{Z(l)}{l(l+1) - (\lambda + 4w - 2w^2)}, \tag{2.18}$$

(where $\eta$ the viscosity of the solute). One can therefore extract the value of $\kappa$ from these measurements:[37,38]

$$\kappa \simeq 3.10^{-13} \div 2.10^{-12}\,\text{erg} \tag{2.19}$$

for phospholipid bilayers. This should be compared with $\kappa \simeq 10^{-14}$ erg for some surfactant films (e.g. in microemulsions or quasi-ternary swollen systems[44]). The fact that different experiments made on similar systems lead to rather different results is not really surprising for the two main reasons:

(i) theories describing the dynamical aspects of flickering are still approximative. Solving the Navier–Stokes equations even within the limit of very low Reynolds numbers $\left(R \sim \frac{\rho\kappa}{\eta^2 r_0} \sim 10^{-4}\right)$ and the deterministic "creeping flow"[39] is still a difficult task because of the non-trivial boundary conditions;

(ii) measurements are based on the projection of three dimensional shapes on a two dimensional focal plane, which implies strong assumptions in analysing the vesicle behavior.

For sake of completeness let us mention some of the main techniques used to measure the bending constant $\kappa$ as well as other properties of vesicle membranes (e.g. viscosity, shear modulus etc.):[40]

(i) non disturbing techniques, such as flicker spectroscopy,[33] reflection interference contrast microscopy[43] or phase contrast microscopy;[37]
(ii) weak disturbation techniques, such as the electric field jump method[44] in which one slightly perturbs the vesicles and observes their relaxation;

(iii) techniques using large distrubations such as the micropipette methods[45] (see Fig. 4.1).

However, the most accurate methods of measuring $\kappa$ for pure, liquid membranes are probably developed in the study of lamellar phases of lyotropic crystals, which we shall discuss in Sec. 4. The interest of the methods mentioned before is thus their applicability to biological systems.

Finally, let us mention that as of now there are no known experiments which estimate the Gaussian rigidity $\bar{\kappa}$[(46)]. Good candidates for the experimental methods to measure $\bar{\kappa}$ will be those which are sensitive to the topological changes of the membranes. In fact, the Gauss-Bonnet theorem[15] tells us that

$$\int \bar{\kappa} K d^2\sigma = 4\pi\bar{\kappa}\chi, \tag{2.20}$$

where $\chi$ is the Euler characteristic of the surface; to measure $\bar{\kappa}$ therefore one has to study the topological changes with $\Delta\chi \neq 0$. Possible experiments would consist in studying the fusion between vesicles[12] or the transformation between lamellar, "sponge-like" and vesicle phases of amphiphilic systems (see next section).

## 3. The Role of Thermal Fluctuations in the Behavior of (Fluid) Membranes and Films

### 3.1. *Fluctuations of a Single Fluid Membrane*

In this section we shall study the role of thermal fluctuations in the behavior of membranes. In particular we shall consider how fluctuations modify the elastic bending constants. It is easy to see that on a classical level, i.e. for the $T = 0$ mechanical problem, the rigidity coefficient $\kappa$ does not depend on the linear size $L$ of the membrane. (In fact, if for instance we calculate energy of a spherical vesicle $E = \int \kappa H^2 d^2\sigma$, we obtain $4\pi\kappa$ independent of its radius $L$.) However, if one takes thermal excitations into account one can show that this result does not hold any longer, and $\kappa$ is decreasing with increasing $L$.[47]

Let us first consider a *fluid* membrane which spans a frame of area $A_0 = L \times L$. Amphiphilic molecules are supposed to exchange freely with an exterior reservoir; all the calculations are thus perfomed in the *grand canonical ensemble*. The Hamiltonian describing the membrane

$$\mathcal{H} = \int d^2\sigma \left[ r_0 + \frac{\kappa_0}{2}(H - H_0)^2 + \bar{\kappa}_0 K \right] \tag{3.1}$$

has a term proportional to the total area, $\int d^2\sigma$, and, since we assume that the membrane is *incompressible*, to the total number of molecules $N$. The coefficient $r_0$ can thus be regarded as the *chemical potential* of the reservoir. (Here, the subscript 0 for $r$, $\kappa$ and $\bar{\kappa}$ denotes the bare values of these quantities.)

Two theoretical approaches permit us to study the fluctuations of such a membrane. One which we shall describe in the next paragraph is a *perturbation theory* for almost flat membranes,[48,49] and can be viewed as a low-temperature expansion for fluid membranes. (The perturbation parameter being $k_B T/\kappa_0$.) The other is an *expansion in* $1/d$,[50–52] where $d$ is the dimension of the embedding space. This expansion is non-perturbative in the sense that it also takes into account convoluted, far-from-flat configurations of the membranes. Both types of calculations neglect the effects of self-avoidance on the membrane.

Figure 3.1 summarizes the results of these calculations. There is an unstable fixed point $W$ at the origin of the plane $\left(\alpha_0 \equiv \frac{k_B T}{\kappa_0}, r_0\right)$ from which a critical line $r_0^{cr}(\kappa_0)$ emerges. The perturbation theory is valid on this locus near $W$; its main results for $d = 3$ are the following:

(i) in the first order in perturbation theory the rigidity $\kappa$ is renormalized by fluctuations as[48]

$$\kappa(L) = \kappa_0 - \frac{3k_B T}{4\pi} \ln\left(\frac{L}{a}\right) + \cdots, \tag{3.2}$$

where $a$ is a microscopic cutoff (e.g. the size of molecules).

(ii) the correlation function of the orientations of the membrane (see Fig. 3.2) decays exponentially:

$$\langle \vec{n}(0)\vec{n}(\vec{R}) \rangle \sim e^{-|\vec{R}|/\xi_p} \tag{3.3}$$

with *the persistence length* $\xi_p$ varying as

$$\xi_p \approx a\exp(4\pi\kappa_0/3k_B T) \tag{3.4}$$

The properties of the model around the fixed point $W$ are important for the behavior of the membrane at scales smaller than $\xi_p$, at which the membrane is effectively

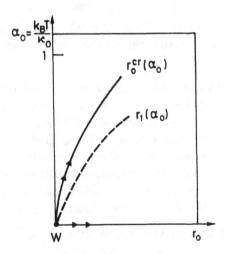

Fig. 3.1.    A possible phase diagram for a single fluid membrane. The perturbation theory described in this section is valid for almost planar membranes near the unstable fixed point W. The broken line corresponds to an instability of a planar solution described in the lectures of F. David.[15] The topology of the membrane is supposed fixed (in contrast with the Fig. 3.5).

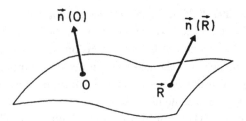

**Fig. 3.2.** The correlation function of the orientation of a fluid membrane $\langle \vec{n}(0)\vec{n}(\vec{R})\rangle$ decays with the characteristic length $\xi_p$.

flat, i.e. its orientations in different points are correlated. Above this length $\xi_p$ the membrane is *crumpled*, i.e. its orientations are decorrelated. The fact that for $T \neq 0$ the persistence length $\xi_p$ is always finite means that (for large enough scales) *noninteracting fluid membranes are always crumpled*.

(iii) for $L \to \infty$ the actual area of the fluctuating membrane diverges and one can evaluate its Hausdorff dimension:

$$\frac{d_F}{2} \equiv \frac{dlnA}{dlnA_0} = 1 + \left(\frac{\alpha_0}{8\pi}\right) - \left(\frac{\alpha_0}{8\pi}\right)^2 + \cdots \qquad (3.5)$$

(iv) one can also formally calculate the renormalization of other elastic constants in this perturbation scheme:[49,52]

$$\bar{\kappa}(L) = \bar{\kappa}_0 - \frac{k_B T}{\pi} \ln\left(\frac{L}{a}\right) + \cdots \qquad (3.6)$$

$$H_0(L) = H_0 \left(1 + \frac{k_B T}{\pi \kappa_0} + \cdots\right) \qquad (3.7)$$

however the physical significance of these formulae has not really been elucidated. Some authors believe for instance that fluctuations do not modify the Gaussian rigidity: $\bar{\kappa}(L) \equiv \bar{\kappa}_0$.[53,54]

We shall return to the perturbation calculation — and in particular to its results for $d > 3$ dimensions — below. It seems useful — for the sake of completeness — to describe the results for the large embedding dimension $d$. The details of these calculations can be found in the lectures by F. David.[15]

In contrast to the perturbation calculation which is valid only in the vicinity of the fixed point $W$, the $d = \infty$ calculation describes the approach to the line $r_0^{cr}(\kappa_0)$(see Fig. 3.1). Under a strong (see below) assumption that the membrane which spans the frame *stays on the average planar*, one can show that:

(i) its total area $A$ diverges when $r$ approaches $r_0^{cr}$ as

$$A \sim (r_0 - r_0^{cr})^{-1/2}. \qquad (3.8)$$

At the critical locus $r_0^{cr}$ the rotational symmetry of the membrane, broken by the existence of the planar frame, is restored.

(ii) for large $L : L \to \infty$ one can define the effective, "physical" quantities $r_{phys}$ and $\kappa_{phys}$ which describe the effective surface tension and the rigidity of the membrane. At the vicinity $r_0^{cr}$ one obtains:

$$r_{phys} \sim \frac{\kappa_0}{a_2} e^{-8\pi\kappa_0/3k_B T}, \quad \kappa_{phys} = 0. \tag{3.9}$$

(iii) The persistence length $\xi_p$ of the fluctuating membrane is

$$\xi_p \sim (k_B T / r_{phys})^{1/2},$$
$$\sim a\kappa_0^{-1/2} e^{4\pi\kappa_0/3k_B T}, \quad (as\, r_0 \to r_0^{cr}) \tag{3.10}$$

a result which agrees quite well with Eq. (3.4).

All these results however should be taken with caution. In fact, it has been shown[51] that at some value $r_1 < r_0^{cr}$ an instability of a planar solution appears, with characteristic wavelengths of order $\xi_p$. For $r_0 \leq r_1$ the membrane does not fluctuate around a planar average configuration. The solution for $r_0 \leq r_1$ is not known; it is believed that the first order transformation at $r_1$ may be connected to the well-known collapse of random surfaces to *branched polymers* (see the lectures by J. Fröhlich[55]). In such a case the membranes would fluctuate around non-selfavoiding branched polymer configurations, with the transverse size of tubular branches of order $\xi_p$.

We will end this description of fluctuating fluid membranes with an example of possible applications of the above results. For this let us consider an ensemble of quasi-spherical vesicles. If this system is in equilibrium, i.e. the vesicles had time to exchange molecules among themselves, then the distribution of sizes can easily be calculated[47,53]

$$P(N) \sim N^p e^{-2N/\bar{N}}, \tag{3.11}$$

where $N$ is the number of amphiphiles in the vesicles, and $p =$ const. The presence of the power-law prefactor is in fact due to the logarithmic renormalization of the bending coefficients $K$ and $\bar{\kappa}$. From Eqs. (3.2) and (3.6) one gets $p = 1$. Such a power-law prefactor can in principle be measured in experiments; however, one should be aware that:

(i) equilibration times can be prohibitively long;
(ii) non-linear, higher order elastic terms such as $H^2 K$, $K^2$ etc.[56] can become important for small $N$.
(iii) non-zero values of $H_0$ can be generated through the exchange of amphiphiles.

## 3.2. *Perturbation Calculations and the Concept of Crumpling Transition*

We shall now present a summary of the perturbation and renormalization group calculations which lead to the results described above.[48,57]

In the Monge representation, a surface is described by a function $u(x, y)$, where $x$ and $y$ are two coordinates of the embedding space. This representation is well suited to describing almost planar configurations of surfaces and will be adopted here. In this representation the Hamiltonian (3.1) can be written as (with $H_0 \equiv 0$, $\bar{\kappa} = 0$)

$$\mathcal{H} \equiv \mathcal{H}_0 + \mathcal{H}_I = \int d^D x \left[ r_0 + \frac{r_0}{2} (\nabla u)^2 + \frac{\kappa_0}{2} (\nabla^2 u)^2 \right] + \mathcal{H}_I, \qquad (3.12)$$

where we have generalized it to D-dimensional surfaces in the $d = D+1$ dimensional space. One can now regard the $\mathcal{H}_I$ term as a perturbation around the Gaussian Hamiltonian $\mathcal{H}_0$ and calculate the physical quantities as a series in $\frac{k_B T}{\kappa_0}$. The lowest order terms of $\mathcal{H}_I$ are:

$$\mathcal{H}_I = \int d^D x \left\{ \frac{r_0}{8} (\nabla u)^4 + \frac{\kappa_0}{4} (\nabla^2 u)^2 (\nabla u)^2 + \frac{K_0}{2} \nabla^2 u \times \nabla u \cdot \nabla (\nabla u)^2 \right\} + \cdots .$$
$$(3.13)$$

The calculations can be summarized schematically in the following way:

(i) one defines

$$Z[\lambda] = \frac{1}{Z[0]} \int \mathcal{D} u \, e^{-\beta[\mathcal{H} + \mathcal{H}_I - \lambda u]} \qquad (3.14)$$

and $W[\lambda] = \ln Z[\lambda]$. Then one can introduce the so-called effective potential $\Gamma(\mathcal{U})$ as:

$$\Gamma[\mathcal{U}] = \int d^D x \lambda(x) u(x) - W[\lambda], \qquad (3.15)$$

where $\mathcal{U} = \frac{\delta W}{\delta \lambda} = \langle u(x) \rangle$. The effective potential can be expanded around $\mathcal{U} = 0$:

$$\Gamma[\mathcal{U}] = \Gamma[0] + \int d^D x \frac{\delta \Gamma}{\delta \mathcal{U}} \bigg|_{\mathcal{U}=0} \mathcal{U}(x) + \int d^D x \, d^D x' \frac{\delta^2 \Gamma}{\delta \mathcal{U}(x) \delta \mathcal{U}(x')} \bigg|_{\mathcal{U}=0} \mathcal{U}(x) \mathcal{U}(x') + \cdots .$$
$$(3.16)$$

The second derivative term

$$\Gamma^{(2)}(\vec{x}, \vec{x}') \equiv \frac{\delta^2 \Gamma}{\delta \mathcal{U}(\vec{x}) \delta \mathcal{U}(x')} \bigg|_{\mathcal{U}=0}$$

and its Fourier transform $\Gamma^{(2)}(q)$ is used to define the effective tension $r_{eff}$ and rigidity $\kappa_{eff}$:

$$\Gamma^{(2)}(q) = k_B T \left( r_{eff} q^2 + \kappa_{eff} q^4 + \cdots \right) \qquad (3.17)$$

(ii) in order to calculate $\Gamma^{(2)}(q)$ (and other quantities) one builds a perturbation scheme, with the free propagator:

$$\longrightarrow \qquad G_0(p) = \frac{1}{r_0 p^2 + \kappa_0 p^4} \qquad (3.18)$$

and the three types of vertex corresponding to three terms of $\mathcal{H}_I$, Eq. (3.13):

and

In the order of *one loop*, i.e. in the first order expansion in $\frac{k_B T}{\kappa_0}$, the following diagrams contribute to $\Gamma^{(2)}(q)$:

$\qquad 4 \cdot q^2 \int \dfrac{p^2}{r_0 p^2 + \kappa_0 p^4} \dfrac{d^D p}{(2\pi)^D}$

$\qquad 8 \cdot \int \dfrac{(p \cdot q)^2}{r_0 p^2 + \kappa_0 p^4} \dfrac{d^D p}{(2\pi)^D}$

$\qquad 2 \cdot q^4 \int \dfrac{p^2}{r_0 p^2 + \kappa_0 p^4} \dfrac{d^D p}{(2\pi)^D}$

$\qquad 2 \cdot q^2 \int \dfrac{p^4}{r_0 p^2 + \kappa_0 p^4} \dfrac{d^D p}{(2\pi)^D}$

$\qquad 2 \cdot q^2 \int \dfrac{(p \cdot q)^2}{r_0 p^2 + \kappa_0 p^4} \dfrac{d^D p}{(2\pi)^D}$

$\qquad 2 \int \dfrac{(p \cdot q)^2 p^2}{r_0 p^2 + \kappa_0 p^4} \dfrac{d^D p}{(2\pi)^D}.$

Note that all these diagrams are UV divergent in $D = 2$. One must therefore introduce some *regularization scheme*. Here we chose the *dimensional regularization*, i.e. we put $D = 2 - \epsilon$ so that the integral

$$I = \frac{1}{(2\pi)^D} \int \frac{d^D p}{r_0 + \kappa_0 p^2}$$

behaves for $\epsilon \approx 0$ as $I(\epsilon) \sim \frac{1}{\epsilon} \left[ \frac{1}{(2\pi)^D} S_{D-1} \frac{1}{\kappa_0} \left( \frac{r_0}{\kappa_0} \right)^{-\epsilon/2} \right]$. The existence of the pole at $\epsilon = 0$ reflects the UV divergence of $I$ in $D = 2$. $\left(\text{Here } S_{D-1} = \frac{(2\pi)^{D/2}}{\Gamma(D/2)}.\right)$ With this choice one can easily evaluate $\Gamma^{(2)}(q)$ and obtain:

$$r_{\text{eff}} = r_0 \left( 1 + \frac{k_B T}{4\pi\kappa_0} \left( \frac{r_0}{\kappa_0} \right)^{-\epsilon/2} \frac{S_{D-1}}{(2\pi)^{D-1}} \frac{1}{\epsilon} \right) \tag{3.19a}$$

$$\kappa_{\text{eff}} = r_0 \left( 1 - \frac{3k_B T}{4\pi\kappa_0} \left( \frac{r_0}{\kappa_0} \right)^{-\epsilon/2} \frac{S_{D-1}}{(2\pi)^{D-1}} \frac{1}{\epsilon} \right) \tag{3.19b}$$

(iii) We now get the equations for the *renormalization flow* in the framework of the so-called *minimal subtraction scheme*. Let $\mu$ be an arbitrary wavevector scale of the system, and $\kappa$ (without dimension) the renormalized bending constant on this scale: $\kappa_{eff} = \mu^{-\varepsilon}\kappa(\mu)$. In particular we want to calculate the $\beta$ function:

$$\beta(\kappa) \equiv \mu \left.\frac{d\kappa}{d\mu}\right|_{\kappa_0} , \tag{3.20}$$

where $\kappa_0$, kept constant, is $\kappa_0 = \mu^{-\varepsilon}\kappa(\mu)\, Z_\kappa(\kappa, \varepsilon)$.

By expanding the renormalization factor $Z_\kappa$ in powers of $\frac{k_B T}{\kappa}$ and $\frac{1}{\varepsilon}$:

$$Z_\kappa(\kappa, \varepsilon) = 1 + A_\kappa \frac{k_B T}{\kappa} \cdot \frac{1}{\varepsilon} + \cdots \tag{3.21}$$

one obtains

$$\beta(\kappa) = \varepsilon\kappa + A_\kappa \cdot k_B T + \cdots . \tag{3.22}$$

From Eq. (3.19) and $\kappa_{eff} = \mu^{-\varepsilon}\kappa(\mu)$ one easily arrives at $A_\kappa = \frac{3 S_{D-1}}{2(2\pi)^D}$. Therefore one finally gets

$$\beta_\kappa \equiv \mu \left.\frac{d\kappa}{d\mu}\right|_{\kappa_0} = \kappa\left(\varepsilon + \frac{3 k_B T}{2\kappa} \frac{S_{D-1}}{(2\pi)^D}\right) \tag{3.23}$$

and similarly

$$\beta_r \equiv \mu \left.\frac{dr}{d\mu}\right|_{r_0} = -D - r\left(\varepsilon + \frac{3 k_B T}{2\kappa} \frac{S_{D-1}}{(2\pi)^D}\right) \tag{3.24}$$

(iv) The renormalization flow defined Eqs. (3.23) and (3.24) for $D = 2$ is in agreement with a general picture described above, e.g. with Fig. 3.1. In particular, from Eq. (3.23) we obtain the important result (3.2) for the dependence of the rigidity $\kappa$ on the size of the membrane $L$, simply by integrating this equation up to the scale $\mu = L^{-1}$.

More interestingly, for $\varepsilon < 0$ and thus for *hypothetical* $2 + |\varepsilon|$ dimensional membranes embedded in $3 + |\varepsilon|$ space Eq. (3.23) implies the existence of a non-trivial fixed point at (see Fig. 3.3):

$$T_c = \frac{2\kappa}{3 k_B} \frac{(2\pi)^D}{S_{D-1}} |\varepsilon|. \tag{3.25}$$

This corresponds to *the crumpling transition*: for $T < T_c$ the rigidity $\kappa(L)$ increases with $L$: the membrane behaves as a flat, rigid object, with $\xi_p = \infty$. It is only for $T > T_c$ that the membrane is crumpled, and $\xi_p$ is finite.

As we have already stressed earlier, two-dimensional fluid membranes are always in the crumpled phase: $T_c = 0$, a situation similar to the case of linear polymers. However, the long-range forces present in the system can lower the critical dimension $D = 2$ for the crumpling transition. In such a case the two-dimensional membranes would have $T_c > 0$ and the *true long-range order in the orientations* of the membrane would exist for low enough temperatures.

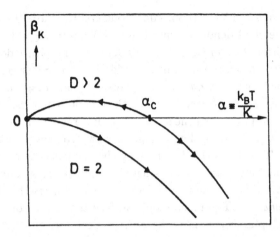

Fig. 3.3.  The $\beta$ function for the bending rigidity $\kappa$ calculated to order $\varepsilon$. For $D = 2$ one obtains an assymptotically free theory corresponding to completely crumpled membranes. For $D > 2$ (or in the presence of long range forces) a crumpling transition fixed point $\alpha_C$ is found.

An important contribution made by L. Peliti and his collaborators[16,58,59] was the observation that *polymerized membranes* can undergo a crumpling transition (ever for $D = 2$). For these systems the internal, phonon-like degrees of freedom induce a non-linear coupling between the thermal undulations of the membrane.[58] This coupling can also be regarded as an effective long-range interaction between the (Gaussian) curvature in distant points, and it can induce the finite temperature transition.

The existence of the crumpling transition for polymerized membranes has been confirmed by Monte-Carlo simulations (see the lectures by Y. Kantor[60]), and the large-$d$ expansion of a continuum model (see the lectures by F. David[15]).

## 3.3. *The Thermodynamic Behavior of an Ensemble of Fluid Membranes*

Up to now we have considered the case of a single, isolated membrane with a fixed topology. Real amphiphilic systems, however, such as lipid/water mixtures or microemulsions, consist of many fluctuating films or membranes which exchange their constituents and therefore easily vary their topology. Recently, a lot of theoretical effort has been devoted to the study of the thermodynamics of such systems.[61] We present here some recent results obtained by D. Huse and the author[61] within the framework of a simple model of fluctuating random surfaces (films).

The model considers fluid films which are non-interesting and the varying topology. It is based on the following *film Hamiltonian*, studied in the grand canonical ensemble:

$$\mathcal{H} = \int d^2\sigma \left[ r_0 + \frac{\kappa_0}{2} H^2 + \bar{\kappa}_0 K \right]. \tag{3.26}$$

The films are assumed to be without edges; one can therefore define the two *sides* of a given film (we shall denote them $S1$ and $S2$). Since $H_0 = 0$ the Hamiltonian (3.26) has an additional symmetry: the two sides of the films are identical (Fig. 3.4). Note that there are no bulk terms which would favor the $S1$ or $S2$ "component".

Let us first consider the case $\bar{\kappa}_0 = 0$. A plausible phase diagram of this model as a function of $r_0$ and $\alpha_0 = \frac{k_B T}{\kappa_0}$ (analogous to the one presented on Fig. 3.1) is shown on Fig. 3.5. For large positive $r_0$ the system does not want to have too much of film present (we remind the reader that $r_0$ can be viewed as the chemical potential of the amphiphiles). It achieves this by breaking the symmetry between the two sides of the film, forming, for instance, an $S1$-rich phase in which some finite unconnected domains of the minority $S2$ component are present. This is the so called droplet phase or *vesicle phase*.[62] The macroscopic surface tension, $\sigma$, of a film separating $S1$-rich and $S2$-rich phases is nonzero here. As $r_0$ decreases the vesicles grow in size and start forming larger connected domains. The domains of minority component percolate before the volume fraction it occupies reaches $1/2$. This leads to a *tense, bicontinuous phase* for intermediate positive values of $r_0$. In this phase the $S1/S2$ symmetry is still spontaneously broken, and a macroscopic surface tension is still nonzero, although infinite percolating domains of both $S1$ and $S2$ components are present.

For a large $\kappa_0$ and sufficiently negative $r_0$ the system prefers to include as much film as possible, but not to bend the film. A possible way to do this is to stack films parallel to one another. This produces the *smectic lamellar phase*, which is characterized long-range order in the orientation of the films, and quasi long-range

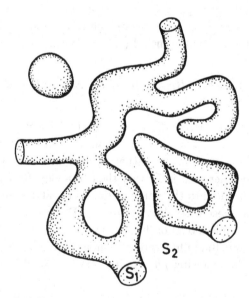

Fig. 3.4.   A fluctuating fluid film which separates two bulk phases. The topology of the film can vary but it is assumed that the film does not have free edges.

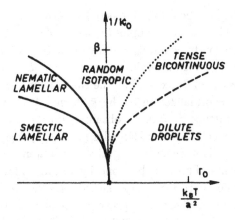

Fig. 3.5. A possible phase diagram for a fluctuating fluid membrane. The topology of the membrane can vary freely. [Ref. 61.]

order in the positions of the film planes. This quasi long-range positional order gives rise to power law divergences of the scattering intensity $S(q)$ (see next section). In the present model the average distance between the lamellar is determined by the competition between negative $r_0$ term and the "steric", fluctuation-induced repulsion[63] (see next section). There is also the possibility that at intermediate negative values of $r_0$ a *lamellar nematic phase* could appear with long-range order in the orientation of the films, but only short-range positional order. Note that the full rotational symmetry of the Hamiltonian (3.26) is spontaneously broken in both lamellar phases. At values of $r_0$ near zero, between the lamellar and tense phases, there should exist a *disordered, random isotropic phase*, where no symmetries of the Hamiltonian are spontaneously broken. The random isotropic phase, also called *sponge phase* has recently become the subject of several theoretical and experimental studies[64] (in lipid/water systems). For $r_0 > 0$ the system can be shown to be analogous to a ferromagnetic Ising model, with $r_0$ the nearest neighbor coupling and $\kappa_0$ a higher-order (4 spins or more) interaction that does not break the global Ising symmetry. The sponge phase is then the high-$T$ paramagnetic phase, while tense bicontinuous and dilute vesicle phases correspond to ferromagnetic phase. Note that in this analogy the magnetic field corresponds to a bulk chemical potential which advantages one of the components ($S1$ or $S2$), while $H_0$ is a higher order (3 spins or more) interaction that breaks the Ising symmetry.

If one allows $\bar{\kappa}_0 \neq 0$ then other phases not shown on Fig. 3.5 can be stabilized. From the Gauss-Bonnet theorem (see lectures by F. David[15]) the integral of the Gaussian curvature is

$$\int d^2\sigma\, K = 4\pi(n_c - n_h) \qquad (3.27)$$

where $n_c$ and $n_h$ are the number of disconnected parts of the film and the number of handles, respectively. Thus $\bar{\kappa}_0$ only couples to the topology of the film $\bar{\kappa}_0 > 0$ favors more handles and fewer parts, while $\bar{\kappa}_0 < 0$ favors more parts and fewer handles.

Figure 3.6 shows a possible phase diagram for our model with $\bar{\kappa}_0 \neq 0$ as a function of $r_0$ and $\bar{\kappa}_0$ at $\kappa_0 > k_B T$ fixed. (For the sake of simplicity we have assumed here that there is no renormalization of the $\bar{\kappa}_0$ term.) If we first move on the $r_0$ axis of this diagram, putting $\bar{\kappa}_0 = 0$, we obtain again the same sequence of phase as in Fig. 3.5, namely lamellar smectic-lamellar nematic-sponge-tense bicontinuous-vesicles. However, as $\bar{\kappa}_0$ is sufficiently negative the system prefers to form many disconnected components. This can be done by forming many roughly spherical vesicles. When the vesicles are very close-packed they should freeze into a *vesicle crystal*, although dynamical effects can prevent this in real systems, where amorphous phases could form instead. Figure 3.7 shows a photograph of a dense phase of niosomes, i.e. multilammelar vesicles made of non-ionic surfactants.[65] Such structures are thermodynamically stable (at least for a period of many years).

If $\bar{\kappa}_0$ is increased from zero we expect that each of the phases discussed above will change itself into a new ordered phase, which we call *"plumbers nightmare"*.[66,67] The ideal plumber's nightmare phases are fully connected, periodic and topologically nontrivial surfaces that have $H = 0$ everywhere. For this last reason they are often called minimal surfaces[68] (see the lectures by F. David[15]), but for a periodic structure its area does not need to be minimal.

Plumber's nightmares have been studied intensively by mathematicians.[69] Figure 3.8 shows a newly obtained example of a surface with tetragonal symmetry:[70]

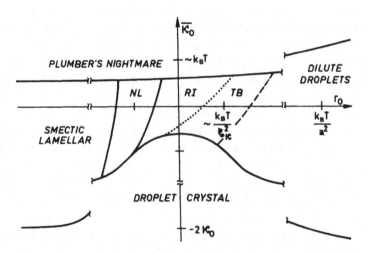

Fig. 3.6.   Proposed phase diagram for a fluctuating fluid membrane the topology of which can vary. For $\bar{\kappa}_0$ one recovers the sequence of phase presented in Fig. 3.5; the abbreviations are: NL — nematic lamellar, RI — random isotropic, TB — tense bicontinuous. For $\bar{\kappa}_0$ positive and large enough the membrane may form a crystalline structure called a "plumber's nightmare". For large negative values of $\bar{\kappa}_0$ the membranes form a crystal of vesicles or droplets. [Ref. 61.]

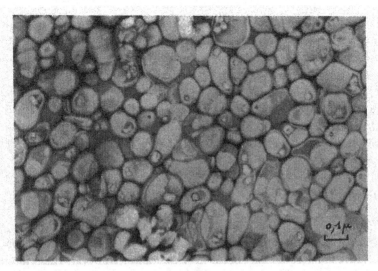

Fig. 3.7.    Electron microphotograph of an ensemble of niosomes [G. Vanlerberghe *et al.*, Ref. 65, (reprinted with permission of Springer-Verlag)]. These multilamellar structures are now currently used in cosmetics.

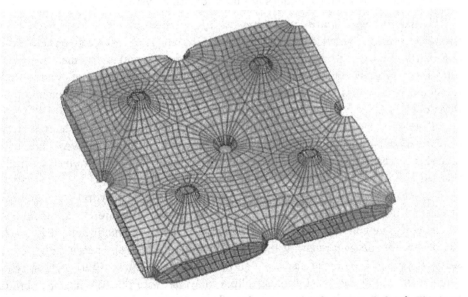

Fig. 3.8.    Unit cell of a tetragonal minimal surface [courtesy of A.C. Maggs, Ref. 70]. This is an example of an ordered plumber's nightmare.

we can see that such structures have many handles (of order one per unit cell) and, ideally, only one component, so for $\bar{\kappa}_0 > 0$ the Gaussian energy term favors them. In principle, many possible plumber's nightmares can be allowed by a given system: there could exist *several structural phase transitions* among them.[70]

The different periodic phases and the transitions between them have indeed been observed in various amphiphilic systems.[71] Note, however, that in real systems the plumber's nightmare phases could appear as the result of competition between the curvature energy term and the energy of stretching of the hydrocarbon chains[72] (the latter being an "internal" degree of freedom not explicitly treated here) or the interactions between membranes (also neglected here). We shall return to this important point in the last section. Before doing this, however, we shall now discuss the simplest and the most studied of the phases described above, which is the lamellar smectic phase.

## 4. Unbinding Transitions and the Swelling of Lamellar Phases

### 4.1. *Molecular Forces between Membranes*

In the studies of fluctuating membranes described above we have completely neglected molecular interactions between membranes. However, these interactions often determine the equilibrium of amphiphilic systems.

Three main experimental methods are used to determine the molecular forces between amphiphilic layers (we restrict here the discussion only to lipid/water systems although the same techniques are used for other systems):

(i) the *micromechanical measurements of interactions between vesicles.*[74] In these experiments two vesicles, partially sucked into micropipettes, are brought close one another, and the force acting between them is measured (e.g. by measuring the pressure in the micropipettes, see Fig. 4.1). One advantage of this method is that it can measure the forces between both fluctuating and non-fluctuating membranes (unswollen and swollen vesicles, respectively).

(ii) *direct force measurements* between absorbed layers.[10] Amphiphilic molecules can be absorbed on mica cylinders and the force can be measured with the great precision. At the same time the distance between layers can be determined with the precision of a few Angströms!

(iii) *measurements of the osmotic pressures* necessary to dehydrate the lamellar phases.[75] Figure 4.2 shows a typical curve of the osmotic pressure $P$, e.g. controlled by changing the concentration of long polymeric chains in equilibrium with lamellae, plotted versus the (mean) distance, $l$, between lamellae (e.g. measured by X-ray scattering). For $P \to 0$ the distance $l$ saturates at some equilibrium value, $l_0$ typically of order 30 Å for neutral phospholipid bilayers. Note that the intermembrane forces measured in this way act between *fluctuating* membranes and thus are *not* purely of molecular origin.[75]

From these and other measurements a simple picture of molecular intermembrane interactions emerges; the effective potential $V_m(l)$ can be written as a sum of three principal terms

$$V_m(l) = V_h(l) + V_{vW}(l) + V_{el}(l), \qquad (4.1)$$

where:

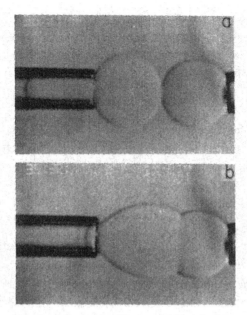

Fig. 4.1. Video micrographs of vesicle-vesicle adhesion. (a) Vesicles in contact, but do not adhere. (b) Spontaneous adhesion allowed to progress by controlling the suction applied to the vesicle on the left. [Ref. 74.] This micromechanical method of measuring the forces between membranes was developed by E. Evans and collaborators.

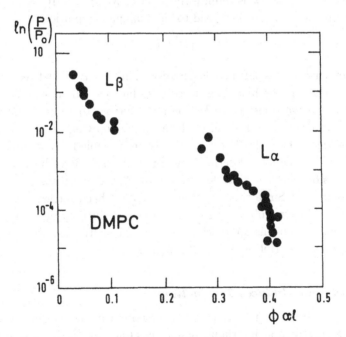

Fig. 4.2. A typical variation of osmotic pressure P with the mean distance between bilayers $\ell$. [Redrawn from R.E. Goldstein, Ph.D. thesis, Cornell U., 1988.] These measurements were done in DMPC/water lamellar phase, and clearly show the existence of a first order phase transition between the fluid-like and solid-like phases (see Sec. 5). [Ref. 100.]

(a) $V_h(l)$ is the so-called *hydration force*, i.e. a repulsive, short range interaction mediated through water layers. It is believed that the ordering of water molecules next to polar heads is the source of this interaction, its from is — in a very good approximation — exponential:

$$V_h(l) \approx A_h e^{-l/\lambda_h} \qquad (4.2)$$

with $A_h$ of order of few hundreds dyne/cm$^2$, $\lambda_h \sim 2 \div 3\,\text{Å}$;

(b) $V_{vW}(l)$ is the attractive *van der Waals interaction*. If retardation effects can be ignored, one has

$$V_{vW}(l) \simeq \begin{cases} -W\delta^2/l^4, & \text{for } l \gg \delta \\ -W/l^2, & \text{for } l \leq \delta \end{cases} , \qquad (4.3)$$

where $W$ is the Hamaker constant, of order of few $10^{-14}$ erg. Retarded van der Walls forces lead to $-1/l^5$ behavior, which is expected to apply for $l \geq 500\,\text{Å}$;

(c) $V_{el}(d)$ denotes the *electrostatic repulsive interactions*, present if membranes are charged. For $l$ compared to the Debye length, $\lambda_E$, of the ionic solution:

$$V_{el}(l) \sim e^{-1/\lambda_E}, \qquad (4.4)$$

where $\lambda_E \sim$ (ion concentration)$^{1/2}$. If there are no ions between the membranes apart from counterions, $\lambda_E$ is infinite (in practice $\lambda_E \geq 10^3 - 10^4\,\text{Å}$ in pure water) and the electrostatic interactions lead to the Langmuir repulsion:

$$V_{el}(l) \sim 1/l, \quad \text{for large } l. \qquad (4.5)$$

Of course for small values of $l$ there is a hard-wall like repulsion between membranes. In practice, however, the hydration force is so big that one does not usually gets to this regime: to overcome the hydration force one has to apply pressures of order of a few $10^4$ atmospheres (this can be done e.g. by applying osmotic stresses). The large amplitude of the hydration forces is also responsible for $l_0 \approx 30\,\text{Å} \gg \lambda_h$.

The direct interactions described by Eq. (4.1) have long been known. More recently, it has been argued that, for $l \simeq 10-100\,\text{Å}$, the intermembrane interactions are more complex.[77] Since this question has not yet been settled satisfactorily we shall assume the intermembrane forces to be well described by Eq. (4.1). As we shall see in the next section this approach — at least for neutral membranes — leads to quite an accurate description of lamellar phases.[78]

## 4.2. Fluctuations-Induced Interactions

Some lamellar phases of the amphiphilic systems can be swollen by adding the solvent to extremely large interlamellar spacings ($d_0 > 10^{-4}\,\text{Å}!$).[79] In such cases the lamellae can again be viewed as *fluctuating, fluid,* (and interacting) *thin membranes,* with the thickness $\delta \ll l_0$. Some years ago, Helfrich[63] pointed out that thermal fluctuations led to an effective *repulsive* interaction between membranes. For $D = 2$ and $d = 3$, he predicted that this effective potential could be expressed as

$$V_{fl}(l) \simeq \frac{(k_B T)^2}{\kappa l^2}. \tag{4.6}$$

This interaction is of entropic origin: a membrane is hindered in its fluctuations by the presence of another, neighboring membrane and therefore is repelled by it.

Equation (4.6) can be rederived from scaling arguments analogous to those presented at this school by M.E. Fisher in his lectures on wetting phenomena.[80] First, let us assume that the membranes are completely separated. Each membrane (which we assume to be perfectly symmetric, so that the spontaneous curvature $H_0 = 0$) is governed then by the Hamiltonian (3.2). We shall omit nonlinear terms (assuming $(\nabla u)^2 \ll 1$), and put $r = 0$, so that we are left with

$$H_0\{u\} = \int d^D\sigma \frac{\kappa_0}{2}(\nabla^2 u)^2. \tag{4.7}$$

This simple form of the Hamiltonian implies that a membrane segment of longitudinal dimension, $L_\parallel$, will typically make transverse excursions $L_\perp \sim L_\parallel^\varsigma$, with[81]

$$\varsigma = \frac{5-d}{2} \quad \text{for } D+1 = d < d_0 \equiv 5. \tag{4.8}$$

Now, if the membranes are bound together, the transverse excursions are limited and the largest "humps"[73] have a typical size $\xi_\perp$. Then, scaling arguments show that these largest humps have an extension, $\xi_\parallel$, with[81]

$$\xi_\perp \approx \left( C_\infty \frac{k_B T}{\kappa} \right)^{1/2} \xi_\parallel^\varsigma, \tag{4.9}$$

where $C_\infty \sim O(1)$. These humps of the confined membrane lead to an increase of the free energy per unit area[81]

$$V_{fl}(\xi_\perp) \approx C_v k_B T \left( \frac{k_B T}{\kappa} \right)^{\tau/2} \xi_\perp^{-\tau}, \quad (d < 5) \tag{4.10}$$

with $C_v \sim O(1)$ and $\tau = 2\frac{d-1}{5-d}$. For $d = 3$, the Helfrich interaction (4.6) is recovered *provided one assumes that $l \sim \xi_\perp$*. This relation is indeed valid for hard-wall intermembrane interactions ($V(l) = V_{HW}(l)$) but does not hold in general: sufficiently long ranged molecular interactions $V_m(l)$ lead to $\xi_\perp \ll l$.[81,82]

Before we go any further, let us stress two important assumptions which have been made implicitly here:

(i) the membranes are supposed to be quasi-planar. For separations $l$ between membranes larger than the persistence length $\xi_p$ the membrane starts to crumple and the above description does not hold;

(ii) nonlinear terms present in the Hamiltonian (3.12) are neglected. If one took such terms into account a renormalization of the rigidity $\kappa_0$ would follow. Thus the coefficient $\kappa$ in Eqs. (4.6) and (4.10) should probably be thought of as $\kappa_0$ renormalized at lengths $\xi_\|$[61]

$$\kappa = \kappa(\xi_\|) \tag{4.11}$$

### 4.3. The Competition between Molecular and Fluctuation-Induced Interactions: Functional Renormalization

In order to study the interplay between direct, molecular interactions (between planar membranes) and fluctuation-induced interactions in a systematic way one cane introduce an effective Hamiltonian:

$$\mathcal{H}\{l\} = \int d^D x \left\{ \frac{\kappa}{2}(\nabla^2 l)^2 + V_m[l(\vec{x})] + Pl \right\}, \tag{4.12}$$

where $l(\vec{x})$ is a local distance between two membranes: $l(\vec{x}) = u_1(\vec{x}) - u_2(\vec{x})$, and $V_m$ is the molecular interaction described above. The linear term $Pl$ represents *the osmotic pressure* effect and/or corresponds to the *constraint* imposed on the average distance between membranes: $\langle l(\vec{x}) \rangle = \vec{l}$. The expression (4.12) implicitly contains a small-distance cutoff, $1/\Lambda \simeq \delta$.

The effective Hamiltonian (4.12) resembles the interface models studied in the context of wetting phenomena (see the lectures by M.E. Fisher[80]). Here we summerize briefly one of the most successful approaches used in this field: the *functional renormalization group* (R.G.) method:[83,84]

(i) the functional R.G. represents an extension of Wilson's approximate recursion relation;[85] the main idea underlying this method is to integrate out the membrane fluctuations of the wavevectors induced between $\Lambda/b$ and $\Lambda$, and then introduce a change of scale: $\Lambda/b \to \Lambda$, $x \to x/b$ and

$$l \to l\,b^\varsigma. \tag{4.13}$$

Such a transformation acts as a *nonlinear map*, $\mathcal{R}$, in the function space of direct interactions, $V(l)$, while it leaves the elastic term $(\nabla^2 l)^2$, unchanged. This implies that the variable $l$ does not acquire an anomalous dimension;[84]

(ii) the initial interaction $V^{(0)}(l) \equiv V(l)$ is renormalized by successive applications of $\mathcal{R}$:

$$V^{(N+1)}(l) = \mathcal{R}[V^{(N)}(l)], \tag{4.14}$$

where

$$\mathcal{R} = [V(l)] = -vb^D \ln \left\{ \int_{-\infty}^{+\infty} \exp \left[ -\frac{1}{2}\left(\frac{z}{a_\perp}\right)^2 - G(l,z) \right] /(2\pi a_\perp^2)^{1/2} \right\} \tag{4.15}$$

with

$$G(l,z) \equiv [V(b^\varsigma l - z) + V(b^\varsigma l + z)]/2z \tag{4.16}$$

and the free energy density scale

$$v \equiv k_B T \int_{\Lambda/b}^{\Lambda} d^D q / (2\pi)^D \qquad (4.17a)$$

and the roughness $a_\perp$ of the membranes arising from excitations of $\Lambda/b \le q \le \Lambda$:

$$a_\perp^2 \equiv \frac{k_B T}{\kappa} \int_{\Lambda/b}^{\Lambda} \frac{d^D q}{(2\pi)^D q^4} \qquad (4.17b)$$

(iii) in the infinitesimal rescaling limit $b = e^{\delta t}$, with $\delta t \to 0$, one obtains the nonlinear flow equation[83]

$$\frac{dV}{dt} = D \cdot V + \varsigma l \frac{\partial V}{\partial l} + \frac{B}{2} \ln \left[ 1 + \frac{A^2}{B} \frac{\partial^2 V}{\partial l^2} \right] \qquad (4.18)$$

with cut-off dependent scale parameters $A$ and $B$.

(iv) comparing the present R.G. scheme with the original Wilson method[85] one can notice three main differences:

(1) normalization in $\mathcal{R}$ is chosen to preserve the form of $V^{(N)}(l)$ for large $l$;
(2) $a_\perp$ is *not* arbitrary but chosen so that the R.G. transformation is exact in linear order in $V$ for all $b$ and $D$;
(3) the "wave-function" renormalization included in (4.13) is exact (the exponent $\eta = 0$);
(4) the fixed points are the potentials $V^*$ such that $\mathcal{R}[V^*(l)] = V^*(l)$. In addition to atrivial Gaussian fixed point $V_G^* \equiv 0$ *the numerical iterations* of the recursion relations (4.15)–(4.17) reveal the existence of two nontrivial fixed points, $V_0^*(l)$ and $V_c^*(l)$, for $d < d_0 = 5$. In $d = d_0 = 5$, these two fixed points do not bifurcate from the *Gaussian fixed point*, $V_R^* = 0$, but rather from a line of drifting fixed points[80,84] (see Fig. 4.3).

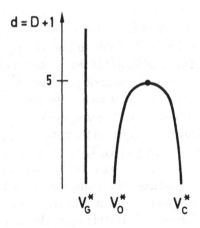

Fig. 4.3.   Unusual bifurcation of fixed point potentials as a function of dimension d. [ Refs. 80, 84.]

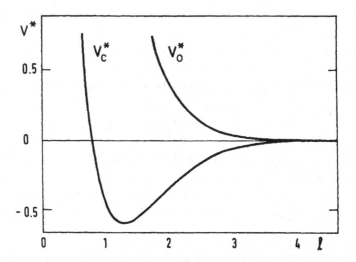

Fig. 4.4. Fixed point potentials obtained through functional renormalization group scheme [redrawn from R. Lipowsky, *Europhys. Lett.* **7**, 255 (1988)].

Within the subspace of molecular interactions, $V_m(l)$, such that $V_m(l) \ll 1/l^\tau$, for large $l$ $\left(\text{where } \tau = \frac{2(d-1)}{5-d}\right)$, there is no relevant perturbation at $V_0^*(l)$, therefore this fixed point has a large domain of attraction. In particular, a hard wall repulsion is mapped onto $V_0^*$. This purely repulsive fixed point (see Fig. 4.4) can then be called the "*hard wall*" fixed-point. In contrast, there is one relevant perturbation at the other fixed point, $V_c^*(l)$; (see Fig. 4.4) we shall thus call it *the "critical" fixed point*. We shall see below that it governs a new critical phenomenon: *the critical unbinding transition.*

## 4.4. *Complete versus Incomplete Unbinding and the Swelling of Lamellar Phases*

Imagine a lamellar phase consisting of a stack of membranes separated by layers of solvent. The addition of solvent results in the *swelling* of the lamellar crystal.[82] In the language of the Hamiltonian (4.12) this process corresponds to a relaxing of the constraint (or a lowering of the osmotic pressure): $P \to 0$. As $P \to 0$, the mean separation, $\bar{l}$, of the membranes can attain a finite limit ($l_0$, see Fig. 4.1), or become arbitrarily large. These two cases correspond to *incomplete* or *complete* unbinding respectively. It is clear that the behavior of the membranes for $P \to 0$ will, in general, depend on their molecular interactions, $V_m(l)$. If these interactions are repulsive, i.e., $V_m(l) \approx V_R(l) \geq 0$ for large $l$, as it is the case for the Langmuir repulsion $V_m(l) \sim 1/l$, then the membranes will completely unbind as $P \to 0$. On the other hand, the direct interaction can also be attractive for large $l$: $V_m(l) \approx V_A(l) + V_R(l)$, with $|V_A(l)| \gg V_R(l)$ for large $l$. In this case, the membranes may *not* unbind completely if $V_A(l)$ is sufficiently long-ranged and satisfies $|V_A(l)| \gg 1/l^2$ for large

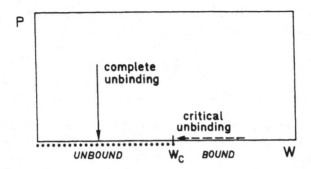

Fig. 4.5. The swelling of a lamellar phase can be viewed as the limit $P \to 0$, where $P$ is a pressurelike variable. For values of Hamaker constant W smaller than the critical value $W_c$ complete unbinding takes place. In this $(P, W)$ diagram the critical unbinding transition corresponds to the approach of $(W, P) = (W_c, 0)$. (Reprinted by permission of *Phys. Rev. B* **35** (1987) 7004.)

$l$ $(d = 3)$. For the systems described above, the van der Walls potential (4.3) satisfies

$$|V_A(l)| \ll 1/l^2, \quad \text{for large } l \tag{4.19}$$

and the membrane can unbind completely or incompletely depending on the strength of the attractive interaction, $V_A(l)$[73,81] (and therefore depending on the value of the Hamaker constant $W$ — see Fig. 4.5). What happens can be determined within the functional $R.G.$ approach: the membranes unbind completely wherever they interact through a molecular potential, $V_m(l)$, such that it is mapped onto the hard wall fixed point, $V_0^*(l)$. Such a *complete unbinding transition* can be charac- terizes by an exponent $\psi$:

$$\bar{l} \sim 1/P^\psi, \quad \text{as } P \to 0. \tag{4.20}$$

This critical exponent is correctly given by *mean-field theory* as long as $V_m(l) = V_R(l) \gg 1/l^2$ $(d = 3)$. This happens, for instance, for the Langmuir repulsion:[82] $\psi = 1/2$ $(d = 3)$. On the other hand for sufficiently short-ranged interactions with $|V_m(l)| \gg 1/l^2$ for large $l$, one finds the *universal* value (unconnected with the nature of microscopic forces!): $\psi = 1/3$ $(d = 3)$. This is a so-called *weak-fluctuation regime* which is characterized by nonclassical critical exponents while the phase boundary is still given by $P = 0$ as in mean-field theory. Not that in this regime one can correctly take the fluctuation effects into account by adding to the term $Pl$ the Helfrich repulsion potential, $V_H(l)$, Eq. (4.6). This is not the case in general, as we shall see later. These scaling regimes can be distinguished experimentally by measuring the X-ray diffraction from the lamellar crystals.[86,87] One can show[82,88] that the effective Ginzburg–Landau Hamiltonian describing the *lamellar smectic phase* can be written as

$$\mathcal{H}_G\{u\} = \int d^2x_\parallel dz \left[ \frac{B}{2}\left(\frac{\partial u}{\partial z}\right)^2 + \frac{K}{2}(\nabla^2 u)^2 \right], \tag{4.21}$$

where $u(x_\parallel, z)$ describes the displacements of the layers (on average perpendicular to the $z$-axis), with

$$K = \frac{\kappa}{\bar{l}}, \quad B = \frac{\kappa \bar{l}^2}{\xi_\parallel^4 \bar{l}}. \tag{4.22}$$

The correlation length $\xi_\parallel$ can be determined[82] from the molecular potential, $V_m(l)$. For instance, for $V_m(l) = V_h(l) + V_{vW}(l)$ (see Eq. 4.1) one obtains for large $\bar{l}B \sim (k_B T)^2 / \kappa \bar{l}^3$.

It is well known[89] that the X-ray diffraction pattern from smectic crystals exhibits Landau-Peierls singularities.

$$I(q_z) \sim (q_z - q_m)^{-(2-X_m)}, \tag{4.23}$$

where $q_m = 2\pi m / \bar{l}$, $(m = 0, 1, 2, \ldots)$ and $q_z$ is the momentum transfer perpendicular to the membranes. Figure 4.6 shows a series of such peaks as the function of dilution (and thus with growing $\bar{l}$). They are well fitted by (4.23) and the exponent $X_m$ can thus be extrated.[87] We expect that the exponent $X_m = \frac{k_B T q_m^2}{8\pi (K \cdot B)^{1/2}}$ is independent of $\bar{l}$ (for large $\bar{l}$) in the weak-fluctuation regime,[81,87] whereas it depends explicitly on $\bar{l}$ within the mean-field regime, e.g. $X_m \sim 1/\bar{l}^{1/2}$ for Langmuir repulsion.[82] This qualitatively different behavior has indeed been observed in recent high-resolution experiments.[42] The weak-fluctuation regime behavior was observed both in lipid/water systems and the quasi-ternary surfactant/oil/brine systems diluted with the oil or the brine.[87] By changing the ion concentration and thus the Debye screening length one gets to the mean-field in the charged systems.[42]

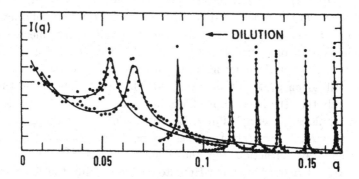

Fig. 4.6.   X-ray diffraction data obtained by high-resolution experiments in lamellar phases of quasi-ternary mixture [Refs. 3, 87]. The system is diluted by adding one component (oil). The shape of the first peak ($m = 1$), which moves with dilution, is well fitted with the power-law Eq. (4.23). (Reprinted by permission of *Phys. Rev. Lett.* **57** (1986) 2718.)

As a conclusion, these experiments constitute:

(i) the first quantitative measurements of the fluctuation-induced repulsion in membrane systems;

(ii) a precise technique for estimating the elastic properties of fluctuating membranes.

### 4.5. *The Critical Unbinding Transition*

Let us now assume that $P = 0$, in other words, that the lamellar crystal is in equilibrium with the bulk solvent. We shall consider for example a molecular inter-action of form $V_m(l) = V_h(l) + V_{oW}(l)$, see Eqs. (4.1)–(4.3). so that the attractive part $|V_A(l)| \ll 1/l^2$, for large $l$. The parameters of such an interaction form a low-dimensional subspace within this parameter space, there is a region where the membranes are completely separated and another region where they are bound together, as it is shown in Fig. 4.7 for instance. These two regions are separated by a phase boundary which consists typically of second order transitions at which $\bar{l}$ diverges as

$$\bar{l} \sim (W - W_{\text{cr}})^{-\psi}. \tag{4.24}$$

These transitions, called *critical unbinding transitions*, are governed by the fixed point $V_c^*(l)$ described above. A numerical study of functional R.G. equations leads to[81] $\psi \simeq 1.0$.

It is important to stress that the prediction of the critical unbinding transition could not be done just by superimposing the molecular forces $V_m(l)$ and the Helfrich interaction $V_H(l)$, Eq. (4.6). One then often talks about the *"strong-fluctuation"* regime. For (nonretarded) van der Waals attractive forces dominating large $l$ behav-ior such a regime governs the critical unbinding for $d < d^* = \frac{11}{3}$,[81] and thus in particular the physical $d = 3$ situation.

Up to now no clear evidence has been obtained of the existence of this non-trivial critical phenomenon.[81] Numerical studies however show that the values of critical parameters (such as $W_{cr}$) lie in the range which can be attained in oreal systems (e.g. for $\kappa \simeq 0.1-2 \times 10^{-12}$ erg, $\delta = 40$ Å, $A_H = 200$ erg/cm$^2$ and $\lambda_H = 3$ Å one obtains[81] $W_{cr} = 6-0.6 \times 10^{-14}$, a *very reasonable value!*). We believe there-fore that critical unbinding *can* be observed in lipid/water systems (by varying the temperature on the bending constant $\kappa$, e.g. through adding "cosurfactants") or in quasiternary systems (by varying the thickness of "membranes", $\delta$).

In any case the study of the stacks of membranes and, in particular, the study of the swelling of the lamellar phases will contribute to our understanding of the inter-actions between the membranes. This is turn can be crucial for the understanding of many fundamental processes in membrane biophysics, such as membrane fusion, exocytosis or endocytosis.[1]

## 5. Membranes with Internal Degrees of Freedom

### 5.1. The "Membranology" of f-Membrane Systems

Up to now we have discussed only the thermodynamics of infinitely thin, or theoretical, "t-membranes". As it was briefly mentioned in Sec. 3, however of freedom can strongly modify the thermal behavior of membranes. For instance, elastic, "internal" degrees of freedom can induce a crumpling transition by coupling to the shape-dependent, "external" variables.[58] In real ("fat" or f-) membranes the internal structure, connected, with amphiphilic molecules, plays an important role. It is one of the three essential aspects of membrane phenomena:

(i) "external" degrees of freedom, such as the curvature of membranes, their area,...

(ii) "internal" degrees of freedom, such as the fluidity of the constituent molecules, the tilt of the hydrocarbon chains, etc.

(iii) interactions between membranes.

Figure 5.1 shows how these three elements interact to produce some of the phenomena we have been describing here lectures. In this last section we would like to deal with the center of this "membranology" diagram, i.e. to consider phases the description of which much be based on all three elements. In order to do so in as progressive manner as possible we shall first study so-called curvature instabilities in a single membrane, with internal degrees of freedom but without interactions.

We shall then consider a stack for such membranes and show that the interactions among them can lead to a well-known phase behavior of the lamellar phases of lyotropic liquid crystals. Since the problem of lamellar phases is already quite complex (or rich, if one prefers to use this term to say so), in order to treat it within a "reasonable" framework, we have to make several simplifying assumptions, e.g. neglect the thermal fluctuations described in previous sections. This theoretical simplification will become even more drastic when we consider the cubic phases of liquid crystals. The geometry of these phases is already rich complex enough for us to forget about their thermodynamics altogether.

### 5.2. Curvature Instability in Fluid Membranes[90,91]

The simplest situation imaginable in which internal degrees of freedom modify the behavior of a fluid membrane is the case when a single, scalar variable $\phi(\sigma)$ is coupled to the local curvature of the membrane.[90] Such a variable can represent (in a more or less simplified way) various internal parameters:

(i) local density of amphiphilic molecules;

(ii) in a membrane made of two types of amphiphiles local concentration of one of the components;

(iii) average local amplitude of tilt of hydrocarbon chains;

(iv) local thickness of the membrane; etc.

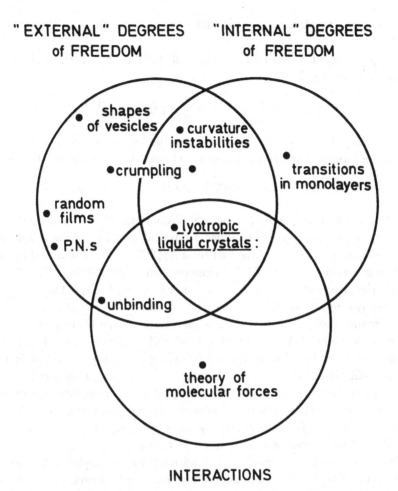

**Fig. 5.1.** "Membranology": the phenomena described in these lectures involve three essential ingredients: "external" degrees of freedom (such as shape, topology etc.), "internal" degrees of freedom and the interactions between membranes.

We shall assume that there is a phase transition associated with the variable $\phi(\sigma)$, such that $\langle\phi(\sigma)\rangle = 0$ in the disordered (high temperature) phase, and $\langle\phi(\sigma)\rangle \neq 0$ in the ordered (low temperature) phase. For instance, the binary mixture of example (ii) can have a consolute point below which the two constituents phase-separate. Near this point, internal order within the membrane can be described by a Landau Hamiltonian:[90]

$$\mathcal{H}[\phi] = \int \left\{ \frac{b}{2}(\nabla\phi)^2 + \frac{A}{2}(\nabla^2\phi)^2 + \frac{a_2}{2}\phi^2 + \frac{a_4}{4}\phi^4 - \mu\phi + \cdots \right\} d^2\sigma, \qquad (5.1)$$

where $b$, $A$, $a_2$, $a_4$ are phenomenological coefficients (assumed positive) and $\mu$ is the chemical potential. Without the gradient terms ($b = A = 0$). Equation (5.1) gives the usual critical point at $\mu = a_2 = 0$, and a line of first-order transitions for: $\mu = 0$, $a_2 < 0$.

Configurations of the membrane itself will be described here by the Gaussian Hamiltonian (see Eq. (3.12))

$$\mathcal{H}[l] = \int \left\{ \frac{1}{2} r_0 (\nabla l)^2 + \frac{1}{2} \kappa (\nabla^2 l)^2 + \cdots \right\} d^2\sigma \qquad (5.2)$$

and the nonlinear terms are neglected. Finally, the two lowest orders in the coupling between the internal variable $\phi$ and the membrane shape can be written as[91]

$$\mathcal{H}_I[\phi, l] = \int \{ \Lambda\phi(\nabla^2 l) + \lambda\phi(\nabla^4 l) + \cdots \} d^2\sigma, \qquad (5.3)$$

where $\Lambda$ and $\lambda$ are two "coupling constants". Equations (5.1)–(5.3) define the model completely. It is interesting to consider a physical meaning of the coupling (5.3). Let us do so on the example of the first term $\Lambda\phi(\nabla^2 l)$. This term is nothing but the spontaneous curvature term of the phenomenological Hamiltonian Eq. (3.1), since $\Lambda^2 l(\sigma)$ is the mean curvature at point $\sigma$. Here, however, the spontaneous curvature $H_0$ is directly proportional to the concentration variable, $\phi(\sigma)$ and thus varies within the membrane. The physical source of such a term is easy to understand: imagine that one constituent of the membrane is smaller than the other, e.g. it has only one hydrophobic chain. As shown on Fig. 5.2 (on a simpler example of a monolayer) such molecules will prefer to sit in curved regions of the membrane for purely "packing" reasons. Thus their concentration $\phi(\sigma)$ is coupled to the local curvature of the film. We expect that the phase separation between the two types of amphiphiles will be followed by the formation of curved regions with a high concentration of one of them. We call this phenomenon *curvature instability*.[90]

This expectation can indeed be confirmed by a simple treatment of our model.[91] If one neglects thermal fluctuations and limits oneself to *one-wavevector approximation* one obtains[91] the phase diagram shown in Fig. 5.3. The low temperature phases $H(IH)$ and $S$ correspond to *curved* and *ordered* ($\langle\phi\rangle \neq 0$) phases of

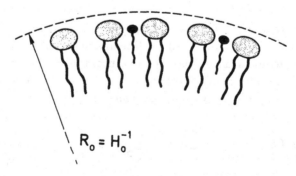

$$R_0 = H_0^{-1}$$

Fig. 5.2. In a binary mixture of amphiphilic molecules the local concentration of one of the constituants can be coupled to the local curvature of the film. This in fact is one of possible sources of the spontaneous curvature $H_0$.

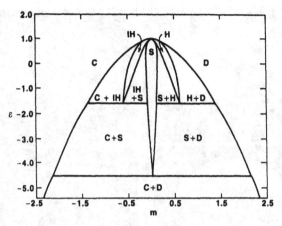

Fig. 5.3.  Phase diagram in the $(m, \epsilon)$ plane where $\epsilon$ is the reduced temperature and $m$ is the reduced order parameter. Here, as an example, this order parameter is chosen to be the local density of amphiphilic molecules. The two homogeneous phases: dilute $(D)$ and condensed $(C)$ are separated by the hexagonal $(H)$, stripe $(S)$ and inverted hexagonal $(IH)$ phases. Two-phase coexistence regions between the various single phases are also shown. This phase diagram was obtained in an approximation valid near the critical point [Ref. 91].

hexagonal and striped symmetries, respectively. The "mountains" and the "valleys" of the curved membrane are at the same time the regions of high concentration of one of the components.

Such ordered phases seem indeed to appear in membranes made of two different phospholipids. Experiments by E. Sackmann *et al.*[92] show the existence of both striped and hexagonally ordered phases. In these experiments, however, one of constituents becomes solid in low temperature phases. The theory presented here should thus be generalized in order to describe these systems.

It is tempting to connect the notion of curvature instability with the phenomenon of *echinocytosis* observed in red blood cells.[90] Figure 5.4 shows a red blood cell in its echinocyte form: it has strongly curved regions of well defined sizes and mutual distances. Although the shape transformation which leads to such structures (echinocytosis) can be induced by various biochemical factors (e.g. change of pH, depletion of ATP, connected to aging, "glass effect" etc.) one of the simplest and cleanest ways to bigger it is to absorb some small drug molecules into the membranes.[93] It can be shown that such molecules (e.g. anaesthetics) then absorb preferentially in one of the two monolayers and that in fact their concentration is higher in the "spikes".[93,94] Moreover the echinocytosis induced in such a way is completely reversible: one can go back to usual forms simply by adding other molecules which preferentially absorb in the other monolayer.[93] Whether this phenomenon really is a kind of curvature instability or only a complex biochemical event (e.g. a modification of the elastic properties of the cytoskeleton) still remains an open question.[90] From a physical point of view it will probably be much more

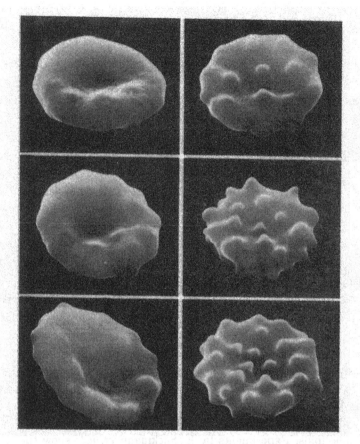

Fig. 5.4.   Transformation from a normal discocyte to an echinocyte (echinocytosis) of a human red blood cell. Note that the spikes have well-defined sizes and mutual spacings. Such transformations can be induced by an asymmetric adsorption of certain drugs. [Electron micrographs from M. Bessis, M. Prenant, *Nouv. Rev. Fran. Hémat.* **12**, 351.]

rewarding to study simpler systems, such as lipid/water mixtures, and look there for the phenomena we have been describing.

## 5.3. *The Polymorphism of Lipid/Water Systems: Different Kinds of f-Membranes*

As we already discussed in Sec. 1 amphiphilic systems present an unusual property of *polymorphism*: even the simplest binary mixtures of e.g. lipids and water show a large variety of thermodynamically distinct phases. In some of these phases lipids form bilayers; this happens for instance in lamellar or cubic structures.[4,71,95] Such *"membrane phases"* seem — in principle — the simplest to explore theoretically. In fact, one could describe them as systems of interacting, "fat"-membranes (f-membranes) and consider the phase transitions between them as some kinds of shape or topology transformations. What we hope is that this description is adequate, and that one does not need to refer to the full bulk theories (which take into account all the microscopic degrees of freedom of the molecules, etc.).

We shall now present an effort in this direction: a phenomenological, mean-field theory of the lamellar phases in (phospho-) lipids/water systems recently developed by R.E. Goldstein and the author.[78]

### 5.4. *Towards a Mean-Field Theory of Lamellar Phases*

Consider lyotropic liquid crystals made for instance of neutral phospholipids and water.[96] The three best known lamellar structures for these systems are the $L_\alpha$, $L_\beta$ and $P_\beta$ phases.[96,97]

In a $L_\alpha$ *phase* the lamellae are fluid (disordered) and, on average, flat. This phase was the one studied in Sec. 4, although we assumed there that the membrane thickness $\delta$ was such smaller than the interlamellar spacing $\bar{l}$. This is not the case in general and one often observes $\delta \geq \bar{l}$.[96] If one lowers the temperature $T$, *or decreases the water content of the mixture*, $1 - \phi$ (as suggested by the adjective *lyotropic!*), one goes to an ordered, "*solid-like*" phase, $L_\beta$, in which the hydrocarbon chains are ordered and molecules do not diffuse freely. In some systems, such as phosphatidylcholines, *PCs*, an intermediate phase appears in a specific range of the $(T, \phi)$-diagram, between the $L_\alpha$ and $L_\beta$ phases. In this "*rippled*" $P_\beta$ phase the lamellae present an undulated structure (Fig. 5.5), and almost solid-like diffusion properties.[97]

In order to build a simple theory of these structural transformations we made a straightfoward synthesis of two theoretical ingredients. The first one is a Landau theory of the intramembrane melting transition.[98,99] Many experiments carried out on lamellar phases *and* on unilamellar vesicles reveal that at the point where this transition takes place the bilayer thickness $\delta$ exhibits jump discontinuities.[75,100] Thus, we base the Landau theory on a *scalar* order parameter $\psi = (\delta(T) - \delta_0)$, where $\delta_0$ is a reference thickness (e.g. that of the fluid phase). We allow $\psi$ to vary with the position vector $\vec{x}$ within the membrane in order to include the possibility of a $P_\beta$ phase. The choice of the order parameter as a simple scalar is an important simplification since the lipid molecules possess many degrees of freedom: hydrocarbon chain conformation, molecular tilt, positional ordering etc. In order to fully account for different symmetries and the details of different phase transitions, more realistic theories must consider elaborate order parameters. Many aspects of the global phase behavior, however, can be explained without referring to the detailed molecular structure of the membranes.

The Landau–Ginzburg Hamiltonian of an isolated membrane is thus taken as

$$\mathcal{H}\left[\psi\right] = \int d^2 x \left\{ \frac{1}{2}\Sigma(\nabla\psi)^2 + \frac{1}{2}K(\nabla^2\psi)^2 + \frac{1}{2}a_2\psi^2 + \frac{1}{3}a_3\psi^3 + \frac{1}{4}a_4\psi^4 \right\} \quad (5.4)$$

The temperature dependence of the coefficients is assumed to reside in $a_2 : a_2 = a_2'(T - T_0)$. For $a_3 \neq 0$ this model leads to a first-order transition between the

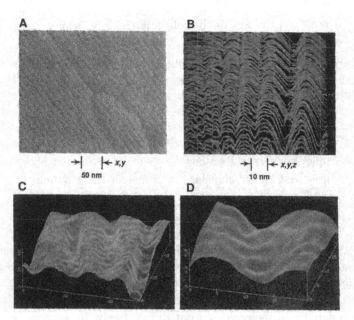

Fig. 5.5. (A) Transmission electron micrograph of a freeze-fracture replica of the $P_\beta$ phase of DMPC. Ripple periodicity is about 12 nm. Ripples are often interrupted by defects, such as the screw dislocation line that passes through this image. (B) Analog STM image of the replica with no $y$ scan. Scan speed is increased from 15 Hz (bottom) to 30 Hz (top). Although detail is washed out in the fastest scans, the essential features are clear: (i) ripple periodicity is 13 nm and average amplitude is 4.5 nm and (ii) the ripples are asymmetric, rising more steeply to the left than to the right. (C) Digital STM image of the replica. The ripple amplitude and configuration are well defined as in (B), although variations occur along the ripple. Note the fine structure that crosses the ripples roughly orthogonal to the ripple direction. The distance scale is in nanometers. (D) Computer zoom of the right-central portion of (C). The bands crossing the ripples are readily apparent. At present, it is difficult to say if the banding reflects an underlying molecular structure of the ripple phase, a structure inherent to the replica itself, or some unknown artifact. [Reprinted from J.A.N. Zasadzinski *et al. Science* **239**, 953 (1988).]

$L_\alpha(\psi = 0)$ and $L_\beta(\psi > 0)$ phases. The modulated $P_\beta$ phase appears if $\Sigma$ becomes smaller than some characteristic value $\Sigma_0(K, \{a_i\}) < 0$. In the *one wave-vector approximation* this modulated phase is characterized by a sinusoidal modulation $\psi = \psi_0 + \eta_0 \cos(q^* x)$ with $q^* = (|\Sigma|/2K)^{1/2}$. The model (5.4) predicts that at the $P_\beta - L_\alpha$ transition $q^* > 0$, except if $\Sigma = a_2 = a_3 = 0$, which corresponds to the so-called *Lifshitz point*. At this point $q^* > \infty$.[101] We do not present here any microscopic explanation why $\Sigma$ becomes negative. One possible scenario could be analogous to the coupling between the internal degrees of freedom and the modulation of the lipid/water interface, similar to the one discussed at the beginning of this section. The Hamiltonian (5.4) could then be considered as the one of Eqs. (5.1) and (5.2) for which the $\phi$ degrees of freedom had already been integrated out.[90]

The second ingredient of the model is a continuum theory of molecules interactions, $V_m$, between the membranes described in Sec. 4. In order to take the $P_\beta$

phase into account and make the theory *self-consistent* we generalize a theory of hydration forces[102] for the case of modulated interspace between membranes.[78]

The model is thus defined by the sum of intra-membrane Hamiltonians (1.4) $\mathcal{H}(\psi_i)$ for each bilayer and the sum of interaction potentials $V_m(\psi_i, \psi_{i+1})$ for the pairs of neighboring membranes. This effective Hamiltonian depends on the temperature $T$ and two phenomenological parameters: $\langle\psi\rangle$ — the average order parameter and $\Phi$ — the water concentration, connected to the average distance $\bar{l}$ between membranes.[78]

The model can be solved in the mean-field approximation, i.e., neglecting altogether the fluctuations within each membrane, as well as the undulations of the membranes (discussed in the previous section). The phase diagrams thus obtained are shown in Figs. (5.6a, b). Each of these diagrams can be applied to the systems in which the rippled $P_\beta$ phase is absent or present, respectively. Both are in semi-quantitative agreement with experiments in the following sense: in principle, there exist enough independent structural and thermodynamic measurements to fix the values of all material constants of the model and obtain those phase diagrams without adjustable parameters. In fact, the diagram in Fig. (5.6a) (with $\Sigma > \Sigma_0$) which has no rippled phase, is like the one found in many phosphatidylethanolamines ($PEs$);[104] while that in Fig. (5.6b) (with $\Sigma > \Sigma_0$) resembles the phase behavior of phosphatidylcholines ($PCs$).[105] In the absence of a *complete* set of thermodynamic measurements on any *single* phospholipid system, we used the values typical of intermediate length phospholipids. As a result we obtained a satisfactory agreement with experimental data — for details see Ref. 78.

The present model not only leads to phase diagrams which are similar to those found in experiments, and at the same time is consistent with the existing osmotic stress measurements, and calorimetric and X-ray diffraction data,[78] but also makes

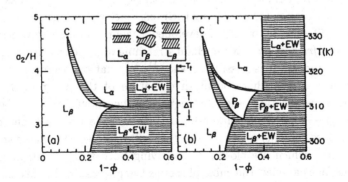

Fig. 5.6. Temperature-composition phase diagram of multilamellar bilayer membranes, with $\Phi$ the volume fraction of lipid. Inset: schematic drawing of the thin ($L_\alpha$), modulated ($P_\beta$), and thick ($L_\beta$) phases. EW denotes a phase of nearly pure ("excess") water. Vertical scales are those of the bare Landau coefficients (left) and in absolute degrees (right) as deduced from typical values of the model parameters. (Reprinted by permission of *Phys. Rev. Lett.* **61** (1988) 2213.)

new predictions which can be checked experimentally. For instance, our generalization of the hydration interaction for the modulated surfaces leads to an additional repulsion term proportional to $q^{*2}$, which may be written as $H_1(\nabla\psi(x))^2$ with $H_1 \sim H\exp(-d/\lambda_h)$. Therefore, the effective gradient-squared term has the coefficient $\Sigma_{eff} = \Sigma + H_1$ and becomes less negative with decreasing $d$. This also implies that $q^*$ decreases exponentially with $d$. In fact, this prediction has recently been verified in X-ray diffraction experiments.[106]

Note that the model presented here needs to be modified if one wants to describe the order within membranes in detail. One knows now[107] that there exist many different $L_\beta$ phases which differ by their in-plane translational and bond-orientational order, as well as the tilt of their ordered molecules. The model should also be further developed to include the effects of thermal fluctuations. We believe, however, that this theory already includes some basic physics governing the phase behavior of interacting membranes.

### 5.5. *Cubic Phases as Crystals of f-Membranes*

Developing the theory described above was facilitated by the simplicity of the geometry of lamellar phases. The planar or sinusoidal interfaces between water and lipid for instance make the calculation of the molecular interactions rather easy. This is not the case for some other phases of lyotropic liquid crystals such as cubic structures. Figure 5.7 shows the micrographs of such structures found in chloroplasts of *Avena* leaves[108] as well as a schematic drawing based on these photographs. Similar structures were found in many pure lipid/water systems.[95,109] We can see that in these equilibrium phases the amphiphilic molecules still form bilayers. One can therefore consider cubic phases as *the crystals of f-membranes*. Precise X-ray data are now available and show that the cubic phases of different crystallographic symmetries can coexist[71,110] in a single system.

Up to now, no simple thermodynamic theory of cubic phases has been developed. The only arguments explaining their existence are of a geometrical nature.[111] Charvolin and Sadoc, for instance, developed a geometrical theory which is based on the concept of *frustration* between chain stretching and the curvature of water/lipid interfaces.[112] Indeed, it is possible to show[111,113] that one cannot curve the membranes in such a way that the mean curvature of the water/lipid interfaces is everywhere constant (and therefore equal to the spontaneous curvature $H_0$ — see Eq. (2.4) — of the monolayers) *and*, at the same time, the thickness of the membranes also remains constant (see, Fig. 5.8). However, the cubic structures are probably those which are closest to satisfying both these constraints simultaneously. Thus, one can view the cubic phases as the phases which relax the frustration between two energy terms:

(i) the stretching energy of the chains, which try to keep the average length of the hydrocarbon chains constant;

(ii) the curvature energy of each water/lipid interface.

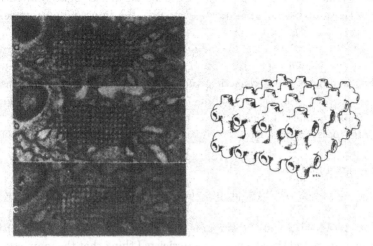

Fig. 5.7. Left: Electron micrographs of serial sections of a prolamellar body in an *Avena* leaf. The sectioning is close to a major lattice plane. Right: the three-dimensional visualization by Gunning [Ref. 108] of the membrane system occurring in the micrographs. Note that this structure corresponds to a simple cubic "plumber's nightmare" discussed in Sec. 3. [Reprinted by permission of *Mol. Cryst. Liq. Cryst.* **63** (1981) 59 (Gordon and Breach Science Publishers S.A.).]

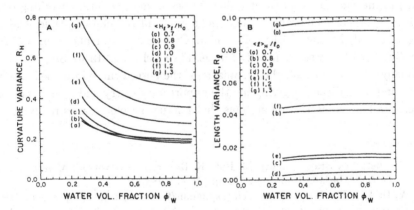

Fig. 5.8. (A): The normalized variance $R_H$ is a measure of the deviation from a constant value, $H_0$, of the mean curvature over the surface, which is at a fixed distance, $\ell$, off the minimal surface. $R_H$ is shown vs. the lipid volume fraction $\phi_\ell$ for different values of the ratio of $\langle H_\ell \rangle_\ell / H_0$ where $\langle H_\ell \rangle_\ell$ is the average curvature of the surface. If $R_H$ were zero, then this parallel surface would have a mean curvature equal to $H_0$ at every point. (B): $R_\ell$ measures the deviation from a constant value $\ell_0$ of the distance between the minimal surface and the surface of constant mean curvature $H$. It is shown vs. the lipid volume fraction for different values of $\langle \ell \rangle_H / \ell_0$, where $\langle \ell \rangle_H$ is the average distance of the constant H surfaces from the minimal surface. The fact that $R_\ell$ is so small means that if the lipid head groups conform to the proper constant curvature surfaces, then the thickness of the bilayer is nearly constant. [Ref. 111.]

Of course it is not obvious whether a simple phenomenological model based on these surface (interface) concepts can indeed be adequate. Moreover, the interactions between the membranes, and in particular the interactions connected with water structure, could play a very important role, as indeed they do for lamellar phases.[78,70] Therefore it is probably too early to claim that the frustration described above provides the *explanation* for the appearance of the cubic structure. The understanding of the geometrical properties of $H = 0$ and $H = $ const triply-periodic surfaces[111,70] however provides a good starting point for developing the future thermodynamic theories of the cubic phases and thus are important for the study of lipid polymorphism.

## Acknowledgments

I hope that this brief description of some equilibrium phenomena taking place in membrane systems has provided the reader with a general idea of the physics of membranes. In fact, the subject of amphiphilic membranes is not new but has only quite recently attracted the attention of physicists I think that the main consequence of the presence of physicists in this field has been the appearance of careful and quantitative experiments. I would therefore like first to acknowledge the experimentalists whose work are largely influenced my thinking. These are among others: Sol Gruner, Adrian Parsegian, Gregoire Porte, Didier Roux, Erich Sackmann, Cyrus Safinya, Marilyn Schneider, Dan Wack, Watt Webb and their collaborators.

I also express my gratitude to all the theoreticians with whom I was fortunate enough to interact and collaborate when I was learning about, or working on, the physical aspects of membranes. These include: David Andelman, David Anderson, François David, Michael Fisher, Raymond Goldstein, Emmanuel Guitter, David Huse, Reinhard Lipowsky, Anthony Maggs, Luca Peliti and Rajiv Singh.

Finally, I wish to thank the organizers of the Jerusalem Winter School: David Nelson and Tsvi Piran for giving me the opportunity to try and share some of my enthusiasm for the fascinating systems that fluctuating membranes really are.

## References

1. B. Alberts, D. Bray, J. Lewis, M. Raff, K. Roberts, J.D. Watson, *Molecular Biology of the Cell* (Garland Publishing Comp., N.Y., 1983).
2. W. Becker, *The World of the Cell* (Benjamin/Cummings, Reading, Mass., 1986).
3. *Physics of Amphiphilic Layers*, eds. J. Meunier, D. Langevin, N. Boccara (Springer-Verlag, 1987).
4. V. Luzzati, in *Biological Membranes*, ed. D. Chapman (Academic Press, London, 1968).
5. S.T. Hyde, S. Andersson, B. Ericsson, K. Larsson, *Z. Krist.* **168**, 213 (1984).
6. B.L. Silver, *The Physical Chemistry of Membranes* (Allen and Unwin, London, 1985).
7. S.J. Singer, G.L. Nicolson, *Science* **175**, 720 (1972).
8. J.D. Robertson, *J. Cell Biol.* **91**, 189 (1981).
9. E. Evans, R. Skalak, *Mechanics and Thermodynamics of Biomembranes* (CRC Press, Boca Raton, Fl., 1980).

10. J.N. Israelachvili, *Intermolecular and Surface Forces* (Academic, Orlando, Fl., 1985).
11. F. Brochard, P.-G. de Gennes, P. Pfeuty, *J. de Physique* **37**, 1099 (1976).
12. J. Zimmerberg, *Bioscience Reports* **7**, 251 (1987).
13. P.B. Canham, *J. Theor. Biol.* **26**, 61 (1970).
14. W. Helfrich, *Z. Naturforsch.* **28c**, 693 (1973).
15. F. David, in *This Volume* (1988).
16. D.R. Nelson, in *This Volume* (1988).
17. A.G. Petrov, J. Bivas, *Progress Surf. Sci.* **18**, 389 (1984).
18. I. Szleifer, D. Kramer, A. Ben-Shaul, W.M. Gelbard, D. Roux, *Phys. Rev. Lett.* **60**, 1966 (1988).
19. J.M. Seddon, G. Cevc, R.D. Kaye, D. Marsh, *Biochemistry* **23**, 2634 (1984).
20. W. Helfrich, in *Physics of Defects (Les Houches Session XXV)*, eds. Balian, R. et al., (North Holland. 1981), p. 716.
21. R. Cantor, *Macromolecules* **14**, 1186 (1981).
22. S. Milner, T.A. Witten, *J. de Physique*, in press (1988).
23. H.J. Deuling, W. Helfrich, *Biophys. J.* **16**, 861 (1976).
24. B.T. Stokke, A. Mikkelsen, A. Elgsaeter, *Eur. Biophys. J.* **13**, 203. *ibid,* 219 (1986).
25. S. Svetina, B. Zeks, *Biomed. Biochim. Acta* **42**, S86 (1983).
26. E. Sackmann, H.-P. Duwe, H. Engelhardt, *Faraday Discuss. Chem. Soc.* **81**, 281 (1986).
27. N. Ostrowsky, J. Peyraud, *J. Chem. Phys.* **77**, 2081 (1982).
28. S. Leibler, R.P.R. Singh, M.E. Fisher, *Phys. Rev. Lett.* **59**, 1989 (1987). Editorial note. Some further studies of this planar vesicle model are reported in (a) C.J. Camacho and M.E. Fisher, *Phys. Rev. Lett.* **65**, 9 (1990); (b) A.C. Maggs, S. Leibler, M.E. Fisher and C.J. Camacho, *Phys. Rev.* **A42**, 691 (1990); (c) C.-J. Camacho, M.E. Fisher and R.K.P. Singh, *J. Chem. Phys.* **94**, 5693 (1991); (d) M.E. Fisher, A.J. Guttmann and S.G. Whittington, *J. Phys.* **A24**, 3095 (1991) ; (e) C.J. Camacho, M.E. Fisher and J.P. Straley, *Phys. Rev.* **A46**, 6300 (1992).
29. Y. Kantor, M. Kardar, D.R. Nelson, *Phys. Rev. Lett.* **57**, 791 (1986).
30. Y. Kantor, M. Kardar, D.R. Nelson, *Phys. Rev.* **A35**, 3056 (1987).
31. G. Parisi, N. Sourlas, *Phys. Rev. Lett.* **46**, 871 (1981).
32. E. Browicz, *Zbl. Med. Wiss.* **28**, 625 (1890).
33. F. Brochard, J.-F. Lennon, *J. de Physique* **36**, 1035 (1975).
34. J.F. Faucon, P. Méléard, M.D. Mitov, I. Bivas, P. Bothorel, unpublished, 1988.
35. G. Belbik, R.-M. Servuss, W. Helfrich, *J. de Physique* **36**, 1773 (1985).
36. M.B. Schneider, J.T. Jenkins, W.W. Webb, *J. de Physique* **45**, 1457 (1984).
37. H. Englehardt, H.P. Duwe, E. Sackmann, *J. de Physique* **46**, L395 (1985).
38. I. Bivas, P. Hanusse, P. Bothorel, J. Lalanne, O. Aguerre-Chariol, *J. de Physique* **48**, 855 (1987).
39. S. Milner, S.A. Safran, *Phys. Rev.* **A36**, 4371 (1987).
40. E. Sackmann, H.-P. Duwe, W. Pfeiffer, in *Proceedings of EPS Meeting (Budapest)* to appear in *Physica Scripta* (1988).
41. J.T. Jenkins, *J. Math. Biol.* **4**, 149 (1977).
42. D. Roux, C.R. Safinya, in Ref. 3., p. 138 (1987).
43. A. Zilker, H. Engelhardt, E. Sackmann, *J. de Physique* **48**, 2139 (1987).
44. H. Engelhardt, H. Gaub, E. Sackmann, *Nature* **307**, 378 (1984).
45. E. Evans, D. Needham, *Faraday Discuss. Chem. Soc.* **81**, 267 (1986).
46. W. Helfrich, *et al. Biophysical J.* (1987).
47. W. Helfrich, *J. de Physique* **46**, 1263 (1985).
48. L. Peliti, S. Leibler, *Phys. Rev. Lett.* **54**, 1690 (1985).

49. D. Förster, *Phys. Lett.* **114A**, 115 (1986).
50. F. David, *Europhys. Lett.* **2**, 577 (1986).
51. F. David, E. Guitter, *Europhys. Lett.* **3**, 1169 (1987).
52. H. Kleinert, *Phys. Lett.* **A114**, 263 (1986).
53. W. Helfrich, *J. de Physique* **47**, 321 (1986).
54. W. Helfrich, W. Harbich, in Ref. 3 (1987).
55. J. Frölich, *This Volume* 1988.
56. M.D. Mitov, *C. R. Acad. Bulg. Sci.* **31**, 513 (1978).
57. P. Miller, T. Baumberger, unpublished 1988.
58. D.R. Nelson, L. Peliti, *J. de Physique* **48**, 1085 (1987).
59. F. David, E. Guitter, *Europhys. Lett.* **5**, 1085 (1988).
60. Y. Kantor, *This Volume*, 1988.
61. D.A. Huse, S. Leibler, *J. de Physique* **49**, 605 (1988).
62. See e.g. H. Hauser, N. Gains, H.-J. Eibl, M. Müller, E. Wehrli, *Biochemistry*, 2126 (1986).
63. W. Helfrich, *Z. Naturförsch.* **33a**, 305 (1978).
64. M.C. Cates, D. Roux, D. Andelman, S.T. Milner, S.A. Safran, *Europhys. Lett.* **5**, 733. Erratum: *ibid.* **7**, 94 (1988).
65. G. Vanderberghe, in Ref. 3., p. 199 (1987).
66. L.E. Scriven, in *Micellization, Solubilization, and Microemulsions*, ed. K.L. Mittal, (Plenum, N.Y., 1977).
67. D.A. Anderson, *Ph.D. Thesis*, 1986.
68. J. Almgren, *Plateau's Problem* (W. Benjamin, N.Y., 1966).
69. J.C.C. Nitsche, 1985 *Vorlesängen über Minimalflüchen*, Springer-Verlag.
70. A.C. Maggs, S. Leibler, unpublished 1988.
71. V. Luzzati, P. Mariani, T. Gulik-Krzywicki, 1987 in Ref. 3.
72. J.-F. Sadoc, J. Charvolin, *J. de Physique* **47**, 683 (1986).
73. R. Lipowsky, S. Leibler, in Ref. 3., p. 98 (1987).
74. E.A. Evans, D. Needham, *J. Phys. Chem.* **91**, 4219 (1987).
75. R.P. Rand, *Ann. Rev. Biophys. Bioeng.* **10**, 277 (1981).
76. J. Israelachvili, P.M. McGuiggan, *Science* **241**, 795 (1988).
77. D.F. Evans, B.W. Ninham, *J. Chem. Phys.* **88**, 2294 (1986).
78. R.E. Goldstein, S. Leibler, *Phys Rev. Lett.* **61**, 2213 (1988).
79. G. Porte, *private communication*, 1988.
80. M.E. Fisher, *This Volume*, 1988.
81. R. Lipowsky, S. Leibler, *Phys. Rev. Lett.* **56**, 2541 (1986).
82. S. Leibler, R. Lipowsky, *Phys. Rev.* **B35**, 7004 (1987).
83. D.S. Fisher, D.A. Huse, *Phys. Rev.* **B32**, 247 (1985).
84. R. Lipowsky, M.E. Fisher, *Phys. Rev. Lett.* **57**, 2411 (1986).
85. K.G. Wilson, *Phys. Rev.* **B4**, 3184 (1971).
86. F. Larche, J. Appell, G. Porte, P. Bassereau, J. Marignan, *Phys. Rev. Lett.* **56**, 1700 (1986).
87. C.R. Safinya, D. Roux, G.S. Smith, S.K. Sinha, P. Dimon, N.A. Clark, A.-M. Bellocq, *Phys. Rev. Lett.* **57**, 2718 (1986).
88. P.-G. de Gennes, *The Physics of Liquid Crystals* (Oxford University Press, 1974).
89. A. Caillé, *C. R. Acad. Sci. Paris* **274B**, 891 (1972).
90. S. Leibler, 1986 *J. de Physique* **47**, 507 (1986).
91. S. Leibler, D. Andelman, *J. de Physique* **48**, 2013 (1987).
92. H. Gaub, E. Sackmann, R. Buschl, H. Ringsdorf, *Biophys. J.* **45**, 725 (1984).
93. B. Chailley, *Ph.D. thesis*, Université Paris VI, 1976.

94. B. Deuticke, *Biochim. Biophys. Acta* **163**, 494 (1968).
95. S.M. Gruner, *Proc. Natl. Acad. Sci. USA* **82**, 3665 (1985).
96. G. Cevc, D. Marsh, *Phospholipid Bilayers: Physical Principles and Models* (Wiley, N.Y., 1987).
97. M.J. Janiak, D.M. Small, G.G. Shipley, *J. Biol. Chem.* **254**, 6068 (1979).
98. J.C. Owicki, H.M. McConnell, *Proc. Natl. Acad. Sci. USA* **76**, 4750 (1979).
99. F. Jahnig, *Biophys. J.* **36**, 329 (1981).
100. L.J. Lis, M. McAlister, N. Fuller, R.P. Rand, V.A. Parsegian, *Biophys. J.* **37**, 657 (1982).
101. R.M. Horneich, M. Luban, S. Shtrikman, *Phys. Rev. Lett.* **35**, 1678.
102. S. Marcelja, N. Radic, *Chem. Phys. Lett.* **42**, 129 (1976).
103. M. Marder, H.L. Frisch, J.S. Langer, H.M. McConnell, *Proc. Natl. Acad. Sci. USA.* **81**, 6559 (1984).
104. G. Cevc, D. Marsh, *Biophys. J.* **47**, 21 (1985).
105. L. Guldbrand, B. Jonsson, H. Wennerstrom, *J. Coll. Int. Sci.* **89**, 532 (1982).
106. D.C. Wack, W.W. Webb, *Phys. Rev. Lett.* **61**, 1210 (1988).
107. G.S. Smith, E.B. Sirota, C.R. Safinya, N.A. Clark, 1988 *Phys. Rev. Lett.* **60**, 813 (1988).
108. B.E.S. Gunning, *Protoplasma* **60**, 11 (1965).
109. K. Fontell, *Mol. Cryst. Liq. Cryst.* **63**, 59 (1981).
110. G.L. Kirk, S.M. Gruner, D.L. Stein, *Biochemistry* **23**, 1093 (1984).
111. D.M. Anderson, S.M. Gruner, S. Leibler, *Proc. Natl. Acad. Sci. USA.* **85**, 5364 (1988).
112. J. Charvolin, *J. de Physique* **46**, C3-173 (1985).
113. J. Charvolin, J.-F. Sadoc, *J. de Physique* **48**, 1559 (1987).

CHAPTER 4

# THE PHYSICS OF MICROEMULSIONS AND
# AMPHIPHILIC MONOLAYERS

David Andelman

*Raymond and Beverly Sackler Faculty of Exact Sciences*
*School of Physics and Astronomy*
*Tel Aviv University, Ramat Aviv*
*Tel Aviv, 69978 Israel*

## Abstract

Surfactants are amphiphiles that combine hydrophobicity with hydrophilicity behavior; namely, they prefer to reside or to create spontaneously liquid/liquid or liquid/gas interfaces. We give here three examples of amphiphilic systems: (i) insoluble monolayers of lipids or fatty acids at the water/air interface — called Langmuir monolayers. (ii) Micellar solutions where the solvent can be either an aqueous solution or a non-polar organic solvent like oil. (iii) Microemulsions which are thermodynamically stable, fluid, oil-water-surfactant mixtures; most microemulsions contain also cosurfactant (alcohol) and/or salt. In this extended abstract we will briefly review some of the main results obtained for Langmuir monolayers and microemulsions.

Insoluble amphiphilic monolayers spread on the water/air interface are of basic interest because of their variety of two-dimensional phase transitions. In addition, they serve as simple models for biological cell membranes. The main experimental technique used to study Langmuir monolayers is a film balance technique which has been invented by Langmuir (1917), (1933) (for a review see, e.g., Adamson (1982), Gaines (1966)). Using this method, surface isotherms measuring surface pressure as function of area per molecule yield a rich variety of surface phases (Gaines (1966)). At very low surface pressures (less than 1 dyn/cm) the monolayer undergoes a two-dimensional liquid-gas transition (Hawkins and Benedek (1974), Kim and Cannell (1975), (1976a)). At higher surface pressure a peculiar "kink" in the isotherms is seen in many experiments. The origin of this singularity is not clear and is a matter of dispute (Pallas and Pethica (1985), Middleton *et al.* (1984), Bell *et al.* (1981), Legre *et al.* (1984) and reference therein). It has been interpreted either as a second-order transition between two liquid phases (termed "liquid-expanded" and "liquid-condensed") or as a first-order transition between liquid and solid phases under poorly controlled conditions: presence of impurities, undersaturated water vapor pressure, retention of the spreading solvent, or non-equilibrium determination of the isotherms (Middleton *et al.* (1984)). At the very compressed state, the monolayer behaves as a solid. It is rather incompressible and will collapse as the surface pressure becomes too high (roughly of the order of 30 dynes/cm).

Recently, other experimental techniques have been developed and applied to the study of structural properties of monolayers. Among others they include electric surface potential (Helm *et al.* (1986), Middleton and Pethica (1981), Kim and Cannell (1976b)), viscoelastic measurements (Abraham *et al.* (1985)), non-linear optics (Rasing *et al.* (1985)), epifluorescence microscopy (McConnell *et al.* (1984), Löesche *et al.* 1983), Löesche and Möhwald (1984), Löesche and Möhwald (1985), Moore *et al.* (1986)), and grazing incident X-ray diffraction from a synchrotron source (Kjaer *et al.* (1987), Dutta *et al.* (1987), Barton *et al.* (1988)). The epifluorescence microscopy, for example, allows direct visualization of monolayers on a length scale of micrometers. In lipid monolayers, an organization of liquid-like and solid-like regions repeating themselves periodically is observed. Those domains can be stripe-like, rounded, or spiral with a definite hardness for monolayers of chiral lipids. The grazing incident X-ray diffraction allowed, for the first time, to obtain microscopic structural information on monolayers directly on the air/water interface. Crystalline order has been examined and positional as well as orientational correlation lengths have been calculated. We expect that these two advanced methods — X-ray scattering and epifluorescence microscopy — will hopefully clarify in the future the nature of the liquid-expanded liquid-condensed transition.

On the theoretical level, a simple Van der Waals equation of state has been proposed to describe the dilute regime of monolayers (Langmuir (1933), Kirkwood (1943), Adamson (1982)). For the more condensed phases the situation is less clear. Several works proposed an explanation of the liquid-expanded liquid-condensed transition by a coupling between the monolayer surface concentration and an additional orientational order parameter of the chains (Legre *et al.* (1984)). Recently, a possible explanation of the modulated structure seen in the epifluorescence experiments has been proposed (Andelman *et al.* (1985), Keller *et al.* (1986), Andelman *et al.* (1987a)). Phase diagrams involving transition between stripe, hexagonal and isotropic phases have been calculated close to a critical point using a Landau-Ginzburg expansion and at low temperatures. The main idea is to consider the effect of dipolar interactions on the creation of modulated phases. The dipoles can be either permanent or induced in a charged monolayer. The competition between the long-range dipolar interaction and the two-dimensional line tension determines the periodicity of the undulations. Other interesting questions which have been recently addressed include the dynamics of growth and spinodal decomposition of dipolar monolayers in the absence of gravity (Brochard *et al.* (1987)), and chirality discrimination in racemic monolayers of lipids which show spiral growth

(McConnell *et al.* (1984), Löesche and Möhwald (1984) and (1985), Andelman and deGennes (1988), Andelman (to be published)).

A different type of an amphiphilic system forming a complex fluid are microemulsions. Like regular emulsions, these are fluid mixtures of oil, water and surfactant but where the oil and water remain separated in coherent domains that are quite small; typically of the order of tens or hundreds of Angstroms in size. Because the surfactant molecules prefer the water-oil interface over bulk oil or water environments, they create an extensive oil-water interface inside the bulk phase. Moreover, due to their small droplet size, microemulsions are believed to be thermodynamically stable.

For many years, microemulsions were studied especially with regards to their phase diagrams (Shinoda and Saito (1968), Robbins (1977)). For a review see Mittal (1977), Mittal and Lindman (1984) and (1987). More recently, experiments probing structural and physical behavior have been performed. For a review on these newer experimental techniques see, e.g., Safran and Clark (1987), Meunier *et al.* (1987). More microscopic investigations involve light, X-ray and neutron scattering (Huang *et al.* (1983), Auvray *et al.* (1984)) and quenched freezing electron microscopy (Jahn and Strey (1987)). Experimentally, it has been observed that the configuration of the oil and water domains varies with the relative composition of water, oil and surfactant. For small fractions of oil in water or water in oil, the structure is that of compact globules (Calje *et al.* (1977), Robbins (1977), Ober and Taupin (1980), Roux *et al.* (1984), Huang *et al.* (1983), Kotlarchyk *et al.* (1984)). However, when the volume fractions of oil and water become comparable, one expect random bicontinuous structure to form (Scriven (1977), Auvray *et al.* (1984), Kaler *et al.* (1983), Talmon and Prager (1978) and (1982), Cazabat *et al.* (1982)). Under other conditions, for example, when the volume fraction of the surfactant is higher than a few percent, various ordered structures reminiscence of liquid-crystalline phases may also arise (Ekwald (1975), Bellocq and Roux (1986), Smith (1984)). Those are phases such as cubic, lamellar or cylindrical and show Bragg peaks in scattering experiments.

Microemulsion phases exist as single phases or coexist with excess water, excess oil or both. We will not discuss here more complicated multiphase coexistences. Of particular interest is the so-called middle-phase microemulsion where the microemulsion phase coexists simultaneously with both excess water and oil. In such situations the liquid/liquid interfacial tension is ultralow: $10^{-3}$–$10^{-5}$ dynes/cm (Saito and Shinoda (1970), Ruckenstein and Chi (1975), Guest and Langevin (1986)). The middle-phase microemulsion can be used in applications like chemically enhanced oil recovery (Shah (1981)). Temperature, salinity and the cosurfactant (alcohol) also play an important role in the relative stability of one phase with respect to the others and the global extent of the coexistence regions (Cazabat and Langevin (1981), Mittal and Lindman (1984), (1987), Safran and Clark (1987)).

Theoretically, the most challenging problems are the ones related to the inter-play between structural properties and macroscopical phase behavior. In addition, it is highly desired both from basic and applied point of views to understand the influence of external controlled parameters such as temperature, salinity, and cosur-factant concentration on micro- and macroscopical properties. Two different theo-retical approaches have been proposed to predict phase diagrams and structure. The first one is the so-called "phenomenological" approach where the surfactant inter-faces are considered as independent entities. Those interfaces separating regions of oil and water are characterized by model dependent energetics. In most of the pro-posed models, each fluctuating interface is characterized by interfacial and curvature energies. Since the characteristic length scales are much bigger than molecular sizes, a continuum approximation that depends on several parameters can be made. For microemulsion phases of compact globules (spheres or cylinders), in the very dilute limit (e.g., a small volume fraction of water in oil), the stability, phase diagram and fluctuations of such objects have been calculated (Huh (1979), (1984), Safran and Turkevich (1983), Safran et al. (1984), Roux and Coulon (1986)). In some cases globule-globule interactions have been considered as well. For cylindrical phases, fluctuations can be treated similarly to the case of semi-flexible polymers in solu-tion (Safran and Turkevich (1983), Safran et al. (1984)).

However, no doubt that the phase which is the most difficult to understand from first principles is the bicontinuous phase, since it is composed of an ensemble of ran-dom fluctuating surfactant interfaces separating coherent regions of oil and water on length scales of many individual molecules. The first attempt to deal with the bicon-tinuous phase from a thermodynamical point of view has been proposed by Talmon and Prager (1978) and (1982). Later works developed more refined models predict-ing phase diagrams of bicontinuous phases (Jouffroy et al. (1982), Widom (1984), Safran et al. (1986), (1987) and Andelman et al. (1987b)). Most of these works have considered the surfactant film as a two-dimensional fluid whose energy is mainly determined by its curvature coefficient and spontaneous radius of curvature. In one case (Widom (1984)), the film has been treated as a compressible two-dimensional fluid of surfactant molecules. The stability of the macroscopic phase depends on the delicate balance between the configurational energy (interfacial and curvature) and the entropy of such fluctuating interfaces. The free-energy that determines the phase behavior is calculated within mean-field approximation and in some works (Andelman et al. (1987b), Safran et al. (1986), (1987)) the size-dependent bending (curvature) coefficient (Helfrich (1973), (1978), (1985) and (1987)) is taken explic-itly into account. Some results exist already on the influence of temperature, salinity and cosurfactant on the phase behavior but only in an indirect way. It is thought that salinity and cosurfactant influence the bending constant and the spontaneous radius of curvature (deGennes and Taupin (1984)). Calculations of phase behavior as function of the latter two parameters exist (Cates et al. (1988)), but the actual dependence of those parameters on the experimentally controlled variables: salinity, cosurfactant and temperature is not presently known in a more quantitative way.

Scattering and quenched freezing microscopy of the random bicontinuous microemulsion support the generally believed picture of the structure of this phase (Auvray *et al.* (1984), Jahn and Strey (1987)). Theoretical calculations of the structure factor of the bicontinuous phase have been done either starting from a purely geometrical construction of the bicontinuous phase (Zemb *et al.* (1987), Teubner and Strey (1987), Berk (1987), Vonk *et al.* (1988)) or by calculating the fluctuations around the mean-field free energy used for the phase diagram (Milner *et al.* (1988), Widom (to be published)).

We briefly mention the other theoretical approach of calculating phase diagrams of microemulsion. The starting point here is to introduce a lattice model for the three component liquid mixture: water, oil, and surfactant (Wheeler and Widom (1968), Widom (1986), Schick and Shih (1986), Chen *et al.* (1987)). Interactions between single molecules are introduced and the amphiphilicity of the surfactant is simulated by introducing a preferred interaction of one surfactant molecule to have as its nearest neighbor an oil molecule from one side and a water molecule from the other side. In this way, the tendency of the surfactant molecule to form an interface between the oil and water is achieved. Further refinements of this microscopic spin model were introduced by considering longer-range interactions. Such an approach is suitable to the study of concentrated surfactant solutions where ordered phases are formed. For surfactants that are partially miscible in the two solvents, thus having a poor amphiphilicity, such an approach is based on a generalized three-component liquid mixture model. However, for bicontinuous phases of surfactants that are quite immiscible in both solvents, it is very difficult to obtain structural information on mesoscopic length scales starting from such a microscopical model although these models give a possible explanation for the ultralow surface tension and predict phase diagrams that capture some of the features of the middle-phase microemulsion.

In this short contribution we summarized current research interests in microemulsions and Langmuir monolayers. Emphasis has been put on the more microscopical and physical approach exploring the connection between phase transitions and structure. We hope that recent advances in experimental techniques will induce more theoretical investigations of these systems. As is discussed in more details by the other contributors to this book, the physics of fluctuating membranes and interfaces is quite interesting and novel from a theoretical point of view. Hopefully, those novel ideas will be applied in the future to more specific, and in some sense more complicated, amphiphilic systems as the ones mentioned here.

### Acknowledgments

The author wishes to acknowledge support from the U.S.-Israel Binational Science Foundation under grant No. 87-00338, the Bat-Sheva de Rothschild Foundation, and Exxon Research and Engineering Company.

## References

1. Abraham, B.M., Miyano, K., Xu, S.Q. and Ketterson, J.B., *Phys. Rev. Lett.* **49**, 1643 (1985).
2. Adamson, A.W., *Physical Chemistry of Interfaces* (Wiley, New York, 1982).
3. Andelman, D., Brochard, F., deGennes, P.G. and Joanny, J.F. *C.R. Acad. Sci. (Paris)* **301**, 675 (1985).
4. Andelman, D., Brochard, F. and Joanny, J.F., *Chem. Phys.* **86**, 3673 (1987a).
5. Andelman, D., Cates, M.E., Roux, D. and Safran, S.A., *J. Chem. Phys.* **87**, 7229 (1987b).
6. Andelman, D. and deGennes, P.G., *C.R. Acad. Sci. (Paris)* **307**, 233 (1988).
7. Auvray, L., Cotton, J.P., Ober, R. and Taupin, C., *J. Phys. (Paris)* **45**, 913 (1984).
8. Barton, S.W., Thomas, B.N., Flom, E.B., Rice, S.A., Lin, B., Peng, J.B., Ketterson, J.B. and Dutta, P., *J. Chem. Phys.* **89**, 2257 (1988).
9. Bell, G.M., Combs, L.L. and Dunne, L.J., *Chem. Rev.* **81**, 15 (1981).
10. Bellocq, A.M. and Roux, D., in *Microemulsions*, eds. S. Friberg and P. Bothorel, Chemical Rubber, New York, 1986.
11. Berk, N.F., *Phys. Rev. Lett.* **58**, 2718 (1987).
12. Brochard, F., Joanny, J.F. and Andelman, D., in *Physics of Amphiphilic Layers*, eds. J. Meunier, D. Langevin and N. Boccara, Springer Verlag, New York, 1987.
13. Calje, A., Agerof, W.G.M. and Vrij, A., in *Micellization, solubilization and Microemulsions*, ed. K. Mittal (Plenum, New York, 1977).
14. Cates, M.E., Andelman, D., Safran, S.A. and Roux, D., *Langmuir* **4**, 802 (1988).
15. Cazabat, A.M. and Langevin, D., *J. Chem. Phys.* **74**, 3148 (1981).
16. Cazabat, A.M., Langevin, D., Meunier, J. and Pouchelon, A., *J. Adv. Colloid Interface Sci.* **16**, 175 (1982).
17. Chen, K., Ebner, C., Jayaprakash, C. and Pandit, R., *J. Phys. C* **20**, L361 (1987).
18. DeGennes, P.G. and Taupin, C., *J. Phys. Chem.* **86**, 2294 (1982).
19. Dutta, P., Peng, J.B., Lin, B., Ketterson, J.B., Prakash, M., Georgopoulous, P. and Erlich, S., *Phys. Rev. Lett.* **58**, 2228 (1987).
20. Ekwald, P., in *Advances in Liquids Crystals I*, ed. G. H. Brown (Academic, New York, 1975).
21. Gaines, G.A., *Insoluble Monolayers at Liquid/Gas Interfaces* (Wiley, New York, 1966).
22. Guest, D. and Langevin, D., *J. Colloid Interface Sci.* **112**, 208 (1986).
23. Hawkins, G.A. and Benedek, G.B., *Phys. Rev. Lett.* **32**, 524 (1974).
24. Helfrich, W. *Naturforsch. Teil* **A28**, 693 (1973).
25. Helfrich, W. *Naturforsch. Teil* **A33**, 305 (1978).
26. Helfrich, W. *J. Phys. (Paris)* **46**, 1263 (1985).
27. Helfrich, W. *J. Phys. (Paris)* **48**, 285 (1987).
28. Helm, C.A., Laxhauber, L., Löesche, M. and Möhwald, M., *J. Colloid Polym. Sci.* **264**, 46 (1986).
29. Huang, J.S., Safran, S.A., Kim, M.W., Grest, G.S., Kotlarchyk, M. and Quirke, N., *Phys. Rev. Lett.* **53**, 592 (1983).
30. Huh, C., *J. Colloid Interface Sci.* **71** (1979).
31. Huh, C., *J. Colloid Interface Sci.* **97**, 201 (1984).
32. Jahn, W. and Strey, R., *in Physics of Amphiphilic Layers*, eds. J. Meunier, D. Langevin and N. Boccara (Springer-Verlag, New York, 1987).
33. Jouffroy, J., Levinson, P. and deGennes, P.G., *J. Phys. (Paris)* **43**, 1241 (1982).
34. Kaler, E.W., Bennett, K.E., Davis, H.T. and Scriven, L.E., *J. Chem. Phys.* **79**, 5673 and 5685 (1983).

35. Keller, D.J., McConnell, H.M. and Moy, V.T., 1986 *J. Phys. Chem.* **90**, 2311 (1986).
36. Kim, M.W. and Cannell, D.S., *Phys. Rev. Lett.* **33**, 889 (1975).
37. Kim, M.W. and Cannell, D.S., *Phys. Rev.* **A13**, 411 (1976a).
38. Kim, M.W. and Cannell, D.S., *Phys. Rev.* **A14**, 1299 (1976b).
39. Kirkwood, J.G., *Publ. Am. Assoc. Advmt. Sci.* **21**, 157 (1943).
40. Kjaer, K., Als-Nielsen, J., Helm, C.A., Laxhauber, L.A. and Möhwald, M., *Phys. Rev. Lett.* **58**, 2224 (1987).
41. Kotlarchyk, M., Chen, S.H., Huang, J.S. and Kim, M.W., *Phys. Rev.* **A29**, 2054 (1984).
42. Langmuir, I., *J. Am. Chem. Soc.* **39**, 354 (1917).
43. Langmuir, I., *J. Chem. Phys.* **1**, 756 (1933).
44. Legre, J.P., Albinet, G., Firpo, J.L. and Tremblay, A.M.S., *Phys. Rev.* **A30**, 2720 (1984).
45. Löesche, M., Sackmann, E. and Möhwald, M., *Ber. Bunsenges. Phys. Chem.* **87**, 848 (1983).
46. Löesche, M. and Möhwald, M., *J. Phys. Lett. (Paris)* **45**, L785 (1984).
47. Löesche, M. and Möhwald, M., *Eur. Biophys.* **11**, 35 (1985).
48. McConnell, M.H., Tamm, L.K. and Weis, R.M., *Proc. Natl. Acad. Sci. (USA)* **81**, 3249 (1984).
49. Meunier, J., Langevin, D. and Boccara, N., *Physics of Amphiphilic Layers* (Springer-Verlag, New York, 1987).
50. Middleton, S.R. and Pethica, B.A., *J. Chem. Soc. Faraday Symp.* **16**, l09 (1981).
51. Middleton, S.R., Iwasaki, M., Pallas, N.R. and Pethica, B.A., *Proc. Roy. Soc. London Ser. A* **396**, 143 (1984).
52. Milner, S.T., Safran, S.A., Andelman, D., Cates, M.E. and Roux, D., *J. Phys. France* **49**, 1065 (1988).
53. Mittal, K., *Micellization, Solubilization and Microemulsions* (Plenum, New York, 1977).
54. Mittal, K. and Lindman, B., *Surfactants in Solution* (Plenum, New York, 1984).
55. Mittal, K. and Lindman, B., *Surfactants in Solution* (Plenum, New York, 1987).
56. Moore, B., Knobler, C.M., Broseta, D. and Rondelez, F., *J. Chem. Soc. Faraday Trans. 2* **86**, 1753 (1986).
57. Ober, R. and Taupin, C., *J. Phys. Chem.* **84**, 2418 (1980).
58. Pallas, N.R. and Pethica, B.A., *Langmuir* **1**, 509 (1985).
59. Rasing, Th., Sen, Y.N., Kim, M.W. and Grubb, S., *Phys. Rev. Lett.* **55**, 2903 (1985).
60. Robbins, M.L., in *Micellization, Solubilization and Microemulsions*, ed. K. Mittal (Plenum, New York, 1977).
61. Roux, D., Bellocq, A.M., and Bothorel, P., in *Surfactants in Solution*, eds. K. Mittal and B. Lindman (Plenum, New York, 1984).
62. Roux, D. and Coulon, C., *J. Phys. (Paris)* **47**, 1257 (1986).
63. Ruckenstein, E. and Chi, J., *J. Chem. Phys. Soc. Faraday Trans. 2* **71**, 1690 (1975).
64. Safran, S.A. and Turkevich, L.A., *Phys. Rev. Lett.* **50**, 1930 (1983).
65. Safran, S.A., Turkevich, L.A. and Pincus, P.A., *J. Phys. (Paris) Lett.* **45**, L69 (1984).
66. Safran, S.A., Roux, D., Cates, M.E. and Andelman, D., *Phys. Rev. Lett.* **57**, 491 (1986).
67. Safran, S.A., Roux, D., Cates, M.E. and Andelman, D., in *Physics of Amphiphilic Layers*, eds. J. Meunier, D. Langevin and N. Boccara (Springer Verlag, New York, 1987).
68. Safran, S.A. and Clark, N.A., *Physics of Complex and Supermolecular Fluids*, (Wiley, New York, 1987).

69. Saito, H. and Shinoda, K., *J. Colloid Interface Sci.* **32**, 647 (1970).
70. Schick, M. and Shih, W.H., *Phys. Rev.* **B34**, 1797 (1986).
71. Schick, M. and Shih, W.H., *Phys. Rev. Lett.* **59**, 1205 (1987).
72. Scriven, L.E., in *Micellization, Solubilization and Microemulsions*, ed. K. Mittal (Plenum, New York, 1977).
73. Shah, D.O., *Surface Phenomena in Enhanced Oil Recovery* (Plenum, New York, 1981).
74. Shinoda, K. and Saito, H., *J. Colloid Interface Sci.* **26**, 70 (1968).
75. Smith, D.H., *J. Colloid Interface Sci.* **102**, 435 (1984).
76. Talmon, Y. and Prager, S., *J. Chem. Phys.* **69**, 2984 (1978).
77. Talmon, Y. and Prager, S., *J. Chem. Phys.* **76**, 1535 (1982).
78. Teubner, M. and Strey, R., *J. Chem. Phys.* **87**, 3195 (1987).
79. Vonk, C.G., Billman, J.F. and Kaler, E.W., *J. Chem. Phys.* **88**, 3970 (1988).
80. Wheeler, J.C. and Widom, B., *J. Am. Chem. Soc.* **90**, 3064 (1968).
81. Widom, B., *J. Chem. Phys.* **81**, 1030 (1984).
82. Widom, B., *J. Chem. Phys.* **84**, 6943 (1986).
83. Zemb, T.N., Hyde, S.T., Derian, P.-J., Barnes, I.S. and Ninham, B.W., *J. Phys. Chem.* **91**, 3814 (1987).

# CHAPTER 5

# PROPERTIES OF TETHERED SURFACES

Yacov Kantor

*School of Physics and Astronomy,*
*Tel Aviv University, Tel Aviv, Israel*

The statistical mechanics of polymerized surfaces is discussed. The radius of gyration, $R_g$, of a model system, representing a flexible surface without excluded volume interactions, increases as $\sqrt{\ln L}$, where $L$ is the linear size of uncrumpled surface, i.e., the surface overfills the embedding space. With excluded volume interactions the surface expands but remains very crumpled and its $R_g$ increases as $L^\nu$, with $\nu \approx 0.8$, in a three-dimensional space. Very rigid surfaces are asymptotically flat. As the rigidity of the surface varies, it undergoes a second order phase transition with diverging specific heat from a crumpled to a flat state. These lecture notes stress the analogies and relations between the surfaces and $D$-dimensional manifolds and branched polymers.

## 1. Introduction

### 1.1. *What is a "Tethered Surface"?*

Recently there have been many studies of random two-dimensional surfaces. (For a review, see Frölich (1985).) Some studies focused on random surfaces related to high-temperature plaquette expansions of lattice gauge theories (see, e.g., Parisi (1979), Drouffe *et al.* (1979)), while others stressed the properties related to condensed matter physics, such as the behavior of membranes (see, e.g., Helfrich (1987), and references therein). However, there is no single universality class encompassing all surfaces (Cates, 1985a). It is, therefore, important to clearly define the type of surface one is considering.

In these lecture notes, I consider a system of particles (atoms or monomers) that are connected to form a regular two-dimensional array embedded in $d$-dimensional space. The precise type of the two-dimensional lattice is not important. Figure 1 depicts a triangular network of particles in a three-dimensional space. The precise form of the "bonding potential" between the neighboring particles of the network is also unimportant. However, it is essential that the bonds between the adjacent atoms or monomers of the array cannot be broken. This feature is typical for many polymeric structures, where at the experimentally relevant temperatures the

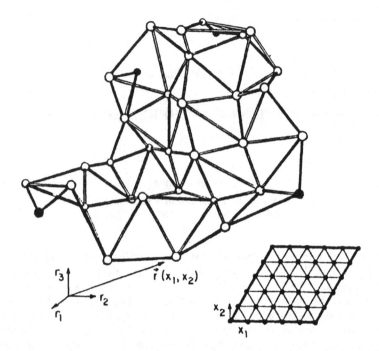

Fig. 1. Triangular tethered surface embedded in a three-dimensional space. The picture in the lower right corner depicts the topology (connectivity) of the surface. Positions of the particles in the internal (two-dimensional) coordinate space are determined by vector **x**. The actual positions of the particles as shown in the main figure are the three-dimensional external coordinates **r(x)**. Boundary bonds and corner particles are shown in black.

bonds are practically unbreakable. Thus, the surfaces preserve their *connectivity* and will be denoted "tethered (or polymeric) surfaces" (Kantor *et al.*, 1986). Such objects can be created by cross-linking of monomers on liquid-liquid, or liquid-gas, or solid-gas interfaces. A typical example of such a structure is cross-linked poly(methyl methacrylate) extracted from the surface of sodium montmorillonite clays (Blumstein *et al.*, 1969). Alternatively, polymeric surfaces can be obtained by cross-polymerization of lipid bilayers (see, e.g., Fendler and Tundo (1984)).

The physical properties of tethered surfaces differ significantly from the properties of *liquid* surfaces. In particular we shall see that the tethered surfaces may be in both crumpled and flat states, while the liquid surfaces are always crumpled beyond a certain persistence length. One can understand the origin of the differences by considering a simple example: Suppose we are trying to "fold" a flat surface of size $L \times L$ into a spherical shell of the same area. Clearly, one cannot perform such folding with a piece of paper (without cutting it) since it requires stretching and compression of various areas of the sheet. On the other hand, if we allow a free motion of the molecules along the surface (i.e., remove the fixed connectivity constraint), such "folding" will become possible. Thus, the fixed connectivity is an essential part of the model described in these lecture notes.

## 1.2. *The Tethered Surface as a Polymer*

Fixed connectivity of tethered surfaces provides some conceptual and technical advantages:

(a) The infinite strength of the bonds within the surface ensures the integrity of that object. One can, at least in principle, consider a *single* surface with definite microscopic interactions, and deduce the long wavelength behavior of the surface. Unlike its liquid counterparts, a single polymeric surface cannot evaporate even in the absence of external pressure or external reservoir of monomers, i.e. a "freely suspended" tethered surface is in a true thermodynamic equilibrium.

(b) Fixed connectivity provides a convenient and natural way to describe the surface. In this sense, the tethered surface closely resembles a linear polymer. The spatial conformation of a linear polymer can be described by a set of position vectors $r(x)$, where $x$ is the internal index (number of a monomer along the chain), while $r$ is the position of that monomer. A natural generalization for a two-dimensional surface, is made by using a *two-dimensional* vector $x$ which denotes the *internal* position of the atom (i.e., its location within the array), as depicted in Fig. 1. Such use of a fixed (flat) internal space, and its clear separation from the external variable $r$, on which the Hamiltonian of the system depends, enables an unambiguous definition of the statistical measure, thus avoiding some problems regarding a proper measure, which may occur in, say, liquid surfaces.

One should view a tethered surface not only as a member in the large family of two-dimensional objects, but also, as a particular case of a polymeric structure. It is very useful to consider the two-dimensional polymeric surface as a particular case of $D$-dimensional polymeric manifolds, or as a case of a branched polymer. Throughout these lecture notes, I will stress the analogies between these surfaces and other polymeric structures.

These notes are based on work which was done in collaboration with M. Kardar and D. R. Nelson. I have attempted to provide a self-contained description of the main features of these surfaces, and have stressed the main problems which have yet to be resolved. Since the subject is undergoing rapid development, I did not attempt to make its review exhaustive, but rather tried to supply the reader with the main references, where he can find a more detailed description.

The following section describes the behavior of oversimplified models for polymeric structures, which are both flexible and have no excluded volume interactions. Section 3 analyzes the effects of the excluded volume. Finally, bending rigidity is added to surfaces in Sec. 4, bringing the model close to realistic structures. In Sec. 5, I summarize the properties of the tethered surfaces and discuss the possible extensions and further directions in the development of this subject.

*Y. Kantor*

## 2. Phantom Chains and Networks

### 2.1. *Linear Polymers*

Linear polymers are formed by interconnection of a large amount of chemical (monomeric) units into a (topologically) linear structure, where each monomer is connected only to two other monomers. Despite the complexity of macromolecular structures, some of their properties are remarkably simple, and independent of their detailed chemical composition. (For a review, see, Flory (1979), de Gennes (1979).) The origin of such "universality" can be understood from the following idealized model: Consider a linear chain consisting of $L$ freely-jointed rods, embedded in $d$ dimensions. The length of each rod $a$ is fixed, while the angle between the neighboring rods is not restricted (see, e.g., Weiner (1983)). Each spatial conformation of such a chain is a particular example of an $L$-step random walk. Since the vector connecting the end-points of the chain $\mathbf{r}$ is a sum of $L$ independent individual step vectors, its probability distribution $p(\mathbf{r})$ approaches a Gaussian form

$$p(\mathbf{r}) \sim e^{(-d/2La^2)r^2}, \qquad (2.1)$$

as $L \to \infty$. Actually the probability distribution approaches its limiting form very fast, and for $L \sim 10$, it is already barely distinguishable from the Gaussian. This limiting behavior is a direct consequence of the central limit theorem, and does not depend on the binding potential between the neighboring monomers of the chain. For an arbitrary binding (central force) potential, (2.1) is the ultimate ($L \to \infty$) result, provided that $a^2$ in that expression is replaced by the mean-squared distance between a pair of neighboring monomers. Introduction of bending forces or restriction of the angles between the successive bonds of the chain complicates the treatment (see, e.g., Weiner (1983)), but results only in the replacement of $a$ in (2.1) by an effective length ("Kuhn statistical segment"), and replacement of the actual number of monomers $L$ by the number of Kuhn segments.

One may think of (2.1) as a statistical weight generated by a Hamiltonian

$$H = \frac{1}{2} k_B T \frac{d}{La^2} r^2 \qquad (2.2)$$

of a "Gaussian spring", with a temperature dependent force constant. Notice, that it differs from a regular "Hookean spring", by the fact that it has a vanishing equilibrium length. It is important to realize the entropic origin of the behavior described by (2.2): The "spring-like" Hamiltonian (2.2) simply indicates, that when two ends of the molecule are brought closely together, the phase space available for the intermediate points of the chain is larger. This $H$ essentially represents the $-TS$ term in the expression for the free energy $F = U - TS$, where $U$ is the energy and $S$ is the entropy.

It is convenient to subdivide a very long molecule into submolecules, each of which is large enough to be described by (2.2). The "energy" of the entire molecule will now become a sum of the contributions of several Gaussian springs:

$$\frac{H}{k_BT} = \frac{1}{2}K_0\sum_x[\mathbf{r}(x+a) - \mathbf{r}(x)]^2,$$  (2.3)

where $x$ is the *internal* coordinate of a unit (monomer or submolecule), measuring its position along the chain, and $a$ is the distance between the successive units. In the continuum limit (2.3) is replaced by

$$\frac{H}{k_BT} = \frac{1}{2}K\int\left(\frac{d\mathbf{r}}{dx}\right)^2 dx.$$  (2.4)

One should keep in mind, that here, and through the entire Sec. 2, we limit the interaction of a monomer to few neighboring monomers along the chain and completely disregard the interactions between the monomers located at the remote parts of the chain. An important omission of the model is the absence of the steric (or excluded volume, or self-avoiding) effect: In reality two monomers cannot occupy the same position in space, and must repel each other, when they come close together. (Of course, the details of the interaction also depend on the solvent in which the polymer is placed.)

### 2.2. Gaussian Networks and Surfaces

A natural generalization of a linear polymer is a network of Gaussian chains, such as depicted in Fig. 2. It is defined by a set of nodes $\{i\}$ connected by Gaussian springs with force constants $K_{ij}$, i.e., the Hamiltonian of the network is

$$\frac{H}{k_BT} = \frac{1}{2}\sum_{i,j} K_{ij}(\mathbf{r}_i - \mathbf{r}_j)^2.$$  (2.5)

Notice that (2.5) already *assumes* that the nodes are interconnected via *long* chains, described by "energies" proportional to the squared end-to-end distance. In general, we *cannot* prove, that a network of monomers with more realistic microscopic potential indeed approaches the form (2.5) on, say, sufficiently long length-scales.

The geometrical properties of Gaussian networks are extremely simple. Usually, one can calculate the mean-squared distance between two nodes, say $l$ and $m$,

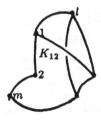

Fig. 2.   Example of a network. Full circles (•) denote the nodes. For a Gaussian network, the lines depict the connecting springs. For a resistor network, the lines represent conductances $K_{ij}$.

since the thermal averaging of $(\mathbf{r}_l - \mathbf{r}_m)^2$ for the Hamiltonian (2.5) only requires calculation of the Gaussian averages:

$$\langle (\mathbf{r}_l - \mathbf{r}_m)^2 \rangle = \frac{\int (\mathbf{r}_l - \mathbf{r}_m)^2 e^{-H/k_B T} \prod_i d\mathbf{r}_i}{\int e^{-H/k_B T} \prod_i d\mathbf{r}_i}. \tag{2.6}$$

This type of average can be easily performed for simple networks (see, e.g., Kardar and Nelson, 1988), such as periodic arrays of monomers. However, for more complicated networks, it is useful to notice a simple relation between this statistical-mechanical problem, and the conductivity problem in a resistor network. Such a relation has been known (in various forms) in polymer physics for quite some time, and has been used, in conjunction with various *approximations* to discuss entropic elasticity (see, e.g., Stauffer *et al.* (1982)). However, only recently it has been used (Cates, 1984, 1985b) to apply our understanding of the resistor networks to the field of polymer physics. In the absence of a convenient reference, I present a somewhat detailed proof of the *exact* relation.

Consider a resistor network with a connectivity defined in Fig. 2, where $K_{ij}$ is the conductance of an elementary resistor connecting the nodes $i$ and $j$. The two-point resistance $\mathcal{R}_{lm}$ between the nodes $l$ and $m$ equals the potential difference $\phi_l - \phi_m$ between the nodes divided by the current $I$ which is injected in $l$ and extracted at $m$. The potential difference can be found from the solution of Kirchhoff's circuit rules, which is equivalent to minimization of heat dissipation. Thus, the conductance problem is reduced to the minimization of $S$ defined by

$$S \equiv S_0 - I(\phi_l - \phi_m), \tag{2.7a}$$

$$S_0 \equiv \frac{1}{2} \sum_{i,j} K_{i,j}(\phi_i - \phi_j)^2, \tag{2.7b}$$

where the second term on the r.h.s. of (2.7a) accounts for the fact that the minimization is performed under the restriction, that the current $I$ is injected and extracted at $l$ and $m$ respectively, by an external current source. One can further replace the minimization of the quadratic form $S$ by averaging with the Gaussian weight $e^{-S}$ to obtain

$$\mathcal{R}_{lm} = \left[ \frac{\phi_l - \phi_m}{I} \right]_{\min\{S\}} = \frac{\int \left[ \frac{\phi_l - \phi_m}{I} \right] e^{-S} \prod_i d\phi_i}{\int e^{-S} \prod_i d\phi_i}$$

$$= \frac{\int (\phi_l - \phi_m)^2 e^{-S_0} \prod_i d\phi_i}{\int e^{-S_0} \prod_i d\phi_i} \tag{2.8}$$

(cf., Stephen, 1978). We now notice that (2.6) and (2.8) almost coincide. Although (2.6) involves integrals over $d$-dimensional *vectors* $\mathbf{r}_i$, each component of the vectors can be integrated separately, and we essentially have $d$ identical *scalar* integrals.

Thus we obtain a simple relation between the mean-squared distance in the statistical mechanics and the resistance:

$$\langle (\mathbf{r}_l - \mathbf{r}_m)^2 \rangle = d\mathcal{R}_{lm}. \tag{2.9}$$

In polymer physics, one usually characterizes the spatial extent of a network by its squared radius of gyration

$$R_g^2 \equiv \frac{1}{2N^2} \sum_{i,j} \langle (\mathbf{r}_i - \mathbf{r}_j)^2 \rangle. \tag{2.10}$$

In the case of a Gaussian network one may use (2.9) to relate this radius to the resistance of an analogous resistor network

$$R_{go}^2 = d\mathcal{R}. \tag{2.11}$$

Here, $\mathcal{R}$ is the two-point resistance averaged over all possible pairs of the nodes of the network. It is important to notice, that $R_{go}$ strongly depends on the connectivity (topology) of the network, and, except for a trivial multiplicative factor, is independent of the dimension $d$ of the embedding space.

The simplest generalization of a *linear* polymer is a $D$-dimensional manifold. $D = 1$ corresponds to a linear polymer, while $D = 2$ is a "regular" surface. We consider a manifold of internal dimensions $L \times L \times \cdots \times L$ ($D$ times). It is convenient to index the monomers of the manifold in the internal coordinate space by a $D$-dimensional vector $\mathbf{x}$ (see Fig. 1). The general expression (2.5) now simplifies, since the summation now is performed only over the pairs of neighboring monomers of the manifold. In the continuum limit, a simple generalization of (2.4) is

$$\frac{H}{k_B T} = \frac{1}{2} K \int (\nabla \mathbf{r})^2 d^D \mathbf{x} \tag{2.12a}$$

$$(\nabla \mathbf{r})^2 \equiv \sum_{i=1}^{D} \left( \frac{\partial \mathbf{r}}{\partial x_i} \right)^2. \tag{2.12b}$$

The extent to which a manifold is crumpled in the embedding space can be characterized by the critical exponent $\nu$, which relates the spatial extent of the manifold to its internal size $L$

$$R_g \sim L^\nu. \tag{2.13}$$

Alternatively, one can use the fractal dimension $d_f$ (Mandelbrot, 1977, 1982) to relate the mass (number of monomers) of the manifold $N$ to its size in the embedding space

$$N \sim R_g^{d_f}. \tag{2.14}$$

Since for a $D$-dimensional manifold $N = L^D$, these two indices are related by

$$\nu = D/d_f. \tag{2.15}$$

One can find the critical exponents for Gaussian manifolds either by a direct calculation of (2.6) with the Hamiltonian (2.12) (Kardar and Nelson, 1988) or by a simple recollection of the distance $(L)$ dependence of the potential produced by a point charge in electrostatics (or produced by a current inserted at a point in conducting medium) in $D$-dimensional space, and taking advantage of the relation (2.11). Either way one finds, for $D \leq 2$:

$$\nu_o = \frac{2 - D}{2},\tag{2.16a}$$

$$d_{fo} = \frac{2D}{2 - D},\tag{2.16b}$$

where the additional subscript $o$ indicates that the exponents are related to the *Gaussian* manifolds.

For a two-dimensional surface $(D = 2)$, the exponent $\nu_o$ vanishes and the fractal dimension is infinite, since for such manifolds

$$R_{go}^2 = \frac{d}{\pi K}\ln L.\tag{2.17}$$

One can visualize a surface as, say, a triangular network of Gaussian springs embedded in $d$ dimensions (Kantor *et al.*, 1986). [A somewhat related, but not equivalent, model has been considered by Billoire *et al.* (1984) and Gross (1984).]

Networks with irregular connectivity usually cannot be solved analytically. However, one may take advantage of the fact that the *conductivity* of the numerous types of fractal structures, such as lattice animals, percolation clusters or cluster aggregates, has been investigated numerically. Typically, numerical investigations of fractals on the lattices are concerned with the establishment of a relation between the mean resistance $\mathcal{R}$ of such a fractal and its linear size (on the lattice) $L$, i.e. calculation of the critical exponent $\tilde{\zeta}$ defined by

$$R \sim L^{\tilde{\zeta}}.\tag{2.18}$$

The fractal dimension $d_{fl}$, which relates the mass of the fractal $N$ to its linear size $L$ *on the lattice* is also measured. These two exponents combined with the relation (2.11) suffice for the calculation of the radius of gyration $R_{go}$ of a Gaussian network, which has the same *connectivity* as the lattice fractal. Thus for fractals we find

$$\nu_o = \tilde{\zeta}/2,\tag{2.19a}$$

$$d_{fo} = 2d_{fl}/\tilde{\zeta}.\tag{2.19b}$$

The exponent $\nu_o$ is not a very useful quantity for fractal networks, since it relates $R_{go}$ to the *linear size of lattice fractal*, which depends on the choice of embedding of the network on a lattice: E.g., the same linear polymer can be represented on a lattice as a straight line and as a random walk, thus leading to different values of the linear size of the fractal, and to different values of $\nu_o$. On the other hand, $d_{fo}$ relates $R_{go}$ to a physically well-defined mass $N$ of the fractal. Since $d_{fo}$ may depend only on the connectivity of the structure, while both $d_{fl}$ and $\tilde{\zeta}$ depend on the particular embedding, it is not surprising that $d_{fo}$ depends only on the ratio between the two. Actually, it can be reexpressed in terms of spectral (or fracton) dimension (Alexander and Orbach, 1982; Rammal and Toulouse, 1983) $\tilde{d} \equiv 2d_{fl}/(\zeta + d_{fl})$:

$$d_{fo} = \frac{2\tilde{d}}{2 - \tilde{d}}. \tag{2.20}$$

Notice the relation between (2.16b) and (2.20) — the spectral dimension plays the role of the dimension of the manifold in the case of fractals.

### 2.3. *Properties of Phantom Tethered Surfaces*

In the previous subsection, I did not address the question, whether a tethered (two-dimensional) surface with realistic interatomic interactions approaches the Gaussian form (2.12) in the long wavelength limit. As an example, let us consider a triangular array of monomers embedded in a $d$-dimensional space, such as depicted in Fig. 1. The Hamiltonian with pairwise nearest-neighbor interactions is

$$\frac{H}{k_B T} = \sum_{\langle \mathbf{x}, \mathbf{x}' \rangle} V[\mathbf{r}(\mathbf{x}) - \mathbf{r}(\mathbf{x}')], \tag{2.21}$$

where $\mathbf{x}$ and $\mathbf{x}'$ are the *internal* coordinates of a pair of neighboring monomers, while $\mathbf{r}$'s are their positions in the *external* (embedding) space. Since the self-avoiding interactions between the distant parts of the surface are ignored, i.e., the surface can freely cross itself, it will be denoted "*phantom* surface".

In the previous subsection, I have shown that the model is solvable for $V(\mathbf{r}) = \frac{1}{2} K_0 r^2$. In particular, the radius of gyration of a finite surface is given by (2.17). Unfortunately, such a potential prefers the neighboring monomers to be located at the same point, and poorly represents any realistic microscopic interaction. Since other choices of $V(\mathbf{r})$ produce analytically unsolvable problems, we must resort to more approximate techniques. One such method is to construct an approximate renormalization group via Migdal–Kadanoff bond-moving approximation for integrating out the intermediate particles (Kadanoff, 1976). Since the high connectivity of a two-dimensional array prevents an exact rescaling, the Migdal–Kadanoff procedure replaces the actual connectivity by its approximation, i.e., the interactions are moved as shown in Fig. 3 to produce an isolated one-dimensional set of degrees of freedom. This *approximate* step conserves the number of bonds in the original problem and is followed by an *exact* decimation of a subset of particles.

It can be verified analytically that any Gaussian spring potential is exactly invariant under such transformation. We confirmed numerically (Kantor *et al.*, 1987) that several simple potentials converge to a Gaussian spring potential under the repeated application of Migdal–Kadanoff approximation.

Going beyond the approximate rescaling of the potential, the asymptotic Gaussian behavior was confirmed numerically by a Monte Carlo (MC) simulation

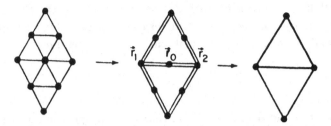

Fig. 3.  Migdal–Kadánoff rescaling procedure for a triangular lattice: the first approximate step changes the connectivity of the lattice, and is followed by exact decimation.

(Kantor *et al.*, 1986, 1987). We considered a hard-sphere-and-string model with

$$V(\mathbf{r}) = \begin{cases} 0, & a < r < b \\ \infty, & \text{otherwise} \end{cases}. \tag{2.22}$$

Although the abrupt changes in the potential with increasing $r$ are not realistic, it has, at least, the realistic feature of not allowing the neighboring monomers of the array to come too close together. Our model closely resembles models used to simulate linear polymers (see, e.g., Baumgärtner (1984)). Since our potentials do not introduce an energy scale into the problem, the results are independent of temperature and the free energy is due solely to entropic effects. We used the MC procedure to numerically equilibrate $L \times L$ parallelograms excised from a triangular lattice with free edges. The procedure was performed for $L$ ranging from 2 to 16. The $R_{go}$ has been measured as a function of $L$ and has been found to approach a simple logarithmic behavior as $L$ increased, i.e., the $L$-dependence coincided with the prediction (2.17) obtained for a Gaussian surface.

   Although the numerical proof of the Gaussian behavior has been made for a particular type of potential $V(\mathbf{r})$ in $d = 3$, it is plausible to assume that it will be valid for any binding (central force) potential in an arbitrary $d$. Thus we shall use (2.12) as a starting point for the investigation of more realistic surfaces.

## 3. Excluded Volume Effects

### 3.1. *Bounds on the Exponent $\nu$*

The next step towards a realistic description of surfaces is to consider the effects of self-avoiding interactions. One expects that introduction of such interactions will create larger (more open) structures. In particular, the exponent $\nu$ of such a structure should be larger than $\nu_o$. Indeed, for linear polymers, $\nu \approx 0.59 > \nu_o = \frac{1}{2}$ in $d = 3$. The effect should be even more significant for two-dimensional surfaces, since in the absence of the excluded volume effects, their $R_g$ barely depends on $L$ (see Eq. (2.17)), and they overfill the embedding space ($d_{fo} = \infty$ for any $d$).

   The physical limits on the values of $\nu$ are determined from a simple argument: Consider a two-dimensional surface of size $L \times L$ of finite thickness $w$ in a three-dimensional space. When the surface is stretched, its $R_g \sim L$. Since the volume of the surface is $L^2 w$, it can be "compactified" into an object of linear dimensions $\sim L^{2/3} w$. Thus, for such conformation, $R_g \sim L^{2/3}$. These two cases identify the physical bounds on the exponent $\nu$: $\frac{2}{3} \leq \nu \leq 1$. Generally, for a $D$-dimensional surface in $d$-dimensional space:

$$\frac{D}{d} \leq \nu \leq 1, \tag{3.1a}$$

$$d \geq d_f \geq D. \tag{3.1b}$$

Are these, indeed, the best possible bounds, which can be obtained from simple geometric considerations? There are many ways to compactify a linear ($D = 1$) polymer in any $d$. However, the high connectivity of a two-dimensional surface restricts its possible spatial conformations, and it is not obvious that the compact conformation ($\nu = \frac{2}{3}$ in $d = 3$) can actually be attained. A "table-top experiment" (Kantor *et al.*, 1986, 1987; Gomes, 1987) in which sheets of foil have been "randomly" crumpled, showed that the diameter of the crumpled ball increased with the linear (uncrumpled) size of the sheet $L$ as $L^{0.8}$, i.e., the resulting structure is *not* compact. Actually, it is quite difficult to find an "intelligent" (non-random) folding procedure, which will allow compactification of an arbitrarily large surface. (We assume, that the surface is elastic, and require that the elastic stretching energy required to compactify the surface increase slower than the surface area.) The reader is invited to try some simple folding procedures on a *very large* piece of paper, and to convince himself, that the task is not trivial. Nevertheless, at least one folding procedure (suggested by R. C. Ball), depicted in Fig. 4, succeeds in achieving that goal. Thus, the connectivity of the surface does not narrow down the bounds $\frac{2}{3} \leq \nu \leq 1$ for $D = 2$ and $d = 3$. We can only hope that similar foldings also exist for manifolds with $D > 2$ (for $d \geq D$), and, thus, (3.10) are the best possible bounds.

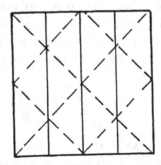

Fig. 4.   Compactification of a two-dimensional sheet. The lines depict the creases created by the folds (of 180°) which can be seen if the sheet is unfolded back to a flat state. The solid and dashed lines indicate different directions of folding ("upwards" and "downwards"). Distances between the neighboring parallel creases are of order $L^{2/3} w^{1/3}$.

### 3.2. Analytic Estimates of $\nu$

Excluded volume effects can be introduced into the continuum description of manifolds by generalizing Edwards' (1965) model for polymers (Kantor et al., 1986). The full Hamiltonian now becomes

$$\frac{H}{k_B T} = \frac{1}{2} K \int (\nabla \mathbf{r})^2 d^D x + \frac{1}{2} v \int \int \delta^d[\mathbf{r}(\mathbf{x}) - \mathbf{r}(\mathbf{x}')] d^D x \, d^D x', \qquad (3.2)$$

where the interaction $\nu$ measures the excluded volume effect. Dimensional analysis of (3.2) shows that for $D < 2$ there exists an upper critical dimension

$$d_c(D) = \frac{4D}{2-D}, \qquad (3.3)$$

such that for $d > d_c$ the excluded volume interactions are irrelevant, and $\nu$ and $d_f$ are given by (2.16). For $D \geq 2$ self-avoidance is relevant in any $d$. The critical exponent at any point in the $(D, d)$-plane can (in principle) be evaluated by a systematic expansion from any point $(D^*, d_c(D^*))$ on the line of critical dimensions (3.3) (Kardar and Nelson, 1987, 1988; Aronovitz and Lubensky, 1987; Duplantier, 1987). To the lowest order in the "distance" from the line of critical dimensions, one finds

$$\nu(D, d) = \frac{2-D}{2} + \frac{2-D^*}{8[D^* + 2C(D^*)]} \epsilon, \qquad (3.4a)$$

$$\epsilon \equiv 4D - (2-D)d, \qquad (3.4b)$$

$$C(D^*) \equiv \frac{\sqrt{\pi} \Gamma(2/(2-D^*))}{2^{2D^*/(2-D^*)} \Gamma((2+D^*)/2(2-D^*))}. \qquad (3.4c)$$

Unfortunately, this expansion is too short to be useful for $D = 2$ in $d = 3$ since it predicts $\nu = 0.536$ (expansion around $d_c(D^*) = 3$), which is certainly incorrect since it violates the bound (3.1). Thus, for the time being, we must rely on more approximate estimates.

The scaling behavior of self-avoiding polymers can be studied by Flory-type approximation (see, e.g., de Gennes (1979)). Consider a $D$-dimensional manifold of internal size $L$ (and mass $N = L^D$), occupying a region of size $R_g$ in the $d$-dimensional space. According to Flory, we may approximate the free energy $F$ of the manifold by

$$\frac{F}{k_B T} = \frac{1}{2} \left( \frac{R_g}{R_{go}} \right)^2 + \frac{1}{2} v \left( \frac{N}{R_g^d} \right)^2 R_g^d. \qquad (3.5)$$

The first term on the r.h.s. of (3.5) is the elastic (entropic) free energy of a *phantom* manifold ($R_{go}$ is the radius of gyration of the manifold without self-avoidance). The second term is the mean-field estimate of the repulsive interaction energy (the squared density of the monomers $(N/R_g)^2$ is a mean-field-type estimate of the number of pair of monomers coming into close contact with each other in a unit volume). By minimizing (3.5) and using (2.16) to relate $R_{g0}$ to $L$, we find $R_g \sim L^{\nu_F}$, with

$$\nu_F = \frac{D+2}{d+2}.$$ (3.6a)

Thus, the Flory estimate of the fractal dimension of the manifold is

$$d_{fF} = \frac{d+2}{D+2}D.$$ (3.6b)

In particular, for $D = 2$ and $d = 3$, we have $\nu_F = 4/5$ and $d_{fF} = 2.5$. Somewhat more formally, we can determine $\nu_F$ from (3.2) by requiring (Oono, 1981) that the rescaling factors of the internal coordinate $\mathbf{x}$ and the external coordinate $\mathbf{r}$ be related in such a way that the ratio between the two terms in (3.2) will remain unchanged under the rescaling.

For a more general case of a network with fractal connectivity (Cates 1984, 1985b), one can use (2.18) and (2.19), as well as the relation between $N$ and $L$ for a fractal, to obtain

$$d_{fF} = \frac{d+2}{\tilde{d}+2}\tilde{d}.$$ (3.7)

As in the then case of phantom networks, the expression (3.7) is obtained from (3.6b) by a substitution $D \to \tilde{d}$.

Despite its numerous deficiencies, the Flory-type theory produces remarkably good estimates of $\nu$ for linear polymers. Note that in the trivial cases of $d = D$ and $d = d_c(D)$, it produces exact answers. However, since it is an uncontrolled approximation, it is difficult *a priori* to assess its accuracy.

### 3.3. *Monte Carlo Investigation of Tethered Surfaces*

We performed a numerical (MC) equilibration of tethered surfaces with self-avoidance (Kantor *et al.*, 1986, 1987). As in the case of phantom surfaces, we equilibrated $L \times L$ parallelograms, with nearest neighbor interaction described by (2.22). However, now the repulsive (hard-core) interaction was "switched on" for any pair of monomers, i.e. no two monomers could come closer together than a distance $a$. The parameters in (2.22) were chosen to be $b/a = \sqrt{3}$, since such a choice ensures a complete impenetrability of the surface.

The size of equilibrated parallelograms $L$ ranged from 2 to 11. Figure 5 depicts an equilibrium conformation of a self-avoiding surface. To study scaling properties of these surfaces, the $R_g$ was calculated as a function of $L$. We found a nice power law dependence with $\nu \approx 0.83$.

The main source of possible errors in the simulation of such small surfaces are not the statistical errors, but rather "systematic errors" which appear since we are not in the asymptotic regime. Thus, it is useful to have two distinct ways to estimate $\nu$, e.g. directly from the two-point (density-density) correlation function and from the "mass versus radius of gyration" curve. Differences in the exponents obtained by

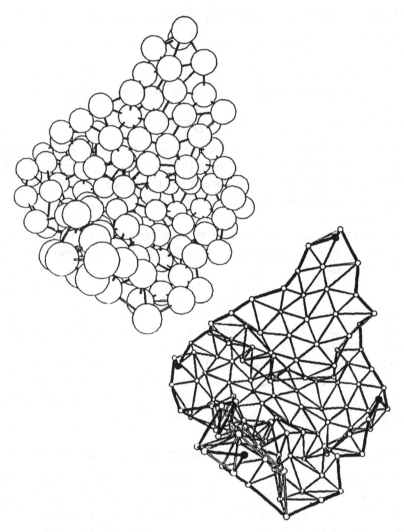

Fig. 5. Shape of the self-avoiding surface for $L = 11$ (top). Sphere sizes indicate the range of the repulsive potential. "Skeleton" of the same surface (bottom): For clarity, the sizes of the monomers (spheres) were taken to be $\frac{1}{5}$ of the actual range of the hard-core potential. Bonds indicate the nearest-neighbor atoms between which the "string" attractive potential acts. Boundary bonds and corner monomers are shown in black.

those methods provide an estimate of the "systematic error." We examined the Fourier transform of the two-point correlation function (or the structure factor)

$$S(\mathbf{k}, L) \equiv \frac{1}{L^4} \sum_{\mathbf{x}, \mathbf{x}'} \langle e^{i\mathbf{k} \cdot [\mathbf{r}(\mathbf{x}) - \mathbf{r}(\mathbf{x}')]} \rangle. \tag{3.8}$$

In analogy with polymeric systems, we assume, that the structure factor satisfies the scaling form $S(\mathbf{k}, L) = S(kR_g) = S(kL^\nu)$. The lack of sphericity of the self-avoiding surfaces is evident from Fig. 5, while our scaling relation for $S$ assumed an isotropic $k$-dependence. The deviation from sphericity can be characterized by the ratio between the minimal and the maximal principal moments of inertia of the surface. This ratio was found to be $\approx 0.16$, and *independent* (within the accuracy of the measurement) of $L$, thus justifying the use of spherically-symmetric scaling form. $S(0) \approx 1$ for $k < 1/R_g$, while for very large $k$ ($k > 1$) we expect to have $S = 1/L^2$. To satisfy these conditions, as well as the scaling relation we must require $S \sim k^{-2/\nu}$ for $1 < k < 1/R_g$. From the $k$-dependence of $S$ in this intermediate range of $k$'s, we find $\nu \approx 0.77$. Thus, our overall estimate of $\nu$ is $\sim 0.8$. The corresponding fractal dimension is 2.5. This result is in good agreement with the Flory estimate $\nu_F = 4/5$, thus confirming the accuracy of the estimate and the physics embodied in (3.2).

## 4. Crumpling Transition in Tethered Surfaces

### 4.1. *Very Rigid and Very Flexible Surfaces*

What happens when we introduce bending rigidity into the tethered surfaces? In the previous sections, we ignored the rigidity and were led to the continuum descriptions (2.12) and (3.2) of the, respectively, phantom and self-avoiding surfaces. The $(\nabla r)^2$-term in those expressions does not represent the microscopic interactions, but results from a "coarse-graining" of the problem and is generated by the entropy in a crumpled surface. We might expect, that introduction of small rigidity will modify the persistence length, but will cause no change in the asymptotic behavior of the surfaces.

Investigations of the properties of very rigid surfaces usually take a point of view diametrically opposed to the crumpled surface ideas: One assumes that the surface is (at least locally) flat and describes its fluctuations using the Monge parametrization in terms of normal dispacement $f, \mathbf{r}(x_1, x_2) = (x_1, x_2, f)$. To the lowest order in $f$ and its gradients, the surface energy may be written (Landau and Lifshitz, 1970)

$$F = \frac{1}{2}\tilde{\kappa} \int d^2x (\nabla^2 f)^2 + \frac{1}{2} \int d^2x (2\mu u_{ij}^2 + \lambda u_{kk}^2), \qquad (3.9)$$

where the strain matrix $u_{ij}$ is related to $f$ and the in-plane displacements $u_i$ by $u_{ij} = \frac{1}{2}(\partial_i u_j + \partial_j u_i + \partial_i f \partial_j f)$, $\tilde{\kappa}$ is the bending rigidity, and $\mu$ and $\lambda$ are the in-plane Lamé constants. In *liquid* membranes the second term of (3.9) vanishes, and this leads to a crumpled surface, since the short wavelength transverse oscillations (undulations) reduce the effective rigidity of liquid surfaces on large length scales (Helfrich, 1985). However, it has been shown by an approximate (self-consistent) treatment of the vibrations, that in the presence of that term the transverse fluctuations are suppressed ($\langle f^2 \rangle$ increases slower than $L^2$), and the surface remains asymptotically flat (Nelson and Peliti, 1987).

Clearly, the results for very rigid and very flexible surfaces are not compatible, and one may expect a phase transition from a crumpled to a flat phase. We applied Monte Carlo methods to investigate the possibility of such a transition (Kantor and

Nelson, 1986, 1987). The model consisted of a hexagonal surface $L$ monomers across, excised from a triangular lattice. The energy assigned to a particular conformation contained the nearest-neighbor interactions (2.22) and a bending energy term

$$\frac{H_b}{k_B T} = -\kappa \sum_{\langle \alpha, \beta \rangle} (\mathbf{n}_\alpha \cdot \mathbf{n}_\beta - 1), \qquad (3.10)$$

where the sum is performed over pairs $\langle \alpha, \beta \rangle$ of adjacent unit normals $\{\mathbf{n}_\alpha\}$ erected perpendicular to each elementary triangle. Due to extremely long MC equilibration times, we did not attempt to include the excluded volume interactions into the model.

Figure 6 depicts the equilibrium conformations of the surface for several values of $\kappa$. We observe a dramatic change in the shape of the surface: For small values of $\kappa$, the surface overfills the space, as expected from flexible phantom surfaces. For large $\kappa$, the surface is quite "flat". We obtained quantitative evidence, supporting the claim that the observed effect is not a mere crossover. In particular, investigation of the $L$- and $\kappa$-dependence of $R_g$ leads to the conclusion that for $L \to \infty$,

$$R_g(L) = \begin{cases} \xi \sqrt{\ln L} & \text{for } \kappa < \kappa_c \\ \zeta L & \text{for } \kappa > \kappa_c, \end{cases} \qquad (3.11)$$

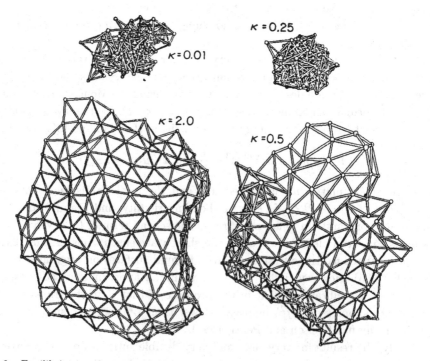

Fig. 6. Equilibrium conformations of hexagons $L = 15$ monomers across in the crumpled phase ($\kappa = 0.01$ and $\kappa = 0.25$), in flat phase ($\kappa = 2.0$), and close to the transition point ($\kappa = 0.5$).

where $\kappa_c \approx 0.33$. The values of $\xi$ and $\zeta$ in (3.11) depend on $\kappa$: When $\kappa$ approaches $\kappa_c$ from below, the persistence length $\xi$ increases and diverges at the transition point. The parameter $\zeta$, which measures the possible shrinkage of the flat phase, decreases as $\kappa$ approaches $\kappa_c$ from above, and vanishes at the transition point. Precisely at the transition point $\kappa_c$ (as well as away from $\kappa_c$, on length scales shorter than the correlation length), one may expect an intermediate "semicrumpled" regime, with $R_g \sim L^{\nu'}$. We indeed obtain such behavior with $\nu' \approx 0.8$.

However, the best evidence of the presence of the phase transition is provided by the specific heat measurements. Figure 7 depicts the $\kappa$-dependence of the specific heat per monomer $C$. For $\kappa = 0$ the fluctuations are purely entropic since the potential (2.22) allows only conformations with zero potential energy. Consequently, $C = 0$ at this point. (We suppress the trivial kinetic part $\frac{3}{2}k_B$ of the specific heat.) For sufficiently large $\kappa$, one can neglect the coupling between the transverse oscillations. [The two-dimensional (in-plane) degrees of freedom are entropic and do not contribute to the specific heat.] Therefore, for large $\kappa$ we expect to obtain $C = \frac{1}{2}k_B$, in accordance with the Dulong–Petit law. Indeed all curves in Fig. 7 approach that limit for $\kappa \rightarrow \infty$. In the absence of a phase transition, one might expect a smooth interpolation of $C$ as $\kappa$ changes from 0 to $\infty$. On the other hand, we see a peak, which sharpens for large $L$. The position of the peak drifts towards small $\kappa$'s as $L$ increases and tends to a constant (positive) value $\kappa_c$.

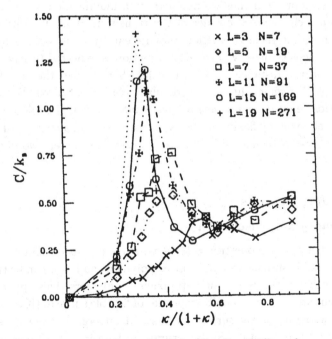

Fig. 7. Specific heat per monomer $C$ as a function of bending rigidity $\kappa$ for several values of $L$. The peak in the specific heat becomes more pronounced for larger $L$.

These MC results have been obtained for systems of quite limited size, and no attempt has been made to determine the critical exponents of the specific heat. The data is, however, sufficient to practically exclude the possibility of the first order phase transition (or, at least, to limit the possible latent heat to a very small value). Recently David and Guitter (1988) studied the behavior of rigid two-dimensional surfaces in large embedding dimension $d$, and recovered (to the first order in $1/d$) the second order phase crumpling transition. There is a reasonable quantitative agreement between those predictions and our MC results regarding the position of the critical point and the scaling behavior in the "semicrumpled" phase. On the other hand, a critical fluctuation analysis by an $\epsilon = 4 - D$ expansion, within Landau–Ginzburg theory (Paczuski *et al.*, 1988) predicts a first-order transition for $d < d_c = 219$. Such predictions, however, are not always reliable in the lowest order in $\epsilon$.

## 4.2. *Excluded Volume Effects*

Excluded volume interactions may modify the details of the crumpling transition. In the flat phase, self-avoidance should be unimportant at long wavelengths. The main effect should be to augment the bending forces, thus delaying the transition to the crumpled state. In the crumpled phase, on the other hand, self-avoidance plays a major role of swelling the surface. The available numerical data (Kantor and Nelson, 1986, 1987) indicates that the excluded volume effects *are relevant* at the transition point, and thus we may expect a modification of the critical exponent, or even a change in the order of the transition. If the transition remains of second order, an interesting situation may appear: Just below the transition point ($\kappa \leq \kappa_c$), the density-density correlation function is expected to have a power-law dependence on the distance on length scales shorter than the correlation length ("semicrumpled" regime), just as in "regular" phase transitions. However, beyond the correlation length, we will *again* have a power-law regime determined by the behavior of self-avoiding flexible surfaces. Thus, instead of a crossover from a power-law to homogeneous regime, we will have a crossover between two different power-law regimes.

## 5. Concluding Remarks

### 5.1. *Summary and Discussion*

Our knowledge of the properties of tethered surfaces can be summarized in a few sentences: Flexible surfaces resemble linear polymers, and they can be treated by the methods which generalize the usual approach to polymers. This applies not only to the static, but also to the dynamic properties of flexible surfaces (Kantor *et al.*, 1986, 1987). As the rigidity of the surface increases, it undergoes a remarkable transition from a crumpled to flat state. Such a crumpling transition has no equivalent in the case of linear polymers.

The list of gaps in our knowledge is significantly longer: The problems begin at the earliest stage of transition from a microscopic Hamiltonian to its continuum version. Even in the case of a flexible phantom surface, we had to rely on MC results to justify the continuum form (2.12), and the subsequent use of the generalized Edwards' Hamiltonian (3.2). Similar problems prevent the proof of the validity of the recently suggested Hamiltonian (Paczuski *et al.*, 1988), which has been used to analyze the nature of crumpling transition within the framework of the Landau–Ginzburg theory. Although we understand the important qualitative features of the physics of tethered surfaces, our quantitative knowledge is quite limited: e.g., the value of the exponent $\nu$ has been established by MC investigations of very small systems, while the $\epsilon$-expansion to the lowest order in $\epsilon$ did not produce a useful estimate of $\nu$ for a two-dimensional surface. We do not know the critical exponents, characterizing the crumpling transition, and can only speculate regarding the changes in the transition which will be caused by the excluded volume effects.

## 5.2. *What Next?*

Investigation of the behavior of the tethered surfaces presents an interesting and very rich problem. Presently, we only begin to understand the behavior of a *single* surface. Many questions of practical importance have yet to be answered. For example, it would be useful to know the physical properties of melts of surfaces, because, unlike linear polymers, tethered surfaces will not become ideal in a dense melt, since in their ideal (Gaussian) state they overfill the embedding space. Also, it would be interesting to find out the behavior of the surfaces near an adsorbing wall.

Two-dimensional membranes often exist as closed vesicles containing a volume of space, and this is a possibly important application of polymerized surfaces. Properties of such vesicles depend not only on the intrinsic properties of the membrane, but also on the pressure difference between their interior and the exterior. Behavior of a *linear* "vesicle" in $d = 2$ (i.e., a closed loop) has been recently investigated by Leibler *et al.* (1987). It would be very useful to extend those results to the more realistic case of two-dimensional tethered surface in $d = 3$.

Recently, it has been shown (Seung and Nelson, 1987) that a two-dimensional "solid" (i.e., a surface with a *finite* bond strength between the neighboring atoms or monomers) embedded in three dimensions, will melt, since the dislocation energy is finite in a surface, which is allowed to buckle in three dimensions. It is not known whether it will melt to a liquid or hexatic phase. Since the behavior of a surface depends on its in-plane order (in particular, hexatic membranes may undergo a crumpling transition (Nelson and Peliti, 1987)), it would be useful to have some way to compare the free energies of these two in-plane phases. Tethered surfaces provide a convenient "reference point" and may, probably, be used to compare these two phases. Melting of a two-dimensional "solid" in $d = 3$, has been analyzed under the assumption that in the absence of dislocations it is in the flat phase

(Nelson and Peliti, 1987; Seung and Nelson, 1988). It is not clear, what are the effects of the dislocations in the crumpled phase.

## References

1. Alexander, S. and Orbach, R., *J. Physique Lett.* **43**, L625 (1982).
2. Aronovitz, J. A. and Lubensky, T. C., *Europhys. Lett.* **4**, 395 (1987).
3. Baumgärtner, A., in *Application of Monte Carlo Method in Statistical Physics*, ed. K. Binder (Springer, Berlin, 1984), p. 145.
4. Billoire, A., Gross, D. J. and Marinari, E., *Phys. Lett.* **139B**, 75 (1984).
5. Blumstein, A., Blumstein, R. and Vanderspurt, T. H., *J. Colloid Interface Sci.* **31**, 236 (1969).
6. Cates, M. E., *Phys. Rev. Lett.* **53**, 926 (1984).
7. Cates, M. E., *Phys. Lett.* **161B**, 363 (1985a).
8. Cates, M. E., *J. Physique* **46**, 1059 (1985b).
9. David, F. and Guitter, E., 1988 *preprint*.
10. Drouffe, J.-M., Parisi, G. and Sourlas, N., *Nucl. Phys.* **B161**, 397 (1979).
11. Duplantier, B., *Phys. Rev. Lett.* **58**, 2733 (1987).
12. Edwards, S. F., *Proc. Phys. Soc. London* **85**, 613 (1965).
13. Fendler, J. H. and Tundo, P., *Acc. Chem. Res.* **17**, 3 (1984).
14. Flory, D. J., *Statistical Mechanics of Chain Molecules* (Wiley, New York, 1979).
15. Frölich, J., in *Applications of Field Theory to Statistical Mechanics*, Vol. 216, ed. L. Garido (Springer, Berlin, 1985).
16. de Gennes, P. G., *Scaling Concepts in Polymer Physics* (Cornell University Press, Ithaca, New York, 1979).
17. Gomes, M. A. F., *J. Phys.* **A20**, L283 (1987).
18. Gross, D. J., *Phys. Lett.* **139B**, 187 (1984).
19. Helfrich, W., *J. Physique* **46**, 1263 (1985).
20. Helfrich, W., *J. Physique* **48**, 285 (1987).
21. Kadanoff, L. F., *Ann. Phys. (New York)* **100**, 359 (1976).
22. Kantor, Y., Kardar, M. and Nelson, D. R., *Phys. Rev. Lett.* **57**, 791 (1986).
23. Kantor, Y., Kardar, M. and Nelson, D. R., *Phys. Rev.* **A35**, 3056 (1987).
24. Kantor, Y. and Nelson, D. R., *Phys. Rev. Lett.* **58**, 2774 (1987a).
25. Kantor, Y. and Nelson, D. R., *Phys. Rev.* **A36**, 4020 (1987b).
26. Kardar, M. and Nelson, D. R., *Phys. Rev. Lett.* **58**, 1298 (1987).
27. Kardar, M. and Nelson, D. R., *Phys. Rev.* **A**, in press (1988).
28. Landau, L. D. and Lifshitz, E. M., *Theory of Elasticity* (Pergamon Press, New York, 1970).
29. Leibler, S., Singh, R. R. P. and Fisher, M. E., *Phys. Rev. Lett.* **59**, 1989 (1987).
30. Nelson, D. R. and Peliti L., *J. Physique* **48**, 1085 (1987).
31. Mandelbrot, B. B., *Fractals: Form Chance and Dimension* (Freeman, San Francisco, 1977).
32. Mandelbrot, B. B., *The Fractal Geometry of Nature* (Freeman, San Francisco, 1982).
33. Oono, Y., in *Adv. Chem. Phys.*, eds. I. Prigogine and S. A. Rice, Vol. LXI (Wiley, New York, 1981), p. 301.
34. Paczuski, M., Kardar, M. and Nelson, D. R., 1988, preprint.
35. Parisi, G., *Phys. Lett.* **81B**, 357 (1979).
36. Rammal, R. and Toulouse, G., *J. Physique Lett.* **44**, L13 (1983).
37. Seung, H. S. and Nelson, D. R., 1987, preprint.
38. Stephen, M. J., *Phys. Rev.* **B17**, 4444 (1978).
39. Weiner, J. H., *Statistical Mechanics of Elasticity* (Wiley, New York, 1983).

## CHAPTER 6

## THEORY OF THE CRUMPLING TRANSITION

David R. Nelson

*Department of Physics*
*Harvard University*
*Cambridge, Massachusetts 02138*

The large distance behaviors of membranes fall into a variety of universality classes, depending, for example, on whether the local order is liquid or crystalline. Here we show that membranes with a nonzero shear modulus differ from their liquid counterparts in that they exhibit a flat phase with long-range order in the normals at sufficiently low temperatures.[1] Because entropy favors crumpled surfaces with decorrelated normals, there must be a transition to a crumpled phase at sufficiently high temperatures. The numerical evidence for such a crumpling transition in the absence of self-avoidance[2] is reviewed in the lectures of Kantor. We describe here a simple Landau theory of the crumpling transition which shows how it is affected by self-avoidance.[3] Finally, we discuss the energies of disclinations and dislocations in flexible membranes with local crystalline order.[1,4] Unlike crystalline films forced to be flat by a surface tension, it is energetically favorable for membranes to screen out elastic stresses by buckling into the third dimension. Dislocations, in particular, are predicted to have a finite energy. We conclude that a finite density of dislocations must exist at all nonzero temperatures in nominally crystalline but unpolymerized membranes. The result macroscopically is a hexatic membrane, with zero shear modulus, but extended bond orientational order.[5] The elastic energy which controls undulations in hexatic membranes is discussed briefly.

## 1. Normal-Normal Correlation in Liquid Membranes

Before proceeding to flexible membranes with a shear modulus, we review expectations for normal correlations in liquid membranes.

Figure 1 shows a fragment of a liquid membrane which we assume is approximately parallel to the $(x_1, x_2)$-plane so that we can use a Monge representation for its position,

$$\vec{r}(x_1, x_2) = (x_1, x_2, f(x_1, x_2)). \tag{1.1}$$

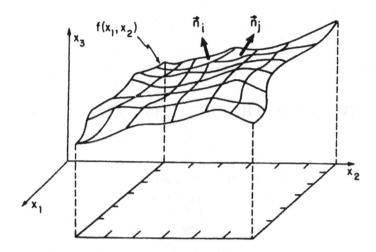

Fig. 1.   Liquid membrane broken up into plaquettes each characterized by a unit normal.

The unit normal at any point is easily shown to be

$$\vec{n}(x_1, x_2) = (-\partial_1 f, -\partial_2 f, 1)/\sqrt{1 + [\vec{\nabla} f]^2}. \tag{1.2}$$

If we think of the membrane as composed of rod-shaped amphiphillic molecules, it is natural to associate these normals with the local rod axis. Upon partitioning the membrane into segments as shown in Fig. 1, we can write down a lattice model of the bending energy,

$$F_b = -\tilde{\kappa} \sum_{\langle i,j \rangle} \vec{n}_i \cdot \vec{n}_j, \tag{1.3}$$

in analogy with a one-Frank-constant elastic energy for nematic liquid crystals.[6] Here the sum is over nearest-neighbor segments, $\vec{n}_i$ is the normal associated with the $i$-th segment, and $\tilde{\kappa}$ is a microscopic bending rigidity.

Equation (1.3) resembles the energy of a classical Heisenberg ferromagnet on a two-dimensional lattice.[7] The normals are like spin vectors, and the rigidity $\tilde{\kappa}$ is like a Heisenberg exchange constant. Rotational symmetry is broken at $T = 0$ by a "ferromagnetic" flat surface, with a uniformly aligned normal field. It is well-known, however, that long-range order is destroyed at any *finite* temperature in the two-dimensional Heisenberg model by spin wave fluctuations.[8] In surfaces, we might also expect long range-order to be destroyed, in this case by surface undulations. The analogy with spin waves is not perfect, however, because the normals are constrained to be part of a surface, which forces them to be expressible as in Eq. (1.2). For small undulations, this restriction means that "spin waves" in the normals must be purely longitudinal; transverse "spin waves" would tear the surface.

To determine how undulations affect correlations in the normals it is useful to take the continuum limit of Eq. (1.3). To leading order in a expansion in gradients of $f(x_1, x_2)$, we can neglect the factor $\sqrt{g} = \sqrt{1 + |\vec{\nabla} f|^2}$ in the measure as well as the $|\vec{\nabla} f|^2$ term in the denominator of (1.2) and find

$$F_b \approx \frac{1}{2} \kappa \int d^2x |\nabla \vec{n}|^2 \approx \frac{1}{2} \kappa \int d^2x [(\partial_1^2 f)^2 + 2(\partial_1 \partial_2 f)^2 + (\partial_2 f)^2]$$

$$\approx \frac{1}{2} \kappa \int d^2x [(\nabla^2 f)^2 - 2 \det(\partial_i \partial_j f)] \tag{1.4}$$

where $\kappa$ is proportional to $\tilde{\kappa}$. The last two terms of (1.4) are just the mean curvature and Gaussian curvature pieces of the Helfrich[9] bending energy of a liquid membrane. The Gaussian curvature is a perfect derivative, which we can see by writing the second term as

$$2 \det(\partial_i \partial_j f) = -\epsilon_{im} \epsilon_{jn} \partial_m \partial_n [(\partial_i f)(\partial_j f)]. \tag{1.5}$$

Upon neglecting the contribution from this surface term, we can write

$$F_b \approx \frac{1}{2} \kappa \int d^2x (\nabla^2 f)^2. \tag{1.6}$$

Following de Gennes and Taupin,[10] we can now estimate fluctuations in the normals. The angle $\theta(x_1, x_2)$ which the normal $\vec{n}(x_1, x_2)$ makes with respect to the $\hat{x}_3$ axis is given by

$$\vec{n} \cdot \hat{x}_3 = \cos \theta = 1/\sqrt{1 + |\vec{\nabla} f|^2}. \tag{1.7}$$

If there is a broken symmetry such that the normals point on average along the $\hat{x}_3$ axis, fluctuations in $\theta^2 = |\vec{\nabla} f|^2$ should be small at low temperatures. Because Eq. (1.6) is a quadratic form, we can calculate $\langle \theta^2 \rangle$ by passing to Fourier space and using the equipartition theorem,

$$\langle \theta^2(x_1, x_2) \rangle \approx k_B T \int \frac{d^2q}{(2\pi)^2} \frac{1}{\kappa q^2} \approx \frac{k_B T}{\kappa} \ln(L/a). \tag{1.8}$$

Just as in many other systems with continuous symmetries in two dimensions,[11] there is a logarithmic divergence with system size $L$, signalling the breakdown of long-range order in the normals. More sophisticated calculations by Peliti and Leibler[12] show that the renormalized wave-vector-dependent rigidity $\kappa_R(q)$ is softened by these fluctuations,

$$\kappa_R(q) = \kappa - \frac{3k_B T}{4\pi} \ln(1/qa). \tag{1.9}$$

Note that if we replace $\kappa$ by $\kappa_R(q)$ in Eq. (1.8), this only makes the divergence worse. The renormalization group calculations of Ref. 12 are consistent with exponential decay of the normal-normal correlation function.

$$\langle \hat{n}(\mathbf{x}) \cdot \hat{n}(0) \rangle \propto e^{-x/\xi} \tag{1.10}$$

with a correlation length which diverges at low temperatures

$$\xi \approx a e^{4\pi\kappa/3k_B T}. \tag{1.11}$$

The low temperature behavior of the two-dimensional Heisenberg model[8] is very similar.

## 2. Tethered Surfaces with Bending Energy

Figure 2 shows a surface in which bending energy and tethering are present simultaneously. If $\vec{r}_i$ denotes the position of the $i$-th vertex, and $\vec{n}_\alpha$ is the normal to the $\alpha$-th triangular plaquette, a microscopic model Hamiltonian would be

$$H = -\tilde{\kappa} \sum_{\langle \alpha, \beta \rangle} \vec{n}_\alpha \cdot \vec{n}_\beta + \sum_{\langle i,j \rangle} V(|\vec{r}_i - \vec{r}_j|), \tag{2.1}$$

where $V(r)$ is a tethering potential between nearest neighbor vertices. We have just seen that liquid membranes, subject only to bending energy, crumple at finite temperatures, in the sense that long-range order in the normals is destroyed. As discussed in the chapter by Kantor, tethered surfaces, subject only to the constraint of fixed bonding connectivity, also crumple, like polymers in a good solvent.[13] We shall now argue that there is a fundamental incompatibility when these two energies are simultaneously present which stabilizes a flat phase at sufficiently low temperatures.

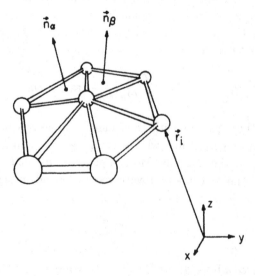

Fig. 2.   Unit normals defined on the triangular plaquettes of a tethered surface.

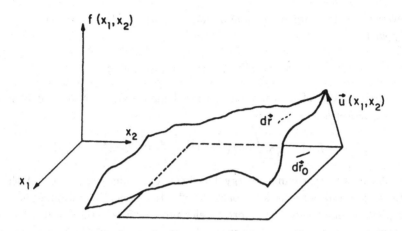

Fig. 3.   Displacement field $\vec{u}(x_1, x_2)$ of a deformed membrane.

As in our discussion of liquid membranes, we start with a locally flat surface and ask if a state with long-range order in the normals is stable to thermal fluctuations. We assume that the tethering potential induces nonzero elastic constants in approximately planar membranes, so that there is elastic stretching energy, as well as the bending energy characteristic of liquid membranes.

As shown in Fig. 3, a three-component displacement field is necessary to describe the deformation of an initially flat membrane. If $\vec{r}_0(x_1, x_2) = (x_1, x_2, 0)$ describes the undistorted membrane at $T = 0$, an arbitrary membrane configuration for $T > 0$ is given by

$$\vec{r}(x_1, x_2) = \vec{r}_0 + \begin{pmatrix} u_1(x_1, x_2) \\ u_2(x_1, x_2) \\ f(x_1, x_3) \end{pmatrix} . \tag{2.2}$$

A small line element $d\vec{r}_0 = (dx_1, dx_2, 0)$ in the undistorted membrane is mapped by this transformation into a line element,

$$d\vec{r} = \begin{pmatrix} (1 + \partial_1 u_1)dx_1 + (\partial_2 u_1)dx_2 \\ (\partial_1 u_2)dx_1 + (1 + \partial_2 u_2)dx_2 \\ (\partial_1 f)dx_1 + (\partial_2 f)dx_2 \end{pmatrix} . \tag{2.3}$$

As usual in discussions of continuum elastic theory,[14] we describe the stretching of this line element by a strain matrix $u_{ij}(x_1, x_2)$,

$$d^2r = d^2r_0 + 2u_{ij}dx_i dx_j, \tag{2.4}$$

where using Eq. (2.3) we find

$$u_{ij} = \frac{1}{2}[\partial_i u_j + \partial_i u_i] + \frac{1}{2}(\partial_i f)(\partial_j f) + \frac{1}{2}(\partial_i u_k)(\partial_j u_k). \tag{2.5}$$

To lowest order in gradients of $u$ and $f$ we can neglect the term $\frac{1}{2}(\partial_i u_k)(\partial_j u_k)$ and simply write

$$u_{ij} \approx \frac{1}{2}[\partial_i u_j + \partial_j u_i + (\partial_i f)(\partial_j f)]. \tag{2.6}$$

The free energy of a nearly flat tethered membrane is a sum of bending the stretching energies,

$$F[f,u] = \frac{1}{2}\kappa \int d^2x (\nabla^2 f)^2 + \frac{1}{2}\int d^2x [2\mu u_{ij}^2 + \lambda u_{kk}^2] \tag{2.7}$$

where the elastic stretching energy has been expanded in powers of the strain matrix, and $\mu$ and $\lambda$ are elastic constants.[14] The vertical membrane displacement $f$ in Eq. (2.6) introduces an important element of "frustration" into (2.7). To see this, imagine that we are given a vertical displacement field $f(x_1, x_2)$ which has a particularly low bending energy, and hence would lead to a low overall energy if the membrane were a liquid. If this is to be a low energy configuration for tethered membranes as well, it must be possible to choose phonon displacements such that $u_{ij}$ vanishes for the given $f$. Note that the term $(\partial_i f)(\partial_j f)$ acts like a matrix vector potential in Eq. (2.6). It will in general be impossible to choose the two independent phonon displacement fields $u_1(x_1, x_2)$ and $u_2(x_1, x_2)$ to cancel all *three* distinct components of this symmetric matrix. We conclude that there must be many low energy configurations of a liquid membrane which will be energetically unfavorable when we introduce stretching energy.

To treat stretching energy more quantitatively, it is useful to eliminate the quadratic phonon field in (2.7) and define

$$\tilde{F}(f) = -k_B T \ln\left\{ \int \mathcal{D}\vec{u}(x_1, x_2)e^{-F[f,u]/k_B T} \right\}. \tag{2.8}$$

To carry out the functional integral in (2.8), it is essential to separate $u_{ij}$ into its $\mathbf{q} = 0$ and $\mathbf{q} \neq 0$ Fourier components

$$u_{ij}(\mathbf{x}) = u_{ij}^0 + A_{ij}^0 + \sum_{\mathbf{q}\neq 0}\left\{ \frac{1}{2}i[q_i u_j(\vec{q}) + q_i u_i(\vec{q})] + A_{ij}(\vec{q}) \right\}e^{i\mathbf{q}\cdot\mathbf{x}}. \tag{2.9}$$

Here, $A_{ij}(q)$ is the $q$th Fourier component of the "vector potential" $A_{ij}(x) = (\partial_i f)(\partial_j f)$,

$$A_{ij}(\mathbf{q}) = \int d^2x e^{-i\mathbf{q}\cdot\mathbf{x}}(\partial_i f)(\partial_j f), \tag{2.10}$$

while $A_{ij}^0$ is the corresponding component for $\vec{q} = 0$. Although there are only two independent phonon degrees of freedom $\vec{u}_i(\mathbf{q})$ in the in-plane strain matrix at nonzero wavevectors, the uniform part of the in-plane strain matrix $u_{ij}^0$ has in fact *three* independent components, reflecting the three independent ways of macroscopically distorting a flat two-dimensional crystal.[15] The $\mathbf{q} \neq 0$ part of

$A_{ij}$ can be decomposed, as can any two-dimensional symmetric matrix, into transverse and longitudinal parts,[16]

$$A_{ij}(\mathbf{x}) = \frac{1}{2}[\partial_i\phi_j(\mathbf{x}) + \partial_j\phi_i(\mathbf{x})] + P_{ij}^T\Phi(\mathbf{x}), \tag{2.11}$$

where $P_{ij}^T$ is the transverse projection operator, $P_{ij}^T = \delta_{ij} - \partial_i\partial_j/\nabla^2$. By applying the transverse projector to both sides of (2.11), we find that

$$\Phi(\mathbf{x}) = \frac{1}{2}P_{ij}^T(\partial_i f)(\partial_i f). \tag{2.12}$$

The functional integral can now be efficiently performed by integrating over the shifted variables

$$\tilde{u}_{ij}^0 = u_{ij}^0 + A_{ij}^0 \tag{2.13a}$$

$$\tilde{u}_i = u_i + \phi_i, \tag{2.13b}$$

which leads to an effective free energy

$$F_{\text{eff}} = \frac{1}{2}\kappa \int d^2x (\nabla^2 f)^2 + \frac{1}{2}K_0 \int{}' d^2x \left[\frac{1}{2}P_{ij}^T(\partial_i f)(\partial_i f)\right]^2 \tag{2.14}$$

where the prime on the integral means that the $q = 0$ part of $P_{ij}$ has been integrated out, and

$$K_0 = \frac{4\mu(\mu + \lambda)}{2\mu + \lambda}. \tag{2.15}$$

To obtain a physical interpretation of the peculiar form assumed by the stretching energy in (2.14), we note first that the Laplacian of the square root of the integrand is just the Gaussian curvature,

$$-\nabla^2\left[\frac{1}{2}P_{ij}(\partial_i f)(\partial_i f)\right] = \det\left(\frac{\partial^2 f}{\partial x_i \partial x_j}\right) = S(\mathbf{x}). \tag{2.16}$$

Thus, the parts of a membrane with a nonzero Gaussian curvature act as source terms in the Laplace equation (2.16), leading inevitably to a large positive contribution to the stretching energy. The elastic coupling $K_0$ penalizes all membrane distortions which are not "isometric," *i.e.*, those with a nonzero Gaussian curvature. There are still many low energy configurations available to the membrane, however, as one can verify by crumpling a piece of paper, which has essentially infinite in-plane elastic constants.

The effect of the nonlinear stretching energy on the renormalized wave-vector-dependent rigidity

$$\kappa_R^{-1}(\mathbf{q}) \equiv q^4\langle|f(\mathbf{q})|^2\rangle \tag{2.17}$$

can be calculated perturbatively in $K_0$.[1] Figure 4 summarizes the relevant Feynman graphs. The slashes on the interaction vertex denote derivatives of $f(x_1, x_2)$. Note that the "tadpole" graphs vanish identically because we have integrated out the

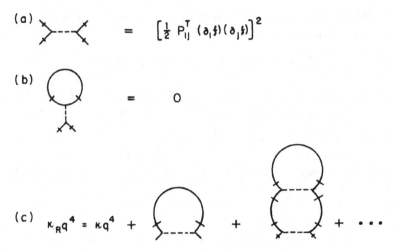

Fig. 4. Graphical rules for calculating the renormalized rigidity: (a) interaction vertex, (b) a graph which vanishes because the $\vec{q} = 0$ part of the interaction has eliminated by integrating out the in-plane strain field, and (c) most divergent terms in the perturbation series for $\kappa_R$.

$\mathbf{q} = 0$ part of the interaction. The first two terms in the perturbation series for $\kappa_R$ are

$$\kappa_R(\mathbf{q}) = \kappa + k_B T K_0 \int \frac{d^2k}{(2\pi)^2} \frac{\left[\hat{q}_i P_{ij}^T(\mathbf{k})\hat{q}_j\right]^2}{\kappa|\mathbf{q} + \mathbf{k}|^4}. \tag{2.18}$$

In contrast to the weak logarithmic singularity for liquid membranes displayed in Eq. (1.9), the integral in Eq. (2.18) exhibits a strong $1/q^2$ divergence for small $q$. The correction to the bare rigidity is positive, showing that the stretching energy *stiffens* the resistance of the membrane to undulations. Summing up the series of badly diverging diagrams displayed in Fig. 4 leads to a self-consistent equation for $\kappa_R(q)$,

$$\kappa_R(\mathbf{q}) = \kappa + k_B T K_0 \int \frac{d^2k}{(2\pi)^2} \frac{\left[\hat{q}_i P_{ij}^T(\mathbf{k})\hat{q}_j\right]^2}{\kappa_R(\mathbf{q} + \mathbf{k})|\mathbf{q} + \mathbf{k}|^4} \tag{2.19}$$

which has the solution, valid for small $q$,

$$\kappa_R(\mathbf{q}) \sim \sqrt{k_B T K_0}\, q^{-1}. \tag{2.20}$$

We can now repeat the analysis of fluctuations of the surface normals carried out for liquid membranes in Sec. 1. Upon inserting the renormalized rigidity (2.20) into Eq. (1.8), we find that the fluctuations in $\theta^2$ are now finite,

$$\langle\theta^2(\mathbf{x})\rangle = k_B T \int \frac{d^2q}{(2\pi)^2} \frac{1}{\kappa_R(q)q^2} \simeq \sqrt{\frac{k_B T}{K_0}} \int \frac{d^2q}{(2\pi)^2} \frac{1}{q} < \infty. \tag{2.21}$$

The mode-coupling analysis which led to this important result *assumed* no significant renormalization of the elastic coupling $K_0$.[1] A recent renormalization group calculation of Aronovitz and Lubensky[17] suggests that the elastic constants will in fact exhibit weak singularities as $q \to 0$, but that $\kappa_R$ still diverges strongly enough to make (1.8) finite. Equation (2.21) shows that tethering stabilizes long-range order in the normals at low temperatures, and suggests the existence of the finite temperature crumpling transition[1,2] discussed in the chapter by Kantor. The possibility of a crumpling transition due to *long-range forces* was suggested in the paper on liquid membranes by Peliti and Leibler.[12] Although it is possible to rewrite the stretching energy in Eq. (2.14) terms of a long-range interaction between *Gaussian* curvatures,[1] this interaction is rather different from the long-range forces considered in Ref. 12. There are, moreover, no true microscopic long-range forces in a tethered surface: they arise here only to compensate for integrating out the underlying physical phonon field.

## 3. Landau Theory of the Crumpling Transition

In this section we discuss an order parameter theory of the crumpling transition.[18] The model reduces to Eq. (2.7) in a low temperature flat phase and leads to the Gaussian crumpled surfaces discussed by Kantor at high temperatures. How self-avoidance modifies this second-order crumpling transition can be determined within the Flory approximation. This approximation should be sufficient for laboratory experiments on flexible membranes, except for very large surfaces close to the critical point.

We focus on the quantity $\vec{r}(\mathbf{x})$, where $\mathbf{x}$ is an internal $D$-dimensional vector which labels the particles, and $\vec{r}$ is an external $d$-dimensional position vector representing the embedding of the particles. In the simplest case, the internal connectivity of the particles corresponds to a $D$-dimensional lattice with nearest neighbor interactions. A statistical description is developed by coarse graining this lattice so that the vector $\mathbf{x}$ becomes a continuous variable which labels a "block" of lattice points.

Symmetries of the microscopic Hamiltonian delimit the form of the free energy functional for the coarse grained variable $\vec{r}(\mathbf{x})$. For a uniform network, overall translational invariance requires that this functional depends on gradients such as the coarse grained tangent vectors $\vec{t}_\alpha = \partial \vec{r}(\mathbf{x})/\partial x_\alpha, \alpha = 1, \ldots, D$. In the crumpled phase, these tangent vectors scale as $\vec{t}_\alpha \sim \ell^{\nu-1}$, where $\ell$ is the coarse graining size. A discussed in the chapter of Kantor, $\nu$ is generically less than one, so the tangents diminish under scaling. In the rigid phase, close to the transition, the $\vec{t}_\alpha$ are also small. Hence, an expansion in powers of the $\vec{t}_\alpha$ and their derivatives is justified. In this simplest case of an isotropic network, overall rotational invariance leads to a "Landau–Ginzburg" expansion[19]

$$\beta F\{r_i(x_\alpha)\} = \int d^D x \left[ \frac{\kappa}{2}(\partial_\alpha \partial_\alpha r_i)^2 + \frac{t}{2}(\partial_\alpha r_i)^2 + u(\partial_\alpha r_i \partial_\beta r_i)^2 + \tilde{v}(\partial_\alpha r_i \partial_\alpha r_i)^2 \right]$$
$$+ \frac{b}{2} \int d^D x d^D x' \delta^d(\vec{r}(\mathbf{x}) - \vec{r}(\mathbf{x}')) \quad (i = 1, 2, \ldots, d), \qquad (3.1)$$

where $\beta = 1/k_B T$. The last term is a nonlocal excluded volume term which represents the effects of self-avoidance at large length scales. Self-avoiding interactions account for interparticle hard core repulsions which mitigate collapse and also prevent the manifold from folding through itself to access unphysical configurations. The local terms represent elastic free energies: the coefficients $t$, $u$, and $\tilde{v}$ can be interpreted as harmonic and anharmonic stretching energies, while the coefficient $\kappa$ is a measure of bending rigidity. Upon identifying the tangents $\vec{t}_\alpha = \partial_\alpha \vec{r}$ with a set of order parameters $\vec{\phi}_\alpha$, an analogy with the usual $\phi^4$ theories of critical phenomena[20] becomes apparent. Given the free energy (3.1), the probability for a configuration $\{\vec{r}(\mathbf{x})\}$ is proportional to the Boltzmann weight $e^{-\beta F}$, and the partition function is a functional integral over all surfaces $\{\vec{r}(\mathbf{x})\}$.

At high temperatures we expect a crumpled phase where $t$ is positive for entropic reasons. Indeed, if we neglect $\kappa$, $u$, and $\tilde{v}$, we then obtain the model of self-avoiding crumpled surfaces discussed in the chapter by Kantor. At low temperatures, however, the microscopic surface tangents tend to align in order to minimize the microscopic bending energy, driving the manifold to form a flat phase described by a negative value of $t$. For $t < 0$, the manifold is stabilized by the anharmonic terms, provided that $u > 0$ and $v = \tilde{v} + u/D > 0$. We anticipate a continuous transition between these phases when $t \equiv \alpha(T - T_c^0) \approx 0$.

As an illustration, we first discuss the mean-field solution in the absence of self-avoidance using the ansatz $\vec{r}(x_\alpha) = \zeta x_\alpha \vec{e}_\alpha$, where the $\{\vec{e}_\alpha\}$ are a set of orthogonal unit vectors specifying the orientation of the manifold in $\mathbf{R}_d$, and the $x_\alpha$ range from 0 to $L$. The prefactor $\zeta$, which has been studied numerically in computer simulations for $D = 2$, $d = 3$, is an order parameter which measures the shrinkage of the manifold in the flat phase due to undulations, and vanishes for $t > 0$. Minimizing Eq. (3.1) for $t \lesssim 0$ leads to $\zeta = \frac{1}{2}\sqrt{-t/Dv}$, which shows that the radius of gyration scales as $R_G = \zeta L \sim |t|^{1/2} L$ when the crumpling transition is approached from below. Fluctuations within the ordered phase can be studied by introducing in-plane phonon modes $u_\alpha$ ($\alpha = 1, \ldots, D$) and out-of-plane undulations $h_\beta$ ($\beta = D+1, \ldots, d$), and setting

$$\vec{r}(x_\alpha) = \zeta[(x_\alpha + u_\alpha)\vec{e}_\alpha + h_\beta \vec{e}\beta], \tag{3.2}$$

where the orthonormal vectors $\{\vec{e}_\beta\}$ are orthogonal to the $\{\vec{e}_\alpha\}$. To leading order in gradients of the $u_\alpha$ and $h_\beta$, the free energy (3.1) reduces to a generalization of Eq. (2.7),

$$\beta F = \int d^D x \left[ \frac{1}{2}\kappa(\nabla^2 h_\beta)^2 + \mu u_{ij}^2 + \frac{1}{2}\lambda u_{kk}^2 \right] \tag{3.3}$$

where the strain matrix is $u_{ij} = \frac{1}{2}[\partial_i u_j + \partial_i u_j + \partial_i h_\beta \partial_j h_\beta]$ and the elastic constants are $\mu = 4u\zeta^4$ and $\lambda = 8\tilde{v}\zeta^4$. The $d = 3$, $D = 2$ version of this low temperature model was studied in Sec. 2, where it was argued that the renormalized bending rigidity grows at large distances and stabilizes the stretched phase against undulations. Note that the elastic constants in this treatment are predicted to vanish like $(T_c - T)^2$ near

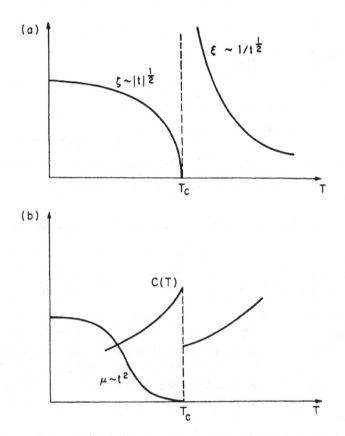

Fig. 5. Behavior of (a) the stretching factor $\zeta$ and correlation length $\xi$ as well as (b) the shear modulus $\mu$ and specific heat $C(T)$ of a tethered surface near the crumpling transition.

the crumpling transition. Mean field theory leads in the usual way to a discontinuity in the specific heat at $t = 0$ and a tangent-tangent correlation length $\xi \sim |t|^{-1/2}$ close to $T_c$. Above $T_c$, $R_G \sim L^{1-D/2}$, while precisely at $T_c$, $R_G \sim L^{\nu_c}$ with $\nu_c = 1 - D/4$. Although the Monte Carlo data on $D = 2$ tethered surfaces in $d = 3$ dimensions without self-avoidance are not yet good enough to obtain precise exponents, all measured quantities in Ref. 2 behave qualitatively as predicted above. Differences, such as the apparently *diverging* specific heat and somewhat larger exponent $\nu_c$ are probably due to critical fluctuations, which will be discussed later. Some of the mean field predictions of the model are summarized in Fig. 5.

The possibility of laboratory experiments on crumpled surfaces requires that we also understand self-avoiding interactions. The Landau expansion (3.1) allows us to treat such effects (within mean field theory) using the Flory approximation.[21] For a network of size $R_G$, the Flory estimates for the individual terms of (3.1) can be obtained as discussed in the chapter by Kantor. The results are

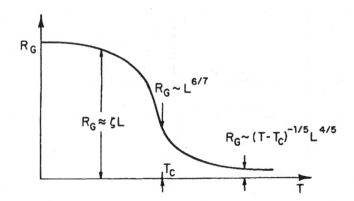

Fig. 6.   Radius of gyration $R_G$ above, below and precisely at the crumpling temperature $T_c$.

summarized by

$$\frac{2\beta F}{D} \approx \kappa R_G^2 L^{D-4} + t R_G^2 L^{D-2} + D v R_G^4 L^{D-4} + b R_G^{-d} L^{2D}/D. \qquad (3.4)$$

Using these estimates, it is straightforward to show that self-avoidance is irrelevant in the flat phase, $i.e.$, for $t < 0$, where $R_G \sim |t|^{1/2} L$. For $t > 0$, the anharmonic terms $u$ and $v$ are asymptotically irrelevant, and balancing the entropic and self-avoiding energies leads to $R_G \sim t^{-\omega} L^{\nu_F}$ where $\omega = 1/(d+2)$, and $\nu_F = (D+2)/(d+2)$ is the Flory exponent for self-avoiding manifolds. At $t = 0$, $R \sim L^{\nu_c}$ where $\nu_c = (D+4)/(d+4)$. The behavior of $R_G$ near the crumpling transition for $D = 2$, $d = 3$ is sketched in Fig. 6.

   Thus far, we have neglected critical fluctuations. As the crumpling temperature is approached from the flat phase, self-avoiding effects are always irrelevant when $L \to \infty$. A simple application of the Ginzburg criterion shows that mean field theory for Eq. (3.1) breaks down for $t^2(\kappa/-t)^{D/2}/8Dv \lesssim 1$. Provided the transition remains continuous, the analogy with $\phi^4$ theory for $T < T_c$ suggests that $\zeta = \frac{R_G}{L} \sim t^\beta$ with a nontrivial $\beta$, which could, $e.g.$, be obtained from careful simulations on surfaces without self-avoidance, since self-avoidance is irrelevant for $T < T_c$. We have just seen that, for $d = 3$, $D = 2$, $R_G \sim t^{-\omega} L^{\nu_F}$ for $T > T_c$, with $\omega \approx 1/5$ and $\nu \approx 4/5$. Assuming that $R_G$ assumes the scaling form $R_G \approx L^{\nu_c} \Psi(tL^y)$ with a common value of $y$ for $t > 0$ and $t < 0$, we find $\nu_c = (\omega + \beta\nu)/(\omega + \beta)$. Upon using the Flory estimates $\nu = 4/5$ and $\omega = 1/5$, and assuming that $\beta$ decreases with fluctuations to a typical $\phi^4$ value of $\beta \approx 1/3$, we find $\nu_c = 7/8$, which is greater than the mean field Flory estimate $\nu_c = 6/7$ displayed in Fig. 6.

   The phenomenological treatment sketched above $assumes$ that the transition remains continuous in the presence of fluctuations, with a well-defined critical exponent $\beta$. When fluctuations are treated to lowest order in an expansion in $\epsilon = 4 - D$ (with $b = 0$ in Eq. (3.1)), one finds in fact a weak fluctuation-driven $first\text{-}order$ phase transition for $d < d_c \approx 219$.[3] Such predictions are not always reliable for $\epsilon = O(1)$ in conventional critical phenomena,[22] so we cannot be certain of what happens here for the interesting case $\epsilon = 2$, $d = 3$. A weak fluctuation-driven first order phase transition certainly cannot be ruled out in the simulations of Ref. 2, however.

In the $d \to \infty$ limit, exact saddle point techniques show an isomorphism to the $n \to \infty$ limit of $O(n)|\vec{\phi}|^4$ models.[3] In this limit, there is a continuous transition with a specific heat singularity $C \sim |t|^{(4-D)/(D-2)}$ and a diverging persistence length $\xi \sim |t|^{-1/(D-2)}$. In contrast to the case $d = 3$, a $D = 2$ surface is always crumpled in the limit $d \to \infty$, while a higher dimensional manifold exhibits a finite temperature rigid phase. Recently, David and Guitter,[23] have solved a model similar to this one, but with infinite bare elastic constants. These authors show that a finite temperature crumpling transition is recovered for $D = 2$ within a $1/d$ expansion for $d < \infty$. In other related work, Aronovitz and Lubensky[17] have studied the singular renormalization of $\kappa$, $\mu$ and $\lambda$ in the low temperature phase described by (3.3) to lowest order in $\epsilon = 4 - D$.

## 4. Defects and Hexatic Order in Membranes

In our discussion of tethered surfaces, it has been convenient to consider polymerized membranes, tied together with permanent covalent bonds. It is interesting to consider instead nominally crystalline membranes bound together by weaker, van der Waals forces. This situation arises, for example, in unpolymerized lipid bilayers at sufficiently low temperatures. If these materials were constrained to be flat, their low temperature crystalline phase would eventually become unstable to a proliferation of unbound dislocations. As first elucidated in a famous argument by Kosterlitz and Thouless,[24] a dislocation with Burger's vector $\vec{b}$ (see Fig. 7) in a flat two-dimensional crystal of radius $R$ has an energy of order $K_0 b^2 \ln(R/a)$, where $K_0$ is given by Eq. (2.15) and $a$ is the lattice spacing. This energy cost suppresses the formation of dislocations at low temperatures. However, there is also an entropy of roughly $2k_B \ln(R/a)$ associated with a dislocation, since it can be located at $(R/a)^2$ possible positions. Above the critical melting temperature $k_B T_M \sim K_0 b^2$, the entropy term dominates, dislocations proliferate, and the crystal melts into a hexatic phase.[25] Figure 7 also shows another type of defect, the disclination, which has an energy of order $K_0 R^2 s^2$. Here $s$ is the disclination charge, defined as the angle in radians of the wedge which must be removed or added to a perfect crystal to make the defect. Disclination energies diverge so rapidly with system size that isolated defects of this kind are extremely unlikely in equilibrated 2d crystals.

In this section we sketch what happens when crystals containing defects are allowed to buckle into the third dimension.[1,4] As shown in Fig. 8a, an initially flat disclination in a triangular lattice (obtained by removing a 60° wedge of material from a perfect crystal) will prefer to buckle into an approximately conical shape in a large enough crystal. A similar, hyperbolic buckling (Fig. 8b) occurs in a sufficiently large crystal containing a negative disclination, obtained by adding a 60° wedge of material. This buckling occurs because the system finds it energetically preferable

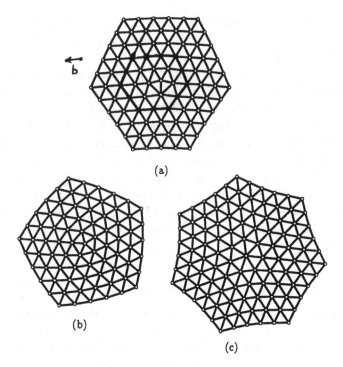

Fig. 7. Defects of interest in polymerized crystalline membranes: (a) dislocation with Burger's vector $\vec{b}$, (b) $+2\pi/6$ disclination and (c) $-2\pi/6$ dislcination.

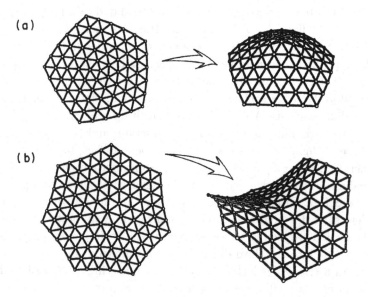

Fig. 8.   Buckling of (a) positive and (b) negative disclinations.

to trade in-plane elastic energy for bending energy. Suppose for simplicity that the in-plane elastic constants are very large, as in an ordinary piece of writing paper. It is easy to check by inserting disclinations into pieces of paper that there are then essentially no elastic distortions in the buckled state. The only remaining energy is now the bending energy, *i.e.*, the first term of Eq. (2.7). It is not hard to show that, for small $s$, the vertical displacement in polar coordinates $(r, \theta)$ is,[1]

$$f(r, \theta) \approx \sqrt{\frac{s}{\pi}} r, \tag{4.1}$$

and

$$f(r, \theta) \approx \sqrt{\frac{2|s|}{3\pi}} r \cos 2\theta, \tag{4.2}$$

for positive and negative disclinations respectively. In both cases $\nabla^2 f \propto 1/r$, and the bending energy in Eq. (2.7) diverges logarithmically, $F \sim \kappa \ln(R/a)$. The disclination energy still diverges with system size, but it has been reduced considerably from the quadratic divergence characteristic of flat membranes. The logarithmic dependence on system size is not restricted to the large in-plane elastic constant, small disclination charge approximation considered here.[4]

The dislocation in Fig. 7 can be regarded as a tightly bound pair of oppositely charged disclinations, in the sense that its core consists of two displaced points of local five- and seven-fold symmetry. When the dislocation is allowed to buckle, as in Fig. 9, we might expect the two (logarithmically divergent) strain fields to cancel at large distances leading to a *finite* dislocation energy.[1] The underlying elasticity equations for buckled membranes are nonlinear,[14] however, so we cannot really apply the superposition principle in this way. Numerical calculations have recently

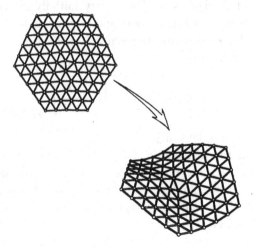

Fig. 9.  Buckling of a dislocation. Note the $\pm 2\pi/6$ disclination pair in the dislocation core.

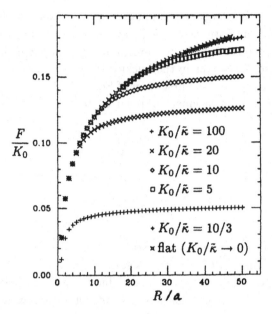

Fig. 10.  Dislocation energy as a function of membrane radius R. The energy ceases to increase logarithmically beyond a critical buckling radius.

been carried out[4] using the model Hamiltonian (2.1), with

$$V(\vec{r}) = \frac{1}{2}[|\vec{r}|^2 - a^2], \qquad (4.3)$$

where $a$ is the lattice constant. The energy of an isolated dislocation at the origin in a system of size of $R$ is shown in Fig. 10 for a variety of ratios of the elastic parameter $K_0$ to the bending rigidity $\tilde{\kappa}$. The energy initially increases logarithmically with distance, but eventually breaks away and increases more slowly after a critical buckling radius $R_c(K_0/\tilde{\kappa})$. Arguments given in Ref. 4 suggest that the energy $E(R)$ approaches a constant in large $R$,

$$E(R) \underset{R \to \infty}{\approx} E_D \left[ 1 - \frac{cR_c}{R} \right] \qquad (4.4)$$

where $c$ is a constant of order unity.

The finiteness of the dislocation energy has important consequences for the statistical mechanics of these membranes.[1] The entropy term in the Kosterlitz–Thouless argument now dominates at all nonzero temperatures. Instead of logarithmically bound dislocation pairs, one now has a finite density of unbound dislocations, with density

$$n_D \approx a^{-2} e^{-E_D/k_B T}. \qquad (4.5)$$

Translational order will be broken up at length scales greater than the translational correlation length

$$\xi_T \approx n_D^{-1/2} \approx a e^{E_D/2k_B T}. \tag{4.6}$$

The resulting phase will be a hexatic, with extended bond orientational order.[25] Because disclinations retain a logarithmically divergent energy even after buckling, hexatic membranes should be stable against disclination unbinding into an isotropic liquid over a range of temperatures. Note that hexatics replace crystals as the inevitable low temperature phase in equilibrated membranes. Similar conclusions would apply even if the energy increased indefinitely for $R > R_c$ in Fig. 10, provided only that the rate of increase is less than logarithmic.

The properties of undulating hexatic membranes are intermediate between the liquid and tethered membranes discussed so far in this chapter. The free energy Eq. (2.7) must now be replaced by[1]

$$F_H = \frac{1}{2}\kappa \int d^2x (\nabla^2 f)^2 + \frac{1}{2} K_A \int d^2x \left[ \partial_i \theta + \frac{1}{2}\epsilon_{jk}\partial_k((\partial_i f)(\partial_j f)) \right]^2 \tag{4.7}$$

where $\theta(x_1, x_2)$ is the local bond angle field defined within the membrane, and $K_A$ is the hexatic stiffness constant.[25] Gradients of $\theta$ are accompanied by a "vector potential"

$$A_i = \frac{-1}{2}\epsilon_{jk}\partial_k((\partial_i f)(\partial_j f)) \tag{4.8}$$

in Eq. (4.7). Just as in Eq. (2.6), this vector potential reflects frustration: the bond angle field cannot return to itself when parallel transported around a region of nonzero Gaussian curvature (The Gaussian curvature is given by the curl of (4.8)). The statistical mechanics associated with Eq. (4.8) was worked out by David et al.[26] Remarkably, there is a low temperature "crinkled" phase with a radius of gyration $R_G$ controlled by a *continuously variable* Flory exponent

$$R_G \sim L^{\nu(K_A)} \tag{4.9}$$

where $L$ is a characteristic membrane linear dimension and

$$\nu(K_A) = 1 - \frac{k_B T}{2\pi K_A} + \mathcal{O}(k_B T)^2. \tag{4.10}$$

Both disclination unbinding transitions and finite crumpling transitions are possible in hexatic membranes. For more information, see the lectures of F. David.

## Acknowledgements

The work described here was carried out in collaboration with L. Peliti, Y. Kantor, M. Kardar, S. Seung, and M. Paczuski. It is a pleasure to acknowledge these fruitful collaborations, as well as stimulating interactions with F. David and S. Leibler. This work was supported by the National Science Foundation, through grant DMR85-14638 and through the Harvard Materials Research Laboratory.

## References

1. D.R. Nelson and L. Peliti, *J. Physique* **48**, 1085 (1987).
2. Y. Kantor and D.R. Nelson, *Phys. Rev. Lett.* **58**, 2774 (1987); *Phys. Rev.* **A36**, 4020 (1987).
3. M. Paczuski, M. Kardar, and D.R. Nelson, *Phys. Rev. Lett.* **60**, 2638 (1988).
4. S. Seung and D.R. Nelson, *Phys. Rev.* **A38**, 1055 (1988).
5. D.R. Nelson and B.I. Halperin, *Phys. Rev.* **B19**, 2457 (1979).
6. P.G. de Gennes, *The Physics of Liquid Crystals* (Clarendon, Oxford, 1974).
7. A.M. Polyakov, *Nucl. Phys.* **B268**, 406 (1986).
8. A.M. Polyakov, *Phys. Rev. Lett.* **59B**, 79 (1975).
9. W. Helfrich, *Z. Naturforsch.* **28C**, 693 (1973).
10. P.G. de Gennes and C. Taupin, *J. Phys. Chem.* **86**, 2294 (1982).
11. D.R. Nelson, in *Phase Transitions and Critical Phenomena*, Vol. 7, edited by C. Domb and J. Lebowitz (Academic, New York).
12. L. Peliti and S. Leibler, *Phys. Rev. Lett.* **54**, 690 (1985); see also W. Helfrich, *J. Physique* **46**, 1263 (1985).
13. Y. Kantor, M. Karkar, and D.R. Nelson, *Phys. Rev. Lett.* **57**, 791 (1986); *Phys. Rev.* **A35**, 3056 (1987).
14. L.D. Landau and E.M. Lifshitz, *Theory of Elasticity* (Pergammon, New York, 1970).
15. For an analogous treatment of compressible spin models in d-dimensions, see J. Sak, *Phys. Rev.* **B10**, 3957 (1974).
16. See, *e.g.*, S. Sachdev and D.R. Nelson, *J. Phys.* **C17**, 5473 (1984).
17. J.A. Aronovitz and T.C. Lubensky, *Phys. Rev. Lett.* **60**, 2634 (1988).
18. This section closely follows the presentation by Paczuski *et al.* in Ref. 3.
19. M. Kardar and D.R. Nelson, *Phys. Rev.* **A38**, 966 (1988).
20. K.G. Wilson and J. Kogut, *Phys. Reports* **12C**, 75 (1974).
21. P.G. de Gennes, *Scaling Concepts in Polymer Physics* (Cornell University Press, Ithaca, NY, 1979).
22. C. Dasgupta and B.I. Halperin, *Phys. Rev. Lett.* **47**, 1556 (1981).
23. F. David and E. Guitter, *Europhys. Lett.* **5**, 709 (1988).
24. J.M. Kosterlitz and D.J. Thouless, *J. Phys.* **C5**, 124 (1972); *J. Phys.* **C6**, 1181 (1973).
25. D.R. Nelson and B.I. Halperin, *Phys. Rev.* **B19**, 2457 (1979).
26. F. David, E. Guitter, and L. Peliti, *J. Phys. (Paris)* **48**, 2059 (1987).

CHAPTER 7

GEOMETRY AND FIELD THEORY
OF RANDOM SURFACES AND MEMBRANES

François David

Service de Physique Théorique de Saclay,
Institut de Recherche Fondamentale du Commissariat à l'Energie Atomique,
91191 Gif-sur-Yvette Cedex, France

## 1. Introduction

Random surfaces (two-dimensional objects fluctuating in d-dimensional Euclidean
space) and their Minkovskian counterpart, relativistic strings ($1 + 1$ dimensional
objects in $1 + (d - 1)$ Minkovskian space) have been considered in various areas of
physics, spanning from biophysics and chemical physics to high energy physics. Let
us mention a few examples. The fluctuations of the interface between two phases
(for instance between a liquid and a gas) are governed by the surface tension. Transitions may occur if the interface interacts with some underlying crystalline structure (roughening transition) or with an external wall or other interfaces (wetting
transition). Near the critical point where the two phases become undiscernable,
the surface tension is very small and large fluctuations of the interface take place.
Membrane-like surfaces (such as phospholipid mono- or bilayers), may have a very
large total area, in this case large fluctuations may also take place, governed by the
bending rigidity energy, and the effective surface tension is very small.

Random surfaces appear also in the strong coupling expansion for lattice gauge
theories, which involves plaquette ensembles. Many efforts have also been devoted
to formulate a theory of strong interactions as a theory of strings. Although this
approach has not really been successful, it led to the tremendous development of
string theories as quantum theories of gravity.

In all these developments the technics of classical and quantum field theory
are essential. The purpose of these lectures will be to give an introduction to those
technics. A special emphasis will be put on surfaces with curvature energy, which are
expected to be "smooth" at small scales and therefore susceptible of a description
in terms of differential geometry. Discretized models of random surfaces are treated
in J. Fröhlich's lectures.

Section 2 is devoted to an introduction to differential and Reimannian geometry
for surfaces. I have tried to make this introduction accessible to readers with no

extensive background in the language and technics of modern differential geometry. I have used the "old-fashioned" presentation with tensors, indices, (rather than the modern one involving forms, vector bundles ...) which proves to be useful when precise field theoretical calculations have to be done.

Section 3 deals with the free field on a curved background as the simplest example of how the curvature properties of the surface may modify the properties of the fluctuations of a field living on that surface.

Section 4 deals with models for fluid surfaces with bending energy. Particular emphasis is put on the choice of the measure for the fluctuations and on the calculation of the renormalization of the rigidity moduli by thermal fluctuations.

Section 5 is devoted to a non-perturbative approach to fluid membranes, which consists in taking the dimension of bulk space to be large, and to a comparison with the perturbative results of Sec. 4.

Section 6 discusses briefly the possible connection between the behavior of fluid membranes at scales larger than the persistence length $\xi_p$ and the properties of other models of random surfaces, including string models and the models discussed in J. Fröhlich's lectures.

The last two sections deal with surfaces with internal structure and bending energy. Section 7 discusses hexatic membranes with orientational order and Sec. 8 discusses elastic membranes with crystalline order. These systems are also discussed in D. Nelson's and Y. Kantor's lectures.

Owing to my slowness in writing these lecture notes, some interesting developments occurred since this school took place, particularly on elastic membranes and the relationship between some strings models and discrete models of surfaces. I have refered to those results when this was useful.

It is a pleasure for me to thank D. Nelson and T. Piran for organizing a very pleasant and stimulating school and for giving me the opportunity to give those lectures. I am very grateful to my collaborators A. Billoire, J.M. Drouffe, E. Guitter, J. Jurkiewicz, A. Krzywicki, L. Peliti and B. Petersson, and to J. Ambjørn, B. Durhuus, J. Fröhlich, I. Kostov, F. Koukiou, T. Jonsson, S. Leibler and D. Petritis, for their interest and discussions on the subject. Finally I thank the Niels Bohr Institute, where those notes were started, and the Aspen Center of Physics, where they were finished, for their hospitality.

## 2. Differential Geometry for Surfaces

In this section we shall try to give a elementary, but hopefully self-consistent introduction to differential and Riemannian geometry, with special emphasis on the specific applications for two-dimensional surfaces. Those mathematics are classic and may be found in many textbooks. A list of useful references is Refs. 1 and 4 (this is by no means an exclusive or exhaustive list).

## 2.1. *Surfaces, Tangent Vectors, Tensors*

A surface S is defined as a two-dimensional (smooth) object embedded in $d$-dimensional Euclidean flat space $\mathbf{R}^d$. If not specified we consider only *compact* surfaces, namely bounded surfaces with no boundary. To describe the surface one can cover it by "patches" $U_\alpha$ such that every point of some patch $U_\alpha$ may be labelled by two coordinates

$$\underline{\sigma} = (\sigma^i; i = 1, 2) \tag{2.1}$$

in some subset $V$ of $\mathbf{R}^2$.

The position in $\mathbf{R}^d$ of a point P in patch $U_\alpha$ with coordinate $\underline{\sigma}$ is represented by a $d$ component vector

$$\vec{X}(\underline{\sigma}) = \{X^\mu(\underline{\sigma}); \ \mu = 1, \ldots, d\}. \tag{2.2}$$

The surface will be assumed to be "smooth" enough, in most cases $\vec{X}^\mu(\sigma)$ will be twice differentiable.

*Tangent vectors*

A basis for the plane tangent to S at point P is given by the two tangent vectors

$$\vec{t}_i = \partial_i \vec{X}(\underline{\sigma}), \qquad i = 1, 2, \tag{2.3}$$

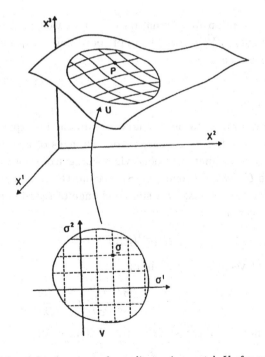

Fig. 2.1.  A local system of coordinates in a patch $U$ of a surface.

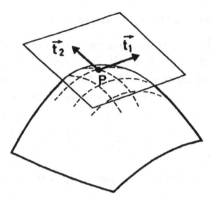

Fig. 2.2.   The tangent plane at point $P$.

where $\partial_i$ denotes the partial derivative with respect to coordinate $\sigma^i$

$$\partial_i = \frac{\partial}{\partial \sigma^i}. \tag{2.4}$$

Any vector $\vec{V}$ tangent to $S$ at $P$ may be written as

$$\vec{V} = V^i \vec{t}_i = \sum_{i=1}^{2} V^i \vec{t}_i \tag{2.5}$$

with Einstein's convention of summation over repeated indices. The two $V^i$ are called the components of $\vec{V}$ in the system of coordinate $\{\sigma^i\}$ and obviously depend on the coordinate system.

*Change of coordinates*

It is very important for two reasons to know how quantities expressed in a coordinate system $\underline{\sigma}$ (in some patch $U$ of $S$) vary as one changes of coordinate system. First we need to change coordinate systems when going from some patch $U_\alpha$ to some neighboring patch $U_\beta$ (when moving on S). Second the same patch may be labelled by various coordinate systems, by a simple change of coordinates. Let us consider a new coordinate system

$$\underline{\sigma}' = \{\sigma'^j(\underline{\sigma}); j = 1, 2\}$$

A basis for tangent vectors in $\underline{\sigma}'$ is simply

$$\vec{t}_i{}' = \frac{\partial \vec{X}}{\partial \sigma'^i}(\underline{\sigma}') = \frac{\partial \sigma^j}{\partial \sigma'^i} \frac{\partial \vec{X}}{\partial \sigma^j}(\underline{\sigma}) = \frac{\partial \sigma^j}{\partial \sigma'^i} \vec{t}_j \tag{2.6}$$

and the components $V'^j$ of the *same* vector $V$ than in (2.5)

$$\vec{V} = V'^j \vec{t}_j{}' \tag{2.7}$$

are simply given in the new coordinate system by

$$V'^j = \frac{\partial \sigma'^j}{\partial \sigma^i} V^i. \tag{2.8}$$

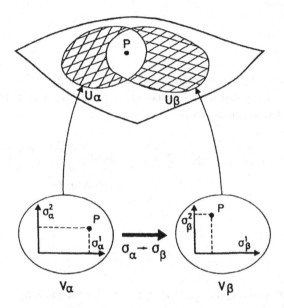

Fig. 2.3. Changes of coordinates allow us to connect two different patches and to get a global description of the surface.

*Tensors*

The $\vec{t}_i$ and the $V^i$ are the components of the simplest examples of objects called *tensors*, or more exactly *tensor fields*, since they depend on the point $P$ on $S$. A type $\binom{P}{Q}$ tensor $T$ is an object defined by its $2^{P+Q}$ components in a coordinate system $\underline{\sigma}$

$$T^{i_1 \cdots i_P}_{j_1 \cdots j_Q}$$

labelled by $P$ upper indices and $Q$ lower indices, which are such that under a change of coordinates $\underline{\sigma} \to \underline{\sigma}'$, the components $T$ transform into $T'$ according to

$$T'^{k_1 \cdots k_P}_{\ell_1 \cdots \ell_Q} = \frac{\partial \sigma'^{k_1}}{\partial \sigma^{i_1}} \cdots \frac{\partial \sigma'^{k_P}}{\partial \sigma^{i_P}} \cdot \frac{\partial \sigma^{\ell_1}}{\partial \sigma'^{j_1}} \cdots \frac{\partial \sigma^{\ell_Q}}{\partial \sigma'^{j_Q}} \cdot T^{i_1 \cdots i_P}_{j_1 \cdots j_Q}. \tag{2.9}$$

This transformation rule may easily be obtained by writing the tensor $T$ in terms of its components as

$$T = T^{i_1 \cdots i_P}_{j_1 \cdots j_Q} \cdot d\sigma^{j_1} \cdots d\sigma^{j_Q} \cdot \frac{\partial}{\partial \sigma^{i_1}} \cdots \frac{\partial}{\partial \sigma^{i_P}}. \tag{2.10}$$

This is in fact the mathematical definition of a tensor field, once a rigorous definition of the "$\frac{\partial}{\partial \sigma^i}$" (differential operator), of the "$d\sigma^i$" (differential form) and of the product "·" (tensorial product) has been given. One sees immediately that in the case of tangent vectors considered above, the $V^i$ are the components of a type $\binom{1}{0}$ tensor, often called a *contravariant vector*, and the $\mu$-components (in bulk space) $t_i^\mu$ of $\vec{t}_i$ are the components of a type $\binom{0}{1}$ tensor $t^\mu$. A type $\binom{0}{1}$ tensor is called a *contravariant* vector.

### The metric tensor

The simplest more elaborate example of tensor is obtained by considering the infinitesimal Euclidean distance between two points with respective coordinates $\underline{\sigma}$ and $\underline{\sigma} + d\underline{\sigma}$. Their distance $dS$ is

$$dS^2 = \left[\vec{X}(\underline{\sigma} + d\underline{\sigma}) - \vec{X}(\underline{\sigma})\right]^2 = d\sigma^i d\sigma^j \frac{\partial \vec{X}}{\partial \sigma^i}(\underline{\sigma}) \frac{d\vec{X}}{\partial \sigma^j}(\underline{\sigma})$$

$$= d\sigma^i d\sigma^j g_{ij}(\underline{\sigma}), \tag{2.11}$$

$$g_{ij}(\underline{\sigma}) = \frac{\partial \vec{X}}{\partial \sigma^i} \frac{\partial \vec{X}}{\partial \sigma^j}, \tag{2.12}$$

are the components of a type $\binom{0}{2}$ tensor, called the *metric tensor*, or the *first fundamental form*. It allows to a define a scalar product for two contravariant vectors $V$ and $W$ by

$$V \cdot W = V^i g_{ij} V^j = |V||W| \cos\theta. \tag{2.13}$$

Obvious examples of type $\binom{1}{1}$ tensors are the Kronecker's symbols $\delta^i{}_j$ and $\delta^j{}_i$, whose components do not depend on the coordinate system. From the metric tensor $g$ we can construct a type $\binom{2}{0}$ tensor $g^{-1}$ by considering the matrix elements of the inverse of the metric tensor

$$g^{ij} = \left(g_{k\ell}^{-1}\right)_{ij}. \tag{2.14}$$

Finally it is obvious that both $g$ and $g^{-1}$ are symmetric, namely

$$g_{ij} = g_{ji}; \quad g^{ij} = g^{ji}. \tag{2.15}$$

### The antisymmetric tensor

Another important type $\binom{0}{2}$ tensor is obtained by considering the area $A$ of the polygon generated by two tangent vectors $V$ and $W$. It is easy to show that it may be written

$$A = |V||W| \sin \theta = W^i V^j \gamma_{ij}, \tag{2.16}$$

with

$$\gamma_{ij} = [\det g]^{1/2} \epsilon_{ij}. \tag{2.17}$$

$\epsilon_{ij}$ is the totally antisymmetric matrix

$$\epsilon_{ij} = \begin{pmatrix} 0 & -1 \\ 1 & 0 \end{pmatrix} \quad \text{or} \quad \epsilon_{11} = \epsilon_{22} = 0, \qquad \epsilon_{21} = -\epsilon_{12} = 1 \tag{2.18}$$

and $\det g$ is the determinant of the metric tensor

$$\det g = \epsilon_{ij} \epsilon_{k\ell} g_{ik} g_{j\ell} = g_{11} g_{22} - g_{12} g_{21}. \tag{2.19}$$

$\gamma_{ij}$ are the components of a $\binom{0}{2}$ antisymmetric tensor which allows us to define the elements of area.

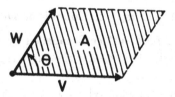

Fig. 2.4.   Area $A(V, W)$.

*Infinitesimal element of area*

For instance the element of area, generated by the two infinitesimal vectors $(d\sigma^1, 0)$ and $(0, d\sigma^2)$ is simply

$$d^2 S = d\sigma^1 d\sigma^2 \sqrt{\det g}. \tag{2.20}$$

*Orientable surfaces*

In fact it is not always possible to define globally the antisymmetric tensor $\gamma_{ij}$, since it requires a choice of orientation on the surface. This is possible only for orientable surfaces. For a non-orientable surface, for instance the Moebius strip or the Klein bottle, if we start from $\gamma_{ij}$ at some point and "transport" it along the surface (in a way which will be defined rigorously later as the parallel transport), we may come back to the origin and find that we have now $-\gamma_{ij}$, thus the orientation is reversed and the sign of elements of area is changed. In practice we shall always consider in the following orientable surfaces.

*Raising and lowering of indices*

$g_{ij}$ and $g^{ij}$ allow to raise the lower indices. For instance from a contravariant vector with components $V^i$ one constructs a covariant vector with components $V_i$ by "contraction"

$$V_i = g_{ij}V^j. \tag{2.21}$$

In particular, from $\gamma_{ij}$ one constructs

$$\gamma^i_j = g^{ik}\gamma_{kj} \tag{2.22}$$

and

$$\gamma^{ij} = g^{ik}\gamma_{k\ell}g^{\ell j} = [\det g]^{-1/2}\epsilon_{ij}. \tag{2.23}$$

$\gamma^i_j$ performs rotation by $\pi/2$ over tangent vector, indeed the tangent vector $V_\perp$

$$V^i_\perp = \gamma^i_j V^j \tag{2.24}$$

has some length and is orthogonal to $V$, $V_\perp \cdot V = 0$.

## 2.2. Geodesics, Parallel Transport, Covariant Derivatives

Let us consider on curve $\mathbf{C}$ on a surface S, labelled by a curvilinear coordinate $s$ going from 0 to 1. In a coordinate system $\underline{\sigma}$ the curve is given by $\sigma^i(s)$. The length of the curve is

$$L(\mathbf{C}) = \int_0^1 ds \left| \frac{d\vec{X}(\sigma(s))}{ds} \right| = \int_0^1 ds \sqrt{\frac{d\sigma^i}{ds} g_{ij}(\underline{\sigma}) \frac{d\sigma^j}{ds}}. \tag{2.25}$$

*Geodesics*

The generalization of the notion of "straight lines" on a surface is geodesics: A geodesic is a curve with extremal length with respect to small variations. If one changes the curve by an infinitestinal amount $\delta\sigma^i(s)$ (with $\delta\sigma^i(0) = \delta\sigma^i(1) = 0$), the variation of the length is $\delta L = 0$. In a parametrization $s$ such that

$$\left| \frac{d\vec{X}}{ds} \right| = \text{cst} \tag{2.26}$$

(constant speed), one can show (by writing the Euler–Lagrange equations for the action $L$) that any geodesic satisfies the following second order differential equation

$$\frac{d^2\sigma^i(s)}{ds^2} + \Gamma^i_{jk}(\underline{\sigma}(s)) \frac{d\sigma^j(s)}{ds} \frac{d\sigma^k(s)}{ds} = 0, \tag{2.27}$$

where the quantity $\Gamma^i_{jk}$, which is not a tensor, is

$$\Gamma^i_{jk}(\underline{\sigma}) = \frac{1}{2}g^{i\ell}(\underline{\sigma})[\partial_k g_{j\ell}(\underline{\sigma}) + \partial_j g_{\ell k}(\underline{\sigma}) - \partial_\ell g_{jk}(\underline{\sigma})] \tag{2.28}$$

and is called the *affine connection*.

## Parallel transport

The physical meaning of $\Gamma$ is quite clear when one generalizes the usual notion of translation of vectors in flat space to surfaces. Let us consider two points $P$ and $P'$, close to each other, with coordinates $\underline{\sigma}$ and $\underline{\sigma} + \delta\underline{\sigma}$ respectively, and $V$ a tangent vector to the surface at $P$. To move the vector $V$ to $P'$ in a geometrical way which does not depend on the coordinate system, let us consider the geodesic $\underline{\sigma}(s)$ joining $P$ to $P'$ (it is unique if the distance between $P$ and $P'$ is small enough) and let us consider the vector $V'$ tangent to $S$ at $P'$ which has the following two properties (which determine it in a unique way)

– $V'$ has same length than $V$
– the angle between $V'$ and the geodesic at $P'$ is the same than the angle between $V$ and the geodesic at $P$.

From the equation of the geodesic it is easy to show that in the coordinate system considered

$$V'^i = V^i - \Gamma^i_{jk}(\underline{\sigma})V^j \delta\sigma^k. \tag{2.29}$$

In other words, $-\Gamma^i_{jk}$ represents the element of matrix $\binom{i}{j}$ of the linear variation undergone by a vector $V$ when parallel transported along direction $k$.

An important property of the affine connection $\Gamma^i_{jk}$ is that it is symmetric in $j$ and $k$. One may define more general affine connections which are not symmetric, in such a situation the connection is said to be "with torsion".

Parallel transport along a curve C is easily defined by decomposing the curve into infinitesimal segments. The parallel transported vector $V^i(s)$ along the curve $\underline{\sigma}(s)$ satisfies the differential equation

$$\frac{dV^i(s)}{ds} + \Gamma^i_{jk}(\underline{\sigma}(s))V^j(s)\frac{d\sigma^k(s)}{ds} = \frac{DV^i}{Ds} = 0. \tag{2.30}$$

## Covariant derivative

One can now easily extend to vectors, and to tensors, the notion of partial derivatives. Let $V^i(\underline{\sigma})$ be a tangent vector field. The derivative of $V(\underline{\sigma})$ in direction $\delta\underline{\sigma}$ is

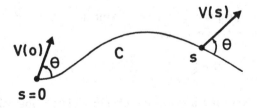

Fig. 2.5.   Parallel transport of a vector $V$ along the curve C.

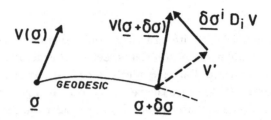

Fig. 2.6.  Geometrical representation of Eq. (2.31).

obtained by comparing the vector $V(\underline{\sigma}+\delta\underline{\sigma})$ with the vector $V'$ obtained by parallel transport of $V(\underline{\sigma})$ at point $\underline{\sigma}+\delta\underline{\sigma}$, namely

$$\delta\sigma^j D_j V^i(\underline{\sigma}) = V^i(\underline{\sigma}+\delta\underline{\sigma}) - \left[ V^i(\underline{\sigma}) - \Gamma^i_{jk}(\underline{\sigma})V^k(\underline{\sigma})\delta\sigma^j \right]. \tag{2.31}$$

Thus the covariant derivative is

$$D_j V^i = \partial_j V^i + \Gamma^i_{jk} V^k \tag{2.32}$$

and one can check that it is a type $\binom{1}{1}$ tensor. More generally, the covariant derivative of a type $\binom{P}{Q}$ tensor is defined as

$$D_k T^{i_1\cdots i_P}_{j_1\cdots j_Q} = \partial_k T^{i_1\cdots i_P}_{j_1\cdots j_Q} + \sum_{\alpha=1}^{P} \Gamma^{i_\alpha}_{k\ell} T^{i_1\cdots i_{\alpha-1}\ell i_{\alpha+1}\cdots i_P}_{j_1\cdots\cdots\cdots\cdots\cdots\cdots j_Q} - \sum_{\beta=1}^{Q} \Gamma^{\ell}_{kj_\beta} T^{i_1\cdots\cdots\cdots\cdots\cdots\cdots i_P}_{j_1\cdots j_{\beta-1}\ell j_{\beta+1}\cdots i_Q} \tag{2.33}$$

and is a type $\binom{P}{Q+1}$ tensor. By definition the covariant derivative of a scalar function is the ordinary derivative. The covariant derivative shares all usual properties of the ordinary derivative, with one very important exception: *Covariant derivatives do not commute,*

$$D_i D_j \neq D_j D_i, \tag{2.34}$$

except when applied to scalar functions

$$D_i D_j f = D_j D_i f. \tag{2.35}$$

Another important property is that covariant derivatives of the metric tensor and of the antisymmetric tensor vanish:

$$D_i g_{jk} = 0, \quad D_i g^{jk} = 0, \quad D_i \gamma_{jk} = 0. \tag{2.36}$$

This implies that lowering and raising of indices is an operation which commutes with covariant derivatives (this is quite useful for practical calculations).

## 2.3. *Integration, Stokes Formula*

If $D$ is a (disk shaped) domain bounded by a clockwise oriented contour C and if $V_i$ is a contravariant vector field (1-form) in $D$, Stokes formula states that

$$\int_D d^2\sigma\epsilon_{ij}\partial_i V_j = \int_C d\sigma^i V_i. \qquad (2.37)$$

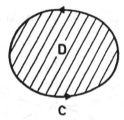

Fig. 2.7.   Domain $D$ and its oriented boundary $C$.

Expressed in terms of tangent (covariant) vector field and of the element of area $d^2S = d^2\sigma\sqrt{\det g}$ this becomes

$$\int_D d^2S\gamma^j{}_i D_j V^i = \int_C d\sigma^i V^f g_{ij} \qquad (2.38)$$

or equivalently

$$\int_D d^2S D_i V^i = \int_C d\sigma^i V^j \gamma_{ij}. \qquad (2.39)$$

## 2.4. *Extrinsic Curvature*

Up to now, no result really depends on the fact that the metric tensor $g_{ij}$ derives from the embedding of the surface in $\mathbf{R}^d$, or that the surface is two-dimensional (except for results involving $\gamma_{ij}$). However, different surfaces, embedded in different ways in $\mathbf{R}^d$, may have the same metric tensor $g_{ij}$, but differ by the way they are curved in bulk space. To measure such a curvature, one generalizes the notion of curvature of a curve by considering the extrinsic curvature tensor

$$\vec{K}_{ij} = D_i \vec{t}_j = D_i D_j \vec{X}. \qquad (2.40)$$

It is a symmetric $\binom{0}{2}$ tensor and each component is a vector in bulk space.

Let us show that each component $\vec{K}_{ij}$ is normal to the surface: indeed

$$\vec{K}_{ij}\vec{t}_k = \left(D_i\vec{t}_j\right)\vec{t}_k = D_i(\vec{t}_j \cdot \vec{t}_k) - \vec{t}_j(D_i\vec{t}_k) = -\vec{K}_{ki}\vec{t}_j \qquad (2.41)$$

$$\text{since } D_i(\vec{t}_j\vec{t}_k) = D_i g_{jk} = 0. \qquad (2.42)$$

Repeating this operation twice, we get $\vec{K}_{ij}\vec{t}_k = 0$. $\vec{K}_{ij}$ has a simple geometrical interpretation. Obviously

$$\partial_i\vec{t}_j = \Gamma^k_{ij}\vec{t}_k + \vec{K}_{ij}. \qquad (2.43)$$

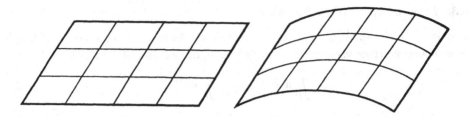

Fig. 2.8.   Two different surfaces with the same intrinsic metric.

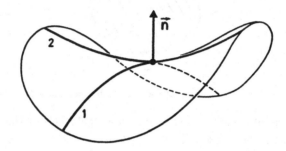

Fig. 2.9.   The two principal directions 1 and 2 at a saddle point.

Thus $\Gamma_{ij}^k$ are the tangential components of $\partial_i \vec{t}_j$ and $\vec{K}_{ij}$ is the component of $\partial_i \vec{t}_j$ normal to the surface, which measures the amount of rotation of tangent vectors when parallel transported along the surface.

In the particular case of a surface in $\mathbf{R}^3$, in $\vec{K}_{ij}$ is proportional to the unit normal vector to the surface $\vec{n}$, it writes

$$\vec{K}_{ij} = K_{ij}\vec{n}, \qquad (2.44)$$

where $K_{ij}$ is a symmetric tensor (the second fundamental form of Gauss), which may also be defined from

$$\partial_i \vec{n} = K_{ij}\vec{t}^{\,j}. \qquad (2.45)$$

At every point $\underline{\sigma}\, K_{ij}(\underline{\sigma})$ may be diagonalized in a tangent frame which defines *the principal directions of curvature* at this point. In this frame it writes

$$K_{ij} = \begin{pmatrix} K_1 & 0 \\ 0 & K_2 \end{pmatrix}, \qquad (2.46)$$

where $K_1, K_2$ are the two principal curvatures and encode the local curvature properties of the surface.

$$r_i = 1/K_i, \qquad i = 1, 2, \tag{2.47}$$

are called the *principal curvature radii*. The signs of $K_1$ and $K_2$ correspond to the following geometrical situations.

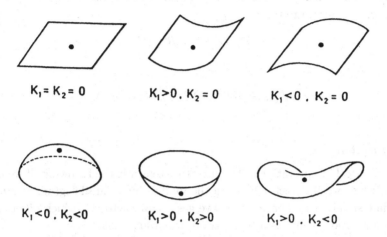

$$K_1 = K_2 = 0 \qquad\qquad K_1 > 0, \ K_2 = 0 \qquad\qquad K_1 < 0, \ K_2 = 0$$

$$K_1 < 0, \ K_2 < 0 \qquad\qquad K_1 > 0, \ K_2 > 0 \qquad\qquad K_1 > 0, \ K_2 < 0$$

Fig. 2.10.   Different cases for the two principal curvatures.

Rather than dealing with $K_1$ and $K_2$ one prefers often to deal with the two following scalar quantities. The Gauss curvature

$$K = \det(K_{ij}) = \frac{1}{r_1 r_2} \tag{2.48}$$

and the mean curvature

$$H = \frac{1}{2}\mathrm{Tr}(K_{ij}) = \frac{1}{2}\left(\frac{1}{r_1} + \frac{1}{r_2}\right) \tag{2.49}$$

which generalize in $\mathbf{R}^d$ to

$$K = \frac{1}{4}\gamma^{ij}\gamma^{k\ell}\vec{K}_{ik}\vec{K}_{j\ell},$$

$$= \frac{1}{2}(\vec{K}_i^i \vec{K}_j^j - \vec{K}_i^j \vec{K}_j^i) \tag{2.50}$$

and to the vector normal to the surface (and which has therefore $d-2$ independent components)

$$\vec{H} = \frac{1}{2}g^{ij}\vec{K}_{ij}. \tag{2.51}$$

## 2.5. *The Riemann Curvature Tensor*

*Riemann curvature*

Let us come back to the difference between ordinary derivatives in flat space and covariant derivatives in a general curved surface: covariant derivatives do not commute. In fact an explicit calculation shows that the commutator

$$[D_i, D_j] = D_i D_j - D_j D_i \tag{2.52}$$

acting on a vector field $V^k$ acts only as a *linear transformation* which defines a new $\binom{1}{3}$ tensor, the curvature tensor $R^k{}_{\ell ij}$

$$[D_i, D_j]V^k = R^k{}_{\ell ij} V^\ell \tag{2.53}$$

explicitly

$$R^k{}_{\ell ij} = \left(\partial_i \Gamma^k_{\ell j} + \Gamma^k_{im}\Gamma^m_{j\ell}\right) - (i \leftrightarrow j). \tag{2.54}$$

*Scalar curvature*

The geometrical intepretation of the curvature tensor is the following. If we perform parallel transport of a vector $V$ along the infinitesimal parallelogram generated by two infinitesimal vectors $\delta\sigma$ and $\delta\sigma'$, we get a new vector $V'$ which differs from $V$ only by a rotation, which is a linear transformation. In fact

$$(V' - V)^k = d\sigma'^i d\sigma^j R^k{}_{\ell ij} V^\ell \tag{2.55}$$

and $\delta^k_\ell - d\sigma'^i d\sigma^j R^k{}_{\ell ij}$ is the $\binom{k}{\ell}$ matrix element corresponding to a rotation by some infinitesimal angle $\delta\theta = \theta_{ij}\delta\sigma'^i d\sigma^j$. Thus $R^k{}_{\ell ij}$ may be written as

$$R^k{}_{\ell ij} = \gamma^k{}_\ell \theta_{ij} \tag{2.56}$$

(we have seen that $\gamma^k_\ell$ performs a $\pi/2$ rotation on vectors). But since $R^k{}_{\ell ij}$ is obviously antisymmetric in $(i, j)$, we may write $\theta_{ij}$ as $\gamma_{ij}\frac{1}{2}R$ and

$$R^k{}_{\ell ij} = \gamma^k{}_\ell \gamma_{ij}\frac{1}{2}R \tag{2.57}$$

where $R$ is a scalar function, *the scalar curvature*, and is the only independent component of the curvature tensor.

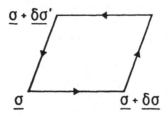

Fig. 2.11. Parallel transport along an infinitesimal circuit defines the Riemann curvature tensor.

The geometrical meaning of the scalar curvature $R$ is the following. If $D$ is a (disk-shaped) domain bounded by a clockwise-oriented closed curve C and if a vector $V$ is parallel transported along C, after one turn, it is rotated by an angle $\theta$

proportional to the integral of $R$ inside $D$:

$$\theta = \frac{1}{2} \int_D d^2 S R. \tag{2.58}$$

**Proof:** This may be shown either by decomposing the domain $D$ into infinitesimal elementary parallelograms, or by considering a smooth vector field $N^a(\sigma)$ in $D$ with unit length

$$|N|^2 = N^i N^j g_{ij} = 1 \tag{2.59}$$

(it is always possible to construct such a vector field in a domain $D$ with the shape of a disk). When parallel transported in direction $d\sigma^i$, the vector $N(\sigma)$ is rotated with respect to $N(\sigma + d\sigma)$ by an elementary angle

$$d\Omega = \gamma_{jk} N^j D_i N^k d\sigma^i = \Omega_i d\sigma^i, \tag{2.60}$$

integrating $d\Omega$ over the boundary $C = \partial D$ of $D$ gives the angle $\theta$ (whatever the vector field $N$ is)

$$\theta = \int_C d\sigma^i \Omega_i = \int_C \gamma_{jk} N^j D_i N^k d\sigma^i, \tag{2.61}$$

using Stokes' formula

$$\begin{aligned}
\theta &= \int_D d^2\sigma \sqrt{|g|} \gamma^{ij} D_j \Omega_i \\
&= \int_D d^2\sigma \sqrt{|g|} \gamma^{ij} \gamma_{k\ell} D_j (N^k D_i N^\ell) \\
&= \int_D d^2\sigma \sqrt{|g|} \gamma^{ij} \gamma_{k\ell} [N^k (D_j D_i N^\ell) + (D_j N^k)(D_i N^\ell)],
\end{aligned} \tag{2.62}$$

the first term is, using antisymmetry of $\gamma^{ij}$,

$$\begin{aligned}
\int_D d^2\sigma \sqrt{g} \frac{1}{2} \gamma^{ij} \gamma_{k\ell} N^k [D_j, D_i] N^\ell \\
= \int_D d^2\sigma \sqrt{g} \frac{1}{2} \gamma^{ij} \gamma_{k\ell} N^k R^\ell{}_{mji} N^m = \int_D d^2\sigma \sqrt{g} \frac{R}{2},
\end{aligned} \tag{2.63}$$

while the second term vanishes also by antisymmetry.

There is a remarkable connection between Gauss curvature $K$ (obtained from the extrinsic curvature tensor) and the scalar curvature (obtained from the Riemann

curvature tensor). It is given by

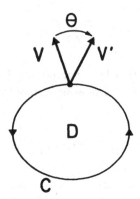

Fig. 2.12.  Parallel transport of $V$ along $C$ induces a rotation by an angle $\theta$.

*Gauss' Theorema egregium*

$$K = \frac{1}{2}R \tag{2.64}$$

**Proof:** Starting from the definition of $K$

$$K = \frac{1}{4}\gamma^{ij}\gamma^{k\ell}(D_i\vec{t}_k)(D_j\vec{t}_\ell) \tag{2.65}$$

using the fact that $\vec{t}_k \cdot \vec{t}_\ell = g_{k\ell}$ and that the covariant derivative of $g$ vanishes, we can rewrite $K$ as

$$K = -\frac{1}{8}\gamma^{ij}\gamma^{k\ell}\vec{t}_k[D_i, D_j]\vec{t}_\ell, \tag{2.66}$$

which gives

$$K = \frac{1}{2}R. \tag{2.67}$$

## 2.6. *The Gauss–Bonnet Theorem*

We have seen that the scalar curvature $R$ is an "intrinsic" quantity of the surface, which depends only on the metric tensor $g_{ij}$ on $S$ and not on the embedding. In fact a much stronger result is valid for the integral of $R$ over a closed surface. It is a topological invariant, which depends on the global shape of $S$ and which does not change under any smooth deformation of $S$.

*Genus and the Euler characteristic*

Let us consider an *orientable closed* surface $S$ with *no boundary*. One can show that its shape (topology) is only characterized by an integer $g \geq 0$, called the *genus of $S$*, which represents the "number" of "handles" of $S$. More precisely, two orientable closed surfaces with no boundary with the same genus are homeotopic, namely may be mapped into one another.

Let us introduce another concept. The *Euler characteristic* $\chi$ which may be defined as follows.

Let us consider a smooth vector field $V$ on a closed orientable surface $S$, which vanishes only at some isolated points $P_1, \ldots, P_N$ (this is the generic case). At each point $P_I$ we may associate an integer $n_I$, called the *index* of the vector field $V$ at point $P_I$, which is the *winding number* of $V$ around $P_I$, which shall be defined more rigorously later. Figure 2.13 represents vector fields with winding numbers $+1$, $0$ and $-1$.

$$n = 0 \qquad n = 1 \qquad n = -1$$

Fig. 2.13.  The index of a vector field at a point is the number of turns that it performs around this point.

The Euler characteristic is defined as

$$\chi = \sum_{I=1}^{N} n_I \tag{2.68}$$

and shall be shown to depend only on $S$ and to be the same for any vector field $V$ on $S$.

*Gauss–Bonnet theorem*

$$\int \sqrt{g} R = 4\pi \chi = 8\pi(1 - g). \tag{2.69}$$

**Proof:** Taking the vector field $V$ introduced before, let us consider infinitesimal anti-clockwise oriented circles $C_I$ with infinitesimal radius $\epsilon$ around the points $P_I$ $(I = 1, N)$ which bound small disks $D_I$ containing $P_I$. Away from the $P_I$'s, we can construct a unit norm vector field $N$ by taking

$$N^i = \frac{V^i}{|V|} = (V^i g_{ij} V^j)^{-1/2} \cdot V^i. \tag{2.70}$$

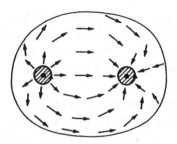

Fig. 2.14.  Construction of the unit vector field $N$ away from the disks $D_I$ where $V$ is small.

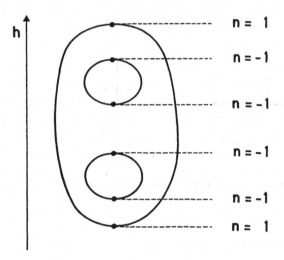

Fig. 2.15.  An example of surface with $g = 2$ and $\chi = 2$.

We may decompose $_I^{S-\underset{I}{\cup}D_I}$ (the exterior of the disks) into disk-shaped domains and apply the same reasoning used in the derivation of (2.68) to show that

$$\int_S d^2\sigma\sqrt{g}R = \int_{S-\underset{I}{\cup}D_I} d^2\sigma\sqrt{g}R + 0(\epsilon) \simeq -2\sum_I \int_{C_I} d\sigma^i\Omega_i$$

$$\simeq -2\sum_I \theta_I, \qquad (2.71)$$

where

$$\Omega_i = \gamma_{jk}N^j D_i N^k. \qquad (2.72)$$

There is a $-$ sign because the circles $C_I$'s, which circle anti-clockwise around $D_I$, circle clockwise around the exterior $_I^{S-\underset{I}{\cup}D_I}$ and because $\theta_I$ is nothing but the angle of rotation of $N$ with respect to itself when parallel transported along $C_I$, after one complete turn. Thus $\theta_I$ must be a multiple of $2\pi$ by some integer which is nothing but minus the index of $V_i$ at $P_I, n_I$, which we introduced previously:

$$\theta_I = -2\pi n_I. \qquad (2.73)$$

Thus

$$\int_S \sqrt{g} R = 4\pi \left( \sum_I n_I \right) = 4\pi \chi \tag{2.74}$$

and this shows that the Euler characteristic $\chi$ does not depend on $V$ but only on the surface $S$.

The second part of the theorem

$$\chi = 2(1 - g) \tag{2.75}$$

is simply obtained by considering a surface with $g$ handles embedded in three-dimensional space $R^3$, by considering the height $h = (X^3)$ of each point on the surface as a smooth function on $S$ and by taking as a vector field $V$ on $S$ the gradient of $h$

$$V^i = g^{ij} \partial_j h. \tag{2.76}$$

The points where $V$ vanishes are extrema of $h$ which may be local maxima or minima (with index $n = +1$), or saddle point, (with index $n = -1$). Counting on a specific example, as the one depicted in Fig. 2.15, shows that

$$\chi = \sum_I n_I = 2 - 2g. \tag{2.77}$$

The Gauss–Bonnet theorem is one of the simplest examples of an index theorem, which relates the integral of some local quantities on a surface to some global topological properties of the surface. There is a version of the Gauss–Bonet theorem valid for surfaces with boundaries, which involves the integral of the extrinsic curvature of the boundaries.

We shall now present some other simple applications of Riemannian geometry which are quite important for the theory of surfaces.

### 2.7. *Minimal Surfaces*

Let us consider a surface $S$ in $d$-dimensional space $\mathbf{R}^d$, bounded by a curve $C$ in $\mathbf{R}^d$. $S$ is said to be a minimal surface (or rather an extremal surface), if its total area $A$ is stationary under infinitesimal changes of shape which leave the boundary $C$ fixed.

Let us represent the surface by $\vec{X}(\underline{\sigma})$, where the coordinates $\underline{\sigma} = (\sigma^i)$ belong to some domain $D$ bounded by a boundary $\partial D$ mapped on $C$. Then it is quite easy to write the differential equation satisfied by $\vec{X}(\sigma)$ if $S$ is minimal. The area is

$$A = \int_d d^2\sigma \sqrt{g}. \tag{2.78}$$

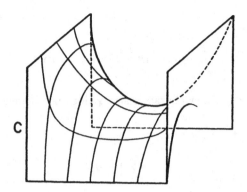

Fig. 2.16.   A minimal surface bounded by a contour C.

Under a small change $\vec{X}(\sigma) \to \vec{X}(\sigma) + \delta\vec{X}(\sigma)$ such that $\delta\vec{X} = 0$ on the boundary $\partial D$ of $D$, the metric tensor changes as

$$\delta(g_{ij}) = \delta(\partial_i\vec{X}\partial_j\vec{X}) = \vec{t}_i\partial_j\delta\vec{X} + \vec{t}_j\partial_i\delta\vec{X} \tag{2.79}$$

and the variation of the area is

$$\delta A \int_d d^2\sigma\delta\sqrt{g} = \frac{1}{2}\int_d d^2\sigma\sqrt{g}g^{ij}\delta g_{ij}. \tag{2.80}$$

Integrating by part by using Stokes formula and the fact that $\delta\vec{X}$ vanishes on the boundary, we get

$$\delta A = -\int_d d^2\sigma\sqrt{g}(d^i\vec{t}_i)\delta\vec{X}. \tag{2.81}$$

If $S$ is minimal, $\delta A = 0$ for any variation $\delta\vec{X}$ and therefore

$$d^i\vec{t}_i = \vec{K}^i_i = 0. \tag{2.82}$$

Therefore, a surface is minimal if and only if the mean curvature $\vec{H}$ vanishes.

### 2.8. *Conformal (or Isothermal) Coordinates*

Another important result, valid only for two-dimensional surfaces, and which goes back to Gauss, is the following. Locally, it is always possible to construct a coordinate system $\underline{\sigma}$ on a surface such that the metric tensor $g_{ij}(\underline{\sigma})$ is proportional to the unit matrix

$$g_{ij}(\underline{\sigma}) = \rho(\underline{\sigma})\delta_{ij}. \tag{2.83}$$

**Proof:** Let us start from an arbitrary coordinate system $\underline{\tau} = (\tau^1, \tau^2)$ and let us take us $\sigma^2(\underline{\tau})$ a non-constant solution of Laplace equation (the heat diffusion equation)

$$\Delta\sigma^2 = d^i d_i\sigma^2 \frac{1}{\sqrt{g(\underline{\tau})}}\frac{\partial}{\partial\tau^i}\sqrt{g(\underline{\tau})}g^{ij}(\underline{\tau})\frac{\partial}{\partial\tau^j}\sigma^2 = 0 \tag{2.84}$$

such that $\partial_i\sigma^2 \neq 0$ ($\sigma^2 \neq$ cst). It is always possible to find such a $\sigma^2$ locally. Then one can take as other coordinate a function $\sigma^1$ such that the lines of constant $\sigma^1$

are normal to the lines of constant $\sigma^2$. This is satisfied if the gradient vector field $\partial_i \sigma^1$ is orthonormal to $\partial_i \sigma^2$:

$$\partial_i \sigma^1 = -\gamma_i^j(\underline{\tau}) \partial_j \sigma^2, \qquad (2.85)$$

which gives two first order differential equations. These two equations are compatible because $\sigma^2$ satisfies (2.84) and therefore there is a solution $\sigma^1$ of (2.85).

Then one can check that $(\sigma^1, \sigma^2)$ provides a conformal coordinate system. The metric tensor $\tilde{g}$ in $(\underline{\sigma})$ satisfies

$$\begin{aligned}
\tilde{g}^{12}(\underline{\sigma}) &= \tilde{g}^{21}(\underline{\sigma}) = 0, \\
\tilde{g}^{11}(\underline{\sigma}) &= \tilde{g}^{22}(\underline{\sigma}) = \rho(\underline{\sigma}),
\end{aligned} \qquad (2.86)$$

with

$$\rho(\underline{\sigma}) = \frac{1}{\sqrt{g(\underline{\tau})}} \left| \frac{\partial(\sigma^1, \sigma^2)}{\partial(\tau^1, \tau^2)} \right|. \qquad (2.87)$$

This result has very important consequences.

(i) Locally, any metric may be written as

$$g_{ij}(\sigma) = \rho(\sigma) g_{ij}^o(\sigma), \qquad (2.88)$$

where $g_{ij}^o(\underline{\sigma})$ is a *flat* metric such that the curvature tensor vanishes ($R^o = 0$). One says that any metric is locally conformally equivalent to a flat metric in two dimensions.

(ii) In a conformal coordinate system, the scalar curvature reads simply

$$R(\underline{\sigma}) = -\frac{1}{\rho(\underline{\sigma})} \left( \frac{\partial^2}{\partial \sigma_1^2} + \frac{\partial^2}{\partial \sigma_2^2} \right) \ln \rho(\underline{\sigma}) = -\frac{1}{\rho} \partial^2(\ln \rho). \qquad (2.89)$$

(iii) There is a natural analytic structure on two-dimensional surfaces. Indeed, let us consider two different conformal coordinate systems $\underline{\sigma} = (\sigma^1, \sigma^2)$ and $\underline{\tau} = (\tau^1, \tau^2)$ on the same patch. It is easy to show that if $\underline{\sigma}$ and $\underline{\tau}$ are conformal.

$$\sigma^1 + i\sigma^2 = f(\tau^1 + i\tau^2) \qquad (2.90)$$

where $f$ in an analytic (or anti-analytic if the change of coordinates reverses orientation) function of one complex variable. Thus conformal changes of coordinates are simply analytic changes of the complex coordinate

$$z = \sigma^1 + i\sigma^2.$$

Two-dimensional Riemannian manifolds may thus be viewed as one-dimensional complex manifolds (or complex curves).

Note also that a conformal transformation $z \to f(z)$ changes the length of tangent vectors but *not the angles*.

(iv) (1) does not mean that *globally* any metric is conformally equivalent to a flat metric. However, (1) has a simple but non-trivial global extension:

Any metric on a closed compact surface with no boundary is conformally equivalent to a metric with *constant* curvature.

Since the integral of the curvature is a topological invariant depending on the genus $g$ of the surface, this curvature $R^{(o)}$ may be chosen to be

$$
\begin{aligned}
R^{(o)} &= 1 && \text{if } g = 0 \text{ sphere,} \\
R^{(o)} &= 0 && \text{if } g = 1 \text{ torus,} \\
R^{(o)} &= -1 && \text{if } g \geq 2 \text{ higher genus surface.}
\end{aligned}
\tag{2.91}
$$

## 3. Fields on Surfaces

For various reasons, we shall have to consider statistical systems living on a curved surface. In the vicinity of a critical point, where correlation lengths *and* typical scales for variations of curvature of the surface are large, this reduces in the continuum limit to consider Euclidean quantum field theories on a curved two-dimensional space.[5] The simplest example of such fields in provided by the massless free field, that we consider now.

### 3.1. Free Field

Consider a free field $\phi(\underline{\sigma})$ living on a 2-$d$. Riemannian manifold $M$ with metric tensor $g_{ij}$. The natural generalization of the Gaussian action which is reparametrization invariant is

$$
S = \frac{1}{2} \int_M d^2\sigma \sqrt{g} g^{ij} \partial_i \phi \partial_j \phi.
\tag{3.1}
$$

If $M$ has no boundaries, integrating by parts gives

$$
S = \frac{1}{2} \int_M d^2\sigma \sqrt{g} \phi(-\Delta)\phi,
\tag{3.2}
$$

where $\Delta$ is the scalar Laplacian (2.84).

*Partition function*

The partition function for such a system is formally

$$
Z = \int \mathbf{D}[\phi] e^{-S[\phi]} \backsimeq [\det(-\Delta)]^{-1/2}.
\tag{3.3}
$$

In order to define this functional integral in a more rigorous way, we stress that we require that the partition function must be reparametrization invariant. Therefore the integration measure $\mathbf{D}[\phi]$ has also to be reparametrization invarariant. One simple way to achieve this is to decompose $\phi$ into a basis of eigenvectors of the scalar Laplacian[6]

$$\phi = \sum_{k=0}^{\infty} a_k \phi_k \quad -\Delta \phi_k = \lambda_k \phi_k,$$

$$\int_M d^2\sigma \sqrt{g} \phi_k \phi_l = \delta_{kl}, \tag{3.4}$$

and to write

$$\mathbf{D}[\phi] = \prod_k da_k, \quad \det(-\Delta) = \prod_k \lambda_k = \exp\left(\sum_k \ln \lambda_k\right). \tag{3.5}$$

Two problems arise in the calculation of this determinant.

(1) There is a zero mode $\lambda_o = 0$ corresponding to the constant function $\phi_o = \left[\int_M d^2\sigma \sqrt{g}\right]^{-1/2}$, which necessitates some care.

(2) Large eigenvalues $\lambda_k$ give a divergent contribution. Indeed, the corresponding eigenfunctions $\varphi_k$ may be written locally within the WK B approximation as plane waves with wavevector $k_i$

$$\phi_k \propto \exp(-ik_i\sigma^i) \qquad \lambda_k = |k^2| = k_i k_j g^{ij} \tag{3.6}$$

and as for the Laplacian in flat space, they give a contribution

$$\sum_{k \text{ large}} \ln \lambda_k \sim \int \frac{d^2k}{(2\pi)^2} \ln k^2, \tag{3.7}$$

which diverges like $k^2$. We need to introduce a short distance regulator $a$, or a high momentum cut-off $\Lambda \sim 1/a$, to suppress the contribution of eigenmodes with wave vectors $k^2 > \Lambda^2$ in order to define in a proper way the partition function. Such a regulator has to be introduced in a reparametrization invariant way and we expect that in the limit of large cut-off $\Lambda$, the logarithm of the partition function will have an expansion as

$$\ln Z = B\Lambda^2 + C \ln \Lambda^2 + D + 0\left(\frac{1}{\Lambda^2}\right). \tag{3.8}$$

As we shall see in the next section, for two-dimensional surfaces, these coefficients may be computed as a function of the geometry of the surface. As a consequence, for "smooth" surfaces such that the curvature $R$ is (at any point) much smaller that the regulator $\Lambda^2$, the partition function may be calculated (almost) exactly!

## 3.2. *The Heat Kernel Regularization*

A simple way to give a sense of $\det'(-\Delta)$ (the $'$ means that the contribution of the zero mode is always omitted) by suppressing the contribution of the large eigenmodes is to write its logarithm as

$$\text{Tr}'\log(-\Delta\epsilon) = \log\det'(-\Delta\epsilon) = -\text{Tr}'\left[\int_\epsilon^\infty \frac{dt}{t}e^{t\Delta}\right] = -\int_\epsilon^\infty \frac{dt}{t}\text{Tr}'(e^{t\Delta}). \quad (3.9)$$

The parameter $\epsilon \sim \Lambda^2$ acts as a cut-off. Indeed, for small eigenvalues $\lambda_k \ll \epsilon$ the $t$ integral gives a contribution of order $\ln(\lambda_k\epsilon)$, while for large eigenvalues $\lambda_k \gg \epsilon$ the $t$ integral gives a contribution of order $\exp(-\epsilon\lambda_k)$. $\text{Tr}'$ means that the sum over eigenvalues $\lambda_k$ starts at $k = 1$. Now $\text{Tr}'(e^{t\Delta})$ may be written as

$$\text{Tr}'(e^{t\Delta}) = \left[\int_M d^2\sigma\sqrt{g}G(\sigma',\sigma;t)\right] - 1, \quad (3.10)$$

where the "heat-kernel" $G(\sigma,\sigma';t)$ is the integral kernel of the operator $e^{t\Delta}$, which is defined as the solution of the differential equation,

$$\frac{d}{dt}G(\sigma,\sigma';t) = \Delta_\sigma G(\sigma,\sigma';t) \quad (3.11)$$

with initial condition

$$G(\sigma,\sigma';t = 0) = \frac{1}{\sqrt{g(\sigma)}}\delta^2(\sigma - \sigma') = \delta^2_{\text{cov.}}\sigma,\sigma' \quad (3.12)$$

Thus $G$ describes the diffusion (according to Laplace equation) in a curved space of a distribution (for instance of heat) initially localized at the point $\sigma'$ at $t = 0$. The short time behavior of this heat kernel has been extensively studied in mathematics. We shall need only its short time behavior at coinciding points, which reads[5-7]

$$G(\sigma,\sigma;t) \simeq \frac{1}{4\pi}\left[\frac{1}{t} + \frac{R(\sigma)}{6} + 0(t)\right], \quad (3.13)$$

where $R(\sigma)$ is the scalar curvature at point $\sigma$. This result has a simple physical significance. The leading term $\frac{1}{t}$ corresponds to diffusion in flat space. The correction $R/6$ corresponds to the fact that diffusion will be slower on a space with positive curvature than on a space with negative curvature. Indeed, the area of a ball in this space grows more slowly with its radius if $R > 0$ than if $R < 0$. Hence the "available space" for diffusion is smaller and diffusion is slower for positive curvature.

From this formula we obtain immediately the small $\epsilon$ expansion of the logarithm of the partition function by integrating over $t$,

$$\text{Tr}'\log(-\Delta\epsilon) \simeq -\frac{1}{4\pi\epsilon}\int_M d^2\sigma\sqrt{g} - \frac{1}{24\pi}\int_M d^2\sigma\sqrt{g}R\log\epsilon + 0(1). \quad (3.14)$$

Hence the quadratic divergent part $B$ is proportional to the area of the surface and the logarithmically divergent part $C$ is proportional to the Euler character. Such a formula may be extended to the case of surfaces with boundaries. Let us stress that

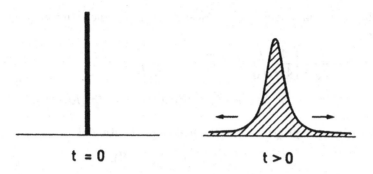

Fig. 3.1. Diffusion of a distribution localized at $t = 0$.

if the coefficient of the quadratic divergence is not universal and should depend on the regulator used, one can show that the coefficient of the logarithmic divergence is universal, provided that the regulator respects reparametrization invariance and is "smooth enough" in suppressing the contribution of higher modes (it is worth mentioning that a sharp cut-off giving zero weight to all modes such that $\lambda_k > 1/\epsilon$ and weight 1 to the others does not satisfy this smoothness assumption).

### 3.3. The Conformal Anomaly and the Liouville Action

In fact the method described above allows also to extract very strong information about the finite part $D$ of $\mathrm{Tr}\log(-\Delta)$. More precisely, one can compute exactly how this finite part changes when an arbitrary local conformal transformation

$$g_{ab}(\underline{\sigma}) \rightarrow g_{ab}(\underline{\sigma})e^{\varphi(\underline{\sigma})} = g'_{ab}(\underline{\sigma}) \tag{3.15}$$

is performed. Indeed under such a transformation, the scalar Laplacian changes as

$$\Delta_g \rightarrow \Delta_{g'} = e^{-\varphi(\underline{\sigma})}\Delta_g. \tag{3.16}$$

Therefore, under an infinitesimal change $\delta\varphi$

$$
\begin{aligned}
\frac{\delta}{\delta\varphi(\sigma)}\mathrm{Tr}'\log(-\Delta\epsilon) &= -\int_\epsilon^\infty \frac{dt}{t}\mathrm{Tr}\left[t\frac{\delta\Delta}{\delta\varphi}e^{t\Delta}\right] \\
&= \int_\epsilon^\infty \frac{dt}{t}\sqrt{g(\sigma)}[t\Delta e^{t\Delta}]_{\sigma\sigma} \\
&= \int_\epsilon^\infty dt\frac{d}{dt}[e^{t\Delta}]_{\sigma\sigma}\sqrt{g(\sigma)} \\
&= -[e^{\epsilon\Delta}]_{\sigma\sigma}\sqrt{g(\sigma)} = -\sqrt{g(\sigma)}G(\sigma,\sigma;\epsilon) \\
&= -\sqrt{g(\sigma)}\left[\frac{1}{4\pi\epsilon} + \frac{R(\sigma)}{24\pi} + 0(\epsilon)\right].
\end{aligned}
\tag{3.17}
$$

This is the conformal anomaly.[5,7-9] Most remarkably, the result of (3.17) may be written as

$$
\frac{\delta}{\delta\varphi(\sigma)}\left[\frac{1}{4\pi\epsilon}\int_M d^2\sigma\sqrt{g(\sigma)} - \frac{1}{48\pi}\int_M d^2\sigma \right.
$$
$$
\left. \times \int_M d^2\sigma'\sqrt{g(\sigma)}R(\sigma)G(\sigma,\sigma')\sqrt{g(\sigma')}R(\sigma')\right],
\tag{3.18}
$$

where $G(\sigma,\sigma')$ is the propagator in the metric $g_{ij}$ solution of

$$
(-\Delta_\sigma)G(\sigma,\sigma') = \delta_{\mathrm{cov}}(\delta,\delta').
\tag{3.19}
$$

Therefore one can integrate out the variation over $\varphi$, starting for instance from a constant curvature metric $g_{ij}^o$ conformally equivalent to the metric $g_{ij}$, to obtain the finite part of the trace of the logarithm as

$$
\mathrm{Tr}\log(-\Delta_g\epsilon) = \frac{1}{4\pi\epsilon}\int\sqrt{g} - \frac{1}{48\pi}\iint\sqrt{g}R\left(\frac{1}{-\Delta_g}\right)\sqrt{g}R + f[g^o] + 0(\epsilon),
\tag{3.20}
$$

where $f[g^o]$ depends only on the conformal class of the metric $g$, characterized by the metric $g^o$, and contains the logarithmic divergent part obtained in (3.14).

The finite part in (3.20) is called the *Liouville action*.[8] In a conformal system of coordinates where $g_{ij}(\sigma) = \delta_{ij}e^{\varphi(\sigma)}$ it takes the simple form

$$
S_{\mathrm{Liouville}}(\varphi) = -\frac{1}{48\pi}\int d^2\sigma[(\partial\varphi)^2 + \mu^2 e^\varphi]; \quad \mu^2 = -\frac{1}{6\epsilon}.
\tag{3.21}
$$

It plays a very important role in the study of surfaces with constant curvature (where the Liouville equation $-\partial^2\varphi + e^\varphi = 0$ was originally introduced by Liouville), in string theory and, as we shall see, in the study of membranes with hexatic order.

Finally let us mention that the free field coupled to a 2-d metric is the simplest example of 2-d conformal field theories,[5,10] a subject which has seen a tremendous development during the last years.

## 4. Fluid Membranes Models

In this section, we shall apply the notions of geometry and of field theory to models describing the statistics of fluid membranes, namely models of random surfaces with bending rigidity but where elements of the membrane are free to move in the plane of the membrane and to adjust themself to the shape deformations. Such an hypothesis is valid if the characteristic time for in-plane diffusion of molecules is much shorter than the characteristic time for thermal undulations corresponding to shape deformations. In such a case one can assume that the energy of a configuration depends only on the geometry of this configuration and is therefore independent of any particular coordinate system chosen to describe the membrane, (which could correspond physically to a labeling of distinct constitutive elements of the membrane). In doing so, one also assumes that internal degrees of freedom (for instance sound waves associated to the finite compressibility of the fluid) are irrelevant for the large distance properties. This assumption is in fact valid, since such degrees of freedom do not exhibit in general critical fluctuations. At the end of this section we shall briefly discuss the effect of global tangential flows in the membrane.

## 4.1. *Continuous Model for Fluid Membranes*

To construct an effective Hamiltonian for fluid membranes, one proceeds in the following way. As argued before, the Hamiltonian must be only a function of the field $\vec{X}(\underline{\sigma})$ (describing the position in $d$-dimensional Euclidean space of points of the membrane), invariant under displacements and rotations in $\mathbf{R}^d$, and reparametrization invariant. Expanding in local terms involving more and more derivatives and keeping only the terms relevant by naïve power counting, the most general Hamiltonian has only three terms, and is given by[11,12]

$$H = r_0 \int d^2\sigma \sqrt{g} + \frac{\kappa_0}{2} \int d^2\sigma \sqrt{g}(\Delta \vec{X})^2 + \frac{\bar{\kappa}_0}{2} \int d^2\sigma \sqrt{g}R, \qquad (4.1)$$

where

$$g_{ij} = \partial_i \vec{X} \partial_j \vec{X} \qquad (4.2)$$

is the induced metric, $g$ its determinant, $\Delta$ the scalar laplacian in metric $g_{ij}$ and $R$ the scalar curvature.

In this formulation the total area $\int d^2\sigma \sqrt{g}$ is allowed to vary. The conjugate parameter $r_0$ has engineering dimension (length)$^{-2}$. It is strongly relevant and will often be called the *microscopic surface tension*. Physically it corresponds to a chemical potential for unit elements of area for the membrane.

$\kappa_0$ and $\bar{\kappa}_0$ are respectively the *bending rigidity* and the *Gaussian rigidity*, they have engineering dimension 0 and correspond to marginal parameters. $\bar{\kappa}_0$ plays no role if the topology of the membrane is fixed but is important if topological fluctuations occur, as this is the case in microemulsions. The microscopic origin of $\kappa_0$ and $\bar{\kappa}_0$ is described in S. Leibler's lectures. Other terms involving for instance higher powers of the extrinsic curvature are irrelevant. Indeed, if one rescale the whole surface by a factor $\lambda$

$$\vec{X}(\underline{\sigma}) \to \lambda \vec{X}(\underline{\sigma})$$

the area energy scale obviously as $\lambda^2$, the curvature energy terms are unchanged and higher terms scale as negative powers of $\lambda$. Hence they are negligible if one considers large wavelength deformations of the surface.

There are two important special cases where additional relevant terms can be included in the Hamiltonian; they correspond to space dimension $d = 3$ and 4.

*d = 3: spontaneous curvature*

If $d = 3$ the membrane has an "interior" and an "exterior" and may have a spontaneous curvature.[11] The corresponding energy term is

$$H_{\text{sp. curv.}} = S_0 \int d^2\sigma \sqrt{g}(\Delta \vec{X} \cdot \vec{n}), \qquad S_0 = -\kappa_0 H_0, \tag{4.3}$$

where $\vec{n}$ is the vector normal to the surface and $H_0$ the spontaneous mean curvature. This requires the choice of an orientation on the surface.

*d = 4: "winding number"*

If $d = 4$ another term may be considered.[5,12] It is obtained from the quantity

$$Q = \frac{1}{4\pi} \int d^2\sqrt{g}\epsilon_{\mu\nu\rho\sigma} g^{ij} \partial_i Q^{\mu\nu} \partial_j Q^{\rho\sigma}, \qquad Q^{\mu\nu} = \gamma^{ij} \partial_i X^\mu \partial_j X^\nu, \tag{4.4}$$

where $\epsilon_{\mu\nu\rho\sigma}$ is the totally antisymmetric rank 4 tensor.

$Q$, which makes sense only in dimension 4, is in fact, like the Euler character $\chi$, an integer quantity of topological origin. It is nothing but the *winding number* of the surface in four-dimensional space, and generalizes to two-dimensional surfaces the notion of winding number of a one-dimensional curve in two-dimensional space,[5]

$$Q_{1D} = \frac{1}{2\pi} \int ds \epsilon_{\mu\nu} \frac{\partial^2 X^\mu}{\partial s^2} \frac{\partial X^\nu}{\partial s} \qquad s \text{ curvilinear coordinate.} \tag{4.5}$$

Since $Q$ may be negative or positive, one can add to the Hamiltonian a topological term of the form

$$H_\theta = i\theta Q, \qquad 0 \le \theta < 2\pi, \tag{4.6}$$

depending on an angle $\theta$ and still get a real partition function.

However if $\theta \neq 0$ (modulo $2\pi$) the Boltzmann weights in the partition function are no more real and positive. Nevertheless, it has been suggested that the case $\theta = \pi$ could correspond to some kind of "fermionic" surfaces, in analogy with the case of $1 - D$ random walks in two dimensions with a topological term (4.5), which for $\theta = \pi$ describes in fact free fermions, while random walks at $\theta = 0$ describes free bosons.[13] This topological term with $\theta = \pi$ could perhaps describe some kind of self interaction between surfaces.

## 4.2. Partition Function, Gauge Fixing

Our purpose is to study the statistics of a membrane in thermodynamic equilibrium at temperature $T$. The partition function $Z$ should write

$$Z = \int \mathbf{D}[\vec{X}]e^{-\beta H}, \qquad \beta = (k_B T)^{-1}, \tag{4.7}$$

where the measure $\mathbf{D}[\vec{X}]$ is defined in the space of all membranes configurations. This measure should satisfy two properties, whose physical significance is quite natural.

(i) *Reparametrization invariance*

The weight of a configuration should not depend on the coordinate system used to describe this configuration.

(ii) *Locality*

Deformations of the surface at far away points must be treated independently and the measure should factorize (there are no long distance correlations on the surface).

If (ii) is not satisfied, this means that additional degrees of freedom should be introduced to describe these correlations and a different physical problem is studied.

Finally, to define the partition function, a *short distance cut-off* has to be introduced. It represents the minimal wave length possible for shape deformations.

At that point we immediately run into a problem. Many different configurations $\vec{X}(\sigma)$ are equivalent under some change of coordinates and describe the same membrane. We should choose a particular configuration in each "equivalent classe" to really sum over configuration. But doing so in an arbitrary way in general violates (ii). The way to deal with this problem is similar to the one used in the functional quantization of gauge theories.

(1) One first sums in a "dumming" way over all possible configurations $\vec{X}(\sigma)$ with a measure which is local and reparameterization invariant. In our case such a measure is provided by Fujikawa's measure which writes (somewhat formally)[5−9]

$$\mathbf{D}[\vec{X}] = \prod_{\sigma} \prod_{\mu=1}^{d} \left[ dX^{\mu}(\sigma)[g(\sigma)]^{1/4} \right]. \tag{4.8}$$

This measure satisfies (i) and (ii) because it is the covariant measure associated with a Riemannian metric on the space of all configurations $\{\vec{X}(\sigma)\}$, which is itself local and reparametrization invariant.

(2) Since we sum over physically equivalent configurations, we pick up a set of inequivalent configurations ("gauge slice") by a "gauge fixing" condition. Then we write the functional integral as an integral over the gauge slice weighted by the volume of each equivalent class of all configurations physically equivalent to their representant in the gauge slice.

Before dealing with the particular case of surface, let us recall briefly the general formulation of this procedure.[14] Let $X$ be a space labelled by coordinates $x = (x^i)$ $(i = 1, M)$.

A symmetry group $G$ acts on $X$ with generators $t_\alpha$ $(\alpha = 1, N = \dim G)$. We want to compute an integral

$$I = \int d\mu(x) f(x), \tag{4.9}$$

where the measure $d\mu$ and the function $f$ are *invariant* under the action of elements of $g$ of $G$

$$x \to x_g = \exp\left[\sum_\alpha \theta^\alpha t_\alpha\right] x. \tag{4.10}$$

Then it is enough to pick one configuration $x$ in the set $\tilde{x}$ of all configurations equivalent to $x$ by $G$

$$\tilde{x} = \{x_g; g \in G\} \tag{4.11}$$

by choosing $N$ constraints $F_\alpha$ in an appropriate way

$$F_\alpha(x) = 0, \qquad \alpha = 1, N. \tag{4.12}$$

Then, using the invariance of $f$ and $d\mu$, we can rewrite $I$ as

$$I = \int_X d\mu(x)f(x) \prod_{\alpha=1}^N \delta(F_\alpha(x)) \cdot |\det J_{\alpha\beta}(x)| \cdot \text{vol } G, \tag{4.13}$$

where $J_{\alpha\beta}(x)$ is the matrix of the derivatives of the constraints,

$$J_{\alpha\beta}(x) = \left.\frac{\partial F_\alpha(x_g)}{\partial \theta^\beta}\right|_{\sigma=0}. \tag{4.14}$$

$\det J$ is called the Faddeev–Popov determinant. As a trivial example one can consider a rotationally invariant two-dimensional integral

$$I = \int_{\mathbf{R}^2} dx\, dy\, f(x^2 + y^2). \tag{4.15}$$

A rotation writes $e^{\theta t}, t = \begin{bmatrix} 0 & -1 \\ 1 & 0 \end{bmatrix}$. Choosing as constraint $F(x,y) = y\,(x > 0)$ the matrix $J$ is $J = \left.\frac{\partial y}{\partial \theta}\right|_{\substack{y=0 \\ \theta=0}} = x$ and one gets the well-known result

$$I = \int dx\, dy\, \theta(x)\, \delta(y) J f(x^2 + y^2)\, \text{vol}(0(2)) = \int_0^\infty dx\, x f(x^2) 2\pi. \tag{4.16}$$

*Gauge fixing for reparametrization invariance*

In the case of reparametrization invariance, the generators of infinitesimal diffeomorphisms are vector fields $\varepsilon^a$,

$$\sigma^a \to \sigma'^a = \sigma^a + \varepsilon^a(\sigma), \tag{4.17}$$

which act on $\vec{X}$ as

$$\vec{X}(\sigma) \to \vec{X}_\varepsilon(\sigma) = \vec{X}(\sigma) + \varepsilon^a(\sigma)\frac{\partial}{\partial\sigma^a}\vec{X}(\sigma). \tag{4.18}$$

A general class of gauge constraints, which includes most of the cases considered in the literature, is the *normal gauge*.[15]

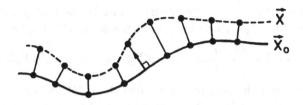

Fig. 4.1.   The normal gauge.

We choose a particular configuration of the surface and a particular system of coordinate, in which this configuration (which is called the background configuration) is $\vec{X}_0(\sigma)$. Configuration "close enough" to the background configuration can be written as

$$\vec{X}(\sigma) = \vec{X}_0(\sigma) + \vec{x}(\sigma). \tag{4.19}$$

The normal gauge consists in taking $\vec{x}$ orthogonal to the background configurations $\vec{X}_0$. This is achieved by the two sets of constraints $F_a(\sigma)$ $(a = 1, 2)$:

$$F_a(\sigma) = \vec{x}(\sigma)\frac{\partial}{\partial\sigma_a}\vec{X}_0(\sigma) = 0. \tag{4.20}$$

Such a gauge is valid only for small deformations around $\vec{X}_0$. It cannot deal with the possibility of "overhangs", neither of course with the possibility of changes of topology. However this is sufficient for the perturbative calculations that we shall perform. Let us note that some global gauges such as the conformal gauge (choose a system of coordinates in which the induced metric is conformally equivalent to a fixed reference metric), although more complicated to deal with, allow to consider arbitrarily large deformations.

The particular case of the Monge representation for a surface corresponds to take as a background configuration a plane parametrized by a Cartesian system of coordinates.[16]

The Faddeev–Popov determinant is obtained by calculating how the constraints $F_a$ are affected by an infinitesimal diffeomorphism generated by a vector field $\epsilon^a$:

$$\delta F_a(\sigma) = \delta[\vec{X}(\sigma) - \vec{X}_o(\sigma)]\frac{\partial}{\partial\sigma^a}\vec{X}_o(\sigma) = \delta\vec{X}(\sigma)\frac{\partial}{\partial\sigma^a}\vec{X}_o(\sigma)$$

$$= \epsilon^b(\sigma)\frac{\partial}{\partial\sigma^b}\vec{X}(\sigma)\frac{\partial}{\partial\sigma^a}\vec{X}_o(\sigma)$$

$$= \epsilon^b(\sigma)\left[\frac{\partial}{\partial\sigma^b}\vec{X}_o(\sigma) + \frac{\partial}{\partial\sigma^b}\vec{x}(0)\right]\frac{\partial}{\partial\sigma^a}\vec{X}_o(\sigma)$$

$$= \epsilon^b(\sigma)\left[g_{ab}^o(\sigma) + \frac{\partial}{\partial\sigma^b}\vec{x}(\sigma)\frac{\partial}{\partial\sigma^a}\vec{X}_o(\sigma)\right]. \tag{4.21}$$

Hence, from the general result, the corresponding Faddeev–Popov determinant is the determinant of the operator acting on vector fields with the kernel

$$J_{ab}(\sigma, \sigma') = \frac{\delta F_a(\sigma)}{\delta \epsilon^b(\sigma')} = \delta(\sigma - \sigma')[g^o_{ab}(\sigma) + \partial_b \vec{x}(\sigma)\partial_a \vec{X}_o(\sigma)] \qquad (4.22)$$

(both the index $a$ and the coordinates $\sigma$ label the constraints and the generators of infinitesimal transformations). One notices that for the normal gauge $J$ is purely local, so that its determinant factorizes into

$$\det(J) = \prod_\sigma \det[g^o_{ab} + \partial_b \vec{x}\partial_a \vec{X}_o(\sigma)] \qquad (4.23)$$

so that the partition function writes

$$Z = \text{cst} \int \mathbf{D}[\vec{X}] \prod_{\sigma, a} \delta(F_a(\sigma)) \det[J] e^{-\beta \cdot H}. \qquad (4.24)$$

In the particular case of the Monge representation, $\partial_b \vec{x}\partial_a \vec{X}_o(\sigma) = 0$ and $g^o_{ab} = \delta_{ab}$ so that the Faddeev–Popov determinant is trivial. As we shall see later, the fact that $\det J$ is purely local for normal gauges means that it will play no role in the calculation of the renormalization of the bending constants. This is not the case in general gauges such as the conformal gauge where the Faddeev–Popov determinant is not local and must be taken into account.[12]

## 4.3. *Effective Action and the Background Field Method*

In order to compute the renormalization of the bending constants, one must calculate the free energy of surface configurations which are not equilibrium configurations but close to such configurations. This is done in the following standard way.[15,17] We assume that some external force is applied to the surface by adding to the Hamiltonian $H$ a "source term" of the form

$$H_{\text{source}} = \int d^2\sigma \sqrt{g} \vec{X}(\sigma)\vec{J}(\sigma). \qquad (4.25)$$

The partition function depends now on the external source term via

$$Z[J] = \int \mathbf{D}[\vec{X}] e^{-\beta H + H_{\text{source}}}. \qquad (4.26)$$

The mean position of some points of the surface is

$$\vec{X}_{c\ell} = \langle \vec{X} \rangle = \frac{1}{\sqrt{g}} \frac{\delta}{\delta \vec{J}} \ln Z[J] \qquad (4.27)$$

and one defines the effective action (that is the free energy) of the classical configuration $\vec{X}_{c\ell}$ via the Legendre transform

$$\Gamma(\vec{X}_{c\ell}) = \left[ \int \sqrt{g} \vec{X}_{c\ell} \vec{J} - \ln Z \right] \frac{1}{\beta} \qquad (4.28)$$

so that the minimum of the effective action corresponds to the expectation value of the field $\vec{X}$ in the absence of source term $J$. Classically, that is at zero temperature

$(\beta = \infty), \Gamma$ coincides with $H$. To first order in the temperature (one loop) one can show that this standard method to compute the effective action is equivalent to the following "background method". Write the configuration $\vec{X}$ as the classical part $\vec{X}_{c\ell}$ plus fluctuations $\vec{x}$:

$$\vec{X} = \vec{X}_{c\ell} + \vec{x}. \tag{4.29}$$

Expand the Hamiltonian $H(\vec{X})$ in terms of $\vec{x}$, disregard the linear term (this is in fact what the source term does) and integrate over the fluctuations. The leading contribution comes from the quadratic part of $H$ in $\vec{x}$ and the Gaussian integration gives for the free energy the logarithm of a determinant

$$\Gamma(\vec{X}_{c\ell}) = H(\vec{X}_{c\ell}) + \frac{1}{2}\beta^{-1}\log\cdot\det\left[\frac{\partial^2 H(\vec{X}_{c\ell} + \vec{x})}{\partial\vec{x}(\sigma)\partial\vec{x}(\sigma')}\right]_{\vec{x}=0} + 0(\beta^{-2}). \tag{4.30}$$

This procedure is slightly complicated by the gauge fixing procedure. One has to consider only fluctuations $\vec{x}$ satisfying the gauge constraints and add to (4.30) $-\beta^{-1}\log\det(J)$ to obtain $\Gamma$.

### 4.4. *Renormalization of the Bending and Gaussian Rigidity*

We are now in a position to study the renormalization of the bending rigidity $\kappa$ by thermal fluctuations. We shall use the above described background method[15,18] and limit ourself to the special case $d = 3$. The general case has been considered in Refs. 12 and 19.

In $d = 3$ we can write

$$\vec{X}(\sigma) = \vec{X}_{c\ell}(\sigma) + \vec{x}(\sigma) = \vec{X}_{c\ell}(\sigma) + \nu(\sigma)\vec{n}(\sigma), \tag{4.31}$$

where $\vec{n}(\sigma)$ is the normal vector. The extrinsic curvature writes $\vec{K}_{ij} = K_{ij}\vec{n}$ and the mean curvature and Riemann curvature are

$$H = \frac{1}{2}C = \frac{1}{2}\text{Tr}(K_{ij}), \quad R = 2\det(K_{ij}).$$

We start from the action

$$H = \frac{\kappa}{2}\int d^2\sigma\sqrt{g}(\sigma)C^2. \tag{4.32}$$

Expanding $H$ to second order in $\nu$ is lengthy but presents no conceptual difficulty. We refer to Refs. 15 and 18. One has for instance

$$g_{ij} = g_{ij}^{c\ell} - 2\nu K_{ij}^{c\ell} + K_{ik}^{c\ell}K_j^{kc\ell}\nu^2 + \partial_i\nu\partial_j\nu + 0(\nu^3). \tag{4.33}$$

The term quadratic in $\nu$ of $H$ is finally

$$H^{(2)} = \int d^2\sigma\sqrt{g^{c\ell}}\left[\Delta\nu\Delta\nu + \nu\frac{1}{2}(C^2 + 2R)\Delta\nu + 2CC^{ij}D_i\nu D_j\nu \right.$$
$$\left. + \frac{(C^2 - 2R)(5C^2 - R)}{2}\nu^2\right], \tag{4.34}$$

where the covariant derivatives $D_i$ and the Laplacian $\Delta$ refer to the background metric $g_{ij}^{c\ell} = \partial_i \vec{X}_{c\ell} \partial_j \vec{X}_{c\ell}$. Thus the effective potential is at one loop.

$$\Gamma = H + \frac{1}{2}k_B T \operatorname{Tr}\log\left[\Delta^2 + \frac{1}{2}(C^2 - 2R)\Delta + 2CC^{ij}D_iD_j + \text{scalar function of } (\sigma)\right]$$

$$+ \text{Faddeev–Popov term.} \qquad (4.35)$$

The issue is to determine how short wavelength fluctuations of $\nu$, with wavevector $k$ satisfying

$$k_{\min} < |k| < k_{\max}, \qquad (4.36)$$

where $k_{\max}$ represents the ultraviolent regulator to the theory ($k_{\max} \sim \frac{\pi}{a}$, where $a$ is some short distance cut-off) contribute to $\Gamma$. For that purpose, let us consider a background configuration which is "almost flat", namely such that

$$C^2, R \ll k_{\min}^2. \qquad (4.37)$$

Then one can expand the $\operatorname{Tr}\log[\ ]$ in powers of the curvature as

$$\operatorname{Tr}\log[\ ] = 2\operatorname{Tr}\log(-\Delta) + \operatorname{Tr}\left[\Delta^{-2}\left[\frac{1}{2}(C^2 - 2R)\Delta + 2CC^{ij}D_iD_j\right]\right],$$

$$+ \text{higher order terms}, \qquad (4.38)$$

where the trace of some operator $O$ means (see Sec. 3)

$$\operatorname{Tr} O = \int d^2\sigma \sqrt{g}(\sigma)O(\sigma,\sigma), \qquad (4.39)$$

where $O(\sigma,\sigma')$ is the kernel associated to $O$. Therefore the first term in the trace means

$$\operatorname{Tr}\left[\Delta^{-1}\left(\frac{1}{2}C^2 - 2R\right)\right] = \int d^2\sigma \sqrt{g}(\sigma)\left[2R(\sigma) - \frac{1}{2}C^2(\sigma)\right]\left(\frac{1}{-\Delta}\right)_{\sigma\sigma}, \qquad (4.40)$$

where $\left(\frac{1}{-\Delta}\right)_{\sigma\sigma'}$ is nothing but the propagator $G(\sigma,\sigma')$ considered in Sec. 3. Using for instance the heat kernel regularization, one shows easily that at short distances the propagator behaves as

$$G(\sigma,\sigma') \simeq -\frac{1}{4\pi}\ln\left[(\sigma - \sigma')^i(\sigma - \sigma])^j g_{ij}(\sigma)\right] \qquad (4.41)$$

and that the contributions of short wavelength modes in the trace gives

$$\int d^2\sigma\left[\frac{1}{2}C^2 - 2R\right]\cdot\frac{1}{4\pi}\ln\left[\frac{k_{\max}^2}{k_{\min}^2}\right] + \text{subdominant terms.} \qquad (4.42)$$

Similarly, the second term gives

$$\operatorname{Tr}[\Delta^{-2}[2CC^{ij}D_iD_j]] \simeq -\int d^2\sigma\sqrt{g}C^2\cdot\frac{1}{4\pi}\ln\left[\frac{k_{\max}}{k_{\min}^2}\right]. \qquad (4.43)$$

The first term $\operatorname{Tr}\log(-\Delta)$ depends only on the intrinsic metric and therefore cannot give a contribution involving the mean curvature $C$. It has been calculated in Sec. 3

and was shown to give

$$\text{Tr}\log(-\Delta) = \text{cst}\left(k_{\max}^2 - k_{\min}^2\right) \int d^2\sqrt{g} - \frac{1}{24\pi} \int d^2\sigma\sqrt{gR} \ln\left(\frac{k_{\max}^2}{k_{\min}^2}\right). \quad (4.44)$$

The other terms in the expansion of the $\text{Tr}\log[\ ]$ can easily be shown not to contribute to any divergent contribution for large momenta. Finally let us briefly discuss the contribution of the Faddeev–Popov determinant. As seen previously $\det(J)$ is a purely local object which factorizes at different points. Thus

$$\text{Tr}\log(J) \simeq \int d^2\sigma\sqrt{g} \ [\text{divergent contribution}] \quad (4.45)$$

and cannot give terms proportional to $C^2$ or $R$.

The calculation is now complete. Starting from the Hamiltonian

$$H = \int d^2\sigma\sqrt{g}\left(\frac{\kappa}{2}C^2 + \frac{\bar{\kappa}}{2}R\right) \quad (4.46)$$

the contribution of the short wavelength thermal undulations to the effective action can be absorbed into a redefinition of the coupling constants

$$\Gamma = \int d^2\sigma\sqrt{g}\left[r_{\text{eff}} + \frac{\kappa_{\text{eff}}}{2}C^2 + \frac{\bar{\kappa}_{\text{eff}}}{2}R\right]. \quad (4.47)$$

$r_{\text{eff}}$ diverges quadratically with $|k|$. $\kappa_{\text{eff}}$ behaves as[16]

$$\kappa_{\text{eff}} = \kappa - \frac{3}{8\pi}k_BT \ln\left|\frac{k_{\max}^2}{k_{\min}^2}\right| + \cdots \quad (4.48)$$

and $\bar{\kappa}_{\text{eff}}$ as[18]

$$\bar{\kappa}_{\text{eff}} = \bar{\kappa} + \frac{5}{12\pi}k_BT \ln\left|\frac{k_{\max}^2}{k_{\min}^2}\right| + \cdots \quad (4.49)$$

while

$$r_{\text{eff}} \propto k_BT(k_{\max}^2 - k_{\min}^2). \quad (4.50)$$

The calculation can be extended to $d$-dimensional bulk space with only technical difficulties. The result for $\kappa_{\text{eff}}$ is[12]

$$\kappa_{\text{eff}} = \kappa - \frac{d}{8\pi}k_BT \ln\left|\frac{k_{\max}^2}{k_{\min}^2}\right| + \cdots \quad (4.51)$$

and for $\bar{\kappa}_{\text{eff}}$[19]

$$\bar{\kappa}_{\text{eff}} = \bar{\kappa} - \frac{d-8}{12\pi}k_BT \ln\left|\frac{k_{\max}^2}{k_{\min}^2}\right| + \cdots. \quad (4.52)$$

The physical consequences of the renormalization of $\kappa$ are discussed at length in S. Leibler's and D. Nelson's lectures. Because of short distance thermal undulations, there is a decrease of the effective rigidity. The length scale $\xi_p$ at which the effective rigidity is of the order of $k_B T$ corresponds to the persistence length[20] and from the above result scales *for large $\kappa$* as

$$\xi_p \sim a \exp\left(\frac{4\pi}{d}\frac{\kappa}{k_B T}\right), \tag{4.53}$$

where $a$ is the short distance cut-off.

The meaning of the renormalization of the gaussian rigidity $\bar{\kappa}$ is less clear since $\bar{\kappa}$ can be "measured" experimentally only when topological fluctuations occur. One expects that such fluctuations will take place only at scales larger than the persistence length, where the perturbative results becomes unapplicable.

## 4.5. *Renormalization of the Surface Tension*

Up to now, the effect of the surface tension $r$ has been neglected. Since an effective surface tension $r_{\mathrm{eff}}$ is generated by thermal fluctuation, in fact a microscopic surface tension has to be introduced in order to counterbalance this effect and to obtain a small effective surface tension and large surfaces. If one adds to the Hamiltonian $H$ a term $r \int d^2\sigma \sqrt{g}$, from the expansion of $g_{ij}$ to second order in $\nu$, $\sqrt{g}$ expands as

$$\sqrt{g} = \sqrt{g^{cl}}\left[1 - \nu C + \frac{1}{2}\nu(-\Delta + R)\nu + \cdots\right] \tag{4.54}$$

and we must add in the $\mathrm{Tr}\log[\;]$ a term of the form

$$\mathrm{Tr}\log[r(-\Delta + R)]. \tag{4.55}$$

When expanding in powers of curvatures this gives a correction to the effective action proportional to

$$-r\,\mathrm{Tr}\left[\frac{1}{\Delta}\right] \simeq r \int d^2\sigma \sqrt{g}\,\frac{1}{4\pi}\ln\left(\frac{k_{\max}^2}{k_{\min}^2}\right). \tag{4.56}$$

Thus the surface tension $r$ does not affect the renormalization of $\kappa$ and $\bar{\kappa}$ (as expected) but $r$ gets a new renormalization

$$r_{\mathrm{eff}} = r\left[1 + \frac{1}{8\pi}\ln(k_{\max}^2/k_{\min}^2)\right] - \mathrm{cst}(k_{\max}^2 - k_{\min}^2). \tag{4.57}$$

## 4.6. *Effect of Tangential Flows*

In the previous derivation, a crucial assumption is that the energy of the surface depends only on its shape and that in-plane fluctuations of the elements of fluid can be integrated out. In fact, as pointed out by W. Helfrich[21] and D. Förster,[22] things are more subtle because of the following effect. The 2-dimensional fluid constituent of the membrane may be considered as incompressible at large scales. As seen from (4.54), a normal displacement at some point will induce a change of local area proportional to the mean curvature $C$ at that point. To keep the density constant, a global in-plane fluid displacement in needed, its amplitude decreasing like one over the distance to the point, when some normal displacement is performed. As a consequence, the kinetic energy of such a displacement diverges logarithmically with the size of the membrane. Such a phenomenon unduces long distance correlations between mean curvatures and hypothesis (ii) (locality) may not be valid.[22] In fact it has been shown recently that because of transverse undulations, such large distance correlations are in fact screened, so that the locality hypothesis holds.[23]

## 5. Fluid Membranes: Non-Perturbative Issues and the Large $d$ Limit

Of course a major issue is to understand the properties of membrane systems at scales larger that the persistence length. Indeed beyond that scale topological fluctuations will occur and steric repulsion effects will have to be taken into account. Those effects are of crucial importance to understand the structure of microemulsions systems. Two approaches are possible:

One can consider effective discrete models with a short distance cut-off which will play the role of the persistence length. Such models are described in J. Fröhlich's lectures.

One can develop non-perturbative schemes to study the continuous model. A possible approach is the large $d$ limit,[24,25] that we now describe.

### 5.1. *The Large $d$ Limit*

The large $d$ limit, where $d$ is the dimension of bulk space, is very similar to the large $N$ limit for $N$ component spin systems. One way to construct it is to treat the internal metric $g_{ij}$ and the position field $\vec{X}$ as independent variables and to enforce the constraint $g_{ij} = \partial_i \vec{X} \partial_j \vec{X}$ with a Lagrange multiplier field $\lambda^{ij}$. Then the partition function writes

$$ Z = \int \mathbf{D}[\vec{X}]\mathbf{D}[g_{ij}]\mathbf{D}[\lambda^{ij}]e^{-\beta H}, \tag{5.1} $$

$$ H = r_0 \int \sqrt{g} + \frac{\kappa_0}{2} \int \sqrt{g}(\Delta\vec{X})^2 + \frac{d}{2} \int \sqrt{g}\lambda^{ij}(\partial_i\vec{X}\partial_j\vec{X} - g_{ij}), \tag{5.2} $$

where the $\lambda^{ij}$ field is integrated over a complex contour going from $-i\infty$ to $+i\infty$. The integration over the $d$-component field $\vec{X}$ is Gaussian and can be performed explicitly. The final result for the effective action $\Gamma(\vec{X}, g_{ij}, \lambda^{ij})$, where $\vec{X}, g_{ij}$ and $\lambda^{ij}$ are now classical backgrounds fields, is

$$ \Gamma(\vec{X}, g_{ij}, \lambda^{ij}) = H(\vec{X}, g_{ij}, \lambda^{ij}) + \frac{d}{2}k_BT \operatorname{Tr} \log\left(\frac{\kappa_0'\Delta^2 - D_i\lambda^{ij}D_j}{k_BT}\right) + o(d), \tag{5.3} $$

where we have rescaled

$$ \kappa_0 = d\kappa_0', \qquad r_0 = dr_0'. \tag{5.4} $$

In the large $d$ limit ($\kappa_0', r_0'$ fixed) only the saddle point of the effective action $\Gamma$ contributes and the free energy for a classical configuration $\vec{X}$ is obtained by extremizing $\Gamma$ with respect to the auxiliary fields $g_{ij}$ and $\lambda^{ij}$:

$$\Gamma(\vec{X}) = \text{Extremum over } (g_{ij}, \lambda^{ij}) \text{ of } \Gamma(\vec{X}, g_{ij}, \lambda^{ij}). \qquad (5.5)$$

In particular, the constraint $g_{ij} \propto \partial_i \vec{X} \partial_j \vec{X}$ should be recovered by extremizing $\Gamma$ only with respect to $\lambda^{ij}$.

Of course one has to compute the $\text{Tr} \log[\;]$. This is possible only for quite simple background configurations. The simplest one is the case of the planar configuration, that we now discuss.

## 5.2. Planar Configuration

Let us consider a membrane which is enforced *classically* to stay in a reference plane.[25] This may be achieved by the boundary conditions, for instance by limiting the membrane by a square frame with size $L$, but by allowing its area to fluctuate (grand canonical formulation) by taking the surface tension to be small enough (Fig. 5.1), and then by taking the thermodynamic limit $L \to \infty$.

As an ansatz for the background configuration which minimizes the effective action we take a planar configuration

$$\vec{X}(\sigma) = \sigma^1 \vec{u}_1 + \sigma^2 \vec{u}_2 \qquad (5.6)$$

where $\vec{u}_1, \vec{u}_2$ are 2 orthonormal vectors in the reference plane ($\vec{u}_i \vec{u}_j = \delta_{ij}$), and we take constant auxiliary fields

$$\begin{aligned} g_{ij}(\sigma) &= \rho \delta_{ij}, \\ \delta^{ij}(\sigma) &= \lambda/\rho \delta^{ij}. \end{aligned} \qquad (5.7)$$

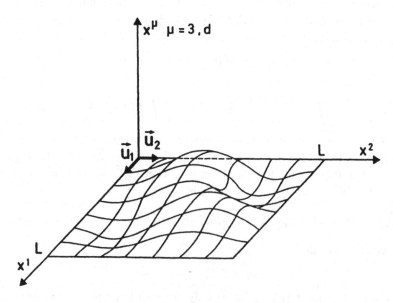

Fig. 5.1.  The boundary conditions keep the surface in the $(x^1, x^2)$ plane.

In this background the $\mathrm{Tr}\log(\ )$ may be computed exactly and the final result for the effective action density $f$ is

$$
f = \frac{\Gamma}{L^2} = d\Big\{ r'_0\rho + \gamma(1-\rho)
$$
$$
+ \frac{k_BT}{8\pi}\Big[ \Lambda^2 \log \frac{1}{k_BT}\Big(\kappa'_0 + \frac{\lambda}{\Lambda^2}\Big) + \frac{\lambda}{\kappa'_0}\log\Big(1 + \frac{\Lambda^2}{\lambda}\kappa'_0\Big)\Big]\Big\}, \quad (5.8)
$$

where $\Lambda$ is a sharp regulator (in estimating the $\mathrm{Tr}\log$ we are only summing over eigenvalues $\lambda$ of the Laplacian $\Lambda = \frac{1}{\rho}\partial^2$ such that $|\lambda| < \Lambda^2$). From its definition $f$ is nothing but the physical surface tension $r_{\mathrm{physical}}$ of the system. At the extremum of $\Gamma$ the equation $\frac{\partial f}{\partial \lambda} = 0$ implies that $f = \lambda d$ and thus

$$
r_{\mathrm{physical}} = \lambda d. \quad (5.9)
$$

The mean total area of the surface is nothing but

$$
\langle \mathrm{Area}\rangle = \frac{\partial \Gamma}{\partial r_0} = L^2\rho. \quad (5.10)
$$

Thus

$$
\rho = \frac{\langle \mathrm{Area}\rangle}{L^2} = \frac{\mathrm{Total\ Area}}{\mathrm{Frame\ Area}}. \quad (5.11)
$$

This provides a physical interpretation of the parameters $\lambda$ and $\rho$.

Extremizing $f$ with respect to $\lambda$ and $\rho$ leads to the following phase diagram in the $(r'_0, \kappa'_0)$ plane depicted in Fig. 5.2.

There is a critical line $L$ where the area of the surface diverges ($\rho = \infty$). Above that line, that is for a bare surface tension $r'_0$ larger than the critical value $r'_{\mathrm{cr}}(\kappa'_0)$,

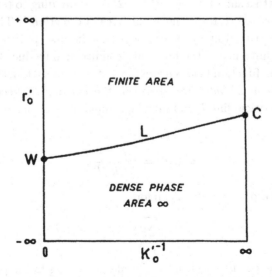

Fig. 5.2.   Phase diagram at $d = \infty$ (assuming translation invariance in the plane).

the mean area of the surface is finite and the surface tension $\lambda$ is positive. As one goes to the critical line from above, the mean area diverges as

$$\rho \simeq (r_0' - r_{\text{crit}}'(\kappa_0'))^{-1/2} \tag{5.12}$$

but the surface tension stays finite.

$$\lambda_{\text{crit}} = \Lambda^2 \kappa_0' \left[ \exp\left(\frac{8\pi\kappa_0'}{k_B T}\right) - 1 \right]^{-1}. \tag{5.13}$$

Below the critical line the system does not exist. If one takes into account selfavoidance this domain should correspond to a dense phase.

The mean orientation of the surface and the correlations between tangents may be studied by associating to a point of the surface the $d \times d$ antisymmetric matrix which characterizes the tangent plane at that point

$$Q^{\mu\nu} = \frac{1}{\sqrt{|g|}} \epsilon^{ij} \partial_i X^\mu \partial_j X^\nu. \tag{5.14}$$

The correlation between two tangent planes at points $\sigma$ and $\sigma'$, which generalizes to $d$ dimensions the scalar product of normal vectors in three dimensions, is

$$\cos\theta_{\sigma\sigma'} = -\frac{1}{2}\text{Tr}[Q(\sigma) \cdot Q(\sigma')]. \tag{5.15}$$

The order parameter $\cos\theta = \langle -\frac{1}{2}\text{Tr}(Q(\sigma)Q_0)\rangle$ (where $Q_0$ is associated to the reference frame plane) decreases from 1 for $r_0' = \infty$ (infinite tension) to zero on the critical line. This means that above the critical line the symmetry group $0(d)$ of rotation invariance in $d$-dimensional bulk space is explicitly broken by the introduction of the frame (to a subgroup $0(2) \times 0(d-2)$ corresponding to rotations within the plane of the frame and to rotations in the normal directions). The $0(d)$ invariance is restored on the critical line $L$, since the area is infinite and the surface does not feel anymore the influence of the boundary conditions introduced by the frame.

The correlation function between two tangent planes as a function of the distance $\ell = |\vec{X}(\sigma) - \vec{X}(\sigma')|$ in *bulk space* between the two points becomes rotationally invariant on the critical line $L$ and may be computed. It behaves as some universal function $f$

$$\langle\cos\theta\rangle = f\left(\frac{\ell^2\lambda}{k_B T}\right), \tag{5.16}$$

where $f$ goes to zero at infinity. Therefore

$$\sqrt{\frac{k_B T}{\lambda}} = \xi_p \tag{5.17}$$

corresponds to the persistence length. $\xi_p$ scales according to the perturbative renormalization group calculation at large $\kappa$.

## 5.3. *Renormalization Group Behavior*

Effective coupling constants behave at large distances in a way which may be studied by perturbing the background configuration around the equilibrium configuration

$$\vec{X} = \vec{X}_{\text{plane}} + \vec{x}_\perp \qquad \vec{X}_{\text{plane}} = \sigma^1 \vec{u}_1 + \sigma^2 \vec{u}_2$$
$$\vec{x}_\perp \cdot \vec{u}_i = 0 \quad i = 1, 2, \tag{5.18}$$

and by calculating the corresponding effective potential to quadratic order in $\vec{x}_\perp$. The result is in fact very simple since from the expression for $\Gamma$ normal fluctuations $\vec{x}_\perp$ are decoupled from the other fluctuations of $g_{ij}$ and $\lambda^{ij}$. Thus even for non-zero but small $\vec{x}_\perp$, the saddle point value for $g_{ij}$ and $\lambda^{ij}$ stay extrema of the effective action which writes

$$\Gamma(\vec{x}_\perp) = d\lambda \int d^2\sigma \left(1 + \frac{1}{2}\partial\vec{x}_\perp\partial\vec{x}_\perp\right) + \frac{d\kappa'_0}{2\rho}\int d^2\sigma(\partial^2\vec{x}_\perp)^2 + 0(x_\perp^4). \tag{5.19}$$

The two terms in (5.19) are simply the first terms of the expansion of the area and of the total extrinsic curvature of the background configuration. Thus the physical surface tension and bending rigidity at large distances are

$$r_{\text{phys}} = d\lambda, \qquad \kappa_{\text{phys}} = \frac{d\kappa'_0}{\rho} \tag{5.20}$$

and the renormalization group functions are obtained in the standard way by computing how the microscopic (bare) couplings depend on the regulator $\Lambda$ for fixed physical couplings

$$\beta(\kappa'_0, r'_0) = \Lambda \frac{\partial}{\partial\Lambda}\kappa'_0 \bigg|_{\text{phys}}, \tag{5.21}$$

$$\gamma(\kappa'_0, r'_0) = \Lambda \frac{\partial}{\partial\Lambda}r'_0 \bigg|_{\text{phys}}. \tag{5.22}$$

In particular, along the critical line we get for $\beta$

$$\beta(\kappa'_0) = \frac{k_B T}{4\pi}\left[1 - \frac{k_B T}{4\pi}\left(1 - \exp\frac{8\pi}{k_B T}(\kappa_{\text{phys}} - \kappa'_0)\right)\right]^{-1}. \tag{5.23}$$

$\beta$ behaves for large rigidity like

$$\beta(\kappa'_0) \simeq \frac{k_B T}{4\pi} + 0\left(\frac{1}{\kappa'_0}\right), \tag{5.24}$$

as predicted by the perturbative renormalization group calculation.

Integrating out the R.G. equation one defines an effective rigidity $\kappa_{\text{eff}}(q)$ depending on the momentum scale $q$ by

$$q\frac{d}{dq}\kappa_{\text{eff}}(q) = \beta(\kappa_{\text{eff}}(q)). \tag{5.25}$$

The $q$ dependence of $\kappa_{\text{eff}}(q)$ is represented in Fig. 5.3 for various values of the surface tension. Starting from the bare bending rigidity $\kappa_0$ at the cut-off scale $\Lambda$, it flows to

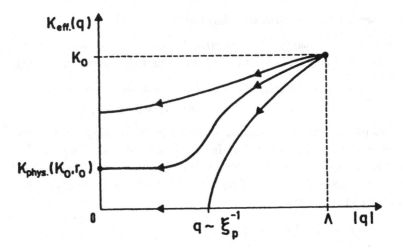

Fig. 5.3.   Behavior of the effective rigidity $\kappa_{\text{eff}}(q)$ for different values of the surface tension $r_0$.

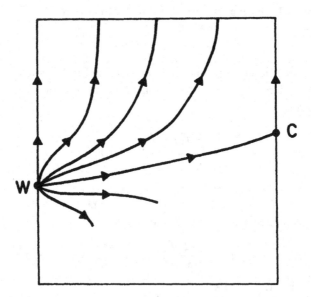

Fig. 5.4.   Renormalization group flow for $r$ and $\kappa$ at large $d$.

the physical rigidity $\kappa_{\text{phys}}$ at zero momentum (large distance). On the critical line where $\rho = \infty$ and thus $\kappa_{\text{phys}} = O$, $\kappa_{\text{eff}}(q)$ vanishes at a finite value of $q$, which is of the order of the inverse of the persistence length $\xi_p$. This discontinuity is probably a consequence of our definition of the effective bending constant and of the large $d$ limit.

The renormalization group flow obtained from the large $d$ calculation is depicted in Fig. 5.4. As expected, if we do not start on the critical line, the effective surface tension becomes large at large distances, since the membrane has a finite area and is under tension. On the critical line, the effective couplings flow from the critical point $W$ at infinite rigidity ($\kappa'_0 = \infty$) with one relevant direction (corresponding to $r_0$) and one marginally relevant one (corresponding to $\kappa_0$), which describes rigid surfaces with small surface tension, to an ordinary critical point $C$ at vanishing rigidity ($\kappa'_0 = 0$) describing the large distance physics of the membrane. According to our large d calculation the effective action at $C$ should be (keeping only the terms relevant at large distance) the action of a Gaussian surface.

$$\Gamma_{\text{eff}}(\vec{x}_\perp) = r_{\text{phys}} \int d^2\sigma \frac{1}{2}(\partial \vec{x}_\perp)^2. \tag{5.26}$$

Such an object should "crumple" with an infinite Hausdorff dimension.

### 5.4. *Conformal Fluctuations and Instabilities*

The previous analysis of the effective action for small fluctuations around the flat background is valid provided that this background is stable under fluctuations of the auxiliary fields $g_{ij}$ and $\lambda^{ij}$ and under longitudinal deformations $x_{||}$. A general fluctuation takes the form

$$\begin{aligned}
\vec{X} &= \vec{X}_{\text{plane}} + \vec{x}_\perp + \tilde{x}^i_{||}\vec{u}_i, \\
g_{ij} &= \rho\delta_{ij} + \tilde{g}_{ij}, \\
\lambda^{ij} &= \lambda g^{ik}\left[\delta^i_k + \tilde{\lambda}^i_k\right]
\end{aligned} \tag{5.27}$$

and the effective action expands to second order in the fluctuations into

$$\begin{aligned}
\Gamma(\vec{X}, g_{ij}, \lambda^{ij}) = \int d^2\sigma &\left[r_{\text{phys}}\left[1 + \frac{1}{2}(\partial\tilde{x}_\perp)^2\right] + \frac{\kappa_{\text{phys}}}{2}(\partial^2\tilde{x}_\perp)^2\right] \\
&+ \Gamma^{(2)}(\tilde{x}^i_{||}, \tilde{g}_{ij}, \tilde{\lambda}^i_j),
\end{aligned} \tag{5.28}$$

where $\Gamma^{(2)}$ is a quadratic form. Since $\Gamma$ is reparametrization invariant we shall fix a guage by choosing a conformal system of coordinates (see Sec. 2.8) where $g_{ij} = \rho\delta_{ij}e^{\phi(\sigma)}$. Thus

$$\tilde{g}_{ij} = \rho\delta_{ij}\phi(\sigma) \tag{5.29}$$

and $\Gamma^{(2)}(\tilde{x}^i_{||}, \phi, \tilde{\lambda}^i_j)$ involves only a $6\times 6$ symmetric matrix. The issue of the positivity of $\Gamma^{(2)}$ is complicated by the fact that $\lambda^i_j$ is an auxiliary field which is integrated from $-i\infty$ to $+i\infty$ in the functional integral. Hence the fluctuations $\tilde{\lambda}^i_j$ are imaginary. The whole discussion of the positivity of $\Gamma^{(2)}$ and of the stability of the flat background, as well as the details of the calculations, are contained in Ref. 25. We shall only give the main conclusions of this analysis here.

It is most convenient to discuss first the stability of $\Gamma^{(2)}$ under small imaginary fluctuations of $\tilde{\lambda}^i_j$, the real fields $\tilde{x}^i_{||}$ and $\phi$ being fixed. It turns out that $\Gamma^{(2)}$ always stays positive under fluctuations of $\tilde{\lambda}^i_j$. Moreover, correlations between $\tilde{\lambda}^i_j$ are always short ranged, with typical correlation length of the order of the persistence length $\xi_p$. The longitudinal fluctuations $\tilde{x}^i_{||}$ share the same properties. Therefore the issue of the stability of the flat background reduces to the study of the positiveness of the effective action for conformal deformations of the metric

$$\Gamma^{(2)}(\phi) = \text{Extremum over}\left(\tilde{x}^i_{||}, \tilde{\lambda}^i_i\right) \text{ of } \Gamma^{(2)}\left(\tilde{x}^i_{||}, \phi, \tilde{\lambda}^i_j\right). \tag{5.30}$$

From translation invariance in the plane it takes the form

$$\Gamma^{(2)}(\phi) = \int d^2p\,\hat{\phi}(p)G(p^2)\hat{\phi}(-p), \tag{5.31}$$

where $\hat{\phi}(p)$ is the 2-dimensional Fourier transform of $\phi(\sigma)$ and $G(p^2)$ the inverse propagator for the field $\phi$.

An estimate of the "mass" of the conformal field $\phi$ is given by $G(0)$. It is found to be positive above the critical line $L$ and to decrease to zero as $r_0 \to r_{0\ \text{crit}}$. Thus on the critical line $L$, the $\phi$ field becomes long ranged and must be taken into account in the effective action at large distance. However at large $d$ we run into a problem. As depicted in Fig. 5.5, where $G(p^2)$ is plotted for various values of the bare surface tension $r'_0, G(p^2)$ becomes negative for a *finite* wave number $p_\perp$, (whose typical scale is given by the inverse of the persistence length $\xi_p$), for a value of the bare surface tension $r_1^{\text{inst}}$ larger than the critical value $r_0^{\text{crit}}$. This means that the flat configuration becomes instable with respect to conformal fluctuations before one reaches the critical point. The new physical ground state which minimizes the full effective action $\Gamma$ will be no more homogeneous, (namely characterized by a constant average internal metric $g_{ij}$), but will be non-homogeneous, with "bumps" or "ripples" with typical wavelength of the order of the persistence

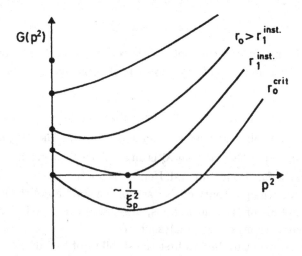

Fig. 5.5.   The energy $G(p^2)$ of a conformal fluctuation with wavevector $p$ for different values for the surface tension $r_0$.

length $\xi_p$. Hence the instability displayed by the large $d$ calculation is a strictly non-perturbative phenomenon, which could not have been predicted by the loop expansion in powers of $\kappa^{-1}$. Indeed, for large momenta $|p| \gg \xi_p^{-1}$, $G(p^2)$ is positive and the flat configuration is always stable with respect to short wavelength conformal fluctuations.

Many issues remain to be understood concerning the physical meaning of this instability. Is the transition between the flat phase and the "bumped" phase continuous or discontinuous as one decreases the tension parameter $r_0$? What is the nature of the "bumped" phase? As one decreases the surface tension $r_0$ below $r_1^{\text{inst}}$ how does the area of the surface diverge and to which new universality class does the system belong? Is such an instability a particularity of the large $d$ limit or does it exist for low dimensions ($d = 3$)? What is the effect of topological fluctuations on such a transition and what is its relationship with the appearance of a bicontinuous phase in real layers systems?

The first question may be studied in the following way. At the instability threshold $r_0 = r_1^{\text{inst}}$, one should compute the effective action for the marginal mode $\phi(\vec{p}_\perp) = \phi_1 e^{i\vec{p}_\perp \cdot \vec{\sigma}}$, $\Gamma(\phi_1)$, at least to third order in $\phi_1$ and check whether $\phi_1 = 0$ is an absolute minimum (continuous transition) or only a relative extremum (first order transition) of $\Gamma$. Such a calculation has not been carried out yet but the most plausible case is the second one (discontinuous transition).

The other issues are non-perturbative and much more difficult to answer. Some partial insights may be gained by considering which effective theory may govern the system at scales much larger than the persistence length and by making comparisons with other random surface models.

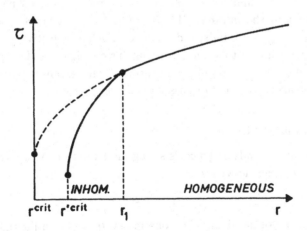

Fig. 5.6.  Conjectured behavior of the physical surface tension $\tau$ as $r_0$ is decreased. At $r = r_1$ a first order transition occurs. For $r_0 < r_1$ the surface is in an inhomogeneous phase (heavy line) where $\tau$ is smaller than in the homogeneous phase (dashed line). The area diverges at a new critical point $r^{\text{crit}}$.

## 6. Effective Models for Fluid Membranes and Strings

### 6.1. *The Polyakov String Model*

From the large $d$ limit, we have seen that on the critical ling $L$, both the position field $\vec{X}$ and the metric conformal field $\phi$ have long range correlations, while the auxiliary field $\lambda^{ij}$ has only short range correlations (of the order of the persistence length $\xi_p$). The existence of an instability is in fact a consequence of the negative sign of the kinetic term for $\phi$ at small momenta on the critical line. We shall assume that the above described features of the model persist for finite dimensions $d$ and shall make the following hypothesis.[5-25]

Starting from the action

$$S = \frac{\kappa}{2} \int \sqrt{g}(\Delta\vec{X})^2 + r_0 \int \sqrt{g} + \int \sqrt{g}\lambda^{ij}(\partial_i\vec{X}\partial_j\vec{X} - g_{ij}), \qquad (6.1)$$

let us assume that there is a critical point $r_0 = r_0^{\mathrm{crit}}$ where a finite persistence length $\xi_p$ is generated so that $\lambda^{ij}$ gets a finite expectation value

$$\langle\lambda^{ij}\rangle = g^{ij}\langle\lambda\rangle \qquad (6.2)$$

and that the fluctuations of $\lambda$ are short ranged (this is reasonable since $\lambda^{ij}$ is a Lagrange multiplier and it should not have observable excitations at large distances). Then at scales $\ell$ larger that $\xi_p$ one can replace $\lambda^{ij}$ by its expectation value in (6.1) and obtain an effective action for the membrane, which depends only on $\vec{X}$ (representing the coarse grained position of the membrane) and $g_{ij}$ (corresponding to some coarse grained density of the membrane)

$$S_{\mathrm{eff}}(X, g_{ij}) = s_0 \int \sqrt{g} + \langle\lambda\rangle \int \sqrt{g}g^{ij}\partial_i\vec{X}\partial_j\vec{X}, \qquad (6.3)$$

where $s_0 \sim r_0 - 2\langle\lambda\rangle$ and the curvature energy term has been neglected since it is now irrelevant in the infrared (this is not the case at short distance with the action (6.1) since $g_{ij}$ is not independent of $\vec{X}$). Such an effective action describes a Gaussian surface coupled to a fluctuating intrinsic metric $g_{ij}$. This model has been first introduced in the context of string theories[8] and extensively studied after the work of A. Polyakov. We shall discuss its main properties.

### 6.2. *The Liouville Model*

The simplest way to study Polyakov's string model is to choose a conformal gauge where the internal metric writes

$$g_{ij} = \delta_{ij}e^{\phi}, \qquad (6.4)$$

where $\phi$ is the conformal field. Then one integrates out the fluctuations of the $\vec{X}$ field and also take into account the Faddeev–Popov determinant introduced by the gauge fixing, which is in this case non trivial. As seen in Sec. 3, the integration of the $\vec{X}$ field gives nothing but $\det(-\Delta)^{d/2}$, where $\Delta$ is the scalar Laplacian in the metric $g$. Provided that *the fluctuations of the conformal field $\phi$ are not too large*, in such a way that the fluctuations of the total curvature are much smaller than the short distance regulator of the effective theory (represented by $\xi_p$)

$$\langle \mathbf{R}^2 \rangle \sim \langle (\Delta \phi)^2 \rangle \ll \xi_p^{-4}, \tag{6.5}$$

this determinant is given by the Liouville term. The Faddeev–Popov determinant may be in fact computed by using the same technics and is found to be also proportional to the Liouville term, but with a different numerical factor. One gets an effective action for $\phi$ which is[8]

$$S_{\text{eff}}(\phi) = \frac{26 - d}{48\pi} \int d^2\sigma \left[ \frac{(\partial \phi)^2}{2} + \mu^2 e^\phi \right], \tag{6.6}$$

where $\mu^2$ is a "mass" term which represents the effect of the surface tension $r_0$. In fact

$$\frac{26 - d}{48\pi} \mu^2 \sim r_0 - r_0^{\text{crit}}, \tag{6.7}$$

where $r_0$ is the bare surface tension we started from and $r_0^{\text{crit}}$ the critical value of the surface tension where the mean area diverges. The action (6.6) has many interesting features:

For $d$ large the kinetic term $(\partial \phi)^2$ has negative sign, with in fact the same coefficient than the one obtained in the large $d$ calculation.

For $r_0 > r_0^{\text{crit}}$ the classical extremum of the action obeys the Liouville equation

$$-\partial^2 \phi + \mu^2 e^\phi = 0, \tag{6.8}$$

which can be solved exactly. For $d$ large ($\mu^2 < 0$) it turns out that the corresponding metric $g_{ij} = \delta_{ij} e^\phi$ has a constant positive curvature $R$ which scales as

$$R \sim \frac{1}{r_0 - r_0^{\text{crit}}}. \tag{6.9}$$

Thus classical solutions describe closed surfaces (with the topology of the sphere) with an area which diverges as $r_0 \to r_0^{\text{crit}}$, as expected if there where no instability associated to the negative sign of the kinetic term. On the other hand for $d < 26$ the solutions describe surfaces with constant negative curvature and the physical interpretation of such solutions is not clear, indicating some trouble.

These problems become more apparent if fluctuations are taken into account. In fact the Liouville model is an integrable model which can be treated exactly.[29,30] The coupling constant $(26 - d)$ plays the role of a Boltzmann factor and the model turns out to make sense only in a low temperature phase[29,30] corresponding to

$$d < 1. \tag{6.10}$$

It is beyond the scope of these lectures to explain how the exact solution is obtained but let us give some results. For instance, the mean area of a closed surface with the topology of the sphere becomes singular at the critical point as

$$\text{Mean Area} = \left\langle \int d^2\sigma\sqrt{g} \right\rangle \simeq (r_0 - r_0^{\text{crit}})^{-\gamma} + \text{regular part}, \qquad (6.11)$$

with[31]

$$\gamma = \frac{1}{12}\left[(d-1) - \sqrt{(1-d)(25-d)}\right]. \qquad (6.12)$$

In the weak coupling phased $d \leq 1, \gamma$ is negative and thus the mean area stays finite (and of the order of $\xi_p^2$) at the critical point. The regime $d > 25$, where formally $1 \leq \gamma \leq 2$, corresponds in fact to the regime where the kinetic term for $\phi$ is negative and where strong instabilities in the conformal factor take place. In the intermediate regime $1 < d < 25$ the real part of $\gamma$ is $> 0$ but $\gamma$ is complex, indicating also some kind of instability. In fact, various arguments indicate[31,32] that for $d > 1$, the assumption that fluctuations of the intrinsic metric at the scale of the short distance cut-off in the Polyakov string model are small breaks down. If this is the case, the effective theory governing fluctuations of the surface at large scales is not the Liouville action but belongs to some different universality class. More insight is provided by the study of discretized models.

## 6.3. Discretized Models for Surfaces

Discretized model of random surfaces are described in J. Fröhlich's lectures and we shall discuss them only briefly.

A first class of models is provided by models of non-self avoiding planar surfaces made of plaquettes on some hypercubic lattice $\mathbf{Z}^d$ in $\mathbf{R}^d$.[33] The lattice size $a$ corresponds then to the persistence length and the action is taken to be equal to the area of the surface (number of plaquettes). (More realistic models of self-avoiding surfaces made out of plaquettes are discussed in Ref. 34.) It has been shown[35] that, under some reasonable scaling assumptions, for planar surfaces, if the exponent $\gamma$ defined by (6.11) is positive (i.e. if the mean area diverges at the critical point), then at the critical point the dominant contributions are "fingered surfaces" corresponding to branched polymers. The corresponding critical exponent $\gamma$ is $\gamma = 1/2$ and the Hausdorff dimension of the surface is $d_H = 4$. Numerical experiments indicate that in 3 and 4 dimensions that is indeed what happens. The inclusion of bending rigidity does not change this result in a large class of models.[36] An intuitive reason for such a "collapse" into branched polymers is the following. If the area diverges ($\gamma > 0$) it is easier, for entropic reasons, for the surface to increase its area by gluing together small blobs (with typical size of the order of the cut-off $a$) than by making larger blobs by some kind of global rescaling.

A second class of models is provided by discretized models of the Polyakov action (6.3), where the functional sum over continuous metrics $g_{ij}$ is replaced by a sum over discretized metrics provided by random triangulations.[37] Then the short distance cut-off $a$ is given by the intrinsic length of the links between neighboring points of the random triangulation, which is kept fixed. The embedding in the continuous $d$-dimensional Euclidean space is realized by a Gaussian action (spring model) which is a discretized version of the Gaussian action $\int d^2\sigma \sqrt{g} g^{ij} \partial_i \vec{X} \partial_j \vec{X}$ for fixed metric $g_{ij}$. Such models may be extended to continuous real values of the space dimension $d$ by simple analytic continuation.

The partition function of these models may be computed exactly for some special values of the dimension ($d = -2, 0, 1/2$ and 1 up to now),[37] using combinatoric methods or an equivalence with some random matrix problems. The results, in particular for the exponent $\gamma$, are in perfect agreement with the analytic predictions of the Liouville model. This is particularly interesting, since this is an explicit case where the field theoretical functional integral over two-dimensional metrics is shown to give the same results than discrete models in their continuum limit.

For $d > 1$, no analytical results have been obtained so far but numerical results (from Monte Carlo simulations and high temperature series) indicate that, at least for not too large values of $d$, $\gamma$ is positive and that the surface may look like a branched polymer (with $\gamma = 1/2$ and $d_4 = 4$), although large finite size effects make the interpretation of the numerical results delicate.

In any case, those results are consistent with the conjecture that for physical dimensions $d > 2$, at scales larger than the persistence length $\xi_p$, and if no change of topology or self-avoidance effects are included, fluid membranes undergo for small enough surface tension (or large enough area) a fingering transition and become some kind of branched objects. The transition observed in the large $d$ limit is probably a precursor of this phenomenon.

When topology changes are allowed, self-avoidance must be included in the models. It is indeed a general feature of the discrete models studied insofar that if no self-avoidance constraint is included, the sum over topologies diverges (proliferation of handles). One expects that when the surface starts to become a branched objects, topological fluctuations also occur. The nature of the resulting phase must depend in particular of the value of the Gaussian rigidity $\bar{\kappa}$. A general and interesting discussion of the possible phases for a system of self avoiding fluid membranes may be found in S. Leibler's lectures.[38]

## 7. Hexatic Membranes

As discussed in D. Nelson's lectures, membranes which sustain an internal order may exhibit a different behavior than fluid membranes. For a *flat* two-dimensional system made of (isotropic) molecules, the KTNHY theory predicts three possible phases: A low temperature crystalline phase, where the molecules form a regular triangular lattice, and characterized by quasi-long range positional order, an intermediate hexatic phase where translational order is lost but where a quasi-long range orientational order of bonds between nearest neighbors exists and a high temperature liquid phase with no translational or orientational order. The KTNHY theory predicts two successive Kosterlitz–Thouless-like transitions separating these three phases, driven respectively by the liberation of dislocations and the liberation of disclinations.

If the two-dimensional membrane is now free to move in the transversal directions, as described in Ref. 39, when a dislocation (defect of translational order) is present, the membrane will "buckle" in order to reduce the elastic free energy of the dislocation. The total elastic bending free energy of a buckled dislocation will be in general lower than that of a flat dislocation and there are good numerical indications that it is in fact finite.[40] This indicates that, if dislocations are allowed to appear in the system, they will be liberated for arbitrary low temperature and that the hexatic phase is the natural low temperature phase for membranes. The situation is different for polymerized membranes, which can sustain a low temperature crystalline phase.[39] We now describe a field-theoretical model for hexatic membranes.

### 7.1. *Hexatic Membranes: Continuous Model*

In order to construct a field theoretical model[41] describing large distance properties of hexatic membranes, one starts from the model for fluid membranes described in Sec. 4. A hexatic membrane, although fluid, will sustain a local orientational order (defined modulo rotations by $\pi/3$), which is described by a local order parameter $\vec{N}(\sigma)$. It is a unit vector tangent to the membrane, which may be written in terms of two components $N^i$ in the frame of the tangent vectors $\partial_i \vec{X}$:

$$\vec{N}(\sigma) = N^i(\sigma)\partial_i \vec{X}(\sigma), \qquad N^2 = N^i N^j g_{ij} = 1. \tag{7.1}$$

For a flat membrane the hexatic free energy corresponds to the usual $XY$ model

$$H_{\text{hex}} = \frac{K_A}{2} \int d^2\sigma \partial_i \vec{N} \partial_i \vec{N}. \tag{7.2}$$

Using the fact that

$$D_i \vec{N} = (\partial_i \vec{N})_{\parallel} = (D_i N^j)\partial_j \vec{X} = \partial_i \vec{N} - N^j \vec{K}_{ij} \tag{7.3}$$

is the tangential component of $\partial_i \vec{N}$, one can show that the only possible term for the hexatic free energy $H_{\text{hex}}$, which cannot be recast into the usual free energy for fluid surfaces

$$H_{\text{fluid}}(\vec{X}) = r \int d^2\sigma\sqrt{g} + \frac{\kappa}{2}\int d^2\sigma\sqrt{g}(\vec{K}_i^i \vec{K}_i^j) + \frac{\bar{\kappa}}{2}\int d^2\sigma\sqrt{g}R \tag{7.4}$$

is

$$H_{\text{hex}}(\vec{X}, \vec{N}) = \frac{K_A}{2}\int d^2\sigma\sqrt{g}D_i \vec{N} D^i \vec{N}. \tag{7.5}$$

The coupling constant $K_A$ is the *hexatic stiffness* and measures the strength of the coupling between the orientations of neighboring bonds.

In deriving this form of $H_{\text{hex}}$ we have neglected all terms which by power counting are irrelevant at large distance. Other terms which couple $\vec{N}$ to the principal

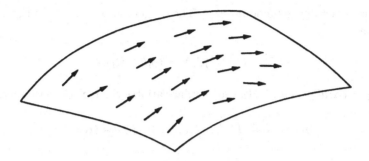

Fig. 7.1. Order parameter $N$ for hexatic membranes.

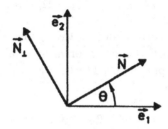

Fig. 7.2. $\vec{N}$ expressed in the frame $(\vec{e}_1, \vec{e}_2)$.

directions of curvature of the surface (such as: $N_a K_b^a K^{bc} N_c$) are not invariant under the hexatic symmetry (global rotation by $\pi/3$ of $\vec{N}$) and are irrelevant at large distance, so that the model has in fact a full $0(2)$ rotational symmetry. This is similar to the fact that a 2-dimensional crystal with hexagonal or triangular structure has isotropic elastic properties at large distances.

To describe $\vec{N}$ in term of an angular variable, it is convenient to associate to every point of the surface two tangent orthonormal vectors $\vec{e}_a$ ($a = 1, 2$). Locally it is possible to construct such a field of two vectors, called a 2-bein, in a continuous way. In components

$$\vec{e}_a = e_a^i \partial_i \vec{X} \tag{7.6}$$

and the orthonormality $\vec{e}_a \vec{e}_b = \delta_{ab}$ implies

$$e_a^i e_b^j g_{ij} = \delta_{ab} \quad \text{and} \quad e_a^i e_a^j = g^{ij}. \tag{7.7}$$

The covarient derivative of $\vec{e}_a$ in direction $i$ defines an angle $\Omega_i$. Under parallel transport in direction $d\sigma^i$, each $\vec{e}_a$ is rotated by an angle $\Omega_i d\sigma^i$. The vector field $\Omega_i$ is defined by

$$D_i \vec{e}_a = (\partial_i \vec{e}_a)_{\parallel} = -\Omega_i \epsilon_{ab} \vec{e}_b \tag{7.8}$$

and $\omega_{iab} = \Omega_i \epsilon_{ab}$ is called the "spin connection" and describes how $\vec{e}_a$ rotates under parallel transport. We can write $\vec{N}$ as

$$\vec{N}(\sigma) = \cos \theta(\sigma) \vec{e}_1 + \sin \theta(\sigma) \vec{e}_2 \tag{7.9}$$

(where $\theta$ is in fact defined modulo $\pi/3$). The tangential component of the derivative of $\vec{N}$ writes

$$D_i\vec{N} = (D_iN^j)\partial_j\vec{X} = (\partial_i\theta - \Omega_i)\vec{N}_\perp, \qquad (7.10)$$

where $\vec{N}_\perp = \cos\theta\vec{e}_2 - \sin\theta\vec{e}_1$. In the angular variable the hexatic action writes then

$$H_{\text{hex}} = \frac{K_A}{2}\int d^2\sigma\sqrt{g}(\partial_i\theta - \Omega_i)(\partial_i\theta - \Omega_b)g^{ij}. \qquad (7.11)$$

In this form, $H_{\text{hex}}$ is invariant under local $0(2)$ rotations

$$\theta \to \theta + \psi, \qquad \Omega_i \to \Omega_i - \partial_i\psi \qquad (7.12)$$

of the vector frame $\vec{e}_a$. $\Omega_i$ corresponds to the "gauge field" considered in Sec. 2 in the derivation of the Gauss–Bonnet theorem. It measures in fact the amount of frustration for parallel transport induced by the intrinsic curvature $R$ of the surface. Indeed, $\Omega_i$ is related to $R$ by

$$R = 2\gamma^{ij}D_i\Omega_j, \qquad (7.13)$$

as seen by computing $[D_i, D_j]\vec{e}_a$. If the surface is flat one can always choose a parallel 2-bein such that $\Omega_i = 0$ and $H_{\text{hex}}$ reduces to the free energy of $XY$ model. In the presence of local curvature $R$, one cannot make the hexatic energy vanish by making the hexatic order parameter $\vec{N}$ parallel. A classical configuration $\theta_o$ minimizing $H_{\text{hex}}$ satisfies the equation

$$-\Delta\theta_o + D^i\Omega_i = 0. \qquad (7.14)$$

One can check that

$$\partial_i\theta_o - \Omega_i = \left[D_i\frac{1}{\Delta_o}D^j - \partial_i^j\right]\Omega_j = \frac{1}{2}\epsilon_{ij}\frac{1}{\Delta}D^jR. \qquad (7.15)$$

It follows that the hexatic free energy of the classical configuration $\theta_o$ is

$$H_{\text{hex}}^o = H_{\text{hex}}(\theta_o) = \frac{K_A}{8}\int d^2\sigma\sqrt{g}(\sigma)\int d^2\sigma'\sqrt{g}(\sigma')R(\sigma)\left(\frac{1}{-\Delta}\right)_{\sigma\sigma'}R(\sigma'), \qquad (7.16)$$

where $(\frac{1}{-\Delta})_{\sigma\sigma'}$ is the massless scalar propagator in the intrinsic metric $g_{ij}$.

Thus the existence of an orientation order and of the hexatic stiffness $K_A$ induces long distance Coulombic interactions between intrinsic curvatures on the surface, analogous to the Coulombic interaction between vortices in the $XY$ mode. This is not a coincidence. Intrinsic curvature introduces frustration for the $\vec{N}$ vectors exactly as defects (disclinations here or vortices in the $XY$ model) do. It is exactly the same mechanism which generates both Coulombic interactions.

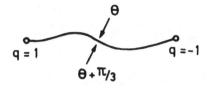

Fig. 7.3. A pair of disclinations may be represented by a line of discontinuity of $\theta$.

Now taking into account the fluctuations around $\theta_o$

$$\theta(\sigma) = \theta_o(\sigma) + \tilde{\theta}(\sigma) \tag{7.17}$$

the full hexatic energy writes

$$H_{\text{hex}}(\theta) = H_{\text{hex}}^o + \frac{K_A}{2} \int d^2\sigma \sqrt{g} g^{ij} \partial_i \tilde{\theta} \partial_j \tilde{\theta}. \tag{7.18}$$

Thus it reduces to the Coulombic interaction between intrinsic curvatures plus an ordinary $XY$ model on the surface for the fluctuations. The first part is nothing but the Liouville action $\frac{K_A}{4} S_{\text{Liouville}}[g_{ij}]$ discussed in Sec. 3.

, In the above derivation, we have been somewhat sketchy on two points. First the angle variable $\theta(\sigma)$ may not in general be defined globally on a surface but only locally. For a general surface with non zero Euler character $\chi$ disclinations must be introduced to define $\vec{N}$ globally, so that

$$\sum_{\text{disclinations}} (\text{charge of the disclination}) = 6\chi. \tag{7.19}$$

Pairs of disclinations may also be present, which will be described by lines of discontinuity of $\theta$ (by multiples of $\pi/3$). In the following renormalization group calculation we shall not treat the effect of disclinations, by assuming that they have a very small fugacity.

### 7.2. Hexatic Membranes: Renormalization Group Behavior

Neglecting disclinations, the fluctuations $\tilde{\theta}$ of $\theta$ may be treated as real field instead of a field with periodicity $\pi/3$. The difference will only give non-perturbative effects which will not affect the one loop calculation. The integration over $\tilde{\theta}$ is then Gaussian and reduces to the calculation of $\det(-\Delta)^{-1/2}$, which has been discussed in Sec. 3. The result, up to a shift in the surface tension $r$, is the Liouville action and is therefore *proportional* to $H_{\text{hex}}^o$. Thus the full intergration over the orientational degrees of freedom has been performed, $\vec{X}(\sigma)$ being fixed, with the final result

$$Z = \int \mathbf{D}[\vec{N}]\mathbf{D}[\vec{X}]e^{-\beta H_{\text{fluid}}(X)+H_{\text{hex}}(N)} = \int \mathbf{D}[\vec{X}]e^{-\beta H_{\text{eff}}(X)}, \qquad (7.20)$$

$$H_{\text{eff}}(\vec{X}) = r' \int \sqrt{g} + \frac{\kappa}{2} \int \sqrt{g}(\Delta\vec{X})^2$$
$$+ \frac{\bar{\kappa}}{2} \int \sqrt{g}R + \left(K_A - \frac{k_B T}{12\pi}\right)\frac{1}{8}S_{\text{Liouville}}\,[g]. \qquad (7.21)$$

We can now study how $\kappa, \bar{\kappa}$ and $K_A$ are renormalized by short wavelength fluctuations. The first result is that there is *no* renormalization of the hexatic stiffness. Indeed the same effective action could have been obtained by introducing on the membrane $N$ independent free fields $\phi_\alpha$ $(\alpha = 1, N)$, provided that $N = \left(1 - \frac{12\pi K_A}{k_B T}\right)$ (with a proper analytic continuation in $N$). The $\sum_\alpha \int \sqrt{g}(\partial\phi^\alpha)^2$ has an obvious $0(N)$ symmetry which cannot be broken by renormalization effects and neither $N$, nor $K_A$ are renormalized. The renormalization of $\kappa$ may be studied along the lines of Sec. 4 by taking into account the effect of $H_{\text{hex}}$ in the background field calculation. On general grounds one expects a new contribution to $\kappa(q)$ proportional to $\frac{K_A}{\kappa}k_B T \ln(q)$. The exact calculation gives[41]

$$\kappa_{\text{eff}} = \kappa + \frac{k_B T}{4\pi}\left(d - \frac{K_A}{\kappa}\frac{3}{4}\right)\ln\left(\frac{k_{\max}}{k_{\min}}\right). \qquad (7.22)$$

The corresponding one loop $\beta$ functions are

$$\beta_{K_A} = 0, \quad \beta_\kappa = \frac{k_B T}{4\pi}\left(d - \frac{3}{4}\frac{K_A}{\kappa}\right). \qquad (7.23)$$

For fixed hexatic stiffness $K_A$ there is an long distance (infra-red) attractive fixed point $\kappa^*$ at

$$\kappa^*(K_A) = \frac{3}{4d}K_A. \qquad (7.24)$$

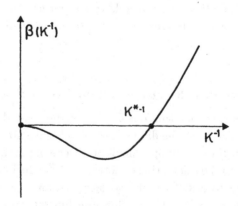

Fig. 7.4.   The $\beta$ function of the inverse bending rigidity $\alpha = \kappa^{-1}$.

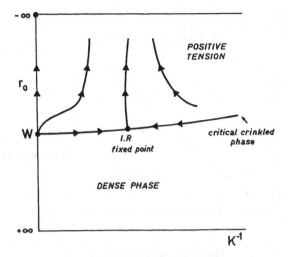

Fig. 7.5. Phase diagram and R.G. flow for hexatic membranes.

For large $\kappa > \kappa^*$, $\kappa_{\text{eff}}(q)$ decreases for small momenta and the membrane becomes less rigid. For small $\kappa < \kappa^*$, $\kappa_{\text{eff}}(q)$ increases and the membrane becomes more rigid. The surface has a finite bending rigidity $\kappa^*$ at large distance, at variance with fluid membranes which have very small rigidity of the order of $k_B T$.

The introduction of the surface tension $r$ does not change the result. The renormalization group flow in the $(r, \kappa^{-1})$ plane for fixed $K_A$ is depicted in Fig. 7.5. The I.R. fixed point has now one unstable direction and governs the critical behavior of hexatic surfaces with infinite area.

The existence of the I.R. stable fixed point has very important consequences for the structure of the surface, which can be obtained by working out the renormalization group equations satisfied by various correlation functions of the model.[41] The membrane is in a "crinkled phase" intermediate between a crumpled phase and a rigid phase. The correlation between normals does not decay exponentially but with a power law

$$\langle \vec{n}(0) \vec{n}(r) \rangle \sim |r|^{-\eta} \tag{7.25}$$

and consequently the persistence length is infinite, while the physical surface tension vanishes. Thus the crinkled phase is characterized by quasi-long range order between normals. The fractal dimension of the surface is larger than two. The one loop renormalization group calculation gives

$$\eta = \frac{2}{3\pi} d(d-2) \frac{k_B T}{K_A}, \tag{7.26}$$

$$d_F = 2 + \frac{d(d-2)}{3\pi} \frac{k_B T}{K_A}, \tag{7.27}$$

and the critical exponents vary continuously with the hexatic stiffness $K_A$. The R.G. flow on the critical surface with zero surface tension is shown on Fig. 7.6. Two

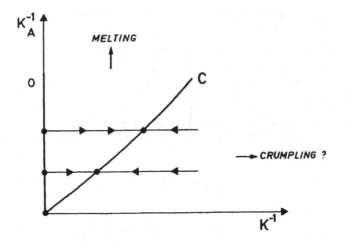

Fig. 7.6.   The lines of fixed points in the $\kappa, K_A$ plane.

lines of fixed points merge at $K_A^{-1} = \kappa^{-1} = 0$. The marginally unstable one 0 at $\kappa^{-1} = 0$ describes $\vec{N}$ vectors on a flat (infinitely rigid) surface and is nothing but the line of Gaussian fixed points of the $XY$ model. The line $L$ of I.R. stable fixed points describes the crinkled phase. The picture is expected to be valid for large bending rigidity *and* hexatic stiffness.

Up to now, we have not considered the effect of disclinations, namely of defects for the orientational order parameter. We expect that at some small value of $K_A$ disclination will be liberated and the hexatic surface will melt. This is known to occur through a Kosterlitz–Thouless transition for $\kappa = \infty$ and the buckling of the membrane should make such a melting easier for finite $\kappa$. On the other hand for large $K_A$ but smaller $\kappa$ one may expect that the membrane will be crumpled. The study of the complete phase diagram and the possible interplay between liberation of disclinations, crumpling and buckling obviously deserve further study.

## 8.  Crystalline Membranes

Crystalline membranes, which are non-zero in plane shear and compression moduli, appear to behave in a very different way than fluid or hexatic membranes.[38] The elastic forces mediated by phonons enhance strongly the correlations between tangent planes and a low temperature flat phase where the membrane has a spontaneous orientation exists, separated from a high temperature crumpled phase by a finite temperature crumpling transition. The theory of this crumpling transition is discussed in detail in D. Nelson's and Y. Kantor's lectures and we shall only discuss here a few more technical aspects.

The basic property of an elastic membrane is that it admits a prefered reference frame associated with the internal crystal structure. The order parameter for an effective Hamiltonian will be simply the $D{\times}d$ matrix $\partial_i X^\mu(\sigma)$ describing the tangent plane to the surface ($D = 2$ is the internal dimension of the surface and $d$ the dimension of bulk space). The simplest isotropic effective Hamiltonian reads[42]

$$H = \int d^D \sigma \left[ \frac{\kappa}{2} (\partial^2 \vec{X})^2 + u (\partial_i \vec{X} \partial_j \vec{X})^2 + v(\partial_i \vec{X} \partial_i \vec{X})^2 + \frac{t}{2} (\partial_i \vec{X} \partial_i \vec{X}) \right] \qquad (8.1)$$

and is similar to the Landau–Ginsburg–Wilson Hamiltonian when identifying $\partial_i \vec{X}$ with the usual $\phi$ order parameter. In mean field, for $t > 0, \partial_i \vec{X} \partial_j \vec{X} = 0$ and the surface is crumpled. For $t < 0, \partial_i \vec{X} \partial_j \vec{X}$ has a spontaneous expectation value

$$\partial_i \vec{X} \partial_j \vec{X} = -\frac{t}{4(u + Dv)} \delta_{ij} \qquad (8.2)$$

and the surface is flat. Rescaling the coordinates $\sigma^i \to \alpha \sigma^i$ in such a way that $\partial_i \vec{X} \partial_j \vec{X} = \delta_{ij}$ we can write $H$ as[43]

$$H = \int d^D \sigma \left[ \frac{\kappa}{2} (\partial^2 \vec{X})^2 + \mu (u_{ij})^2 + \frac{\lambda}{2} (u_{ii})^2 \right], \qquad (8.3)$$

where

$$u_{ij} = \frac{1}{2} (\partial_i \vec{X} \partial_j \vec{X} - \delta_{ij}) \qquad (8.4)$$

appears as the strain tensor, which vanishes at equilibrium. $\mu$ and $\lambda$ are the Lamé elastic coefficients and $\kappa$ the bending rigidity. The rescaled Hamiltonian for the ordered low temperature phase is analogous to the linear sigma model describing the ordered phase of a spin system. The crumpling transition has been studied with the Hamiltonian (8.1) by a $\epsilon = 4 - D$ expansion.[42] One can also start from the Hamiltonian (8.3) describing the low temperature phase and try to construct an effective Hamiltonian at large distances.[43] In analogy with the passage from the linear to the non-linear $\sigma$ model, one expects that the effective elastic constants become larger and larger at large distances and that one may consider only the non-linear version of (8.3) obtained by enforcing the constraint $\partial_i \vec{X} \partial_j \vec{X} = g_{ij}$ with the help of a Lagrange multiplier $\lambda^{ij}$

$$H = \int d^D \sigma \left[ \frac{\kappa}{2} (\partial^2 \vec{X}) + \lambda^{ij} (\partial_i \vec{X} \partial_j \vec{X} - \delta_{ij}) \right]. \qquad (8.5)$$

For the non-linear $\sigma$ mode, one loop perturbation theory tells us that the model is asymptotically free, that the effective spin stiffness decreases at large distance and that in two dimensions a spin system with compact continuous symmetry is always disordered a finite temperature. For the model (8.5) we are in troubles because it does not have a sensible low temperature perturbative expansion. The reason is the following. In the flat phase we expect that (8.5) describes $d-2$ transverse undulation modes, but the constraint leaves only $d - 3$ independent degrees of freedom. This is an indication that additional constraints are present for infinite elastic constants, which could stabilize the flat phase even in $D = 2$ dimensions. This is corroborated if we try to apply the standard argument about the internal energy of a disordered region of size $L$ inside an ordered phase. While for a spin system this energy scales as $L^{D-2}$, for a membrane the total bending and stretching energy scales as $L^D$. Hence even for $D = 2$ entropy does not gain over energy and cannot disorder the system at low temperature.

Although the model (8.5) cannot be studied in perturbation theory, it can be studied in a $1/d$ expansion, which is very similar to the $1/N$ expansion for $N$-components spin systems and to the $1/d$ expansion for fluid surface. Indeed integrating over the $\vec{X}$ fluctuations we get the effective action:

$$S_{\text{eff}}(\vec{X}, \lambda^{ij}) = \int d^2\sigma \left[ \frac{\kappa}{2} (\partial^2 \vec{X})^2 + \lambda^{ij}(\partial_i \vec{X} \partial_j \vec{X} - \delta_{ij}) \right]$$
$$+ \frac{d}{2} k_B T \, \text{Tr} \log \left[ \frac{\kappa}{2} (\partial^2)^2 - \partial_i \lambda^{ij} \partial_j \right]. \qquad (8.6)$$

Extremizing this effective action leads to the $d = \infty$ solution. Then $\lambda^{ij}$ get a non-zero expectation value

$$\langle \lambda^{ij} \rangle = \Lambda^2 \frac{\kappa}{2} \exp \left( -\frac{8\pi\kappa}{dk_B T} \right) \delta_{ij} \qquad (8.7)$$

($\Lambda$ is the momentum regulator) and the membrane is crumbled for any positive temperature. However, things change when the first $1/d$ correction is taken into account. Indeed, the $\langle \vec{X} \cdot \vec{X} \rangle$ propagator involves at one loop the diagram ⌢⌢⌣ where ∿∿∿ represents the $\langle \lambda\lambda \rangle$ propagator, which is found to behave at large momenta as

$$\langle \lambda^{ij}(p) \lambda^{k\ell}(-p) \rangle \sim p^2 \qquad (8.8)$$

instead of $p^2/\ln p^2$ as this is the case for fluid membranes. As a consequence the 2 point function for $\vec{X}$ behaves at one loop as

$$\Gamma_{\vec{X}\vec{X}}(p) \sim \left[ \kappa + \frac{1}{8\pi} \ln(p^2) \right] p^4 - \frac{1}{d}\kappa(\ln p^2)p^4 \qquad (8.9)$$

(we have rescaled $\kappa \to d\kappa$) and the $\frac{1}{d}$ correction tends to increase the effective rigidity at large distance, while the $d = \infty$ contribution tends to decrease it. The final result is that the $\beta$ function for the inverse of the bending rigidity $\alpha = d/\kappa$ is

$$\beta(\alpha) = \frac{2}{d}\alpha - \frac{1}{4\pi}\alpha^2 + 0 \left( \frac{1}{d} \right) \qquad (8.10)$$

and has an ultraviolet fixed point at

$$\alpha^* = \frac{8\pi}{d} \quad \text{or} \quad \kappa^* = \frac{d^2}{8\pi}. \qquad (8.11)$$

See Fig. 8.1. Hence for large enough bare rigidity $\kappa$ the effective rigidity grows at large distance like

$$\kappa_{\text{eff}}(q) \sim q^{-2/d} \qquad (8.12)$$

and the surface is flat, with Hausdorff dimension $d_H = 2$, while for small rigidity $\kappa_{\text{eff}}(q)$ goes to zero and the surface is crumpled with $d_H = \infty$. These two regimes

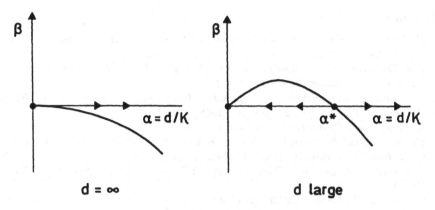

Fig. 8.1.   The $\beta$ function for $\alpha = \kappa^{-1}$ at $d = \infty$ and $d$ large but finite.

are separated by a continuous crumpling transition governed by the non-trivial fixed point $\kappa^*$. At that point the surface is a non-trivial object with Hausdorff dimension

$$d_H = \frac{2d}{d-1}. \tag{8.13}$$

Thus, this calculation provides strong evidence for the existence of a flat phase and of a finite temperature crumpling transition for two-dimensional elastic membranes. This would seem to contradict the Mermin–Wagner theorem,[44] which states that there is no breakdown of a continuous symmetry in two-dimensional systems, but this is not the case. Indeed, (8.12) means that the interactions between phonons and transverse undulations mediate long range forces which enhance correlations between normals to the surface. Indeed those correlations behave as

$$\langle \vec{n}\vec{n} \rangle \sim \frac{1}{\kappa_{\text{eff}}(q)q^2} = \frac{1}{q^{2-\eta}}, \qquad \eta = 2/d \tag{8.14}$$

and the Mermin–Wagner–Coleman theorem, which relies on the fact that for two dimensional spin systems $\eta = 0$ and that

$$\langle \vec{n}\vec{n} \rangle \sim \int d^2 q \frac{1}{q^2} \tag{8.15}$$

diverges at small $q$, does not apply. In fact our calculation provides an estimate of the lower critical dimension $D_{\ell c}$ of the membrane below which the crumpling transition occurs at zero temperature. $D_{\ell c}$ is less than 2 and is given by

$$D_{\ell c} - 2 + \eta = 0 \Rightarrow D_{\ell c} = 2 - 2/d + 0\left(\frac{1}{d}\right) \tag{8.16}$$

(corrections to $\eta$ for $D \neq 2$ may be neglected at the order in $1/d$).

The effective field theory describing the elastic properties of the flat phase has been studied within the $\epsilon = 4 - D$ expansion in Ref. 45. Recently a systematic study of the flat phase and of the crumpling transition for arbitrary $2 < D < 4$ has been carried out to order $1/d$ in Ref. 46. The effect of external stress on elastic membranes raises interesting questions which have been considered in Refs. 46 and 47.

208                                    F. David

# References

1. M. Spivak, *A Comprehensive Introduction to Differential Geometry* (Publish or Perish, Boston, 1970).
2. B.A. Dubrovin, A.T. Fomenko and S.P. Novikov, *Modern Geometry, Methods and Applications* (Springer, New York, 1984).
3. E. Kreyszig, *Introduction to Differential Geometry and Riemannian Geometry* (University of Toronto Press, Toronto, 1968).
4. J.P. Petit, *Le Géométricon* (Belin editor, Paris, 1980).
5. For many issues discussed in Chapters III to VI see: A.M. Polyakov, "Gauge Fields and Strings" (Harwood Academic Publishers, Chur, 1987).
6. S. Hawking, *Comm. Math. Phys.* **55**, 133 (1977). K. Fujikawa, *Phys. Rev.* **D23**, 2262 (1981).
7. O. Alvarez, *Nucl. Phys.* **B216**, 125 (1983).
8. A.M. Polyakov, *Phys. Lett.* **103B**, 207 (1981).
9. D. Friedan, in *Recent Advances in Field Theory and Statistical Mechanics*, Les Houches 1982, eds. J.B. Zuber and R. Stora (North-Holland, Amsterdam, 1984).
10. For recent reviews see J. Cardy, "Conformal Invariance" in Domb and Lebowitz, *Phase Transitions and Critical Phenomena*. Vol. 11 (Academic Press, New York, 1986).
11. P.B. Canham, *J. Theor. Biol.* **26**, 61 (1970). W. Helfrich, *Z. Naturforsch.* **28C**, 693 (1973).
12. A.M. Polyakov, *Nucl. Phys.* **B268**, 406 (1986).
13. L. Brink, P. Di Vecchia and P. Howe, *Phys. Lett.* **B65**, 471 (1976).
14. See for instance L. Fadeev and V. Popov, *Phys. Lett.* **25B**, 29 (1967) and V. Popov, *Functional Integrals in Quantum Field Theory and Statistical Physics* (D. Reidel, Dordrecht, Holland, 1983).
15. D. Förster, *Phys. Lett.* **114A**, 115 (1986).
16. L. Peliti and S. Leibler, *Phys. Rev. Lett.* **54**, 1690 (1985). The Monge representation was used previously in D.J. Wallace and R.K. Zia, *Phys. Rev. Lett.* **43**, 808 (1979).
17. For a general presentation of the effective action and functional tecnics see Ref. 5 and C. Itzykson and J.B. Zuber. "Quantum Field Theory" (McGraw-Hill, New-York, 1980).
18. H. Kleinert, *Phys. Lett.* **114A**, 263 (1986).
19. E. Guitter, unpublished.
20. P.G. de Gennes and C. Taupin, *J. Phys. Chem.* **86**, 2294 (1982).
21. W. Helfrich, *J. de Physique* **46**, 1263 (1985); **47**, 322 (1986).
22. D. Förster, *Europhys. Lett.* **4**, 65 (1987).
23. F. David, *Europhys. Lett.* **6**, 603 (1988).
24. F. David, *Europhys. Lett.* **2**, 577 (1986).
25. F. David and E. Guitter, *Europhys. Lett.* **3**, 1169 (1987) and *Nucl. Phys.* **B295**, 332 (1988).
26. P. Olesen and S.K. Yang, *Nucl. Phys.* **B283**, 73 (1987). F. Alonso and D. Espriu, *Nucl. Phys.* **B283**, 393 (1987). H. Kleinert, *Phys. Rev. Lett.* **58**, 1915 (1987).
27. E. Braaten, R.D. Pisarski and S.M. Tse, *Phys. Rev. Lett.* **58**, 93 (1987).
28. R.D. Pisarski, *Phys. Rev. Lett.* **58**, 1300 (1987), 2608(E) and *Phys. Rev.* **D38**, 578 (1988). The large d calculation is similar to the one of Ref. 25 but the conclusions of the stability analysis differ strongly.
29. T.L. Curthright and C.B. Thorn, *Phys. Rev. Lett.* **48**, 1309 (1982); *Ann. of Phys.* **147**, 365 (1983), **153**, 147 (1983).
30. J.L. Gervais and A. Neveu, *Nucl. Phys.* **B238**, 125 and 396 (1984).

31. V.G. Knizhnik, A.M. Polyakov and A.A. Zamolodchikov, *Mod. Phys. Lett.* **A3**, 819 (1988); F. David, *Mod. Phys. Lett.* **A3**, 1051 (1988).
32. M. Cates, preprint Cavendish Laboratory (1988).
33. For a review see J. Fröhlich in lecture Notes in Physics Vol. 216, Ed. L. Garrido (Springer, 1985).
34. M. Karowski and H.J. Thun, *Phys. Rev. Lett.* **54**, 2556 (1985).
35. B. Durhuus, J. Fröhlich and T. Jonsson, *Nucl. Phys.* **B240**, 453 (1984).
36. J. Fröhlich *et al.*, to appear.
37. For reviews see the contributions of A. Krzywicki, J. Ambjorn, V.A. Kazakov and I.K. Kostov in "Field Theory on the Lattice," proceedings of the Seillac conference 1987. *Nucl. Phys. B (Proc. Suppl.)* 4 (1988).
38. D. Huse and S. Leibler, *J. de Physique*.
39. D. Nelson and L. Peliti, *J. de Phys.* **48**, 1085 (1987).
40. S. Seung and D. Nelson, *Phys. Rev. A.* **38**, 1005 (1988).
41. F. David, E. Guitter and L. Peliti, *J. de Phys.* **48**, 2059 (1987).
42. M. Paczuski, M. Kardar and D.R. Nelson, *Phys. Rev. Lett.* **60**, 2638 (1988).
43. F. David and E. Guitter, *Europhys. Lett.* **5**, 709 (1988).
44. N.D. Mermin and H. Wagner, *Phys. Rev. Lett.* **17**, 1133 (1966).
45. J. Aronovitz and T. Lubensky, *Phys. Rev. Lett.* **60**, 2634 (1988).
46. J. Aronovitz, L. Golubovic and T. Lubensky, University of Pennsylvania preprint (1988).
47. E. Guitter, F. David, S. Leibler and L. Peliti, *Phys. Rev. Lett.* **61**, 2949 (1988) and *J. Physique* **50**, 1787 (1989).

## CHAPTER 8

## STATISTICAL MECHANICS OF SELF-AVOIDING CRUMPLED MANIFOLDS — PART I

Bertrand Duplantier

*Service de Physique Théorique, CEA/Saclay*
*F-91191 Gif-sur-Yvette, France*
*bertrand@spht.saclay.cea.fr*

In this chapter, we address the specific question of self-avoidance effects in membranes and surfaces. The latter are imagined to be reticulated i.e. tethered, and embedded in a solvent. The temperature is high enough so that the membrane is above the crumpling transition, hence in a crumpled state. This lecture studies the minimal Edwards model of self-avoiding crumpled surfaces, which are generalized to manifolds of internal dimension $D$, embedded in space of dimension $d$. An $\varepsilon$-expansion (with $\varepsilon = 4D - (2 - D)d$) can be devised about the upper critical dimension $d^* = 4D/(2 - D)$, above which self-avoidance becomes irrelevant. The specific problems arising in the renormalization of such a model are reviewed.

## 1. Continuum Model of Self-Avoiding Manifolds

### 1.1. *Edwards Model*

Before considering crumpled membranes, let us return for a while to the case of polymers, which are also tethered but one-dimensional objects. Polymers or self-avoiding (SA) walks are the subject of a vast literature.[1,2] A model of SA polymers of great methodological value has been proposed by S.F. Edwards[3] as early as 1965. The statistical weight of a configuration $\{\vec{r}\}$ reads

$$\mathcal{P}\{\vec{r}\} = \exp - \mathcal{A}\{\vec{r}\}$$

with a free energy or "action"

$$\mathcal{A}\{\vec{r}\} = \frac{1}{2} \int_0^S \mathrm{d}s \left(\frac{\mathrm{d}\vec{r}}{\mathrm{d}s}\right)^2 + \frac{1}{2}b \int_0^S \mathrm{d}s \int_0^S \mathrm{d}s' \delta^d[\vec{r}(s) - \vec{r}(s')] \qquad (1.1)$$

where $\vec{r}(s)$ is the configuration in d-dimensional space, with a curvilinear abscissa $0 \leq s \leq S$. $S$ is thus the "length", or "mass", or "Brownian area" of the polymer chain. The first term in (1.1) in the well-known Wiener measure describing a Brownian motion. In particular, for $b = 0$, it is easy to calculate the average end-to-end distance

$$R^2 = \langle [\vec{r}(S) - \vec{r}(0)]^2 \rangle = dS \qquad (1.2)$$

The second (excluded volume) term in (1.1) generates self-avoidance, since one can show (see below) that in the infinite chain limit $S \to \infty$, its effective strength goes to infinity.

This Edwards model is amenable to a rigorous treatment by a *direct* renormalization method,[2,4,5] which leads to a calculable $\varepsilon$-expansion ($\varepsilon = 4 - d$), about the upper critical dimension $d^* = 4$. The direct method deals with polymer partition functions only, which are renormalized in a systematic way, and avoids recourse to the well-known $n = 0$ limit of an $O(n)$ field theory. However, it must be stressed that the two methods are perfectly equivalent and that the correspondence between both is firmly established.[6,7] The interest of the direct method is its very flexibility; in particular it can be extended to polymers of arbitrary topologies.[5] It further paves the way to generalization to crumpled self-avoiding manifolds.[8-10]

The physical arguments leading to the generalization of model (1.1) for crumpled reticulated surfaces or manifolds is described in the first chapter of Nelson's lectures (section IIC). For a two-dimensional ($2D$) tethered surface, the continuum partition function based on a fixed microscopic triangulation can be expressed as the functional integral

$$\mathcal{Z}_0 = \int \mathcal{D}\vec{r}(\underline{\sigma}) \exp -\mathcal{A}_0\{\vec{r}\}$$

$$\mathcal{A}_0\{\vec{r}\} = \frac{1}{2} \int d^2\sigma \left[ \left( \frac{\partial \vec{r}}{\partial \sigma_1} \right)^2 + \left( \frac{\partial \vec{r}}{\partial \sigma_2} \right)^2 \right] \qquad (1.3)$$

For a "phantom" $2D$ continuum elastic network $\underline{\sigma} = (\sigma_1, \sigma_2)$ describes the parametrization of the surface, assumed here to be flat. In presence of self-avoidance the total partition function becomes

$$\mathcal{Z} = \int \mathcal{D}\vec{r}(\underline{\sigma}) \exp - A\{\vec{r}\} \qquad (1.4)$$

with

$$\mathcal{A}\{\vec{r}\} = \mathcal{A}_0\{\vec{r}\} + \frac{b}{2} \int d^2\sigma \, d^2\sigma' \delta^d[\vec{r}(\underline{\sigma}) - \vec{r}(\underline{\sigma}')] \qquad (1.5)$$

In absence of self-avoidance, the elastic free energy (1.3) can be shown to be the universal continuum limit of a large class of random reticulated surfaces.[11-13] However, the analytic proof is performed in the $d \to \infty$ limit, and for finite space dimension $d$, the belief that the same universality holds is based only on numerical simulations results,[14] as described in detail in Kantor's lectures.

When one introduces self-avoidance, the model (1.5) presents a serious mathematical difficulty: the perturbation theory involves the expansion parameter $b A^2$, up to logarithmic terms,[14] where $A$ is the internal area $A = \int d^2\sigma$ of the membrane. The expansion parameter thus grows without bound to infinity when the surface size increases for any embedding space dimension. Hence the upper critical dimension of self-avoidance for $2D$ crumpled membranes is *infinite* and one cannot devise an $\varepsilon$-expansion for treating, e.g., SA effects in $d = 3$. This led Kardar and Nelson,[8] and independently Aronowitz and Lubensky[9] (see also Ball[15]), to consider a modified model which interpolates between polymers and reticulated surfaces, and allows analytic progress.

One considers *manifolds* $\vec{r}(\mathbf{x})$ with a $D$-dimensional flat internal space $\mathcal{V}, \mathbf{x} \in \mathcal{V}$, embedded by $\vec{r}(\mathbf{x})$ in a $d$-dimensional external space. The associated action or free-energy is

$$\mathcal{A}_0\{\vec{r}\} = \frac{1}{2} \int d^D x \sum_{a=1}^{D} \left( \frac{\partial \vec{r}}{\partial x_a} \right)^2 \quad \text{(Gaussian manifold)} \qquad (1.6a)$$

and with self-avoidance

$$\mathcal{A} = \mathcal{A}_0 + \mathcal{A}_I \equiv \frac{1}{2} \int d^D x \sum_{a=1}^{D} \left( \frac{\partial \vec{r}}{\partial x_a} \right)^2 + \frac{b}{2} \int_{\mathcal{V} \times \mathcal{V}} d^D x\, d^D x'\, \delta^d[\vec{r}(\mathbf{x}) - \vec{r}(\mathbf{x}')] \qquad (1.6b)$$

where $\mathcal{V}$ is a bounded flat manifold of volume

$$|\mathcal{V}| \equiv X^D \qquad (1.7)$$

Note that we use units such that the elastic coupling constant $K$ of Nelson's lecture is set equal to one. Our expression (1.6) is *formally* identical to his rescaled free energy (1.29b) with $b \equiv vK^{d/2}$, except that we prefer to consider $\vec{r}$ as the actual distance in embedding space, and rescale the internal coordinates $\mathbf{x}$. Hence for us,[10] $\mathbf{x}$ has dimension $[L]^{2/(2-D)}$, where $L$ is a length scale. For polymers, $D = 1, s \sim [L]^2$ is the "Brownian area" $s$ of Eq. (1.1).[4]

It is very interesting to explore the behaviour of model (1.6) in the full $(d, D)$ plane.

## 1.2. *Gaussian D-Dimensional Manifold*

The action (1.6a) must be dimensionless, hence the distances or sizes $r$ in space of the Gaussian manifold must scale with respect to the internal linear size $X$ as

$$r \sim X^{(2-D)/2} \qquad (1.8)$$

A more precise calculation gives the average distance between any two points $\mathbf{x}_1, \mathbf{x}_2$ on a Gaussian $D$-manifold.[8,9] We introduce the generating function

$$\mathcal{Z}_0(\vec{k}, \mathbf{x}_1; -\vec{k}, \mathbf{x}_2) = \int D\vec{r}(\mathbf{x}) \exp\left(-\mathcal{A}_0\{\vec{r}\} + i\vec{k}.[\vec{r}(\mathbf{x}_1) - \vec{r}(\mathbf{x}_2)]\right) \qquad (1.9)$$

This functional integral in Gaussian variable $\vec{r}$ is immediately evaluated as

$$\mathcal{Z}_0(\vec{k}, \mathbf{x}_1; -\vec{k}, \mathbf{x}_2) = [\det(-\Delta)]^{-d/2} \exp\left(-\frac{1}{2}G(\mathbf{x}_1 - \mathbf{x}_2)k^2\right) \qquad (1.10)$$

where $G(\mathbf{x}_1 - \mathbf{x}_2)$ is the operatorial inverse of the $D$-dimensional Laplacian $\Delta$, namely the Green function

$$\Delta G(\mathbf{x}) = 2\delta^D(\mathbf{x})$$
$$G(\mathbf{x}) = [S_D(2 - D)/2]^{-1}|\mathbf{x}|^{2-D} \tag{1.11}$$

This is nothing but the Coulomb or Newtonian potential in the internal $D$-dimensional space. $S_D = 2\pi^{D/2}/\Gamma(D/2)$ is the unit sphere area in $D$ dimensions. The end-to-end squared distance is deduced from the generating function (1.9) (1.10) in a standard way as

$$\langle[\vec{r}(\mathbf{x}_1) - \vec{r}(\mathbf{x}_2)]^2\rangle = d\,G(\mathbf{x}_1 - \mathbf{x}_2) = \frac{2d}{S_D(2-D)}|\mathbf{x}_1 - \mathbf{x}_2|^{2-D} \tag{1.12}$$

From (1.12) we observe that a Gaussian $D$-manifold is a perfectly well defined *fractal* object, possessing scale invariance and governed by a critical exponent $\nu$

$$\nu = (2 - D)/2 \tag{1.13}$$

such that the average size $R$ between e.g. two points near opposite sides on the boundary scales like $R \sim X^\nu$. The fractal or Hausdorff dimension is accordingly

$$D_H \equiv D/\nu = 2D/(2 - D) \tag{1.14}$$

Note also that a two-dimensional Gaussian membrane has an infinite Hausdorff dimension[11] and accordingly the squared distance (1.12) grows only logarithmically[11,12,14]

$$\langle[\vec{r}(\mathbf{x}_1) - \vec{r}(\mathbf{x}_2)]^2\rangle_{D=2} = \frac{2d}{2\pi}\ln|\mathbf{x}_1 - \mathbf{x}_2| \tag{1.15}$$

It is important to notice, however, that the Green function (1.11), which is the Newtonian potential in infinite $D$-space, corresponds to an infinite Gaussian manifold, or equivalently to the neglect of boundary conditions in a finite one. Inclusion of proper boundary conditions (Dirichlet or Neumann ones corresponding to fixed or free boundary $\delta\mathcal{V}$ of $\mathcal{V}$ in $d$-space) would lead to a more complicated $G$ depending explicitly on the shape of $\partial\mathcal{V}$. We shall return to this question later.

Let us now turn to the interacting manifold as described by (1.6).

### 1.3. *Dimensional Analysis*

Let us perform the rescaling

$$\vec{r}(\mathbf{x}) = X^{(2-D)/2}\vec{\rho}(\mathbf{u})$$
$$\mathbf{x} = X\,\mathbf{u} \tag{1.16}$$

where $X$ is the linear size of the manifold such that $|\mathcal{V}| = X^D$. Due to our special choice of units for $\mathbf{x}$ and $X$, $\vec{\rho}$ is *dimensionless*. The free energy (1.6) can then be rewritten as

$$\mathcal{A}\{\vec{\rho}\} = \frac{1}{2}\int \mathrm{d}^D u \left(\frac{\partial \vec{\rho}}{\partial \mathbf{u}}\right)^2 + \frac{1}{2}bX^{2D-(2-D)d/2}\int_{\mathcal{V}'\times\mathcal{V}'} \mathrm{d}^D u\, \mathrm{d}^D u'\, \delta^d(\vec{\rho}(\mathbf{u}) - \vec{\rho}(\mathbf{u}'))$$

(1.6c)

where all variables are now *dimensionless* and where $\mathcal{V}' = X^{-D}\mathcal{V}$ is the rescaled manifold of unit volume $|\mathcal{V}'| \equiv 1$. We define the *dimensionless interaction parameter* $z$ (generalizing that introduced by Zimm *et al.* long ago in polymer physics[16])

$$z \equiv \left((2-D)\frac{S_D}{4\pi}\right)^{d/2} bX^{2D-(2-D)d/2}$$

(1.17)

We are interested in *large* manifolds $X \to \infty$. Hence a critical line[8,9]

$$d^*(D) = \frac{4D}{2-D}$$

(1.18)

appears in the $(d, D)$ plane (Fig. 1). For $d < d^*(D)$ the effective interaction parameter is $z \to \infty$ for $X \to \infty$, and self-avoidance is fully relevant; while for $d > d^*(D)$, $z \to 0$ when the size $X$ increases, yielding an effective ideal Gaussian manifold at large distances. For polymers $(D = 1)$ one recovers the standard upper critical dimension $d^* = 4$ of second order phase transitions.

A similar dimensional analysis reveals also the Flory approximation[8,9] to the size exponent $\nu$. One assumes positions in $d$-space to scale like $r \sim X^\nu$ in the free energy (1.6), while of course the internal dimensions scale as $x_a \sim X, a = 1, \ldots, D$. Then the Gaussian free energy (1.6a) scales like $\mathcal{A}_0 \sim X^{2\nu+D-2}$, while the interaction free energy in (1.6b) grows as $\mathcal{A}_I \sim bX^{2D-\nu d}$. The Flory result is then obtained

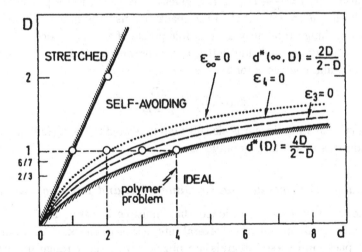

Fig. 1.  The various critical lines in the $(d, D)$ plane. Two-body self-avoidance is relevant for $D$ above the hatched line $d^*(D) = 4D/(2-D)$. The $\varepsilon_n = 0$ lines correspond to the onset of relevance for multicritical $n$-body interactions. For $D$ above the line $d^*(\infty, D) = 2D/(2-D)$, all $n$-body interactions are infrared relevant, but the theory is free from UV divergences.

simply by requiring both terms to be equivalent when $X \to \infty$, yielding

$$\nu_F = (D+2)/(d+2) \tag{1.19}$$

Note that although this approximation has no rigorous foundation, it works quite well numerically (see Kantor's lectures).

### 1.4. Higher Order Interactions

It is also quite interesting to consider possible higher order interaction terms and their relevance to the physics of crumpled manifolds. As in polymer physics,[17,18] one considers a local $n$-point interaction, whose Edwards free energy reads

$$\mathcal{A}_n = \frac{b_n}{n!} \int \prod_{i=1}^{n} d^D x_i \prod_{j=1}^{n-1} \delta^d(\vec{r}(\mathbf{x}_j) - \vec{r}(\mathbf{x}_n)) \tag{1.20}$$

and weights multiple $n$-point contacts in $d$ space. The relevance of such interactions depends crucially on $d$. An analytic way to determine the upper critical dimension $d^*(n)$ of $\mathcal{A}_n$ is to use the rescaling (1.16) in (1.20) to get the dimensionless $n$-point interaction parameter

$$z_n \sim b_n X^{Dn - (n-1)(2-D)d/2} \tag{1.21}$$

leading to a multicritical line $d^*(n, D)$ in the $(d, D)$ plane (Fig. 1), where $z_n \sim X^0$

$$d^*(n, D) = \frac{n}{n-1} \frac{2D}{2-D} \equiv \frac{n}{n-1} D_H \tag{1.22}$$

where $D_H$ is the Gaussian Hausdorff dimension (1.14). Note that a direct *geometric* derivation of $d^*(n) = \frac{n}{n-1} D_H$ can be given.[18] As usual, the $n$-point interaction is relevant for $d \leq d^*(n, D)$ i.e. when the space dimension is sufficiently low and leads to a non vanishing probability for multiple points of order $n$ to occur in a Gaussian manifold. The set of upper multicritical dimensions (1.22) is parallel to the largest one $d^*(2, D)$ and converges for $n \to \infty$ to

$$d^*(\infty, D) = \frac{2D}{2-D} \equiv D_H \tag{1.23}$$

This line is interesting since it is also intimately related to the ultraviolet divergences of the Edwards model for $D$-manifolds.

### 1.5. Analytical Continuation in Dimension and Regularization

Up to now, we have not mentioned that multiple integrals like (1.6b) or (1.20) can actually be ill defined when several interacting points $\mathbf{x}, \mathbf{x}', \ldots$ coincide in the internal space, and these integrals are plagued with short range ultraviolet (UV) divergences. It can be shown that UV divergences occur only for $d \geq \frac{2D}{2-D} \equiv D_H \equiv d^*(\infty)$. Note that this is also the dimension where all $n$-point interactions $\mathcal{A}_n$ (1.21) have the same effective strength (with respect to the Gaussian fixed point) $z_n \sim b_n X^D, \forall n \geq 2$. In the $(d, D)$ plane, the set of upper multicritical lines $(d^*(n, D), D)$ for $n \geq 2$ form a bundle of ordered lines (Fig. 1). Below the lowest one $(d > d^*(2, D) = 2D_H)$, self-avoidance is irrelevant and the manifolds are ideal. Above the highest one $(d < d^*(\infty, D) = D_H)$ the perturbation theory is ultraviolet convergent and requires no cut-off. This allows the use of dimensional regularization: one computes the partition function for $d < d^*(\infty, D)$ and continues the analytic expressions above $d^*(\infty, D) = D_H$ up to the multicritical line of interest e.g. for simple self-avoidance $d^*(2, D)$. As usual in critical phenomena,[19,20] one has to treat infrared divergences in the large size limit $X \to \infty$, and an $\varepsilon$ expansion about $d^*(2, D) = \frac{4D}{2-D}$ is performed[8,9] by setting

$$\varepsilon = 4D - (2 - D)d \qquad (1.24)$$

For a general multicritical point of order $n$, one should similarily expand the theory in

$$\varepsilon_n = 2n\, D - (n - 1)(2 - D)d$$

about the multicritical line $\varepsilon_n = 0$ (Fig. 1).

### 1.6. *Extension to Negative Dimensions*

First, let us notice that the line $d = D$ plays a special role[8,9] (Fig. 1). Indeed for such a small space dimension, one expects the manifold to be fully stretched by self-avoidance. One may wonder what happens in $d < D$. This leads us to leave the $d \geq 0, D \geq 0$ quarter of the plane and investigate the negative dimension behaviour. Note that for $D = 2$ (random surfaces) this is not totally unheard of. Indeed random surfaces have been considered in negative space dimensions $d$, and in particular are exactly solvable in $d = 0$ or $d = -2$.[21-23] However, in contrast to the present case of reticulated self-avoiding manifolds, the above random surfaces have a fluctuating metric and are liquid-like, without self-avoidance (see David's lectures). All the lines (1.18) (1.22) (1.23) extend nicely in the full $(d, D)$ plane (Fig. 2), where the asymptotes $D = 2$ and $d = -2$ play a central role. Notice that self-avoidance seems to be relevant in a hyperbolic strip located between the two branches of $d^* = 4D/(2 - D)$. For $D$ positive and $d$ negative enough and of the same order of magnitude, the manifold becomes ideal again. Of course the physical interpretation of these dimensions and of this resurgence of ideal manifolds remains unclear. Also notice that between points $(d, D) = (0, 0)$ and $(-2, -2)$ there seems to exist a domain where the manifold is stretched but ideal!

We leave these speculations here, and from now on we consider SA manifolds near the positive $d^* = \frac{4D}{2-D}$ line, and work in dimensional regularization.

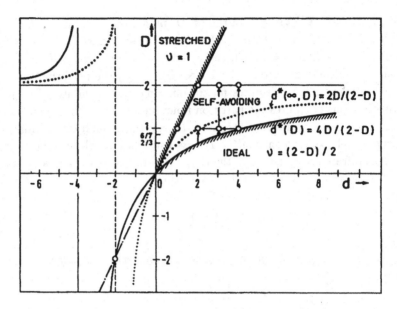

Fig. 2.   The extension of the critical lines of Fig. 1 to negative dimensions.

## 2. Perturbation Expansion

### 2.1. *Rules*

Quantities of interest are the partition functions

$$Z_1\left(\vec{k}, \mathbf{x}_1; -\vec{k}, \mathbf{x}_2\right) = \frac{1}{Z_0} \int D\vec{r}(\mathbf{x}) \exp\left(-\mathcal{A}\{\vec{r}\} + i\vec{k}.[\vec{r}(\mathbf{x}_1) - \vec{r}(\mathbf{x}_2)]\right) \qquad (2.1)$$

normalized by the partition function $Z_0$ of the Gaussian manifold. For $\vec{k} = \vec{0}$, one gets the normalized partition function of the interacting manifold

$$Z_1 \equiv Z/Z_0 = \int D\vec{r}(\mathbf{x})\, e^{-\mathcal{A}\{\vec{r}\}}/Z_0 \qquad (2.2)$$

where $\mathcal{A}$ is the generalized Edwards action (1.6b). Notice that for $b = 0, Z_1 \equiv 1$. As usual, the mean squared end-to-end distance is generated by (2.1) as

$$\left\langle [\vec{r}(\mathbf{x}_1) - \vec{r}(\mathbf{x}_2)]^2 \right\rangle = -\left(\frac{\partial}{\partial \vec{k}}\right)^2 Z_1\left(\vec{k}, \mathbf{x}_1; -\vec{k}, \mathbf{x}_2\right)\Bigg|_{\vec{k}=\vec{0}} /Z_1 \qquad (2.3)$$

Functions (2.1) (2.2) can be calculated by perturbation expansions of the interacting part $\exp\left(-\mathcal{A}_I\right)$ in (1.6b) in power series of $b$. The rules are easily found by Fourier transforming each $\delta^d$ interaction as

$$\delta^d(\vec{r}(\mathbf{x}) - \vec{r}(\mathbf{x}')) = \int \frac{d^d k}{(2\pi)^d} \exp\left(i\vec{k}.[\vec{r}(\mathbf{x}) - \vec{r}(\mathbf{x}')]\right)$$

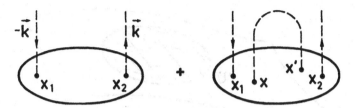

Fig. 3.   Perturbation expansion of $\mathcal{Z}_1(\vec{k}, \mathbf{x}_1; -\vec{k}, \mathbf{x}_2)$ (Eq. (2.5)) in powers of $b$. The dotted line carries an internal momentum, and Gaussian integration (2.4) leads to the interaction integral in (2.5).

as in polymer calculations,[2,4,5] and then averaging over Gaussian configurations (Fig. 3). The useful formula is the Gaussian integral

$$\left\langle \exp\left(i \sum_\alpha \vec{k}_\alpha \cdot \vec{r}(\mathbf{x}_\alpha)\right) \right\rangle_0 = \exp\left[\frac{1}{2} \sum_{\alpha<\beta} \vec{k}_\alpha \cdot \vec{k}_\beta G(\mathbf{x}_\alpha - \mathbf{x}_\beta)\right] \left(\sum_\alpha \vec{k}_\alpha = \vec{0}\right)$$

(2.4)

where the average is performed with weight (1.6a), and where $G$ is Green's function (1.11).

As a result, the single manifold generating function (2.1) reads to first order (Fig. 3).

$$\mathcal{Z}_1(\vec{k}, \mathbf{x}_1; -\vec{k}, \mathbf{x}_2)$$

$$= \exp\left[-\frac{1}{2}k^2 G(\mathbf{x}_1 - \mathbf{x}_2)\right]\left[1 - \frac{b}{2}(2\pi)^{-d/2}\int_{\mathcal{V}\times\mathcal{V}} \mathrm{d}^D x\, \mathrm{d}^D x'\, G^{-d/2}(\mathbf{x} - \mathbf{x}')\right.$$

$$\times \exp\left\{\frac{1}{2}\frac{k^2}{4}G^{-1}(\mathbf{x} - \mathbf{x}')[G(\mathbf{x}_1 - \mathbf{x}) + G(\mathbf{x}_2 - \mathbf{x}')\right.$$

$$\left.\left. - G(\mathbf{x}_1 - \mathbf{x}') - G(\mathbf{x}_2 - \mathbf{x})]^2\right\}\right]$$

(2.5)

From this, we deduce the partition function

$$\mathcal{Z}_1 \equiv \mathcal{Z}_1(\vec{0}, \mathbf{x}_1; \vec{0}, \mathbf{x}_2) = 1 - \frac{b}{2}(2\pi)^{-d/2}\int_{\mathcal{V}\times\mathcal{V}} \mathrm{d}^D x\, \mathrm{d}^D x'[G(\mathbf{x} - \mathbf{x}')]^{-d/2}$$   (2.6)

and the average squared distance (2.3)

$$\langle [\vec{r}(\mathbf{x}_1) - \vec{r}(\mathbf{x}_2)]^2 \rangle = dG(\mathbf{x}_1 - \mathbf{x}_2) + \frac{d}{4}\frac{b}{2}(2\pi)^{-d/2}\mathcal{I}$$   (2.7a)

$$\mathcal{I} = \int \mathrm{d}^D x\, \mathrm{d}^D x'[G(\mathbf{x} - \mathbf{x}')]^{-1-d/2}$$

$$\times [G(\mathbf{x}_1 - \mathbf{x}) - \mathbf{G}(\mathbf{x}_1 - \mathbf{x}') + \mathbf{G}(\mathbf{x}_2 - \mathbf{x}') - G(\mathbf{x}_2 - \mathbf{x})]^2$$

(2.7b)

As expected, we see that self-avoidance ($b > 0$) increases the distance between points on the manifold. As such, the expansions (2.6) (2.7) are not conclusive since

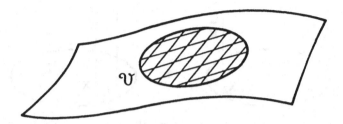

Fig. 4.   Self-avoiding "patch" $\mathcal{V}$ embedded in a larger elastic Gaussian manifold.

the expansion parameter grows like $z$ (1.17). Indeed we may rewrite for instance (2.6) in terms of dimensionless variables, using the explicit form (1.11) of $G$

$$\mathcal{Z}_1 = 1 - \frac{1}{2}z \int_{\mathcal{V}' \times \mathcal{V}'} \mathrm{d}^D x \mathrm{d}^D x' |\mathbf{x} - \mathbf{x}'|^{-(2-D)d/2} \qquad (2.8)$$

where $\mathcal{V}'$ is the rescaled internal space of unit volume, $|\mathcal{V}'| = |\mathcal{V}|X^{-D} \equiv 1$. (Note that here $\mathbf{x}$ and $\mathbf{x}'$ are *dimensionless* and correspond to the $\mathbf{u}$ parameters of Eq. (1.16). For simplicity we nevertheless use the same notation $\mathbf{x}$.) It is worth noticing here again that insisting on the infinite volume form (1.11) of the Coulomb kernel $G$ corresponds to taking a very large manifold, with self-avoidance operating only on a finite part or "patch" $\mathcal{V}$ of it (Fig. 4). This is really the system we are studying here for simplicity, as in Refs. 8, 9, 10.

A power law behaviour can be extracted from expansions like (2.7) (2.8), by performing a direct renormalization[8,9] (or a dimensional renormalization[24]) as in the case of polymers.[4] For this, we first need to extract the polar part of (2.7) (2.8) when $\varepsilon = 4D - (2 - D)d \to 0$.

The simplest integral to evaluate is not the partition function (2.8), as could be believed at first sight, but the squared distance (2.7). Indeed both are superficially logarithmically divergent but the actual UV divergence of $\mathcal{Z}_1$ (2.8) is of order $-D$ in $\mathbf{x}$ units, while that of (2.7) is only logarithmic.

## 2.2. Divergences

### 2.2.1. End-to-end Distance

It is convenient to define in the integrand of (2.7b) $\mathbf{y} = \mathbf{x} - \mathbf{x}'$ and notice that the divergence of (2.7b) occurs for $\mathbf{y} \to 0$. In this limit, we expand the integrand of $\mathcal{I}$ about $\mathbf{y} = \mathbf{0}$ and get

$$\mathcal{I} \cong \int \mathrm{d}^D y \, [G(\mathbf{y})]^{-1-d/2} \int \mathrm{d}^D x \, (\mathbf{y}.\nabla_{\mathbf{x}} U)^2$$

where

$$U \equiv G(\mathbf{x}_1 - \mathbf{x}) - G(\mathbf{x}_2 - \mathbf{x})$$

Taking advantage of the rotational invariance in internal $D$-space, we can average the last factor over the directions of $\mathbf{y}$ as

$$\langle (\mathbf{y} \cdot \nabla_{\mathbf{x}} U)^2 \rangle = \frac{1}{D} y^2 (\nabla_{\mathbf{x}} U)^2$$

Hence the leading divergence of $\mathcal{I}$ factorizes into two integrals

$$\mathcal{I} = \int d^D y \, [G(\mathbf{y})]^{-1-d/2} y^2/D \times \int d^D x \, (\nabla_{\mathbf{x}} U)^2 \qquad (2.9)$$

The second integral is the (finite) electrostatic energy of a moving unit charge located at $\mathbf{x}$, interacting with two $\pm 1$ unit charges located at $\mathbf{x}_1$ and $\mathbf{x}_2$. An elementary integration by parts, where one takes advantage of $\Delta G = 2\delta^D$, gives

$$\int d^D x \, (\nabla_{\mathbf{x}} U)^2 = 4[G(\mathbf{x}_1 - \mathbf{x}_2) - G(\mathbf{0})]$$

For $D < 2$, (and we shall restrict ourselves to this case) $G(\mathbf{0}) = 0$, and the energy is finite. The diverging part for $\varepsilon \to 0$ is the first integral in (2.9). It is easily evaluated using the explicit form (1.11) and the $D$–integral over the manifold

$$\int_{\mathcal{V}} d^D y \, y^{2-(2-D)(1+d/2)} \cong S_D \int_0^X dy \, y^{\varepsilon/2-1} = S_D X^{\varepsilon/2} \frac{2}{\varepsilon}$$

Collecting all these results into (2.7a), we find for any two points on the manifold

$$\langle [\vec{r}(\mathbf{x}_1) - \vec{r}(\mathbf{x}_2)]^2 \rangle = d \, G(\mathbf{x}_1 - \mathbf{x}_2) \mathcal{X}_0 \qquad (2.10)$$

where the swelling factor $\mathcal{X}_0$ reads[8,9]

$$\mathcal{X}_0 = 1 + \frac{z}{2\varepsilon}(2 - D) \frac{S_D^2}{D} \qquad (2.11)$$

Notice that the divergence in $1/\varepsilon$ does not depend on the location of points $\mathbf{x}_1, \mathbf{x}_2$ on the patch $\mathcal{V}$ and at this order is universal, depending only on internal dimension $D$. (For $D = 1, S_1 = 2$, and $\mathcal{X}_0 = 1 + 2z/\varepsilon$, recovering the result of direct renormalization for polymers[4]).

### 2.2.2. *Partition Function*

Let us now consider the first order expansion (2.8) of $\mathcal{Z}_1$. The actual short range ultraviolet divergence is (in $\mathbf{x}$ units) of order $\int d^D y \, y^{-(2-D)d/2} \sim y^{\varepsilon/2-D}$, hence stronger than logarithmic by a $(-D)$ power law. Notice that this UV divergence in (2.8) disappears for $D - (2 - D)d/2 > 0$, i.e. $d < d_H = \frac{2D}{2-D}$, in agreement with the discussion of part I above. This power law divergence of (2.8) requires a *regularization*. A cut-off procedure has been proposed in Refs. 8 and 9. As mentioned above, the dimensional regularization will be preferred here for actual calculations. It is interesting to discuss the relation between both regularizations. As noted in Ref. 8, in a cut-off theory, the most diverging contribution to partition function $\mathcal{Z}_1$ (2.2) is proportional to the volume of the manifold, the next ones come from the boundary, the edges, the corners, etc. For instance, let us rewrite the partition function (2.6) with an ultraviolet cut-off $x_0$ in the dimensional $\mathbf{x}$ variables. We have

$$
\begin{aligned}
\mathcal{Z}_1 &= 1 - \frac{b}{2}\left(\frac{S_D}{4\pi}(2-D)\right)^{d/2}\int_{\mathcal{V}\times\mathcal{V}}\frac{d^D x\, d^D x'}{|\mathbf{x}-\mathbf{x}'|^{(2-D)d/2}}\theta(|\mathbf{x}-\mathbf{x}'|-x_0)\\
&= 1 - \frac{b}{2}(S_D(2-D)/4\pi)^{d/2}S_D X^D\frac{x_0^{D-(2-D)d/2}}{D-(2-D)d/2} + \cdots
\end{aligned}
\tag{2.12}
$$

where we have retained *only the leading ultraviolet divergence*. In particular a sub-leading term proportional to $z$ also exists (as formally written in (2.8)), which we shall calculate later. We can rewrite this expression (2.12) as

$$
\mathcal{Z}_1 = 1 - \frac{1}{2}\frac{S_D}{D-(2-D)d/2}z_0(X/x_0)^D + \cdots
\tag{2.13}
$$

where $z_0 \equiv b x_0^{2D-(2-D)d/2}(S_D(2-D)/4\pi)^{d/2}$ is the dimensionless interaction parameter at the *cut-off length scale*, defined in complete analogy to $z$ (1.17). In (2.13) we observe the expected leading ultraviolet divergence proportional to the volume $X^D$ of the manifold. We can generalize this and write for $X/x_0$ large the dominant ultraviolet behaviour

$$
\mathcal{Z}_1 = \exp\left[\sum_{p\geq 0}f_{D-p}(X/x_0)^{D-p}\right]\times \text{regular terms}\quad X/x_0\to\infty
\tag{2.14}
$$

governed by successive volume, surface ... terms. The $f_{D-p}$ are dimensionless functions of $z_0$ above. For instance (2.13) yields $f_D$ to first order in $z_0$:

$$
f_D = -\frac{1}{2}\frac{S_D}{D-(2-D)d/2}z_0 + O(z_0^2)
$$

The *regular* terms should on the contrary depend on the manifold size through the infrared interaction parameter $z \sim X^{\epsilon/2}$. The upper bound of the sum over integers $p$ in (2.14) depends crucially on whether $D$ is integer or not. If $D$ is integer, a formal term $f_0 X^0$ or $f_0 \ln X$ occurs in the series and one can expect a power law behaviour:

$$
\mathcal{Z}_1 = \exp\left(\sum_{p=0}^{D-1}f_{D-p}X^{D-p}\right)X^{\gamma-1}\times\cdots
\tag{2.15}
$$

where $\gamma - 1 \equiv f_0$. If $D \neq$ integer, the $p$-series does not meet a logarithmic divergence, and for $X/x_0 \to \infty$, it ends at $p_{\max} = E(D)$ (integer part of $D$, $D \leq E(D) < D+1$). So it seems that a power law behaviour $\mathcal{Z}_1 \sim X^{\gamma-1}$ is expected only for a manifold of integer dimension $D$. Notice that some authors may not agree with this.[8,25] We shall return to this question when calculating explicitly $\gamma$ for some manifolds. However all geometrical cases calculated up to now agree with our statement, and give[9,10] $\gamma - 1 \equiv 0$ for $D \notin \mathbf{N}$.

Let us now turn to dimensional regularization, which avoids the use of a cut-off and simply extends analytically the "Feynman integrals" from $d < \frac{2D}{2-D}$, where they are UV convergent, to $d > d_H$. It is interesting to first note the following result of polymer theory $(D = 1)$. One can show[4] that the cut-off and the dimensionally regularized partition functions are related by

$$\mathcal{Z}_1 = \exp[f_1(z_0)X/x_0]\, \mathcal{Z}_{1\,\text{dim.reg.}}(z, d) \tag{2.16}$$

where $f_1(z_0)$ is a function of the dimensionless interaction parameter $z_0 = (2\pi)^{-d/2} \times bx_0^{2-d/2}$, associated with the cut-off. Note that this is in full agreement with (2.14) and that the regular part therein is simply the *dimensionally regularized* partition function, hence the extraction of short range UV divergences is equivalent to dimensional regularization. A similar relationship exists[18] for any polymer theory based on a generalized Edwards model (1.20) with higher order interactions. For a general $D$-dimensional manifold, we can expect a similar formal relation (valid for $X/x_0$ large),

$$\mathcal{Z}_1 = \exp\left[\sum_p f_{D-p}(X/x_0)^{D-p}\right] \mathcal{Z}_{1\,\text{dim.reg.}}(z, d, D) \tag{2.17}$$

where the $p$–sum ends at $E(D)$ for $D \notin \mathbf{N}$, and $D - 1$ for integer $D$. From now on we shall use only dimensional regularization, having in mind the above equivalence to a cut-off theory.


### 2.2.3. *Dimensional Regularization*

We want to calculate the integral in (2.8)

$$I_1 = \int_{\mathcal{V}' \times \mathcal{V}'} d^D x\, d^D x' |\mathbf{x} - \mathbf{x}'|^{-(2-D)d/2} \tag{2.18}$$

as a meromorphic function of $d$ and $D$. Notice that in earlier works, the manifold $\mathcal{V}'$ was taken to be a hypercube[9,10] or a hypersphere.[8,10] Different values of $I_1$ (and thus $\gamma$) were obtained and it was realized[9,10] that the result depends on the shape of the manifold $\mathcal{V}$. The calculation for a hypercube, using a cut-off, can be found in Ref. 9, while that for a hypersphere of *integer* dimension $D$ is performed with a cut-off in Ref. 8. Following Ref. 10, we present here the full calculation of $I_1$ for hyperspheres and hyperellipsoids in the $(d, D)$ plane, as well as for hypercubes.

We introduce the characteristic function of the manifold[10]

$$\phi(\mathbf{q}) = \int_{\mathcal{V}'} d^D x\, \epsilon^{i\mathbf{q}\cdot\mathbf{x}} \tag{2.19}$$

where the integral extends over the bounded domain $\mathcal{V}'$. The distribution

$$|\mathbf{x}|^{-\alpha} = A \int d^D q\, e^{i\mathbf{q}\cdot\mathbf{x}} |\mathbf{q}|^{\alpha-D} \tag{2.20}$$

where

$$A = 2^{-\alpha}\pi^{-D/2}\Gamma[(D-\alpha)/2]/\Gamma(\alpha/2) \tag{2.21}$$

is used, and it should be noted that treating integrals in the sense of distributions is equivalent to dimensional regularization. Setting in integral (2.18)

$$\alpha \equiv (2 - D)d/2 = 2D - \varepsilon/2 \qquad (2.22)$$

we obtain the Fourier representation of (2.18)

$$I_1 = A \int d^D q \phi(\mathbf{q}) \phi(-\mathbf{q}) q^{\alpha-D} \qquad (2.23)$$

The singularities in $I_1$ will come from high $q$ since $\alpha - D = D - \varepsilon/2$, and the asymptotic expansion of $\phi(\mathbf{q})$ for $q \to \infty$ will be relevant.

*Hyperellipsoid*
Take a hyperellipsoid of half-axes $\alpha_i, i = 1, \ldots, D$. The characteristic function is[10]

$$\phi(\mathbf{q}) = \left( \prod_{i=1}^{D} \alpha_i \right) (2\pi)^{D/2} \tilde{q}^{-D/2} J_{D/2}(\tilde{q}) \qquad (2.24)$$

where $\tilde{q}$ is the modulus of dilated vector $\tilde{q}_i = \alpha_i q_i, i = 1, \ldots, D$, and $J_{D/2}$ the standard Bessel function of index $D/2$. Using properties of Bessel functions, integral (2.23) can be performed analytically to give[10]

$$I_1 = \left\langle \left( \sum_i u_i^2 \alpha_i^{-2} \right)^{(\alpha-D)/2} \right\rangle \left( \prod_i \alpha_i \right)$$
$$\times \pi^{D/2} S_D \frac{1}{D-\alpha} \frac{\Gamma(1+D-\alpha)}{\Gamma(1+D-\alpha/2)\Gamma\left(1 - \frac{\alpha-D}{2}\right)} \qquad (2.25)$$

where $\alpha \equiv (2 - D)d/2$, and where $u_i$ describes the components of a unit vector in $\mathbf{R}^D$, the angle brackets denoting the angular average in the internal $D$ space. Notice that the volume of the hyperellipsoid is

$$|\mathcal{V}| = \left( \prod_i \alpha_i \right) \frac{S_D}{D} R^D = \left( \prod_i \alpha_i \right) V_{\text{sphere}}(R) \quad (R \text{ sphere radius})$$

Using identity (2.22) gives

$$I_1|_{\text{ell.}} = \left\langle \left( \sum_i u_i^2 \alpha_i^{-2} \right)^{D/2-\varepsilon/4} \right\rangle \left( \prod_i \alpha_i \right) I_1|_{\text{sph.}} \qquad (2.26)$$

$$I_1|_{\text{sph.}} = \pi^{D/2} S_D \frac{1}{-D+\varepsilon/2} \frac{1}{\Gamma(1+\varepsilon/4)} \frac{\Gamma(1-D+\varepsilon/2)}{\Gamma(1-D/2+\varepsilon/4)} \qquad (2.27)$$

This is an exact result, which can be analytically extended to non integer dimension $D$. In particular the angular average is well defined for integer $D$ and can be continued analytically. The result (2.26) depends *continuously* on the shape of the hyperellipsoid through the parameters $\alpha_i$. The possible poles when $\varepsilon \to 0$ lie in $I_1|_{\text{sph.}}$. Setting $\varepsilon = 0$ formally, we find

$$I_1|_{\text{sph.}}(d^*, D) = \pi^{D/2} S_D(-1/D) \frac{\Gamma(1-D)}{\Gamma(1-D/2)} \qquad (2.27\text{bis})$$

which has poles only at the *odd integers* $D = 1, 3, 5, \ldots = 2k + 1$, $k \in \mathbf{N}$. Notice that for negative integer dimensions $D$, $I_1$ has *no* pole. It is easy to extract the polar part of $I_1$

$$I_1 = 2 \frac{a\mathbf{D}}{\varepsilon} + \text{regular terms} \qquad (2.28)$$

with

$$a_D|_{\text{sph.}} = \begin{cases} \frac{S_D S_{D-1}}{(D-1)D} \frac{(-1)^{(D+1)/2}}{2^{D-1}} & D \in 2\mathbf{N} + 1 \\ 0 & \text{otherwise} \end{cases} \qquad (2.29)$$

and

$$a_D|_{\text{ell.}} = \left( \prod_i \alpha_i \right) \left\langle \left( \sum_i u_i^2 \alpha_i^{-2} \right)^{D/2} \right\rangle a_D|_{\text{sph.}} \qquad (2.30)$$

Since the divergences in $1/\varepsilon$ are intimately related to the existence of a non trivial configuration exponent $\gamma$,[4,8-10] we already observe in (2.27bis) the absence of any singularity for non integer dimensions $D$, and thus check the statement $\gamma \equiv 1$ for non integer $D$. The case of hyperspheres is even more peculiar since for even $D$ the amplitude vanishes. A purely geometrical derivation of (2.29) is actually possible[26] where one clearly sees this phenomenon.

*Hypercube*

Let us consider a hypercubic manifold[9] $\mathcal{V} = [O, X]^D$. A simple calculation of the integral $I_1$ has been given in Ref. 9. The result is

$$I_1 = \frac{2}{\varepsilon} a_D + \cdots$$

with a coefficient

$$a_D = \begin{cases} 2(-1)^D \frac{1}{\Gamma(D)}, & D \in \mathbf{N} \\ 0, & D \in \mathbf{R} \setminus \mathbf{N} \end{cases} \qquad (2.31)$$

This result was later criticized and it has been argued[25] that a complex amplitude $a_D$ could even be found in the case of a hypercube, depending on the way the critical line $d^*(D)$ is approached in the $(d, D)$ plane (see the second Ref. 8). We disagree with this statement, and in view of the possible confusion, we shall give another analytic calculation of $I_1|_{\text{hypercube}}$ in the full $(d, D)$ plane. We use again the

Fourier method[10] introduced above (Eqs. (2.19)–(2.23)). The characteristic function of a hypercube is now in Fourier space

$$|\phi(\mathbf{q})|^2 = \prod_{i=1}^{D} \left( \frac{4\sin^2 q_i X}{q_i^2} \right) \tag{2.32}$$

We now use the integral representation

$$q^{\alpha-D} = \frac{1}{\Gamma\left(\frac{D-\alpha}{2}\right)} \int_0^\infty d\lambda e^{-\lambda q^2} \lambda^{-1+(D-\alpha)/2}$$

to write $I_1 = A\mathcal{J}$, where $A$ is given by (2.21) and where

$$\mathcal{J} \equiv \int d^D q \, q^{\alpha-D} |\phi(\mathbf{q})|^2 = \int_0^\infty \frac{d\lambda}{\Gamma\left(\frac{D-\alpha}{2}\right)} \lambda^{-1+(D-\alpha)/2} \, F^D(\lambda) \tag{2.33}$$

with

$$F(\lambda) \equiv \int_{-\infty}^{+\infty} dq \, e^{-\lambda q^2} \frac{4\sin^2 qX}{q^2} \tag{2.34}$$

Note that (2.33) has now a meaning when $D$ is a continuous parameter. $F'(\lambda)$ is exactly calculable

$$F'(\lambda) = -4 \int_{-\infty}^{+\infty} dq \, e^{-\lambda q^2} \sin^2 qX = -2 \left(\frac{\pi}{\lambda}\right)^{1/2} \left(1 - e^{-X^2/\lambda}\right) \tag{2.35}$$

Note also that $F(0) = 4\pi X$. Hence

$$F(\lambda) = 4\pi X - 2 \int_0^\lambda d\alpha \left(\frac{\pi}{\alpha}\right)^{1/2} \left(1 - e^{-X^2/\alpha}\right) \tag{2.36}$$

We have now a well defined problem of analytical continuation of Mellin integral $\mathcal{J}$ (2.33), in terms of the variable $D - \alpha$, $D$ being considered as a parameter. $\mathcal{J}$ is a priori defined for $D - \alpha > 0$, and we want to continue it analytically to $\alpha \to 2D$. The possible divergences occur at the origin. One has to consider the Taylor expansion or here more generally the asymptotic expansion of $F^D(\lambda)$ for $\lambda \to 0$. We have explicitly

$$F(\lambda) \underset{\lambda \to 0}{=} 4\pi X - 4\pi^{1/2}\lambda^{1/2} + \frac{2}{X^2}\lambda^{3/2}e^{-X^2/\lambda} + O(\lambda^{5/2}e^{-X^2/\lambda}) \tag{2.37}$$

So we see that the Taylor expansion of $F$ ends at $O(\lambda^{1/2})$. This gives

$$F^D(\lambda) = (4\pi X)^D \left\{ 1 + \sum_{n=1}^\infty \frac{D(D-1)\cdots(D-n+1)}{n!} \left[ -\frac{1}{X}\left(\frac{\lambda}{\pi}\right)^{1/2} \right]^n \right\}$$
$$+ O(\lambda^{3/2}e^{-X^2/\lambda}) \tag{2.38}$$

The exponentially small terms play no role for the analytic continuation of $\mathcal{J}$. One has simply to look at the Taylor terms. $\mathcal{J}$ is analytically continued by subtracting from $F^D(\lambda)$ in (2.33) just enough Taylor terms of (2.38), in such a way that the integral converges. The number of these terms of course depends on the value of $D - \alpha$. The poles occur then only for

$$D - \alpha + n = 0, \quad n = 0, 1, 2, \ldots \tag{2.39}$$

On the "critical line" $\alpha = 2D$, this condition gives

$$D = n \in \mathbf{N} \tag{2.40}$$

hence $D$ must be an integer. In the other cases, the integral has poles at $\alpha = D + n$, but not on the "critical line". Therefore the contribution

$$a_D|_{\text{hypercube}} \equiv (2D - \alpha)I_1 = (2D - \alpha)2^{-\alpha}\pi^{-D/2}\frac{\Gamma\left(\frac{D-\alpha}{2}\right)}{\Gamma\left(\frac{\alpha}{2}\right)}\mathcal{J} \tag{2.41}$$

is identically

$$a_D = 0 \quad \text{for } \alpha = 2D, \quad D \notin \mathbf{N} \tag{2.42}$$

since $\mathcal{J}$ does not diverge at $\alpha = 2D$. This contradicts the result[25]

$$a_D = 2\frac{e^{\pi D i}}{\Gamma(D)}$$

supposed to be valid on the whole critical line.

Let us end the calculation of $\mathcal{J}$ for integer $D$, and $\alpha \to 2D$. The first Taylor term in (2.38) which is not subtracted for $\alpha < 2D$ and gives the next pole when $\alpha \to 2D$, is the term $n = D$ in (2.38). Hence the singular part of $\mathcal{J}$ reads

$$\mathcal{J}_{\text{sing}} = \int_0^1 d\lambda \frac{\lambda^{\frac{D-\alpha}{2}} - 1}{\Gamma\left(\frac{D-\alpha}{2}\right)}(4\pi)^D(-1)^D\left(\frac{\lambda}{\pi}\right)^{D/2} = \frac{1}{D - \frac{1}{2}\alpha}\frac{1}{\Gamma\left(\frac{D-\alpha}{2}\right)}4^D\pi^{D/2}(-1)^D \tag{2.43}$$

Inserting this value (2.43) into (2.41), we find for $\alpha = 2D$ the finite value

$$a_D = 2(-1)^D/\Gamma(D) \quad \alpha = 2D, \quad D \in \mathbf{N}$$

which is the original result of Ref. 9, valid for $D \in \mathbf{N}$ only. Otherwise (2.42) holds true.

Let us summarize the perturbation expansions obtained up to now. For the end-to-end distance (2.11) we found a swelling factor

$$\mathcal{X}_0 = 1 + z\frac{1}{\varepsilon}\frac{S_D^2}{2D}(2 - D) \tag{2.44}$$

and for the partition function of a single manifold

$$\mathcal{Z}_1 = \mathcal{Z}/\mathcal{Z}_0 = 1 - z\frac{1}{2}\frac{a_D}{\varepsilon} \tag{2.45}$$

where $a_D$ depends crucially on the *shape* of the internal space of parameters $\mathcal{V}$ and is given by (2.29) (2.30) for a hypersphere and a hyperellipsoid, and by (2.31) for a hypercube. Notice that in all cases $a_D$ *vanishes identically* on the real axis when $D$ *is not a positive integer*. Note finally that the extreme sensitivity of $\mathcal{Z}$ to the boundary $\partial \mathcal{V}$, shows that the choice of the free space Coulomb potential (1.11) is not harmless. If one were using the Coulomb kernel evaluated in a finite box $\mathcal{V}$, its modification by boundaries and edges would probably also reflect in the actual values of $a_D$.

## 3. Direct Renormalization

### 3.1. Scaling Functions

Renormalization amounts, in this direct method,[4,5,8–10] to substituting in scaling functions, a renormalized interaction parameter $g$ for the original dimensionless interaction parameter $z$ (1.17). Scaling functions are functions of $z$ which have a finite limit at the critical point, when the size of the manifold goes to infinity, i.e. $z \to \infty$.

We can take as a measure of the size of the manifold, the distance $R$ between two points separated in the internal metric by the typical length $X : |\mathbf{x}_1 - \mathbf{x}_2| = X$. Hence Eqs. (2.10) (2.11) give

$$R^2 = d\,\mathcal{X}_0(z)G(X) = d\,\mathcal{X}_0(z)\,\frac{2X^{2-D}}{S_D(2-D)} \qquad (3.1)$$

In the asymptotic critical limit, when the size $X$ diverges, we expect $R^2$ to scale like

$$R^2 \sim X^{2\nu}$$

This means that the swelling factor $\mathcal{X}_0(z)$ is itself expected to scale like

$$\mathcal{X}_0(z, d, D) \underset{z \to \infty}{\sim} z^{[2\nu - (2-D)]2/\varepsilon} \sim X^{2\nu - (2-D)} \qquad (3.2)$$

Similarly the normalized partition function $\mathcal{Z}_1$ should behave as

$$\mathcal{Z}_1(z, d, D) \underset{z \to \infty}{\sim} z^{(\gamma-1)2/\varepsilon} \sim X^{\gamma-1} \qquad (3.3)$$

Hence two examples of scaling functions are the logarithmic derivatives i.e. effective exponents

$$\sigma_0(z, d, D) \equiv X \frac{\partial}{\partial X} \ln \mathcal{X}_0 \bigg|_{b,d,D} = \frac{\varepsilon}{2} z \frac{\partial}{\partial z} \ln \mathcal{X}_0 \bigg|_{d,D} \qquad (3.4)$$

and

$$\sigma_1(z, d, D) \equiv X \frac{\partial}{\partial X} \ln \mathcal{Z}_1 \bigg|_{b,d,D} = \frac{\varepsilon}{2} z \frac{\partial}{\partial z} \ln \mathcal{Z}_1 \bigg|_{d,D} \qquad (3.5)$$

such that

$$\begin{aligned} \sigma_0(z \to \infty) &= 2\nu - (2 - D) \\ \sigma_1(z \to \infty) &= \gamma - 1 \end{aligned} \qquad (3.6)$$

## 3.2. *Second Virial Coefficient*

A possibility for the renormalized "coupling constant" is given in terms of the "dimensionless second virial coefficient", paralleling that of polymer theory.[4] One has to consider the two-manifold partition function

$$\mathcal{Z}_2 = \int \mathcal{D}\vec{r}_1(\mathbf{x}) \, \mathcal{D}\vec{r}_2(\mathbf{x}') \, e^{-\mathcal{A}\{\vec{r}_1, \vec{r}_2\}} / \mathcal{Z}_0^2 \tag{3.7}$$

where $\mathcal{A}\{\vec{r}_1, \vec{r}_2\}$ is the action (1.6) generalized to include two-body interactions between the two (identical) manifolds $\mathcal{V}_1$ and $\mathcal{V}_2$, with configurations $\vec{r}_1(\mathbf{x})$ and $\vec{r}_2(\mathbf{x}')$, described by a term

$$-b \int_{\mathcal{V}_1} d^D x \int_{\mathcal{V}_2} d^D x' \delta^d [\vec{r}_1(\mathbf{x}) - \vec{r}_2(\mathbf{x}')].$$

Then a standard exercise in statistical mechanics shows that the osmotic pressure $\Pi$ of a collection of manifolds in solution has the virial expansion[4]

$$\Pi\beta = \mathbf{C} - \frac{1}{2} \frac{\mathcal{Z}_{2,c}}{(\mathcal{Z}_1)^2} \mathbf{C}^2 + \cdots \tag{3.8}$$

where $\beta = 1/k_B T$ is the inverse temperature, $\mathbf{C}$ the number of manifolds per unit volume, and $\mathcal{Z}_{2,c}$ the *connected* part of the two-manifold partition function (3.7). It is then convenient to define the *dimensionless* second virial coefficient

$$g = -\frac{\mathcal{Z}_{2,c}}{(\mathcal{Z}_1)^2} (2\pi R^2/d)^{-d/2} \tag{3.9}$$

where $R^2$ is the manifold typical size (3.1). The parameter $g$ being dimensionless is a function of $z$ (and $d, D$) only.

We need the perturbation expansion of $\mathcal{Z}_{2,c}$, which is given by Feynman-like diagrams in Fig. 5:

$$\mathcal{Z}_{2,c} = -b|\mathcal{V}|^2 + b^2|\mathcal{V}|^2 (2\pi)^{-d/2} \int_{\mathcal{V}\times\mathcal{V}} d^D x \, d^D x' [G(\mathbf{x} - \mathbf{x}')]^{-d/2} + \frac{b^2}{2} (2\pi)^{-d/2}$$

$$\times \int_{\mathcal{V}_1} d^D x_1 d^D x_1' \int_{\mathcal{V}_2} d^D x_2 d^D x_2' \, [G(\mathbf{x}_1 - \mathbf{x}_1') + G(\mathbf{x}_2 - \mathbf{x}_2')]^{-d/2} \tag{3.10}$$

We have taken here for simplicity the two manifolds $\mathcal{V}_1$, $\mathcal{V}_2$ to be identical $\mathcal{V}_1 = \mathcal{V}_2 \equiv \mathcal{V}$, with the same volume $|\mathcal{V}| = X^D$. It is not difficult also to recognize in

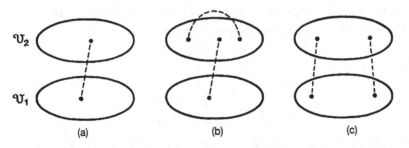

Fig. 5.   Diagrammatic expansion of $\mathcal{Z}_{2,c}$ (Eq. (3.10)).

the second term of (3.10) the self-interaction of a single manifold (compare Figs. 3 and 5, and Eq. (2.6)). This term is the "dangerous" one (2.18), which depends on the shape of the manifolds. But this self-interaction contribution drops out when $\mathcal{Z}_{2,c}$ is normalized by the product of the partition functions of the manifolds as defined in $g$ (3.9)

$$\frac{\mathcal{Z}_{2,c}}{(\mathcal{Z}_1)^2} = -b|\mathcal{V}|^2 \left(1 - \frac{b}{2}(2\pi)^{-d/2} \int_{\mathcal{V}_1} d^D x_1 d^D x_1' \right.$$
$$\left. \times \int_{\mathcal{V}_2} d^D x_2 d^D x_2' \left[G\left(\mathbf{x}_1 - \mathbf{x}_1'\right) + G\left(\mathbf{x}_2 - \mathbf{x}_2'\right)\right]^{-d/2}\right) \qquad (3.11)$$

As before, it is useful to compactify these expressions, by using dimensionless parameter $z$ (1.17) and Eq. (1.11)

$$\mathcal{Z}_{2,c}(\mathcal{Z}_1)^{-2} = -b \, X^{2D} \left(1 - \frac{1}{2}z \, I_2\right) \qquad (3.12)$$

where $I_2$ is the dimensionless integral

$$I_2 \equiv \int_{\mathcal{V}_1'} d^D x_1 d^D x_1' \int_{\mathcal{V}_2'} d^D x_2 d^D x_2' \left(|\mathbf{x}_1 - \mathbf{x}_1'|^{2-D} + |\mathbf{x}_2 - \mathbf{x}_2'|^{2-D}\right)^{-d/2} \qquad (3.13)$$

Using now definition (3.9) of $g$ and expressions (3.1) (3.12), one finds the expansion of $g$ in terms of $z$

$$g = z \left(1 - \frac{1}{2}z \, I_2\right) [\mathcal{X}_0(z)]^{-d/2} \qquad (3.14)$$

where $\mathcal{X}_0$ has the Taylor-Laurent expansion (2.44) in terms of $z$ and $\varepsilon$. It remains to extract the singularity of integral $I_2$ (3.13) near $\varepsilon = 0$. This is easily done for instance by using the integral representation

$$A^{-d/2} = \frac{2-D}{\Gamma(d/2)} \int_0^\infty d\eta \, \eta^{-1-(2-D)d/2} \exp(-A\eta^{D-2}) \qquad (3.15a)$$

and integrating over each manifold 1 and 2 separately. (See in section IV a calculation which contains the present one). The universal result, independent of the shapes of the manifolds, reads[8-10]

$$I_2 = \frac{2}{\varepsilon} \frac{S_D^2}{2-D} \frac{\Gamma^2[D/(2-D)]}{\Gamma(d/2)} \qquad (3.15b)$$

This holds for any *real* $D$, and $2D = (2-D)d/2 + \varepsilon/2$. Inserting this and (2.44) in (3.14) yields the Taylor-Laurent expansion of $g$

$$g = z - \frac{a_D'}{\varepsilon}z^2 + \cdots \qquad (3.16)$$

where

$$a_D' \equiv S_D^2 \left(1 + \frac{1}{2-D} \frac{\Gamma^2[D/(2-D)]}{\Gamma[2D/(2-D)]}\right) \qquad (3.17)$$

As in polymer theory, direct renormalization means two things. First, $g$ (3.9), which is a physically meaningful parameter (see (3.8)) should reach a *finite* fixed point limit when the size of the manifolds goes to infinity, i.e.

$$g(z \to \infty) = g^* < \infty.$$

Second, scaling functions like (3.4) and (3.5), which have a Taylor-Laurent expansion in $z$ and $\varepsilon$, become regular double Taylor series in $g$ and $\varepsilon$, when $g$ is substituted to $z$. This holds true to all orders in polymer theory[4-7] (and also of course in the equivalent field theoretic formalism of critical phenomena[19]). We have to assume here the consistency of this direct renormalization method for self-avoiding manifolds to all orders (see Part II). The Wilson function associated with $g$ is defined as

$$W(z,\varepsilon) \equiv X \frac{\partial}{\partial X}g = \frac{\varepsilon}{2}z\frac{\partial}{\partial z}g \qquad (3.18)$$

and should be also a *regular* function $W[g,\varepsilon]$ of $\varepsilon$, order by order in $g$. To first order, it reads

$$W(z,\varepsilon) = \frac{\varepsilon}{2}z - a_D'z^2 + O(z^3)$$

or in terms of $g$

$$W[g,\varepsilon] = \frac{\varepsilon}{2}g - \frac{1}{2}a_D'g^2 + O(g^3) \qquad (3.19)$$

In the asymptotic limit $X \to \infty$, $g \to g^*$, hence $X\partial g/\partial X = W[g^*,\varepsilon] = 0$. From the regular expansion (3.19) one therefore finds the fixed point value

$$g^* = \frac{\varepsilon}{a_D'} + O(\varepsilon^2) \qquad (3.20)$$

$$\varepsilon = 4D - (2 - D)d, \qquad \varepsilon > 0.$$

It is also interesting to calculate the asymptotic expression of $g$ at the upper critical dimension $d^* = 4D/(2 - D)$. Setting $\varepsilon = 0$ in (3.19) gives

$$X\frac{\partial g}{\partial X} = W[g,0] = -\frac{1}{2}a_D'g^2 + O(g^3)$$

which is readily integrated into

$$g = \frac{2}{a_D' \ln X'}, \qquad d^* = \frac{4D}{2 - D} \qquad (3.21)$$

In particular, for polymers $(D = 1)$, we have $a_1' = 8$, and recover $g^* = \varepsilon/8$,[4] or $g = 1/4\ln X$ in $d = 4$.[27]

### 3.3. *Critical Exponents*

The scaling functions $\sigma_0$ and $\sigma_1$ associated with the swelling of a single manifold by self-avoidance, and with the critical behavior of the partition function, are calculated from Eqs. (3.4) (3.5) and (2.44) (2.45):

$$\sigma_0(z) = \frac{1}{2}\frac{S_D^2}{2D}(2-D)z + \cdots$$

$$\sigma_1(z) = -\frac{1}{2}a_D z + \cdots$$

or in terms of $g$ (3.16)

$$\sigma_0[g,\varepsilon] = \frac{1}{2}\frac{S_D^2}{2D}(2-D)g + O(g^2)$$

$$\sigma_1[g,\varepsilon] = -\frac{1}{2}a_D g + O(g^2)$$

(3.22)

Note that, as always, the renormalization process is a triviality to first order. Only calculations to second order could demonstrate the subtle mechanism of suppressions of all $1/\varepsilon$ multiple poles in the double $(g,\varepsilon)$-expansion. Below $d^* = 2D_H$, i.e. for $\varepsilon > 0$, this gives the $\varepsilon$-expansion of critical exponents $\nu$ and $\gamma$ (Eq. (3.6)) to first order

$$\nu = \frac{2-D}{2} + \frac{1}{4}\frac{S_D^2}{2D}(2-D)g^*$$

$$\gamma - 1 = -\frac{1}{2}a_D g^*$$

(3.23)

The explicit values are

$$\nu = \frac{2-D}{2} + \frac{1}{8D}(2-D)\frac{\varepsilon}{1 + \frac{1}{2-D}\frac{\Gamma^2[D/(2-D)]}{\Gamma[2D/(2-D)]}}$$

(3.24)

$$\gamma - 1 = -\varepsilon a_D \frac{1}{2S_D^2}\left(1 + \frac{1}{2-D}\frac{\Gamma^2[D/(2-D)]}{\Gamma[2D/(2-D)]}\right)^{-1}$$

(3.25)

Recall that the coefficient $a_D$ depends crucially on the *shape* of the manifold $\mathcal{V}$ and is given by (2.29)–(2.31) for hyper-ellipsoids and -cubes respectively.

At the upper critical dimension $d = d^*$, we can also compute the swelling $\mathcal{X}_0$ and partition function $\mathcal{Z}_1$. It is sufficient[27] to set $\varepsilon = 0$ and use the asymptotic value (3.21) of $g$ in (3.22), whence

$$X\frac{\partial}{\partial X}\ln\mathcal{X}_0 = \frac{S_D^2}{2D}(2-D)\frac{1}{a_D'\ln X}$$

$$X\frac{\partial}{\partial X}\ln\mathcal{Z}_1 = \frac{-a_D}{a_D'\ln X}$$

which are integrated trivially into

$$\mathcal{X}_0 \sim (\ln X)^{S_D^2(2-D)/(2Da_D')}$$

(3.26a)

$$\mathcal{Z}_1 \sim (\ln X)^{-a_D/a_D'}$$

(3.26b)

Of special interest is the swelling with respect to the extension of the ideal manifold. We get from (3.1)

$$R^2 \sim \frac{8D}{S_D} \frac{X^{2-D}}{(2-D)^2} (\ln X)^{n(D)}, \quad d^* = 4D/(2-D)$$

$$n(D) = \frac{2-D}{2D} \left(1 + \frac{1}{2-D} \frac{\Gamma^2[D/(2-D)]}{\Gamma[2D/(2-D)]}\right)^{-1}$$
(3.27)

For $D = 1$, at $d^* = 4$ we recover the polymer value[27] $n(1) = 1/4$. For $D = 2$, at $d^* = \infty$, the logarithmic exponent $n(2)$ vanishes like $n(D) \sim (2-D)/4$. Hence a two-dimensional *surface* with self-avoidance, tends to recover its ideal behavior $(R^2 \sim \ln X)$ when embedded in space of infinite dimension, modified as

$$R^2 \sim \ln X \ln(\ln X)$$

## 4. Contact Exponents

An interesting problem in the statistical mechanics of self-avoiding manifolds is the determination of full renormalized correlation functions or probability distributions. Take for instance two points $x_1$ and $x_2$ on a SA manifold and ask about the probability to find them at a relative distance $\vec{r}$ in $d$-space

$$P(\vec{r}; x_1, x_2) = \langle \delta^d[\vec{r} - (\vec{r}(x_1) - \vec{r}(x_2))]\rangle$$
(4.1)

For a Gaussian manifold, this probability is simply the Fourier transform of $Z_0(\vec{k})$ (1.10) or (2.4)

$$P_0(\vec{r}; x_1, x_2) = \int \frac{d^d k}{(2\pi)^d} e^{-i\vec{k}.\vec{r}} \left\langle e^{i\vec{k}.[\vec{r}(x_1)-\vec{r}(x_2)]}\right\rangle_0$$

$$= [2\pi G(x_1 - x_2)]^{-d/2} \exp[-r^2/2G(x_1 - x_2)]$$
(4.1bis)

In the SA limit of a large manifold, one expects similarly this probability to scale as

$$P(\vec{r}; x_1, x_2) = R_{12}^{-d} F(r/R_{12})$$
(4.2)

where

$$R_{12} \equiv \langle [\vec{r}(x_1) - \vec{r}(x_2)]^2\rangle^{1/2}$$
(4.3)

and where $F$ is a universal function. In polymer theory, one knows that $F$ depends on the location of the points $x_1, x_2$ along the chain. Different behaviors exist, (labelled by $a = 0, 1, 2$), depending whether the two points are the extremities of the chain ($a = 0$), or one is an extremity, the other an interior point ($a = 1$), or finally both are interior points.[28-30] This results in three distinct universal functions $F_a, a = 0, 1, 2$. These functions have specific short and long range behaviors.[28-30] When $r \to \infty$, the function $F_a(x)$ decays exponentially as[28,31]

$$F_a(x) \underset{x \longrightarrow \infty}{\sim} x^{\sigma^a} \exp\left(-A_a x^{1/(1-\nu)}\right) \tag{4.4}$$

where the $\sigma^a$ are three universal exponents, and the $A_a$ are amplitudes. For the case $a = 0$, (interior-interior correlation) there is the additional result[32]

$$\sigma^0 = \frac{1 - \gamma + \nu d - d/2}{1 - \nu}$$

At short distance, two parts of a polymer (or of a SA manifold) repel each other and the probability of contact vanishes like

$$F_a(x) \underset{x \longrightarrow 0}{\sim} x^{\theta_a} \tag{4.5}$$

yielding similarly $P_a(r) \sim r^{\theta_a}$. The exponents $\theta_a, a = 0, 1, 2$, are (universal) contact exponents. Note that one can generalize them by considering higher order contacts, where the polymer folds onto itself several times to pass repeatedly at the same point.[29] For the end-to-end correlation again, $\theta_0$ is related to usual exponents $\gamma$ and $\nu$ by[28]

$$\theta_0 = (\gamma - 1)/\nu \tag{4.6}$$

For a self-avoiding *manifold*, we expect a straightforward generalization of these results. However, the Fisher exponent[31] $(1 - \nu)^{-1}$ in the exponential decay for *large* separations, does not extend as such to $D \neq 1$, as can be seen in the exact Gaussian results (4.1bis) and (1.13). At *short* distance, the existence of contact exponents in a manifold $\mathcal{V}$ as in (4.5) seems to be granted on physical grounds.[10] Here again the contact can exist between boundary points of $\mathcal{V}$ (Fig. 6a), a boundary and an interior point (Fig. 6b), or two interior points (Fig. 6c). However the physics is richer for manifolds.[10] When a boundary point $\mathbf{x}_1$ is involved, the contact exponent to another point will depend explicitly on the local shape of the boundary $\partial\mathcal{V}$ of $\mathcal{V}$ around $\mathbf{x}_1$ (and similarly, also of that of $\partial\mathcal{V}$ around $\mathbf{x}_2$ if $\mathbf{x}_2 \in \partial\mathcal{V}$). Let us now turn to the calculation of these $\theta$ exponents for a SA manifold.

We use a simple method which has been introduced in polymer physics.[2,5,29,33] It is not necessary to calculate the full two-point probability distribution of a single manifold, which would be more complicated, but to consider the probability of approach of two points $\mathbf{x}_1, \mathbf{x}_2$ on *distinct* manifolds (Fig. 7). Let us call $\mathcal{Z}_{\mathcal{S}_2}(\mathbf{x}_1, \mathbf{x}_2)$ the star-connected partition function of the manifolds $\mathcal{V}_1, \mathcal{V}_2$ (replicas of $\mathcal{V}$), glued together at the origin $\vec{r}_1(\mathbf{x}_1) = \vec{r}_2(\mathbf{x}_2) = \vec{0}$

$$\mathcal{Z}_{\mathcal{S}_2}(\mathbf{x}_1, \mathbf{x}_2) \equiv \int \mathcal{D}\vec{r}_1(\mathbf{x}) \mathcal{D}\vec{r}_2(\mathbf{x}') \, \delta^d[\vec{r}_1(\mathbf{x}_1)] \delta^d[\vec{r}_2(\mathbf{x}_2)] e^{-\mathcal{A}\{\vec{r}_1, \vec{r}_2\}} / (\mathcal{Z}_0)^2 \tag{4.7}$$

Notice a subtlety in the normalization: the two points $\mathbf{x}_1, \mathbf{x}_2$ are fixed at the origin, but since $\mathcal{Z}_0$ is the Gaussian partition function with one point fixed, $\mathcal{Z}_{\mathcal{S}_2}(\mathbf{x}_1, \mathbf{x}_2)$ is dimensionless, in contrast to $\mathcal{Z}_2$ (3.7) where the relative position of the manifolds was integrated over. $\mathcal{Z}_{\mathcal{S}_2}(\mathbf{x}_1, \mathbf{x}_2)$ is the partition function of a *double-star*

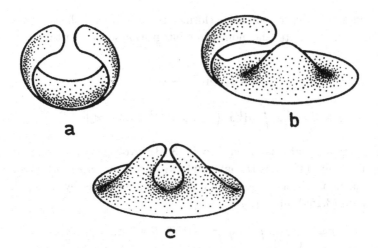

Fig. 6.   The various types of binary contacts of a membrane.

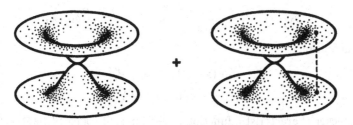

Fig. 7.   Diagrammatic expansion of the double star partition function $\mathcal{Z}_{S_2}(\mathcal{Z}_1)^{-2}$. (Eq. (4.9)).

manifold $S_2$. A simple scaling argument relates the latter to the contact exponent when the two parts of the star manifold are approaching each other. We refer to the polymer theory[5,29] where the relationship between multiple contact exponents and the star partition functions made by contact is established at length (Eq. (1.7) in Ref. 5). The result is simply that the ratio of the double-star partition function $\mathcal{Z}_{S_2}(\mathbf{x}_1, \mathbf{x}_2)$ to that of the two manifolds far away from each other scales with the size $X$ as

$$\frac{\mathcal{Z}_{S_2}(\mathbf{x}_1, \mathbf{x}_2)}{(\mathcal{Z}_1)^2} \sim X^{-\nu\theta} \qquad (4.8)$$

The perturbation expansion of $\mathcal{Z}_{S_2}(\mathcal{Z}_1)^{-2}$ in powers of $b$ is given by Fig. 7. To order $b$, only the interaction diagram between the two manifolds appears, the self-interaction parts of $\mathcal{Z}_{S_2}$ like in Fig. 3 being exactly cancelled by those of $(\mathcal{Z}_1)^2$. Hence from Fig. 7 we find the simple result

$$\mathcal{Z}_{S_2}(\mathbf{x}_1, \mathbf{x}_2)(\mathcal{Z}_1)^{-2}$$
$$= 1 - b \int \frac{\mathrm{d}^d q}{(2\pi)^d} \int_{\mathcal{V}_1} \mathrm{d}^D x \int_{\mathcal{V}_2} \mathrm{d}^D x' \exp\left(-\frac{1}{2} q^2 [G(\mathbf{x} - \mathbf{x}_1) + G(\mathbf{x}' - \mathbf{x}_2)]\right) \quad (4.9)$$

where the internal integrations are performed separately over the two parts $\mathcal{V}_1, \mathcal{V}_2$ of the star manifold $\mathcal{S}_2$. After the momentum integration we can write in *dimensionless* variables

$$\mathcal{Z}_{\mathcal{S}_2}(\mathcal{Z}_1)^{-2} = 1 - z\mathcal{I}$$

$$\mathcal{I} \equiv \int_{\mathcal{V}_1'} \mathrm{d}^D x \int_{\mathcal{V}_2'} \mathrm{d}^D x' \left( |\mathbf{x} - \mathbf{x}_1|^{2-D} + |\mathbf{x}' - \mathbf{x}_2|^{2-D} \right)^{-d/2} \qquad (4.10)$$

Notice the similarity of this integral to $I_2$ (3.13) involved in the second virial coefficient $g$. The only difference is that points $\mathbf{x}_1$ and $\mathbf{x}_2$ are now *not* integrated over. It is convenient to disentangle the $\mathcal{V}_1'$ and $\mathcal{V}_2'$ integrations by using the integral representation (3.15a) and write

$$\mathcal{I} = \frac{2-D}{\Gamma(d/2)} \int_0^\infty \mathrm{d}\eta\, \eta^{-1+2D-(2-D)d/2} \mathcal{J}_{\mathcal{V}_1'}(\mathbf{x}_1) \mathcal{J}_{\mathcal{V}_2'}(\mathbf{x}_2) \qquad (4.11)$$

with

$$\mathcal{J}_{\mathcal{V}_1'}(\mathbf{x}_1) \equiv \frac{1}{\eta^D} \int_{\mathbf{R}^D} \mathrm{d}^D x\, \mathrm{e}^{-(|\mathbf{x}-\mathbf{x}_1|/\eta)^{2-D}} \delta_{\mathcal{V}_1'}(\mathbf{x})$$

where by definition

$$\delta_{\mathcal{V}_1'}(\mathbf{x}) \equiv 1 \quad \text{if } \mathbf{x} \in \mathcal{V}_1'$$
$$\equiv 0 \quad \text{if } \mathbf{x} \notin \mathcal{V}_1'$$

is the Kronecker characteristic function of the internal space $\mathcal{V}_1'$ in $\mathbf{R}^D$. We shift to relative coordinates $\mathbf{u}$

$$\mathbf{x} = \mathbf{x}_1 + \eta\mathbf{u}$$

such that

$$\mathcal{J}_{\mathcal{V}_1'}(\mathbf{x}_1, \eta) = \int_{\mathbf{R}^D} \mathrm{d}^D u\, \mathrm{e}^{-u^{2-D}} \delta_{\mathcal{V}_1'}(\mathbf{x}_1 + \eta\mathbf{u}) \qquad (4.12)$$

In (4.11), recalling that $\varepsilon/2 \equiv 2D - (2-D)d/2$, we see that the divergent part in $1/\varepsilon$ will come from the $\eta \to 0$ region, the large values of $\eta$ being cut-off by the $\delta_{\mathcal{V}'}$ functions in (4.12). In this limit $\eta \to 0$, only the neighbourhood of $\mathbf{x}_1$ (respectively $\mathbf{x}_2$) in manifold $\mathcal{V}_1'$ (resp. $\mathcal{V}_2'$) matters. The function (4.12) is then easily computed as

$$\mathcal{J}_{\mathcal{V}'}(\mathbf{x}, \eta \to 0) = \Omega_D(\mathbf{x}) \int_0^\infty \mathrm{d}u\, u^{D-1} \mathrm{e}^{-u^{2-D}} = \omega(\mathbf{x}) S_D \frac{1}{2-D} \Gamma\left(\frac{D}{2-D}\right) \qquad (4.13)$$

where

$$\omega(\mathbf{x}) \equiv \Omega_D(\mathbf{x})/S_D \equiv \frac{1}{S_D} \int \mathrm{d}^{D-1}\Omega\, \delta_{\mathcal{V}}(\mathbf{x} + 0^+ \mathbf{u}) \qquad (4.14)$$

is the fraction of *solid angle* $\Omega_D$ *spanned by the manifold* $\mathcal{V}$ *around point* $\mathbf{x}$, *in Euclidean internal space* $\mathbf{R}^D$ (Fig. 8). If point $\mathbf{x}$ is *inside* $\mathcal{V}$ (Fig. 8c), $\Omega_D = S_D$, $\omega = 1$; if it belongs to a *smooth* part of the boundary (Fig. 8a), $\Omega_D = S_D/2, \omega = 1/2$.

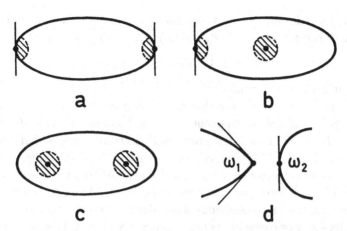

**Fig. 8.** The local domain of $\mathcal{V}$ contributing to the solid angle integral (4.14). The cases $a, b, c$ correspond exactly to those of Fig. 6. The local geometry $d$ corresponds to a boundary-to-boundary contact $a$ in Fig. 6, one of the contact points being a corner.

Any values of $\Omega_D$, hence of $\omega \in [0,1]$ can be obtained for a *corner* point (Fig. 8d). Putting Eqs. (4.11) (4.13) together gives

$$\mathcal{I} = \frac{2}{\varepsilon}\omega_1\omega_2 c_D$$
$$c_D = \frac{S_D^2}{2-D}\frac{\Gamma^2[D/(2-D)]}{\Gamma(d/2)} \tag{4.15}$$

Notice that this result gives us also a calculation of integral $I_2$ (3.13) (3.15b) associated with the second virial coefficient (Fig. 5c), for which one takes $w_{1,2} = 1$. The computation of the contact exponent $\theta$ in (4.8) is performed by direct renormalization as in section III above. The scaling function

$$\nu\theta(z,\varepsilon) = -X\frac{\partial}{\partial X}\ln(\mathcal{Z}_{\mathcal{S}_2}/(\mathcal{Z}_1)^2)$$

has a regular $g$-expansion[10]

$$\nu\theta[g,\varepsilon] = \frac{\varepsilon}{2}z\mathcal{I} + \cdots = g\omega_1\omega_2 c_D + 0(g^2)$$

with a fixed point values $\nu\theta = g^*\omega_1\omega_2 c_D$. To first order in $\varepsilon$, results (3.24) and (3.20) give[10]

$$\theta = \varepsilon\omega_1\omega_2\frac{2}{2-D}\frac{c_D}{a_D'} = \varepsilon\,\omega_1\omega_2\frac{2}{2-D}\left(1+(2-D)\frac{\Gamma[2D/(2-D)]}{\Gamma^2[D/(2-D)]}\right)^{-1} \tag{4.16}$$

As announced, the contact exponent $\theta$ depends on the location of points $x_1, x_2$ on the manifold $\mathcal{V}$ (Fig. 8), through the fractions of solid angle $\omega_1, \omega_2$ spanned by $\mathcal{V}$ around $x_1$ and $x_2$. This is in full agreement with the result of polymer theory $(D = 1)$. Indeed for a point on a linear chain, there are only two possible local geometries: $\omega = 1$ for an interior point and $\omega = 1/2$ for a chain end. Thus for the three polymer contact exponents $\theta_a$ in (4.5), Eq. (4.16) gives for $D = 1$

$$\theta_0 = \varepsilon/4, \quad \theta_1 = \varepsilon/2, \quad \theta_2 = \varepsilon$$

in agreement with known results.[5,28,29]

It is also interesting to remark that the dependence of $\theta$ on $\omega_1$, $\omega_2$ is not similar to that of the configuration exponent $\gamma$ on the boundary's shape $\partial \mathcal{V}$. Indeed the contribution $I_1$ (2.18) of the "dangerous" self-interaction diagram of Fig. 3, which led to the non trivial $\gamma$ (3.25), cancels out in ratio (4.8). So the origin of the shape dependence of contact exponents $\theta$ is quite different and is purely local. Moreover, it was already expected from polymer theory and our solid angle formula (4.16) is its natural generalization. A consequence of this discussion is also that the relation (4.6) of the exterior-exterior contact exponent $\theta_0$ to $\gamma$, $\nu\theta_0 = \gamma - 1$, fails for $D$-dimensional manifolds, if one insists that $\theta_0$ represents the contact of any two exterior points on the boundary. However, this relation should hold true and have another meaning here: $\theta_0 = (\gamma - 1)/\nu$ gives the repulsive potential $-\theta_0 \ln r$ exerted by the boundary onto itself when closing the manifold $\mathcal{V}$ by bringing the lips of $\partial \mathcal{V}$ at a mean distance $r \to 0$.

## 5. On the Nonuniversality of Exponent $\gamma$

As was already mentioned, the use of the bare Coulomb potential $G$ (1.11) in the perturbation expansions means that we are really considering an infinite Gaussian manifold, of which only a part $\mathcal{V}$ is self-avoiding. Even in this case, we have found a configuration exponent $\gamma$ which strongly depends on the boundary's shape (fn integer $D$). If we were insisting on studying a self-avoiding manifold with sharp edges, and with no infinite Gaussian dressing around, we would have to implement in the perturbation expansion a free propagator $G$ which would take into account the boundary conditions in a finite box $\mathcal{V}$. The latter could be Dirichlet ones if the boundary points are fixed in $d$-space, or Neumann if they freely fluctuate. The inclusion of these boundary conditions could modify significantly the physics of the manifold. For instance if the $D$-membrane is rigidly adjusted onto an external frame, one would encounter new relevant operators and possibly buckling or crumpling transitions when approaching the rigid low temperature phase[34] (see David's lecture). In the crumpled phase we are considering here, one could be tempted to imagine the nonuniversality of $\gamma$ as an artifact due to the consideration of a self-avoiding $\mathcal{V}$ inside a larger Gaussian membrane, that could be suppressed by proper physical boundary conditions for the Coulomb kernel $G$ of a sharp-cut manifold. We do not believe this to be the case. Indeed as we shall see, a simple *Gaussian two-dimensional membrane with boundaries, embedded in d-space, has a non trivial and shape dependent exponent* $\gamma$.

Consider a *two-dimensional* membrane $\mathcal{V}(D = 2)$ defined by a space of parameters $\sigma \in \mathcal{V}$, embedded in $d$-space by $\vec{r}(\underline{\sigma})$. The Gaussian partition function associated with action (1.3) reads

$$\mathcal{Z}_0 = \int \mathcal{D}\vec{r}(\underline{\sigma}) \exp\left(-\frac{1}{2}\int d^2\sigma \left(\frac{\partial \vec{r}}{\partial \underline{\sigma}}\right)^2\right) \sim (\det -\Delta|_\mathcal{V})^{-d/2} \qquad (5.1)$$

where $\Delta$ is the two-dimensional Laplacian in space $\mathcal{V}$. Note that here we take for $\mathcal{V}$ an arbitrary two-dimensional manifold with any topology. It can have handles, holes and boundaries. If it is *curved* with an intrinsic non Euclidean metric $g^{ab}$, one has to take a Gaussian weight[35–37] (see David's lectures)

$$\mathcal{A}_0 = \frac{1}{2}\int d^2\sigma\sqrt{g}\, g^{ab} \frac{\partial\vec{r}}{\partial\sigma^a}\cdot\frac{\partial\vec{r}}{\partial\sigma^b} \qquad (5.2)$$

where $g$ is the determinant of $g^{ab}$, and then Eq. (5.1) still holds true. The determinant of the Laplacian has to be taken in (5.1) with appropriate boundary conditions. Fixing the boundary $\partial\mathcal{V}$ of $\mathcal{V}$ in external space $\mathbf{R}^d$ amounts to taking *Dirichlet* conditions at the boundary $\partial\mathcal{V}$ when calculating the spectrum of $-\Delta$ in $\mathcal{V}$. If, on the contrary, one lets the positions in $d$-space of $\partial\mathcal{V}$ fluctuate freely, then det $(-\Delta)$ should be calculated with *Neumann* boundary conditions in internal space $\mathcal{V}$ (the zero mode being eliminated). These two boundary conditions are fortunately related one to another by duality.[38] Before taking the continuum limit, suppose for a moment that the manifold $\mathcal{V}$ is really made of a regular lattice, $\mathcal{L}$, with $N_0(\mathcal{L})$ sites. Then one can show[38] that

$$\det{}'(-\Delta)|_{\mathcal{L},\text{ Neumann}} = N_0(\mathcal{L})\det(-\Delta)|_{\mathcal{L}^*,\text{ Dirichlet}} \qquad (5.3)$$

where the prime in the Neumann determinant stands for the elimination of the zero mode, and where $\mathcal{L}^*$ is the *dual* lattice of $\mathcal{L}$. In the continuum limit, $\mathcal{L}$ and $\mathcal{L}^*$ become identical to manifold $\mathcal{V}$, and we may replace $N_0(\mathcal{L})$ by the area $A$ of $\mathcal{V}$, to get the equivalence

$$\det{}'(-\Delta)|_{\mathcal{V},\text{ Neumann}} = A\det(-\Delta)|_{\mathcal{V},\text{ Dirichlet}} \qquad (5.4)$$

Now the Dirichlet determinant is well known on a two-dimensional manifold.[37–40] It is given by an asymptotic expansion

$$\ln\det(-\Delta)|_{\text{Dir.}} = c_1 A + c_2 L - \zeta(0)\ln A + \cdots \qquad (5.5)$$

where $A$ is the area of $\mathcal{V}$

$$A \equiv \int_\mathcal{V} d^2\sigma\sqrt{g}, \qquad (5.6)$$

$L$ is the length of its boundary

$$L \equiv \int_{\partial\mathcal{V}} ds \qquad (5.7)$$

and $\zeta(0)$ is a *universal coefficient* depending only on the geometry or the topology of $\mathcal{V}$. (The notation $\zeta(0)$ comes here from the so-called $\zeta$-function regularization method for calculating the determinant of the Laplacian[40]). The coefficients

$c_1, c_2$ are not universal and depend on the regularization used. In David's lectures (section III) an ultraviolet cut-off $\varepsilon'$ is used in the parameter space $\underline{\sigma}$, for which $c_1 = -1/4\pi\varepsilon'$ (Eq. (3.14) there, note that the latter is the same as our Eq. (5.5) with no perimeter term $c_2$). If one uses instead a square lattice regularization of $\mathcal{V}$, of mesh size $a$, the value of the coefficients $c_1, c_2$ have been calculated in Ref. 38

$$c_1 = \frac{G}{\pi a^2}, \quad c_2 = -\frac{1}{2a} \ln\left(1 + \sqrt{2}\right) \tag{5.8}$$

where $G$ is Catalan's constant $G = 1 - 1/3^2 + 1/5^2 + \cdots$. The really interesting term is $\zeta(0)$ which is directly related to the configurational exponent $\gamma$ above. Indeed, from Eqs. (5.1) (5.5) we derive the Gaussian partition function (with Dirichlet boundary conditions)

$$\mathcal{Z}_0 \sim \exp\left(c_1' A + c_2' L\right) A^{\zeta(0)d/2} \tag{5.9}$$

with $c_1' \equiv -c_1 d/2$, $c_2' \equiv -c_2 d/2$. Hence for a Gaussian manifold, there exists a non-trivial configuration exponent

$$\gamma_0 - 1\big|_{\text{Gaussian}} = d \left(\zeta(0) \begin{matrix} {}^{+0} \\ {}_{-1} \end{matrix}\right)_{\text{Neumann}}^{\text{Dirichlet}} \tag{5.10}$$

where $\zeta(0)$ is shifted by $-1$ in the case of free Neumann boundary conditions, owing to identity (5.4). Now, this result is quite interesting since $\zeta(0)$ is known exactly! For a *flat* manifold whose boundary is made of smooth arcs $\gamma_j$, meeting at some angles $\alpha_i$ (Fig. 9), the explicit value of $\zeta(0)$ is[39]

$$\zeta(0) = \sum_i \frac{1}{24}\left(\frac{\pi}{\alpha_i} - \frac{\alpha_i}{\pi}\right) + \frac{1}{12\pi} \sum_j \int_{\gamma_j} \frac{ds}{\rho} \tag{5.11}$$

where $\rho$ is the scalar curvature of the arcs. The above corner contribution was found by D.B. Ray (described in Mc Kean and Singer).[39]

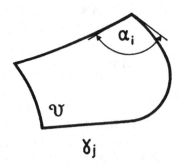

Fig. 9.  A two-dimensional elastic membrane $\mathcal{V}$, with a boundary $\partial\mathcal{V}$ made of smooth arcs $\gamma_j$ joining at edge angles $\alpha_i$.

If the manifold is curved, and has a non-trivial topology, with handles and holes, the general result can be found in O. Alvarez[37]

$$\zeta(0) = \frac{1}{6}\chi(\mathcal{V}) \tag{5.12}$$

where $\chi(\mathcal{V})$ is the Euler characteristic of the manifold

$$\chi(\mathcal{V}) = \frac{1}{4\pi}\int_{\mathcal{V}} d^2\sigma\sqrt{g}R + \frac{1}{2\pi}\int_{\partial\mathcal{V}} \frac{ds}{\rho} \tag{5.13}$$

$R$ being the intrinsic Gauss curvature $R = 1/R_1 R_2$ of $\mathcal{V}$. Notice that the perimeter term in (5.13) coincides with that of (5.11). By the Gauss-Bonnet theorem, $\chi$ is well known to be a topological invariant

$$\chi(\mathcal{V}) = 2 - 2H - B \tag{5.14}$$

where $H$ is the number of handles of $\mathcal{V}$ (the genus of the surface) and $B$ the number of holes (boundaries). It is important to remark also that the corner terms in (5.11) are new and cannot be obtained by simply taking the limit of the second perimeter term in (5.11) with an infinite curvature localized at the corners. A similar non commutative limit phenomenon is well known in all distribution of modes problems,[39] when wedges are present.

Collecting the results (5.10)–(5.14) we get a quite interesting feature of two-dimensional Gaussian membranes: they possess a non-trivial $\gamma_0$ exponent given by a nice explicit formula in terms of the geometrical intrinsic shape of the membrane and of the space dimension. So, when introducing self-avoidance, one should not be so surprised that the shift of $\gamma$ due to excluded volume itself depends strongly on the shape of the tethered manifold. Notice also that the result (3.25) gives the scaling behavior of the ratio $\mathcal{Z}_1 = \mathcal{Z}/\mathcal{Z}_0$ of the partition function of a large manifold with a SA patch to that of the embedding Gaussian manifold. Remarking that $\zeta(0)$ (5.11) (5.12) is purely topological, and does not depend on the size of the embedding manifold, not being metric, we conclude that the total partition function $\mathcal{Z}$ of the complete Gaussian manifold, with its self-avoiding patch of size $X$ (Fig. 4) should scale as

$$\mathcal{Z} \sim X^{\gamma-1} A^{(\gamma_0-1)/2} \tag{5.15}$$

where $X$ is the size of the SA patch and $A$ the total area.

## 6. Conclusion

From this study, it appears that self-avoiding tethered manifolds can be studied at least to first order in $\varepsilon$ by a direct renormalization technique which originated in polymer theory. However, one should remember that $D$-dimensional objects are more complicated that $1D$ ones. This led to some surprises like the dependence of exponent $\gamma$ on the intrinsic manifold shape, which reflects that which actually already exists in absence of self-avoidance for Gaussian membranes. This led to an unexpected application of mathematical results on the Laplacian spectrum in a finite two-dimensional domain.

The next challenge is to study in more detail the renormalizability of the Edwards model for self-avoiding manifolds, which has been assumed implicitly throughout the above.[41] This is the subject of Part II, Chapter 9.[42]

## Acknowledgements

It is a pleasure to thank David Nelson and Tsvi Piran for organizing this enjoyable and exciting Jerusalem Winter School, and offering me the opportunity to give this lecture.

## References

1. P.G. de Gennes, *Scaling Concepts in Polymer Physics* (Cornell University Press, Ithaca, New York, 1979).
2. J. des Cloizeaux, G. Jannink, *Les Polymères en Solution, leur Modélisation et leur Structure* (Editions de Physique, Les Ulis, France, 1987).
3. S.F. Edwards, *Proc. Phys. Soc. London* **85**, 613 (1965).
4. J. des Cloizeaux, *J. Phys. France* **42**, 635 (1981).
5. B. Duplantier, *J. Stat. Phys.* **54**, 581 (1989).
6. M. Benhamou, G. Mahoux, *J. Phys. France* **47**, 559 (1986).
7. See also L. Schäfer and T.A. Witten, *J. Phys. France* **41**, 459 (1980); B. Duplantier, *J. Phys. France* **47**, 569 (1986).
8. M. Kardar, D.R. Nelson, *Phys. Rev. Lett.* **58**, 12 (1987), **58**(E), 2280, *Phys. Rev.* **A 38**, 966 (1988).
9. J.A. Aronowitz, T.C. Lubensky, *Europhys. Lett.* **4**, 395 (1987).
10. B. Duplantier, *Phys. Rev. Lett.* **58**, 2733 (1987).
11. A. Billoire, D.J. Gross, E. Marinari, *Phys. Lett.* **139B**, 75 (1984); D.J. Gross, *ibid.* **139B**, 187 (1984).
12. B. Duplantier, *Phys. Lett.* **141B**, 239 (1984).
13. C. Itzykson, M.C. Bander, *Nucl. Phys.* **B 257** [FS14], 531 (1985).
14. Y. Kantor, M. Kardar, D.R. Nelson, *Phys. Rev. Lett.* **57**, 791 (1986); *Phys. Rev.* **A 35**, 3056 (1987).
15. An earlier attempt was made by R.C. Ball, Cambridge University preprint (unpublished).
16. B.H. Zimm, W.H. Stockmayer, M. Fixman, *J. Chem. Phys.* **21**, 1716 (1953).
17. P.G. de Gennes, *J. Phys. France Lett.* **36**, L55 (1975).
18. B. Duplantier, *J. Chem. Phys.* **86**, 4233 (1987), *Phys. Rev.* **A 38**, 3647 (1988).
19. *Phase Transitions and Critical Phenomena* Vol. 6, C. Domb and M.S. Green editors (Academic, London, 1976).
20. J. Zinn-Justin, *Quantum Field Theory and Critical Phenomena*, Fourth Edition, (Clarendon Press, Oxford, 2002).
21. V. Kazakov, *Phys. Lett.* **B 150**, 282 (1985); F. David, *Nucl. Phys.* **B 257**, 45 (1985).
22. V. Kazakov, I. Kostov, A. Migdal, *Phys. Lett.* **B 157**, 295 (1985); F. David, *Nucl. Phys.* **B 257**, 543 (1985).
23. I. Kostov, M.L. Mehta, *Phys. Lett.* **B 189**, 118 (1987).
24. B. Duplantier, *Europhys. Lett.* **1**, 99 (1986).
25. A. Manohar (unpublished).

26. B. Duplantier (unpublished).
27. B. Duplantier, *Nucl. Phys.* **B 275** [FS17], 319 (1986).
28. J. des Cloizeaux, *Phys. Rev.* **10**, 1655 (1974); *J. Phys. France* **41**, 223 (1980).
29. B. Duplantier, *J. Phys. France* **47**, 1633 (1986); *Phys. Rev.* **B 37**, 5290 (1987).
30. Y. Oono, *Adv. Chem. Phys.* **61** (1985); L. Schäfer, A. Baumgärtner, *J. Phys. France* **47**, 1431 (1986).
31. M.E. Fisher, *J. Chem. Phys.* **44**, 616 (1966).
32. D.S. Mc Kenzie, M.A. Moore, *J. Phys.* **A 4**, L82 (1971).
33. M.E. Cates, T.A. Witten, *Macromolecules* **19**, 732 (1986).
34. E. Guitter, F. David, S. Leibler, L. Peliti, *Phys. Rev. Lett.* **61**, 2949 (1988).
35. A.M Polyakov, *Phys. Lett.* **103B**, 207 (1981); *Gauge Fields and Strings* (Harwood, Chur, 1987).
36. D. Friedan, in Les Houches XXXIX (1982), *Recent Advances in Field Theory and Statistical Mechanics*, ed. by J.B. Zuber and R. Stora (Elsevier, 1984).
37. O. Alvarez, *Nucl. Phys.* **B 216**, 125 (1983).
38. B. Duplantier, F. David, *J. Stat. Phys.* **51**, 327 (1988).
39. M. Kac, *Am. Math. Monthly* **73**, 1 (1966); H.P. Mc Kean, I.M. Singer, *J. Diff. Geom.* **1**, 43 (1967); H.P. Baltes, E.R. Hilf, *Spectra of Finite Systems* (Bibliographisches Institut, Mannheim, 1976).
40. N.L. Balazs, C. Schmit, A. Voros, *J. Stat. Phys.* **46**, 1067 (1987).
41. The next order polar part $O(\varepsilon^{-2})$ in the expansion of the swelling factor $\chi_0(z)$, corresponding to exponent $\nu$, appears to be related to the $O(\varepsilon^{-1})$ one in the expected way (T. Hwa, M. Kardar, private communication, 1988).
42. Part I was written in 1988, and corresponds to the knowledge of that time. Part II below presents the state of the art in 2004.

# CHAPTER 9

# STATISTICAL MECHANICS OF SELF-AVOIDING
# CRUMPLED MANIFOLDS — Part II

Bertrand Duplantier

*Service de Physique Théorique\*, CEA/Saclay*
*F-91191 Gif-sur-Yvette, France*
*bertrand@spht.saclay.cea.fr*

We consider a model of a $D$-dimensional tethered manifold interacting by excluded volume in $\mathbb{R}^d$ with a single point. Use of intrinsic distance geometry provides a rigorous definition of the analytic continuation of the perturbative expansion for arbitrary $D$, $0 < D < 2$. Its one-loop renormalizability is first established by direct resummation. A renormalization operation **R** is then described, which ensures renormalizability to all orders. The similar question of the renormalizability of the self-avoiding manifold (SAM) Edwards model is then considered, first at one-loop, than to all orders. We describe a short-distance multi-local operator product expansion, which extends methods of local field theories to a large class of models with non-local singular interactions. It vindicates the direct renormalization method used earlier in Part I of these lectures, as well as the corresponding scaling laws.

## 1. Interacting Manifold Renormalization: A Brief History

As can be seen in the set of lectures in this volume, which presents an extended version of Ref. 1, the statistical mechanics of random surfaces and membranes, or more generally of extended objects, poses fundamental problems. The study of *polymerized* membranes, which are generalizations of linear polymers[2,3] to two-dimensionally connected networks, is emphasized, with a number of possible experimental realizations,[4-8] or numerical simulations.[9,10] From a theoretical point of view, a clear challenge in the late eighties was to understand self-avoidance (SA) effects in membranes.

The model proposed[a] in Refs. 11 and 12 aimed to incorporate the advances made in polymer theory by renormalization group (RG) methods into the field of polymerized, or tethered, membranes. As we saw in Part I of these lectures, these

---

\*Laboratoire de la Direction des Sciences de la Matière du Commissariat à l'Energie Atomique, URA CNRS 2306.
[a]R.C. Ball was actually the first to propose, while a postdoc in Saclay in 1981, the extension of the Edwards model to $D$-manifolds, with the aim, at that time, to better understand polymers! (unpublished).

extended objects, *a priori* two-dimensional in nature, are generalized for theoretical purposes to intrinsically *D-dimensional manifolds* with internal points $x \in \mathbb{R}^D$, embedded in external *d*-dimensional space with position vector $\vec{r}(x) \in \mathbb{R}^d$. The associated continuum Hamiltonian $\mathcal{H}$ generalizes that of Edwards for polymers[2]:

$$\beta\mathcal{H} = \frac{1}{2} \int d^D x (\nabla_x \vec{r}(x))^2 + \frac{b}{2} \int d^D x \int d^D x' \delta^d(\vec{r}(x) - \vec{r}(x')), \qquad (1.1)$$

with an elastic Gaussian term and a self-avoidance two-body $\delta$-potential with interaction parameter $b > 0$. For $0 < D < 2$, the Gaussian manifold ($b = 0$) is *crumpled* with a Gaussian size exponent:

$$\nu_0 = \frac{2 - D}{2}, \qquad (1.2)$$

and a finite Hausdorff dimension:

$$d_H = D/\nu_0 = 2D/(2 - D); \qquad (1.3)$$

the finiteness of the upper critical dimension $d^* = 2d_H$ for the SA-interaction allows an $\varepsilon$-expansion about $d^{*}$[11-13]:

$$\varepsilon = 4D - 2\nu_0 d \qquad (1.4)$$

performed via the direct renormalization method adapted from that of des Cloizeaux in polymer theory,[14] as we explained in Part I.

Only the polymer case, with an *integer* internal dimension $D = 1$, can be mapped, following de Gennes,[15] onto a standard field theory, namely a $(\Phi^2(\vec{r}))^2$ theory for an *n*-component field $\Phi(\vec{r})$ in external *d*-dimensional space, with $n \to 0$ components. This is instrumental in showing that the direct renormalization method for polymers is mathematically sound,[16] and equivalent to rigorous renormalization schemes in standard local field theory, such as the Bogoliubov–Parasiuk–Hepp–Zimmermann (BPHZ) construction.[17] For manifold theory, we have to deal with *non-integer* internal dimensions $D$, $D \neq 1$, and no such mapping exists. Therefore, two outstanding problems remained in the theory of interacting manifolds: (a) the mathematical meaning of a *continuous* internal dimension $D$; (b) the actual *renormalizability* of the perturbative expansion of a manifold model like (1.1), implying the scaling behavior expected on physical grounds.

In Ref. 18 a simpler model was proposed, of a crumpled manifold interacting by excluded volume with a fixed Euclidean subspace of $\mathbb{R}^d$.[19] The simplified model Hamiltonian introduced there reads:

$$\beta\mathcal{H} = \frac{1}{2} \int d^D x (\nabla_x \vec{r}(x))^2 + b \int d^D x \, \delta^d(\vec{r}(x)), \qquad (1.5)$$

with a pointwise interaction of the Gaussian manifold with an impurity located at the origin (Fig. 1a). Note that this Hamiltonian also represents interactions of a fluctuating (possibly directed) manifold with a nonfluctuating $D'$-Euclidean hyperplane of $\mathbb{R}^{d+D'}$, $\vec{r}$ then standing for the coordinates transverse to this subspace (Fig. 1b). The excluded volume case ($b > 0$) parallels that of the Edwards model

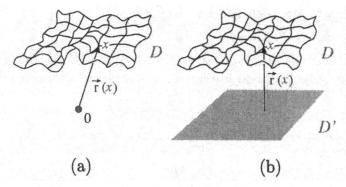

Fig. 1.  (a) A $D$-manifold interacting with an impurity located at point **0** in $\mathbb{R}^d$; (b) interaction with an Euclidean hyperplane of dimension $D'$ in $\mathbb{R}^{d'}$, with $d' = d + D'$.

(1.1) for SA-manifolds, while an attractive interaction ($b < 0$) is also possible, describing pinning phenomena. The (naive) dimensions of $\vec{r}$ and $b$ are respectively $[\vec{r}] = [x^\nu]$ with a Gaussian size exponent

$$\nu = \nu_0 = (2 - D)/2, \tag{1.6}$$

and $[b] = [x^{-\varepsilon}]$ with

$$\varepsilon \equiv D - \nu d. \tag{1.7}$$

For fixed $D$ and $\nu$, the parameter $d$ (or equivalently $\varepsilon$) controls the relevance of the interaction, with the exclusion of a point only effective for $d \leq d^* = D/\nu$. Note that in this model the size exponent $\nu$ is not modified by the local interaction and stays equal to its Gaussian value (1.6), whereas the correlation functions obey (non-Gaussian) universal scaling laws.

For $D = 1$, the model is exactly solvable.[18] For $D \neq 1$, the direct resummation of leading divergences of the perturbation series is possible for model (1.5) and indeed validates *one-loop* renormalization.[18] This result was also extended to the Edwards model (1.1) itself.[20]

A study to all orders of the interaction model (1.5) was later performed in Refs. 21 and 22. A mathematical construction of the $D$-dimensional internal measure $d^D x$ via distance geometry within the elastic manifold was given, with expressions for manifold Feynman integrals which generalize the $\alpha$-parameter representation of field theory. In the case of the manifold model of Ref. 18 the essential properties which make it *renormalizable to all orders* by a renormalization of the coupling constant were established. This led to a direct construction of a renormalization operation, generalizing the BPHZ construction to manifolds (see also Refs. 23 and 24).

Later, the full Edwards model of self-avoiding manifolds (1.1) was studied by the same methods, and its renormalizability established to all orders.[25,26] Effective calculations to second order in $\varepsilon$ ("two-loop" order) were performed in Ref. 27. The large order behavior of the Edwards model (1.1) was finally studied in Ref. 28.

The aim of Part II of these notes is to review some of these developments.

## 2. Manifold Model with Local $\delta$ Interaction

### 2.1. *Perturbative Expansion*

In this chapter, we study the statistical mechanics of the simplified model Hamiltonian (1.5). The model is described by its (connected) partition function:

$$\mathcal{Z} = \mathcal{V}^{-1} \int \mathcal{D}[\vec{r}] \exp(-\beta\mathcal{H}), \qquad (2.1)$$

(here $\mathcal{V}$ is the internal volume of the manifold) and, for instance, by its one-point vertex function

$$\mathcal{Z}^{(0)}(\vec{k})/\mathcal{Z} = \int d^D x_0 \langle e^{i\vec{k}\cdot\vec{r}(x_0)} \rangle, \qquad (2.2)$$

where the (connected) average $\langle \cdots \rangle$ is performed with (1.5):

$$\mathcal{Z}^{(0)}(\vec{k}) = \mathcal{V}^{-1} \int \mathcal{D}[\vec{r}] \exp(-\beta\mathcal{H}) \int d^D x_0 e^{i\vec{k}\cdot\vec{r}(x_0)}. \qquad (2.3)$$

These functions are all formally defined via their perturbative expansions in the coupling constant $b$:

$$\mathcal{Z} = \sum_{N=1}^{\infty} \frac{(-b)^N}{N!} \mathcal{Z}_N, \qquad (2.4)$$

with a similar equation for $\mathcal{Z}^{(0)}$ with coefficients $\mathcal{Z}_N^{(0)}$:

$$\mathcal{Z}^{(0)}(\vec{k}) = \sum_{N=1}^{\infty} \frac{(-b)^N}{N!} \mathcal{Z}_N^{(0)}(\vec{k}). \qquad (2.5)$$

$\mathcal{Z}_N$ has the path integral representation

$$\mathcal{Z}_N = \frac{1}{\mathcal{V}} \int d\mathcal{P}_0 \left[ \int d^D x \delta^d(\vec{r}(x)) \right]^N, \qquad (2.6)$$

where the Gaussian path measure is

$$d\mathcal{P}_0 = \mathcal{D}\vec{r}(x) \exp(-\beta\mathcal{H}_0) \qquad (2.7)$$

with

$$\beta\mathcal{H}_0 = \frac{1}{2} \int d^D x \left( \nabla_x \vec{r}(x) \right)^2. \qquad (2.8)$$

There is no translational invariance in this theory, since the origin is selected by the presence of the impurity. The measure $d\mathcal{P}_0$ thus *includes* integration over global translations of the manifold in $\mathbb{R}^d$. The first term is then simply $\mathcal{Z}_1 \equiv 1$, so that

$$\mathcal{Z} = -b + \mathcal{O}(b^2). \qquad (2.9)$$

The term of order $N$, $\mathcal{Z}_N$, is a Gaussian average involving $N$ interaction points $x_i$ (Fig. 2):

$$\mathcal{Z}_N = \frac{1}{\mathcal{V}} \int d\mathcal{P}_0 \int \prod_{i=1}^{N} d^D x_i \prod_{i=1}^{N} \delta^d(\vec{r}(x_i)). \qquad (2.10)$$

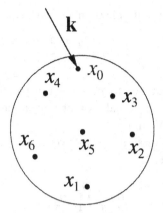

Fig. 2.    Interaction points $x_i$; insertion point $x_0$ for the external momentum $\vec{k}$.

By Fourier transforming the distribution in $d$-space

$$\delta^d\left(\vec{r}(x)\right) = \int \frac{d^d\vec{k}}{(2\pi)^d} \exp\left(i\vec{k}\cdot\vec{r}\right),$$

one gets

$$\mathcal{Z}_N = \frac{1}{\mathcal{V}} \int \prod_{i=1}^{N} d^D x_i \int \prod_{i=1}^{N} \frac{d^d\vec{k}_i}{(2\pi)^d} \int d\mathcal{P}_0 \exp\left[i\sum_{i=1}^{N} \vec{k}_i\cdot\vec{r}_i\right]. \qquad (2.11)$$

For a Gaussian manifold with weight (2.7) (2.8) we have:

$$\int d\mathcal{P}_0 \exp\left[i\sum_{i=1}^{N} \vec{k}_i\cdot\vec{r}_i\right] = (2\pi)^d \delta^d\left(\sum_{i=1}^{N} \vec{k}_i\right)$$

$$\times \exp\left[-\frac{1}{2}\sum_{i,j=1}^{N} \vec{k}_i\cdot\vec{k}_j G(x_i - x_j)\right]. \qquad (2.12)$$

This Gaussian manifold average is expressed solely in terms of the Green function

$$G(x - y) = -\frac{1}{2}A_D|x - y|^{2\nu}, \qquad (2.13)$$

solution[b] of

$$-\Delta_x G(x - y) = \delta^D(x - y), \qquad (2.14)$$

with $2\nu = 2 - D$, and $A_D$ a normalization:

$$A_D = [S_D(2 - D)/2]^{-1} = [S_D\nu]^{-1}, \qquad (2.15)$$

---

[b]In Part I we used the notation $G(x - y) \equiv A_D|x - y|^{2\nu}$ for the (positive) solution of the slightly different equation $\Delta_x G(x - y) = 2\delta^D(x - y)$, while hereafter in II we shall use the proper Newton–Coulomb potential (2.13), in view of the underlying electrostatic representation.

where $S_D$ is the area of the unit sphere in $D$ dimensions

$$S_D = \frac{2\pi^{D/2}}{\Gamma(D)}. \tag{2.16}$$

In the following, it is important to preserve the condition $0 < \nu < 1$ (*i.e.*, $0 < D < 2$), corresponding to the actual case of a crumpled manifold, where $(-G)$ is positive and ultraviolet (UV) finite.

Performing finally the Gaussian integral over the $N-1$ independent real variables $\vec{k}_i$, $(i = 1, \ldots, N-1)$ yields[18]:

$$\mathcal{Z}_N = \mathcal{V}^{-1}(2\pi)^{-(N-1)d/2} \int \prod_{i=1}^{N} d^D x_i \, (\det[\Pi_{ij}]_{1 \le i,j \le N-1})^{-\frac{d}{2}}, \tag{2.17}$$

where the matrix $[\Pi_{ij}]$ is simply defined as

$$\Pi_{ij} \equiv G(x_i - x_j) - G(x_i - x_N) - G(x_j - x_N), \tag{2.18}$$

with respect to the reference point $x_N$, the permutation symmetry between the $N$ points being restored in the determinant.

The integral representation of $\mathcal{Z}_N^{(0)}$ is obtained from that of $\mathcal{Z}_N$ by multiplying the integrand in (2.17) by $\exp(-\frac{1}{2}\vec{k}^2 \Delta^{(0)})$ with:

$$\Delta^{(0)} \equiv \frac{\det[\Pi_{ij}]_{0 \le i,j \le N-1}}{\det[\Pi_{ij}]_{1 \le i,j \le N-1}}, \tag{2.19}$$

and integrating over one more position, $x_0$, (Fig. 2):

$$\mathcal{Z}_N^{(0)}(\vec{k}) = \mathcal{V}^{-1}(2\pi)^{-(N-1)d/2} \int \prod_{i=0}^{N} d^D x_i$$

$$\times \exp\left(-\frac{1}{2}\vec{k}^2 \Delta^{(0)}\right) (\det[\Pi_{ij}]_{1 \le i,j \le N-1})^{-\frac{d}{2}}. \tag{2.20}$$

Notice that the first order term ($N = 1$) specializes to:

$$\mathcal{Z}_1^{(0)}(\vec{k}) = \mathcal{V}^{-1} \int_{\mathcal{V} \times \mathcal{V}} d^D x_0 d^D x_1 \exp\left(-\frac{1}{2}\vec{k}^2 \Pi_{01}\right). \tag{2.21}$$

The resulting expressions are quite similar to those for the Edwards manifold model.[20]

## 2.2. Second Virial Coefficient

In this section, we imagine the manifold to be of finite internal volume $\mathcal{V} = X^D$, and define two dimensionless interaction coefficients, the excluded volume parameter $z$, and the second virial coefficient $g$, as

$$z = (2\pi A_D)^{-d/2} b X^{D-(2-D)d/2},\tag{2.22}$$

$$g = (2\pi A_D)^{-d/2}(-\mathcal{Z}) X^{D-(2-D)d/2}.\tag{2.23}$$

Because of (2.9), the perturbative expansion of the full interaction parameter $g$ starts as:

$$g = z + \mathcal{O}(z^2).\tag{2.24}$$

More precisely we have:

$$g = \sum_{N=1}^{\infty} (-1)^{N-1} z^N I_N \tag{2.25}$$

where we have set

$$\frac{1}{N!}\mathcal{Z}_N \equiv (2\pi A_D)^{-(N-1)d/2} X^{(N-1)\varepsilon} I_N \tag{2.26}$$

in order to get rid of cumbersome factors. Now the dimensionless integral $I_N$ is

$$I_N = \frac{1}{N!} \int_{\mathcal{V}'} \prod_{i=1}^{N} d^D x_i (\det \mathbf{D})^{-d/2},\tag{2.27}$$

with integrations over *rescaled* coordinates, in a unit internal volume $\mathcal{V}' = X^{-D}\mathcal{V} = 1$; $\mathbf{D}$ is the symmetric $(N-1) \times (N-1)$ matrix with elements $(1 \le i, j \le N-1)$

$$\begin{aligned}
D_{ii} &= |x_{iN}|^{2-D} \\
D_{ij} &= \frac{1}{2}\left(|x_{iN}|^{2-D} + |x_{jN}|^{2-D} - |x_{ij}|^{2-D}\right),
\end{aligned}\tag{2.28}$$

where we set $x_{ij} \equiv x_i - x_j$.

## 2.3. *Resummation of Leading Divergences*

In this section we analyse the leading divergence of each $I_N$ for $\varepsilon = D - (2-D)\,d/2 = D - \nu d > 0$. We have $I_1 = 1$, and

$$I_2 = \frac{1}{2} \int_{\mathcal{V}' \times \mathcal{V}'} d^D x_1 d^D x_2 \, |x_1 - x_2|^{-(2-D)d/2}.\tag{2.29}$$

We are interested in evaluating the pole at $\varepsilon = 0$. It is easily extracted as[18]

$$I_2 \simeq \frac{1}{2} \int_{\mathcal{V}'} d^D x_1 \int_0^1 S_D \, dy \, y^{-1+\varepsilon} = \frac{S_D}{2\varepsilon},\tag{2.30}$$

where[c] $y = |x_1 - x_2|$.

---

[c]Note that the precise value of the upper limit for $y, y \lesssim 1$, is immaterial when evaluating the pole part.

The structure of divergences of the generic term $I_N$ will be studied in detail in the next sections. They will be shown to be only *local* divergences, obtained by letting any interaction point subset coalesce. Here, the leading divergence is evaluated as follows.

The determinant in (2.27) is symmetrical with respect to the $N$ points, so we can, for a given $i \in \{1, \ldots, N - 1\}$, and without loss of generality, consider the "Hepp sector" $x_i \rightarrow x_N$, hence $\rho \equiv |x_{iN}| \rightarrow 0$. We then have $D_{ii} = |x_{iN}|^{2-D}$, while for any other $j \in \{1, \ldots, N - 1\}, j \neq i$,

$$D_{ij} \simeq \frac{1}{2} \left( |x_{iN}|^{2-D} - x_{iN} \cdot \nabla |x_{jN}|^{2-D} + \mathcal{O}(\rho^2) \right).$$

Using $\nu = (2 - D)/2$ and the notation $\delta \equiv \min(\nu, 1 - \nu)$, we can write the leading term of this equation, which depends on the position of $D, 0 < D < 2$, with respect to 1, as

$$D_{ij} = |x_{iN}|^\nu \times \mathcal{O}(\rho^\delta).$$

When expanding the determinant det $\mathbf{D}$ with respect to column $i$ and line $i$, we encounter either the diagonal term $D_{ii} = |x_{iN}|^{2\nu} = \mathcal{O}(\rho^{2\nu})$, or non diagonal terms of type $D_{ij}D_{ik} = \mathcal{O}(\rho^{2\nu+2\delta})$. Thus $D_{ii}$ dominates and we can write in the sector $x_i \rightarrow x_N$

$$\det \mathbf{D} \simeq D_{ii} \times \det \mathbf{D}/i = |x_{iN}|^{2-D} \times \det \mathbf{D}/i, \qquad (2.31)$$

where det $\mathbf{D}/i$ is the reduced determinant of order $(N - 2) \times (N - 2)$, in which line $i$ and column $i$ have been removed, hence the point $i$ itself. By symmetry, in any other sector $x_i \rightarrow x_j$, we have similarly,

$$\det \mathbf{D} \simeq |x_{ij}|^{2-D} \times \det \mathbf{D}/i. \qquad (2.32)$$

Among the $N(N-1)/2$ possible pairs $(i, j)$ we define an arbitrary ordered set of $N-1$ pairs $\mathcal{P} = \{(i_\alpha, j_\alpha), \alpha = 1, \ldots, N - 1\}$, such that the distances $|x_{i_\alpha} - x_{j_\alpha}| = y_\alpha \rightarrow 0$ define a *sector* $y_1 \leq y_2 \leq \cdots \leq y_{N-1}$. In this limit, applying the rule (2.32) successively from $\alpha = 1$ to $N - 1$ yields a determinant factorized as

$$\det \mathbf{D} \simeq \prod_{\alpha=1}^{N-1} y_\alpha^{2-D}.$$

The contribution of the sector $\mathcal{P}$ to the integral $I_N$ is given by the iteration of (2.30):

$$I_{N|\mathcal{P}} \simeq \frac{1}{N!} \prod_{\alpha=1}^{N-1} \left[ S_D \int_0^{y_{\alpha+1}} dy_\alpha y_\alpha^{-1+\varepsilon} \right]$$

$$= \frac{1}{N!} \frac{1}{(N-1)!} \left[ \frac{S_D}{\varepsilon} \right]^{N-1}. \qquad (2.33)$$

The number of distinct sectors of $N-1$ ordered pairs $\mathcal{P}$ chosen among $N$ points equals $N!(N-1)!/2^{N-1}$, whence the leading divergence of $I_N$:

$$I_N \simeq \left(\frac{S_D}{2\varepsilon}\right)^{N-1} \tag{2.34}$$

At this order, the dimensionless excluded volume parameter $g$ (2.23) thus reads

$$
\begin{aligned}
g &= \sum_{N=1}^{\infty} (-1)^{N-1} z^N I_N \simeq \sum_{N=1}^{\infty} (-1)^{N-1} z^N \left(\frac{S_D}{2\varepsilon}\right)^{N-1} \\
&= \frac{z}{1+z\frac{S_D}{2\varepsilon}}.
\end{aligned}
\tag{2.35}
$$

### 2.4. *Comparison to One-Loop Renormalization*

The Taylor–Laurent expansion of parameter $g$ to first orders is obtained from (2.25) and (2.30)

$$g = z - z^2 I_2 + \cdots = z - z^2 \frac{S_D}{2\varepsilon} + \cdots . \tag{2.36}$$

It is associated with a Wilson function

$$
\begin{aligned}
W(g,\varepsilon) = X\frac{\partial g}{\partial X} &= \varepsilon z \frac{\partial g}{\partial z} \\
&= \varepsilon z - z^2 S_D + \cdots = \varepsilon g - g^2 \frac{S_D}{2} + \cdots .
\end{aligned}
\tag{2.37}
$$

The fixed point $g^*$ such that $W(g^*,\varepsilon) = 0$ is $g^* = 2\varepsilon/S_D$ and precisely corresponds to the limit of (2.35)

$$g(z \to +\infty) = \frac{2\varepsilon}{S_D} = g^*. \tag{2.38}$$

More interestingly, the (truncated) flow Eq. (2.37)

$$W(g,\varepsilon) = \varepsilon z \frac{\partial g}{\partial z} = \varepsilon g - g^2 \frac{S_D}{2}, \tag{2.39}$$

with boundary condition (2.24), has precisely the solution $g = z/(1+z\frac{S_D}{2\varepsilon})$. So we see that the resummation (2.35) to all orders of leading divergences is exactly equivalent to the one-loop renormalization group equation, as displayed in (2.39). Thus the one-loop renormalizability of the manifold model has been directly established by direct resummation of the perturbation expansion.[18]

This is confirmed by consideration of the vertex function (2.2). The same evaluation of (2.20) gives, after successive contractions of pairs of points in the determinants in (2.19), (2.20), the leading divergence:

$$Z_N^{(0)}(\vec{k}) = \frac{1}{V}(2\pi A_D)^{-(N-1)d/2} X^{(N-1)\varepsilon} \left(\frac{S_D}{2\varepsilon}\right)^{N-1} N!$$

$$\times \int_{V\times V} d^D x_0 \, d^D x_1 \exp\left\{-\frac{1}{2}\vec{k}^2 \Pi_{01}\right\} \qquad (2.40)$$

with the matrix element $\Pi_{01} = -2G(x_0 - x_1)$. The (connected) vertex function (2.3), (2.5) can thus be resummed at this order as

$$Z^{(0)}(\vec{k}) = \sum_{N=1}^{\infty} \frac{(-b)^N}{N!} Z_N^{(0)}(\vec{k})$$

$$= \frac{-b}{1 + z\frac{S_D}{2\varepsilon}} V^{-1} \int_{V\times V} d^D x_0 \, d^D x_1 \exp\left\{-\frac{1}{2}\vec{k}^2(-2G(x_0 - x_1))\right\}. \qquad (2.41)$$

Notice that, at *first order*, $Z^{(0)}$ is determined from (2.21) as

$$Z^{(0)}(\vec{k}) = -b Z_1^{(0)}(\vec{k}) + \mathcal{O}(b^2)$$

$$= -b \, V^{-1} \int_{V\times V} d^D x_0 \, d^D x_1 \exp\left\{-\frac{1}{2}\vec{k}^2(-2G(x_0 - x_1))\right\} + \mathcal{O}(b^2); \qquad (2.42)$$

therefore the resummation of leading divergences in (2.41) amounts exactly to replacing

$$b \to \frac{b}{1 + z\frac{S_D}{2\varepsilon}}$$

in the first order correlation function (2.42). Owing to (2.22), this is indeed equivalent to replacing the bare dimensionless interaction parameter $z$ by the renormalized one $g = z/\left(1 + z\frac{S_D}{2\varepsilon}\right)$, in complete agreement with (2.35) above.

## 2.5. Analytic Continuation in D of the Euclidean Measure

Integrals like (2.17) or (2.20), written with Cartesian coordinates, are *a priori* meaningful only for integer $D$. Up to now, we have only formally extended such integrals to non-integer dimensions. Actually, an analytic continuation in $D$ can be performed by use of *distance geometry*.[22] The key idea is to substitute for the internal Euclidean coordinates $x_i$ the set of all mutual (squared) distances $a_{ij} = (x_i - x_j)^2$ (Fig. 3).

This is possible for integrands invariant under the group of Euclidean motions (as in (2.17) and (2.20)). For $N$ integration points, it also requires, *before* analytic continuation, $D$ to be large enough, *i.e.*, $D \geq N - 1$, such that the $N - 1$ relative vectors spanning these points are linearly independent.

We define the graph $\mathcal{G}$ as the set $\mathcal{G} = \{1, \ldots, N\}$ labeling the interaction points. Vertices $i \in \mathcal{G}$ will be remnants of the original Euclidean points after analytic

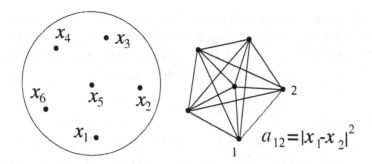

Fig. 3.   Passage from Euclidean coordinates $x_i$ to the complete set of squared distances $a_{ij}$.

continuation, and index the squared distance matrix $[a_{ij}]$. The change of variables $\{x_i\}_{i\in\mathcal{G}} \to a \equiv [a_{ij}]_{\substack{i<j \\ i,j\in\mathcal{G}}}$ reads explicitly[22]:

$$\frac{1}{V}\int_{\mathbb{R}^D}\prod_{i\in\mathcal{G}}d^D x_i \cdots = \int_{A_\mathcal{G}} d\mu_\mathcal{G}^{(D)}(a)\cdots, \tag{2.43}$$

with the measure

$$d\mu_\mathcal{G}^{(D)}(a) \equiv \prod_{\substack{i<j \\ i,j\in\mathcal{G}}} da_{ij}\,\Omega_N^{(D)}\,(P_\mathcal{G}(a))^{\frac{D-N}{2}}, \tag{2.44}$$

where $N = |\mathcal{G}|$, and

$$\Omega_N^{(D)} \equiv \prod_{K=0}^{N-2} \frac{S_{D-K}}{2^{K+1}} \tag{2.45}$$

$\left(S_D = \frac{2\pi^{D/2}}{\Gamma(D/2)}\right.$ is as before the volume of the unit sphere in $\mathbb{R}^D\big)$, and

$$P_\mathcal{G}(a) \equiv \frac{(-1)^N}{2^{N-1}}\begin{vmatrix} 0 & 1 & 1 & \cdots & 1 \\ 1 & 0 & a_{12} & \cdots & a_{1N} \\ 1 & a_{12} & 0 & \cdots & a_{2N} \\ \vdots & \vdots & \vdots & \ddots & \vdots \\ 1 & a_{1N} & a_{2N} & \cdots & 0 \end{vmatrix}. \tag{2.46}$$

The factor $\Omega_N^{(D)}$ (2.45) is the volume of the rotation group of the rigid simplex spanning the points $x_i$. The "Cayley–Menger determinant"[29] $P_\mathcal{G}(a)$ is proportional to the squared Euclidean volume of this simplex, a polynomial of degree $N-1$ in the $a_{ij}$. The set $a$ of squared distances has to fulfil the triangular inequalities and their generalizations: $P_\mathcal{K}(a) \geq 0$ for all subgraphs $\mathcal{K} \subset \mathcal{G}$, which defines the domain of integration $A_\mathcal{G}$ in (2.43).

For real $D > |\mathcal{G}|-2 = N-2$, $d\mu_\mathcal{G}^{(D)}(a)$ is a positive measure on $A_\mathcal{G}$, analytic in $D$. It is remarkable that, as a distribution, it can be extended to $0 \leq D \leq |\mathcal{G}|-2$.[22] For integer $D \leq |\mathcal{G}| - 2$, although the change of variables from $x_i$ to $a_{ij}$ no longer exists, Eq. (2.44) still reconstructs the correct measure, concentrated on $D$-dimensional submanifolds of $\mathbb{R}^{N-1}$, i.e., $P_{\mathcal{K}} = 0$ if $D \leq |\mathcal{K}| - 2$.[22] For example, when $D \to 1$ for $N = 3$ vertices, we have, denoting the distances $|ij| = \sqrt{a_{ij}}$:

$$\frac{d\mu_{\{1,2,3\}}^{(D \to 1)}(a)}{d_{|12|} d_{|13|} d_{|23|}} = 2\delta(|12| + |23| - |13|)$$

$$+ 2\delta(|13| + |32| - |12|) + 2\delta(|21| + |13| - |23|), \qquad (2.47)$$

which indeed describes the 6 possibilities for nested intervals in $\mathbb{R}$, with degeneracy factors 2 corresponding to the reversal of the orientation.

Another nice feature of this formalism is that *the interaction determinants in* (2.17) *and* (2.19) *are also Cayley–Menger determinants!* We have indeed

$$\det[\Pi_{ij}]_{1 \leq i,j \leq N-1} = P_{\mathcal{G}}(a^{\nu}) \qquad (2.48)$$

where $a^{\nu} \equiv [a_{ij}^{\nu}]_{\substack{i<j \\ i,j \in \mathcal{G}}}$ is obtained by simply raising each squared distance to the power $\nu$. We arrive for (2.17) and (2.20) at the representation of "Feynman diagrams" in distance geometry:

$$\mathcal{Z}_N = \int_{A_{\mathcal{G}}} d\mu_{\mathcal{G}}^{(D)} I_{\mathcal{G}}, \quad I_{\mathcal{G}} = (P_{\mathcal{G}}(a^{\nu}))^{-\frac{d}{2}}$$

$$\mathcal{Z}_N^{(0)}(\vec{k}) = \int_{A_{\mathcal{G} \cup \{0\}}} d\mu_{\mathcal{G} \cup \{0\}}^{(D)} I_{\mathcal{G}}^{(0)}(\vec{k}), \qquad (2.49)$$

$$I_{\mathcal{G}}^{(0)}(\vec{k}) = I_{\mathcal{G}} \exp\left(-\frac{1}{2}\vec{k}^2 \frac{P_{\mathcal{G} \cup \{0\}}(a^{\nu})}{P_{\mathcal{G}}(a^{\nu})}\right),$$

which are $D$-dimensional extensions of the Schwinger $\alpha$-parameter representation. We now have to study the actual convergence of these integrals and, possibly, their renormalization.

## 2.6. Analysis of Divergences

Large distance infrared (IR) divergences occur for manifolds of infinite size. One can keep a finite size, preserve symmetries and avoid boundary effects by choosing as a manifold the $D$-dimensional sphere $\mathcal{S}_D$ of radius $R$ in $\mathbb{R}^{D+1}$. This amounts[22] in distance geometry to substituting for $P_{\mathcal{G}}(a)$ the "spherical" polynomial $P_{\mathcal{G}}^{\mathcal{S}}(a) \equiv P_{\mathcal{G}}(a) + \frac{1}{R^2} \det\left(-\frac{1}{2}a\right)$, the second term providing an IR cut-off, such that $a_{ij} \leq 4R^2$. In the following, this IR regularization will simply be ignored when dealing with short-distance properties, for which we can take $P_{\mathcal{G}}^{\mathcal{S}} \sim P_{\mathcal{G}}$. This was also the case when evaluating leading divergences in the sections above.

The complete description of the possible set of divergences is then obtained from the following theorem of distance geometry[29]:

*Schoenberg's theorem.* For $0 < \nu < 1$, the set $a^{\nu} = [a_{ij}^{\nu}]_{\substack{i<j \\ i,j \in \mathcal{G}}}$ can be realized as the set of squared distances of a transformed simplex in $\mathbb{R}^{N-1}$, whose volume $P_{\mathcal{G}}(a^{\nu})$

*is positive, and vanishes if and only if at least one of the mutual original distances itself vanishes, $a_{ij} = 0$.*

This insures that, as in field theory, the only source of divergences in $I_{\mathcal{G}}$ and $I_{\mathcal{G}}^{(0)}$ is at *short distances*. Whether these UV singularities are integrable or not will depend on whether the external space dimension $d < d^* = D/\nu$ or $d > d^*$.

## 2.7. *Factorizations*

The key to convergence and renormalization is the following short-distance *factorization* property of $P_{\mathcal{G}}(a^\nu)$. Let us consider a subgraph $\mathcal{P} \subset \mathcal{G}$, with at least two vertices, in which we distinguish an element, the *root $p$* of $\mathcal{P}$, and let us denote by $\mathcal{G}/_p\mathcal{P} \equiv (\mathcal{G}\backslash\mathcal{P}) \cup \{p\}$ the subgraph obtained by replacing in $\mathcal{G}$ the whole subset $\mathcal{P}$ by its root $p$. In the original Euclidean formulation, the analysis of short-distance properties amounts to that of contractions of points $x_i$, labeled by such a subset $\mathcal{P}$, toward the point $x_p$, according to: $x_i(\rho) = x_p + \rho(x_i - x_p)$ if $i \in \mathcal{P}$, where $\rho \to 0^+$ is the dilation factor, and $x_i(\rho) = x_i$ if $i \notin \mathcal{P}$. This transformation has an immediate resultant in terms of mutual distances: $a_{ij} \to a_{ij}(\rho)$, depending on both $\mathcal{P}$ and $p$. Under this transformation, the interaction polynomial $P_{\mathcal{G}}(a^\nu)$ factorizes into[22]:

$$P_{\mathcal{G}}(a^\nu(\rho)) = P_{\mathcal{P}}(a^\nu(\rho))P_{\mathcal{G}/_p\mathcal{P}}(a^\nu)\{1 + \mathcal{O}(\rho^{2\delta})\} \qquad (2.50)$$

with $\delta = \min(\nu, 1 - \nu) > 0$ and where, by homogeneity, $P_{\mathcal{P}}(a^\nu(\rho)) = \rho^{2\nu}(|\mathcal{P}| - 1)P_{\mathcal{P}}(a^\nu)$.

The geometrical interpretation of (2.50) is quite simple: the contribution of the set $\mathcal{G}$ splits into that of the contracting subgraph $\mathcal{P}$ multiplied by that of the whole set $\mathcal{G}$ where $\mathcal{P}$ has been replaced by its root $p$ (Fig. 4), all correlation distances between these subsets being suppressed. The factorization property (2.50) is the generalization, to an arbitrary set $\mathcal{P}$ of contracting points, of the factorization encountered in (2.32) for the contraction of a pair of points. This is simply, in this interacting manifold model, the rigorous expression of an *operator product expansion*.[22]

The factorization property (2.50) does not hold for $\nu = 1$, preventing a factorization of the measure (2.44) $d\mu_{\mathcal{G}}^{(D)}(a)$ itself. Still, the integral of the measure, when applied to a factorized integrand, does factorize as:

$$\int_{\mathcal{A}_{\mathcal{G}}} d\mu_{\mathcal{G}}^{(D)} \cdots = \int_{\mathcal{A}_{\mathcal{P}}} d\mu_{\mathcal{P}}^{(D)} \cdots \int_{\mathcal{A}_{(\mathcal{G}/_p\mathcal{P})}} d\mu_{(\mathcal{G}/_p\mathcal{P})}^{(D)} \cdots . \qquad (2.51)$$

This fact, explicit for integer $D$ with a readily factorized measure $\Pi_i d^D x_i$, is preserved[22] by analytic continuation only after integration over relative distances between the two "complementary" subsets $\mathcal{P}$ and $\mathcal{G}/_p\mathcal{P}$.

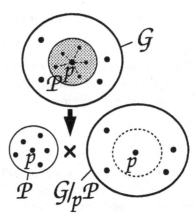

Fig. 4.   Factorization property (2.50).

## 2.8. Renormalization

A first consequence of factorizations (2.50) and (2.51) is the absolute convergence of $\mathcal{Z}_N$ and $\mathcal{Z}_N^0$ for $\varepsilon > 0$. Indeed, the superficial degree of divergence of $\mathcal{Z}_N$ (in distance units) is $(N-1)\varepsilon$, as can be read from (2.49), already ensuring the superficial convergence when $\varepsilon > 0$. The above factorizations ensure that the superficial degree of divergence in $\mathcal{Z}_N$ or $\mathcal{Z}_N^{(0)}$ of any subgraph $\mathcal{P}$ of $\mathcal{G}$ is exactly that of $\mathcal{Z}_{|\mathcal{P}|}$ itself, i.e., $(|\mathcal{P}|-1)\varepsilon > 0$. By recursion, this ensures the absolute convergence of the manifold Feynman integrals. A complete discussion has recourse to a generalized notion of Hepp sectors and is given in Ref. 22. In the proof, it is convenient to first consider $D$ large enough where $d\mu_{\mathcal{G}}^{(D)}$ is a non-singular measure, with a fixed $\nu$ considered as an independent variable $0 < \nu < 1$, and to then continue to $D = 2 - 2\nu, 0 < D < 2$, corresponding to the physical case.

When $\varepsilon = 0$, the integrals giving $\mathcal{Z}_N$ and $\mathcal{Z}_N^{(0)}$ are (logarithmically) divergent. Another consequence of Eqs. (2.50) and (2.51) is thus the possibility to devise a renormalization operation $\mathbf{R}$, as follows. To each contracting rooted subgraph $(\mathcal{P}, p)$ of $\mathcal{G}$, we associate a Taylor operator $\mathcal{T}_{(\mathcal{P},p)}$, performing on interaction integrands the exact factorization corresponding to (2.50):

$$\mathcal{T}_{(\mathcal{P},p)} I_{\mathcal{G}}^{(0)} = I_{\mathcal{P}} I_{\mathcal{G}/_p \mathcal{P}}^{(0)}, \tag{2.52}$$

and similarly $\mathcal{T}_{(\mathcal{P},p)} I_{\mathcal{G}} = I_{\mathcal{P}} I_{\mathcal{G}/_p \mathcal{P}}$. As in standard field theory,[17] the subtraction renormalization operator $\mathbf{R}$ is then organized in terms of forests à la Zimmermann. In manifold theory, we define a *rooted forest* as a set of rooted subgraphs $(\mathcal{P}, p)$ such that any two subgraphs are either disjoint or nested, i.e., never partially overlap. Each of these subgraphs in the forest will be contracted toward its root under the action (2.52) of the corresponding Taylor operator. When two subgraphs $\mathcal{P} \subset \mathcal{P}'$

are nested, the smallest one is contracted first toward its root $p$, the root $p'$ of $\mathcal{P}'$ being itself attracted toward $p$ if $p'$ happened to be in $\mathcal{P}$. This hierarchical structure is anticipated by choosing the roots of the forest as *compatible*: in the case described above, if $p' \in \mathcal{P}$, then $p' \equiv p$. Finally, the renormalization operator is written as a sum over all such compatibly rooted forests of $\mathcal{G}$, denoted by $\mathcal{F}_\oplus$:

$$\mathbf{R} = \sum_{\mathcal{F}_\oplus} W(\mathcal{F}_\oplus) \left[ \prod_{(\mathcal{P},p) \in \mathcal{F}_\oplus} (-\mathcal{T}_{(\mathcal{P},p)}) \right]. \qquad (2.53)$$

Here $W$ is a necessary combinatorial weight associated with the degeneracy of compatible rootings, $W(\mathcal{F}_\oplus) = \prod_{\substack{p \text{ root} \\ \text{of } \mathcal{F}_\oplus}} 1/|\mathcal{P}(p)|$ with $\mathcal{P}(p)$ being the largest subgraph of the forest $\mathcal{F}_\oplus$ whose root is $p$. An important property is that, with compatible roots, the Taylor operators of a given forest now commute.[22] The renormalized amplitudes are defined as

$$\mathcal{Z}_N^{\mathbf{R}\,(0)}(\vec{k}) \equiv \int_{A_{\mathcal{G}\cup\{0\}}} d\mu_{\mathcal{G}\cup\{0\}}^{(D)} \mathbf{R}\left[ I_{\mathcal{G}}^{(0)}(\vec{k}) \right]. \qquad (2.54)$$

The same operation $\mathbf{R}$ acting on $I_{\mathcal{G}}$ leads automatically by homogeneity to $\mathbf{R}[I_{\mathcal{G}}] = 0$ for $|\mathcal{G}| \geq 2$. We state the essential result that now *the renormalized Feynman integral* (2.54) *is convergent*: $\mathcal{Z}_N^{\mathbf{R}\,(0)} < \infty$ for $\varepsilon = 0$. A complete proof of this renormalizability property is given in Ref. 22, the analysis being inspired from the direct proof by Bergère and Lam of the renormalizability in field theory of Feynman amplitudes in the $\alpha$-representation.[30]

The physical interpretation of the renormalized amplitude (2.54) and of (2.53) is simple. Equations (2.51) and (2.52) show that the substitution for the bare amplitudes (2.49) of the renormalized ones (2.54) amounts to a reorganization to all orders of the original perturbation series in $b$, leading to the remarkable identity:

$$\mathcal{Z}^{(0)}(\vec{k}) = \sum_{N=1}^{\infty} \frac{(-b_{\mathbf{R}})^N}{N!} \mathcal{Z}_N^{\mathbf{R}\,(0)}(\vec{k}), \qquad (2.55)$$

where the *renormalized* interaction parameter $b_{\mathbf{R}}$ is simply here (minus) the connected partition function

$$b_{\mathbf{R}} \equiv -\mathcal{Z}. \qquad (2.56)$$

This actually extends to any vertex function, showing that the theory is made perturbatively finite (at $\varepsilon = 0$) by a full renormalization of the coupling constant $b$ into $-\mathcal{Z}$ itself, in agreement with the definition of the second virial coefficient $g$ (2.23) above. From this result, one establishes the existence to all orders of the Wilson function (2.37)

$$W(g,\varepsilon) = X \left. \frac{\partial g}{\partial X} \right|_b,$$

describing the scaling properties of the interacting manifold for $\varepsilon$ close to zero, and which has a *finite limit* up to $\varepsilon = 0$.[22] For $\varepsilon > 0$, an IR fixed point at $b > 0$ yields universal excluded volume exponents; for $\varepsilon < 0$, the associated UV fixed point at $b < 0$ describes a localization transition.

This demonstrated how to define an interacting manifold model with continuous internal dimension, by use of distance geometry, as a natural extension of the Schwinger representation for field theories. Furthermore, in the case of a pointwise interaction, the manifold model is indeed renormalizable to all orders. The main ingredients are Schoenberg's theorem of distance geometry, insuring that divergences occur only at short distances for (finite) manifolds, and the short-distance factorization of the generalized Feynman amplitudes. This provided probably the first example of a perturbative renormalization established for extended geometrical objects.[22] This opens the way to the renormalization theory of self-avoiding manifolds, which we now sketch.

## 3. Self-Avoiding Manifolds and Edwards Models

### 3.1. *Introduction*

In this part, we concentrate on the renormalization theory of the model of tethered self-avoiding manifolds (SAM) of Refs. 11 and 12, directly inspired by the Edwards model for polymers[2]:

$$\mathcal{H}/k_{\mathrm{B}}T = \frac{1}{2} \int d^D x (\nabla_x \vec{\mathbf{r}}(x))^2 + \frac{b}{2} \int d^D x \int d^D x' \delta^d(\vec{\mathbf{r}}(x) - \vec{\mathbf{r}}(x')), \qquad (3.1)$$

with an elastic Gaussian term and a self-avoidance two-body $\delta$-potential with excluded volume parameter $b > 0$. Notice that in contrast with the local $\delta$ interaction model (1.5) studied in Sec. 2, the interaction here is *non-local* in "manifold space" $\mathbb{R}^D$.

The finite upper critical dimension (u.c.d.) $d^*$ for the SA interaction exists only for manifolds with a continuous internal dimension $0 < D < 2$. For $D \to 2$, $d^* \to +\infty$. Phantom manifolds ($b = 0$) are *crumpled* with a finite Hausdorff dimension $d_H = 2D/(2-D)$, and $d^* = 2d_H$. The $\varepsilon$-expansion about $d^*$ performed in Refs. 11–13, and described in Part I above, was directly inspired by the des Cloizeaux *direct renormalization* (DR) method in polymer theory.[14] But the issue of the consistency of the DR method remained unanswered, since for $D \neq 1$, model (3.1) cannot be mapped onto a standard $(\Phi^2(\vec{\mathbf{r}}))^2$ local field theory.

The question of *boundary effects* in relation to the value of the configuration exponent $\gamma$ also requires some study.[13] It caused some confusion in earlier publications.[11-13] In Part I of these lectures, we showed that a finite self-avoiding patch embedded in an infinite Gaussian manifold has exponent $\gamma = 1$ for any $0 < D < 2$, $D \neq 1$. Here the cases of closed or open manifolds with free boundaries will be considered.

## 3.2. *Renormalizability to First Order*

The validity of RG methods and of scaling laws was first justified at leading order in $\varepsilon$ through explicit resummations in Ref. 20, in close analogy to the procedure described in Subsec. 2.3 for the $\delta$-interaction impurity model. We shall not repeat all the arguments here, but comment on some significant results.

Let us consider the spatial correlation function $\langle [\vec{r}(x) - \vec{r}(0)]^2 \rangle$. For a Gaussian (infinite) manifold it equals

$$\langle [\vec{r}(x) - \vec{r}(0)]^2 \rangle_0 = d[-2G(x)] = d\, A_D |x|^{2-D} = d \frac{2}{S_D(2-D)} |x|^{2-D}. \qquad (3.2)$$

In the presence of self-avoidance, it is expected to scale as:

$$\langle [\vec{r}(x) - \vec{r}(0)]^2 \rangle \propto |x|^{2\nu}, \qquad (3.3)$$

with a swelling exponent $\nu \geq \nu_0 = (2-D)/2$ for $d \leq d^*$. It can be directly evaluated by resummation of leading divergences[20]:

$$\langle [\vec{r}(x) - \vec{r}(0)]^2 \rangle = d \frac{2}{S_D(2-D)} |x|^{2-D} \left( 1 + \frac{a}{\varepsilon} b_D |x|^{\varepsilon/2} \right)^{a_0/a}, \qquad (3.4)$$

where $b_D$ is simply the bare interaction parameter $b$ conveniently dressed by coefficients

$$b_D = (2\pi A_D)^{-d/2} b = [4\pi/S_D(2-D)]^{-d/2} b,$$

and where $a_0$ and $a$ are two universal coefficients[20]:

$$a_0 = \frac{S_D^2}{D} \frac{2-D}{2}, \qquad a = S_D^2 \left( 1 + \frac{1}{2-D} \frac{\Gamma^2(D/(2-D))}{\Gamma(2D/(2-D))} \right), \qquad (3.5)$$

The scaling behavior (3.3) is then directly recovered from (3.4) in the large distance or strong self-avoidance limit $b|x|^{\varepsilon/2} \to +\infty$, with a value of the swelling exponent $\nu$ at first order in $\varepsilon$:

$$\nu = \frac{2-D}{2} + \frac{1}{2} \frac{a_0}{a} \frac{\varepsilon}{2}, \qquad (3.6)$$

or explicitly:

$$\nu = \frac{2-D}{2} \left\{ 1 + \frac{\varepsilon}{2} \frac{1}{2D} \left[ 1 + \frac{1}{2-D} \frac{\Gamma^2(D/(2-D))}{\Gamma(2D/(2-D))} \right]^{-1} \right\}, \qquad (3.7)$$

in agreement with the result (3.24) of Part I.

Similarly, for a manifold of finite volume $\mathcal{V} = X^D$, one defines a dimensionless excluded volume parameter $z$, as in Part I of these lectures, by

$$z = b_D X^{2D - (2-D)d/2} = (2\pi A_D)^{-d/2} b X^{\varepsilon/2}. \qquad (3.8)$$

One finds an effective size of the membrane:

$$R^2 = \langle [\vec{r}(X) - \vec{r}(0)]^2 \rangle = \mathcal{X}_0(z, \varepsilon) d \frac{2}{S_D(2-D)} X^{2-D} \qquad (3.9)$$

where $\mathcal{X}_0(z, \varepsilon)$ is the *swelling factor* with respect to the Gaussian size $\langle [\vec{r}(X) - \vec{r}(0)]^2 \rangle_0$ (3.2), as introduced in Part I, Eq. (3.1). The direct resummation of leading divergences to all perturbative orders gives[20]:

$$\mathcal{X}_0(z,\varepsilon) = \left(1 + \frac{a}{\varepsilon}b_D X^{\varepsilon/2}\right)^{a_0/a} = \left(1 + \frac{a}{\varepsilon}z\right)^{a_0/a}. \tag{3.10}$$

At first order in $z$, we recover

$$\mathcal{X}_0(z,\varepsilon) = 1 + \frac{a_0}{\varepsilon}z + \mathcal{O}(z^2), \tag{3.11}$$

which is the perturbative result (2.44) of Part I.

We also intoduced in Part I, Eqs. (3.7)–(3.9), the *dimensionless second virial coefficient* $g$

$$g = -(2\pi R^2/d)^{-d/2}\frac{\mathcal{Z}_{2,c}}{\mathcal{Z}_1^2}, \tag{3.12}$$

where $\mathcal{Z}_1$ and $\mathcal{Z}_{2,c}$ are respectively the (connected) 1-manifold and 2-manifold partition functions. The same direct resummation of leading divergences in perturbation theory gives for $g$

$$g = \frac{z}{1 + az/\varepsilon}, \tag{3.13}$$

with a first order expansion

$$g = z - z^2 a/\varepsilon + \cdots, \tag{3.14}$$

in agreement with I. Eqs. (3.16) and (3.17) [$a$ was noted as $a'_D$ there].

It is interesting to observe the following fact, key to a rigorous approach to renormalizability to first order. The RG flow equations were obtained in Part I. Eqs. (3.4), (3.19) and (3.22) from first order results (here in Part II, (3.11), (3.14)) for the scaling functions

$$W(g,\varepsilon) = X\frac{\partial g}{\partial X} = \frac{\varepsilon}{2}z\frac{\partial g}{\partial z} = \frac{\varepsilon}{2}z - z^2 a + \cdots = \frac{\varepsilon}{2}g - g^2\frac{a}{2} + \mathcal{O}(g^2) \tag{3.15}$$

$$X\frac{\partial}{\partial X}\ln\mathcal{X}_0(z,\varepsilon) = \frac{\varepsilon}{2}z\frac{\partial}{\partial z}\ln\mathcal{X}_0(z,\varepsilon) = \frac{1}{2}a_0 z + \cdots = \frac{1}{2}a_0 g + \mathcal{O}(g^2). \tag{3.16}$$

When truncated to this order, their solutions are exactly the resummed expressions (3.13) and (3.10). Turning things around, the direct resummation of leading poles in $\varepsilon$ indeed establishes one-loop renormalizability.[18,20]

## 3.3. *Renormalizability to All Orders*

We briefly describe below the formalism that allows to prove the validity of the RG approach to self-avoiding manifolds, as well as to a larger class of manifold models with non-local interactions. (See Refs. 22 and 26 for further details.) This formalism is based on an operator product expansion involving *multi-local singular operators*, which allows a systematic analysis of the short-distance ultraviolet singularities of the Edwards model. At the critical dimension $d^*$, one can classify all of the relevant operators and show that the model (3.1) is *renormalizable to all orders*

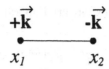

Fig. 5.   The dipole representing the $\delta$ interaction in (3.24).

by renormalizations (i) of the coupling $b$, and (ii) of the position field $\vec{r}$. As a consequence, one establishes the validity of scaling laws for *infinite* membranes, as well as the existence of finite size scaling laws for *finite* membranes. The latter result ensures the consistency of the DR approach.

A peculiar result, which distinguishes manifolds with non-integer $D$ from open linear polymers with $D = 1$, is the absence of *boundary* operator multiplicative renormalization, leading to the general *hyperscaling relation* for the configuration exponent $\gamma$

$$\gamma = 1 - \nu d, \qquad (3.17)$$

valid for finite SAM with $0 < D < 2, D \neq 1$. Note that this hyperscaling value is also valid for *closed* linear polymers (see, *e.g.*, Ref. 31) This result is valid for closed or open manifolds with free boundaries, and has the same origin as the result $\gamma = 1$ obtained in Part I for a finite SA patch embedded in an infinite manifold (see Part I, Subsecs. 2.2.2 and 2.2.3.)

### 3.4. *Perturbation Theory and Dipole Representation*

As in Part I, the partition function is defined by the functional integral:

$$\mathcal{Z} = \int \mathcal{D}[\vec{r}(x)] \exp\left(-\mathcal{H}[\vec{r}]/k_{\mathrm{B}}T\right). \qquad (3.18)$$

It has a perturbative expansion in $b$, formally given by expanding the exponential of the contact interaction

$$\mathcal{Z} = \mathcal{Z}_0 \sum_{N=0}^{\infty} \frac{(-b/2)^N}{N!} \int \prod_{i=1}^{2N} d^D x_i \left\langle \prod_{a=1}^{N} \delta^d(\vec{r}(x_{2a}) - \vec{r}(x_{2a} - 1)) \right\rangle_0$$

$$\equiv \mathcal{Z}_0 \sum_{N=0}^{\infty} \frac{(-b/2)^N}{N!} Z_N, \qquad (3.19)$$

where $\mathcal{Z}_0$ is the partition function of the Gaussian manifold (hence $\mathcal{Z}_0 \equiv 1$), and $\langle \cdots \rangle_0$ denotes the average with respect to the Gaussian manifold ($b = 0$):

$$\langle\!\langle \cdots \rangle\!\rangle_0 = \frac{1}{\mathcal{Z}_0} \int \mathcal{D}[\vec{r}(x)] \exp\left(-\frac{1}{2} \int d^D x (\nabla_x \vec{r}(x))^2\right) (\cdots). \qquad (3.20)$$

Physical observables are provided by average values of operators, which must be invariant under global translations. Using Fourier representation, local operators can always be generated by the exponential operators (or vertex operators), of the form

$$V_{\vec{q}}(z) = e^{i\vec{q}\cdot\vec{r}(z)}. \qquad (3.21)$$

In perturbation theory the field $\vec{r}(x)$ will be treated as a massless free field and the momenta $\vec{q}$ will appear as the "charges" associated with the translations in $\mathbb{R}^d$. Translationally invariant operators are then provided by "neutral" products of such local operators,

$$O_{\vec{q}_1,\ldots,\vec{q}_P}(z_1,\ldots,z_P) = \prod_{l=1}^{P} V_{\vec{q}_l}(z_l), \quad \vec{q}_{\text{total}} = \sum_{l=1}^{P} \vec{q}_l = \vec{0}. \qquad (3.22)$$

The perturbative expansion for these observables is simply

$$\left\langle \prod_{l=1}^{P} e^{i\vec{q}_l\cdot\vec{r}(z_l)} \right\rangle = \frac{1}{\mathcal{Z}} \sum_{N=0}^{\infty} \frac{(-b/2)^N}{N!} \int \prod_{i=1}^{2N} d^D x_i$$

$$\times \left\langle \prod_{l=1}^{P} e^{i\vec{q}_l\cdot\vec{r}(z_l)} \prod_{a=1}^{N} \delta^d\left(\vec{r}(x_{2a}) - \vec{r}(x_{2a-1})\right) \right\rangle_0$$

$$\equiv \frac{1}{\mathcal{Z}} \sum_{N=0}^{\infty} \frac{(-b/2)^N}{N!} Z_N(\{\vec{q}_l\}). \qquad (3.23)$$

Each $\delta$ function in (3.19) and (3.23) can itself be written in terms of two exponential operators as

$$\delta^d(\vec{r}(x_2) - \vec{r}(x_1)) = \int \frac{d^d\vec{k}_1 d^d\vec{k}_2}{(2\pi)^d} \delta^d(\vec{k}_1 + \vec{k}_2) e^{i\vec{k}_1\cdot\vec{r}(x_1)} e^{i\vec{k}_2\cdot\vec{r}(x_2)}. \qquad (3.24)$$

Viewing again the momenta $\vec{k}_1$, $\vec{k}_2$ as charges assigned to the points $x_1$, $x_2$, the bi-local operator (3.24) corresponds to a dipole, with charges $\vec{k}_1 = \vec{k}$, $\vec{k}_2 = -\vec{k}$, integrated over its internal charge $\vec{k}$. We depict graphically each such dipole as in Fig. 5.

Similarly, the product of bi-local operators in (3.19) and (3.23) can be written as an ensemble of $N$ dipoles, that is as the product of $2N$ vertex operators with $N$ "dipolar constraints"

$$C_a\{\vec{k}_i\} = (2\pi)^d \delta^d(\vec{k}_{2a-1} + \vec{k}_{2a}), \qquad (3.25)$$

then integrated over all internal charges $\vec{k}_i$:

$$\prod_{a=1}^{N} \delta^d(\vec{r}(x_{2a}) - \vec{r}(x_{2a-1})) = \int \prod_{i=1}^{2N} \frac{d^d\vec{k}_i}{(2\pi)^d} \prod_{a=1}^{N} C_a\{\vec{k}_i\} \prod_{i=1}^{2N} e^{i\vec{k}_i\cdot\vec{r}(x_i)}. \qquad (3.26)$$

Products of such bi-local operators and of external vertex operators, as in (3.23), are depicted by diagrams such as that of Fig. 6.

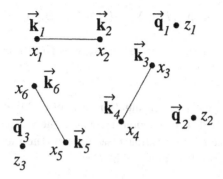

Fig. 6.   Dipole and charges representing bi-local operators and external vertex operators in (3.23).

The Gaussian average in (3.19), (3.26) is easily performed, and with the neutrality condition $\sum_i \vec{k}_i = \vec{0}$, we can rewrite it as

$$\left\langle \prod_i e^{i\vec{k}_i \cdot \vec{r}(x_i)} \right\rangle_0 = \exp\left(-\frac{1}{2}\sum_{i,j}\vec{k}_i \cdot \vec{k}_j G(x_i - x_j)\right), \qquad (3.27)$$

with, as before, the translationally invariant two-point function

$$G(x_i - x_j) = -\frac{1}{2}\langle(\vec{r}(x_i) - \vec{r}(x_j))^2\rangle_0 = -\frac{|x_i - x_j|^{2-D}}{(2-D)S_D}. \qquad (3.28)$$

Integration over the momenta $\vec{k}_i$ then gives for the $N$'th term of the perturbative expansion for the partition function $\mathcal{Z}$ (3.19) the "manifold integral"

$$Z_N = (2\pi)^{-Nd/2}\int \prod_{i=1}^{2N} d^D x_i\,\Delta\{x_i\}^{-\frac{d}{2}}, \qquad (3.29)$$

with $\Delta\{k_i\}$ the determinant associated with the auxiliary quadratic form (now on $\mathbb{R}^{2N}$) $Q\{k_i\} \equiv \sum_{i,j=1}^{2N} k_i k_j \, G(x_i, x_j)$, restricted to the $N$-dimensional vector space defined by the $N$ neutrality constraints $C_a\{k_i\}$, $k_{2a} + k_{2a-1} = 0$. $\Delta\{x_i\}$ is given explicitly by the determinant of the $N \times N$ matrix $\Delta_{ab}$ (with row and columns labeled by the dipole indices $a, b = 1, \dots, N$)

$$\Delta = \det(\Delta_{ab}),$$
$$\Delta_{ab} = G(x_{2a-1}, x_{2b-1}) + G(x_{2a}, x_{2b}) - G(x_{2a-1}, x_{2b}) - G(x_{2a}, x_{2b-1}). \qquad (3.30)$$

Similarly, the $N$'th term in the perturbative expansion of the $P$-point observable (3.23) is

$$Z_N(\{\vec{q}_l\}) = (2\pi)^{-Nd/2}\int \prod_{i=1}^{2N} d^D x_i \Delta\{x_i\}^{-\frac{d}{2}} \exp\left(-\frac{1}{2}\sum_{l,m=1}^{P}\vec{q}_l \cdot \vec{q}_m \frac{\Delta^{lm}}{\Delta}\right). \qquad (3.31)$$

$\Delta^{lm}$ is the $(lm)$ minor of the $(P+N) \times (P+N)$ matrix

$$\begin{bmatrix} G(z_l, z_m) & G(z_l, x_{2b-1}) - G(z_l, x_{2b}) \\ G(x_{2a-1}, z_m) - G(x_{2a}, z_m) & \Delta_{ab} \end{bmatrix}_{\substack{1 \leq l, m \leq P \\ 1 \leq a, b \leq N}} . \qquad (3.32)$$

Note that a proper analytic continuation in $D$ of (3.29) and (3.31) is insured, as in Sec. 2, by the use of distance geometry, where the Euclidean measure over the $x_i$ is understood as the corresponding measure over the mutual squared distances $a_{ij} = |x_i - x_j|^2$, a distribution analytic in $D$.[22]

### 3.5. *Singular Configurations and Electrostatics in* $\mathbb{R}^D$

The integrand in (3.29) is singular when the determinant vanishes, $\Delta\{x_i\} = 0$, or undefined if the latter becomes negative. The associated quadratic form $Q\{k_i\} = \sum_{i,j=1}^{2N} k_i k_j G(x_i, x_j)$, restricted by the $N$ neutrality constraints $\mathcal{C}_a\{k_i\}$: $k_{2a} + k_{2a-1} = 0, a = 1, \ldots, N$, is exactly the *electrostatic energy* of a gas of $2N$ scalar charges $k_i$ located at points $x_i$ in $\mathbb{R}^D$, and constrained to form $N$ neutral pairs $a$ of charges (dipoles). For such a globally neutral gas, the Coulomb energy is *minimal* when the charge density is *zero everywhere*, *i.e.*, when the non-zero charges $k_i$ aggregate into *neutral* "atoms". When $0 < D < 2$, because of the vanishing of the Coulomb potential at the origin, $G(0) = 0$, the corresponding minimal energy is furthermore *zero*, which implies that the quadratic form $Q$ is *non-negative*, and thus its determinant is also non-negative: $\Delta \geq 0$.

Singular $\{x_i\}$ configurations, with $\Delta = 0$, still exist when $Q$ is degenerate, which happens when some dipoles are assembled in such a way that, with appropriate non-zero charges, they still can build neutral atoms. This requires some of the points $x_i$ to coincide *and* the corresponding dipoles to form at least one closed loop (Fig. 7). This ensures that the only sources of divergences are *short-distance singularities*, and extends the Schoenberg theorem used above.

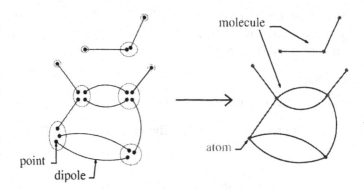

Fig. 7.   The notions of "atoms" and "molecules", built up from dipoles.

### 3.6. *Multi-local Operator Product Expansion*

A singular configuration can thus be viewed as a connected "molecule" (Fig. 7), characterized by a set $\mathcal{M}$ of "atoms" $p$ with assigned positions $x_p$, and by a set $\mathcal{L}$ of links $a$ between these atoms, representing the dipolar constraints $\mathcal{C}_a$ associated with the $\delta_a \equiv \delta^d(\vec{r}(x_{2a}) - \vec{r}(x_{2a-1}))$ interactions. For each $p$, we denote by $\mathcal{P}_p$ the set of charges $i$, at $x_i$, close to point $p$, which build the atom $p$ and define the relative (short) distances $y_i = x_i - x_p$ for $i \in \mathcal{P}_p$ (Fig. 8).

The short-distance singularity of $\Delta^{-d/2}$ is then analyzed by performing a small $y_i$ expansion of the product of the bilocal operators $\delta_a$ for the links $a \in \mathcal{L}$, in the Gaussian manifold theory (Eq. (3.23)). This expansion around $\mathcal{M}$ can be written as a *multi-local operator product expansion* (MOPE)

$$\prod_{a \in \mathcal{L}} \delta^d(\vec{r}(x_{2a}) - \vec{r}(x_{2a-1})) = \sum_{\Phi} \Phi\{x_p\} C \underbrace{\Phi_{\delta \ldots \delta}}_{|\mathcal{L}|} \{y_i\}, \qquad (3.33)$$

where the sum runs over all multi-local operators $\Phi$ of the form:

$$\Phi\{x_p\} = \int d^d\vec{r} \prod_{p \in \mathcal{M}} \left\{ : \{(\nabla_{\vec{r}})^{q_p} \delta^d(\vec{r} - \vec{r}(x_p))\} A_p(x_p) : \right\} \qquad (3.34)$$

Here $A_p(x_p) \equiv A^{(r_p, s_p)}[\nabla_x; \vec{r}(x_p)]$ is a local operator at point $x_p$, which is a combination of powers of $x$-derivatives and field $\vec{r}$, of degree $s_p$ in $\vec{r}(x_p)$ and degree $r_p \geq s_p$ in $\nabla_x$. $(\nabla_{\vec{r}})^{q_p}$ denotes a product of $q_p$ derivatives with respect to $\vec{r}$, acting on $\delta^d(\vec{r} - \vec{r}(x_p))$. The symbol ": :" denotes the *normal product* subtraction prescription at $x_p$ (which, in a Gaussian average, amounts to setting to zero any derivative of the propagator $G_{ij}$ at coinciding points $x_i = x_j = x_p$). For $\mathrm{Card}(\mathcal{M}) \equiv |\mathcal{M}| > 1$,

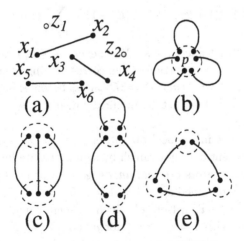

Fig. 8.   A general diagram with two external points and three internal dipoles, representing bilocal interactions $\delta_a$ (a); "molecules" describing singular configurations with one (b), two (c, d) and three (e) "atoms". (b, c, d) give UV divergences, (e) does not.

(3.34) describes the most general $|\mathcal{M}|$-body contact interaction between the points $x_p$, with possible inserted local operators $A_p(x_p)$ at each point $x_p$. For $|\mathcal{M}| = 1$, it reduces to a local operator $A_p(x_p)$.

The coefficient associated with the operator $\Phi$ in the MOPE, $C^{\Phi}_{\delta\ldots\delta}\{y_i\}$, can be written as an integral over the momenta $\vec{k}_i$:

$$C^{\Phi}_{\delta\ldots\delta}\{y_i\} = \int \prod'_{a\in\mathcal{L}} C_a\{\vec{k}_i\} \prod_{p\in\mathcal{M}}$$

$$\times \left\{ \prod_{i\in\mathcal{P}_p} d^d\vec{k}_i \left\{ (\nabla_{\vec{k}})^{q_p}\delta^d\left(\sum_{i\in\mathcal{P}_p}\vec{k}_i\right)\right\} C^{A_p}\{y_i,\vec{k}_i\} e^{-\frac{1}{2}\sum_{i,j\in\mathcal{P}_p}\vec{k}_i\cdot\vec{k}_j G_{ij}}\right\},$$

$$(3.35)$$

where $C^{A_p}\{y_i,\vec{k}_i\}$ is a monomial in the $\{y_i,\vec{k}_i\}$'s, associated with the operator $A_p$, of similar global degree $r_p$ in the $\{y_i\}$'s, and $s_p$ in the $\{\vec{k}_i\}$'s. The product $\prod'$ is over all constraints $a \in \mathcal{L}$ but one.

The MOPE (3.33) follows from the expression (3.24) in terms of free field exponentials plus constraints, and is established in Ref. 26.

### 3.7. Power Counting and Renormalization

The MOPE (3.33) allows us to determine those singular configurations which give rise to actual UV divergences in the manifold integrals (3.29) or (3.31). Indeed, for a given singular configuration $\mathcal{M}$, by integrating over the domain where the relative positions $y_i = x_i - x_p$ are of order $|y_i| \lesssim \rho$, we can use the MOPE (3.33) to obtain an expansion of the integrand in (3.23) in powers of $\rho$. Each coefficient $C^{\Phi}_{\delta\ldots\delta}$ gives a contribution of order $\rho^{\omega_\Phi}$, with degree $\omega_\Phi$ given by power counting as

$$\omega_\Phi = D\{2|\mathcal{L}| - |\mathcal{M}|\} + d\nu_0\{|\mathcal{M}| - |\mathcal{L}| - 1\} + \sum_{p\in\mathcal{M}}\{\nu_0(q_p - s_p) + r_p\} \quad (3.36)$$

with $\nu_0 = (2 - D)/2 < 1$ and $r_p \geq s_p$. Whenever $\omega_\Phi \leq 0$, a UV divergence occurs, as a factor multiplying the insertion of the corresponding operator $\Phi$.

*At the upper critical dimension* $d^* = 2D/\nu_0$, $\omega_\Phi$ becomes *independent* of the number $|\mathcal{L}|$ of dipoles, and is equal to the canonical dimension $\omega_\Phi$ of $\int \prod_{p\in\mathcal{M}} d^D x_p \Phi\{x_p\}$ in the Gaussian theory.

Only three relevant multi-local operators $\Phi$, with $\omega_\Phi \leq 0$ and such that the corresponding coefficient does not vanish by symmetry, are found by simple inspection. Two of these operators are *marginal* ($\omega_\Phi = 0$) at $d^*$: (i) the one-body local elastic term $:(\nabla\vec{r}_p)^2:$, obtained for $|\mathcal{M}| = 1$ ($q = 0, r = s = 2$); (ii) the two-body SA interaction term $\delta^d(\vec{r}_p - \vec{r}_{p'})$ itself, obtained through singular configurations with $|\mathcal{M}| = 2$ atoms (and with $q = r = s = 0$ for $p$ and $p'$) (see Fig. 9).

A third operator is *relevant* with $\omega_\Phi = -D$, *i.e.*, the identity operator $\mathbf{1}$ obtained when $|\mathcal{M}| = 1$ ($q = r = s = 0$). It describes insertions of local "free energy" divergences along the manifold, proportional to the manifold volume, which factor

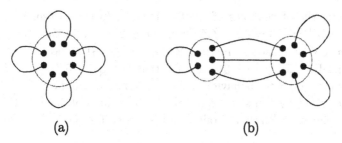

Fig. 9. "Molecules" $\mathcal{M}$ producing (a) the one-body local elastic term $\Phi =: (\nabla \vec{r}_p)^2 :$, and (b) the two-body SA interaction term $\Phi = \delta^d(\vec{r}_p - \vec{r}_{p'})$.

out of partition functions like (3.19), to cancel out in correlation functions (3.23), as already explained in Part I, Subsec. 2.22.

The above analysis deals with *superficial UV divergences* only. A complete analysis of the general UV singularities associated with successive contractions toward "nested" singular configurations can be performed,[26] using the same techniques as in Ref. 22 (Part II, Subsec. 2.8). A basic fact is that an iteration of the MOPE only generates multi-local operators of the same type (3.34).

The results are[26]: (i) that the observables (3.23) are UV finite for $d < d^*(D)$, and are meromorphic functions in $d$ with poles at $d = d^*$; (ii) that a renormalization operation $\mathbf{R}$, similar to the subtraction operation of Subsec. 2.8,[22] can be achieved to remove these poles; (iii) that this operation amounts to a renormalization of the Hamiltonian (3.1).

More explicitly, the renormalized correlation functions $\langle \prod_{l=1}^{P} e^{i\vec{q}_l \cdot \vec{r}_\mathbf{R}(z_l)} \rangle_\mathbf{R}$ have a finite perturbative expansion in the renormalized coupling $b_\mathbf{R}$, when $\langle \cdots \rangle_\mathbf{R}$ is the average w.r.t. the renormalized Hamiltonian

$$\mathcal{H}_\mathbf{R}/k_\mathrm{B}T = \frac{1}{2}Z \int d^D x (\nabla_x \vec{r}_\mathbf{R}(x))^2$$

$$+ \frac{1}{2}b_\mathbf{R}\mu^{\varepsilon/2}Z_b \int d^D x \int d^D x' \delta^d(\vec{r}_\mathbf{R}(x) - \vec{r}_\mathbf{R}(x')). \tag{3.37}$$

Here $\mu$ is a renormalization (internal) momentum scale, necessary for infinite manifolds, $\varepsilon = 4D - 2dv_0$; $Z(b_\mathbf{R})$ and $Z_b(b_\mathbf{R})$ are respectively the field and coupling constant renormalization factors, singular at $\varepsilon = 0$.

At first order, one finds by explicitly calculating $C_\delta^{(\nabla\vec{r})^2}$ and $C_{\delta\delta}^\delta$ that[26]

$$Z = 1 + (2\pi A_D)^{-d/2} \frac{b_\mathbf{R}}{\varepsilon} \frac{S_D^2(2-D)}{2D},$$

$$Z_b = 1 + (2\pi A_D)^{-d/2} \frac{b_\mathbf{R}}{\varepsilon} \frac{S_D^2}{2-D} \frac{\Gamma^2(D/(2-D))}{\Gamma(2D/(2-D))},$$

with $A_D = [S_D(2-D)/2]^{-1}$. For quantities which do not stay finite in the infinite manifold limit $\mathcal{V} \to +\infty$, like partition functions, a shift in the free energy (*i.e.*, an "additive counterterm" in $\mathcal{H}_\mathbf{R}$), proportional to $\mathcal{V}$, is also necessary.

Expressing the observables of the SAM model (3.1) in terms of renormalized variables $\vec{r} = Z^{1/2}\vec{r}_{\mathbf{R}}, b = b_{\mathbf{R}}\mu^{\varepsilon/2}Z_b Z^{d/2}$, one can derive in the standard way RG equations involving Wilson's functions $W(b_{\mathbf{R}}) = \mu\frac{\partial}{\partial\mu}b_{\mathbf{R}}|_b$, $\nu(b_{\mathbf{R}}) = \nu_0 - \frac{1}{2}\mu\frac{\partial}{\partial\mu}\ln Z|_b$. A non-trivial IR fixed point $b_{\mathbf{R}}^* \propto \varepsilon$ such that $W(b_{\mathbf{R}}^*) = 0$ is found for $\varepsilon > 0$. It governs the large distance behavior of the SA infinite manifold, which obeys scaling laws characterized by the size exponent $\nu$. The value obtained in this approach, $\nu = \nu(b_{\mathbf{R}}^*)$, coincides with that obtained at first order in $\varepsilon$ in Eq. (3.7) above.[11–13,20 d]

### 3.8. *Finite Size Scaling and Direct Renormalization*

The direct renormalization formalism considered in Part I, and in Part II, Subsec. 3.2, deals with *finite* manifolds with internal volume $\mathcal{V}$, and expresses scaling functions in terms of a dimensionless second virial coefficient (3.12) $g = -(2\pi R^2/d)^{-d/2}\mathcal{Z}_{2,c}/(\mathcal{Z}_1)^2$, where $\mathcal{Z}_1(\mathcal{V})(= \mathcal{Z}/\mathcal{Z}_0)$ and $\mathcal{Z}_{2,c}(\mathcal{V})$ are respectively the one- and (connected) two-membrane partition functions, and $R$ is the effective radius of the membrane.

When dealing with a finite closed manifold (for instance the $D$-dimensional sphere $\mathcal{S}_D$ (Part II, Subsec. 2.6 and Ref. 22), characterized by its (curved) internal metric, the massless propagator $G$ gets modified. Nevertheless, from the short-distance expansion of $G$ in a general metric,[e] one can show that the short-distance MOPE (3.33) remains valid. The expansion then extends to multi-local operators $\Phi$ of the form (3.34), with local operators $A(x)$ which may involve the Riemann curvature tensor and its derivatives, with appropriate coefficients $C_{\delta\ldots\delta}^{\Phi}$.[26] Still, the coefficients for those operators $\Phi$ that do not involve derivatives of the metric stay the same as in Euclidean flat space.

At $d^*$, UV divergences still come with insertions of relevant multi-local operators with $\omega_\Phi \leq 0$. When $0 < D < 2$, the operators involving curvature are found by power counting to be *all irrelevant*. Thus the *flat infinite membrane* counterterms $Z$ and $Z_b$ still renormalize the (curved) finite membrane theory. Standard arguments parallel to those of Ref. 16 for polymers then help to establish the direct renormalization formalism (see Ref. 26). The second virial coefficient $g(b, \mathcal{V})$ (as any *dimensionless* scaling function) is UV finite once expressed as a function $g_{\mathbf{R}}(b_{\mathbf{R}}, \mathcal{V}\mu^D)$ of $b_{\mathbf{R}}$ (and $\mu$). Then the scaling functions, when expressed in terms of $g$, obey RG flow equations, and stay *finite up to* $\varepsilon = 0$. The existence of a non-trivial IR fixed point $b_{\mathbf{R}}^*$ for $\varepsilon > 0$ implies that in the large volume or strong interaction limit,

---

[d]One can notice the identity between coefficients in the renormalization factors $Z$, $Z_b$ above, and (3.5).

[e]The expansion at the origin of the massless propagator $\tilde{G}$ on a curved manifold reads in Riemann normal coordinates $\tilde{G}(x) \simeq G(x) - \frac{|x|^2}{2D}\langle:(\nabla\mathbf{r})^2:\rangle$, with $G(x) \propto |x|^{2-D}$ the propagator in infinite flat space, and next order terms $\mathcal{O}(|x|^{4-D})$ proportional to the curvature and subdominant for $D < 2$; the normal product $(::)$ is still defined w.r.t. infinite flat space, and gives explicitly for a finite manifold with volume $\mathcal{V}\langle:(\nabla\mathbf{r})^2:\rangle = -1/\mathcal{V}$.

$b\mathcal{V}^{\varepsilon/2D} \to +\infty$, $g$ reaches a finite limit $g^* = g_R(b_R^*)$ (independent of $\mathcal{V}\mu^D$), and so do all scaling functions. This is just direct renormalization, **QED**.

### 3.9. *Hyperscaling*

Let us first consider a *closed* manifold. As mentioned above, the renormalization of partition functions for a (finite) SAM requires a shift $(-f_D X^D)$ of the free energy, proportional to the manifold volume $\mathcal{V}$, and corresponding to the integration of a local contact divergence in the bulk. The configuration exponent $\gamma$ is then defined by the scaling of the partition function[f]

$$\mathcal{Z}_1(\nu) = \mathcal{Z}_0^{-1} \int \mathcal{D}[\vec{r}] \delta^d(\vec{r}(0)) e^{-\beta\mathcal{H} - f_D \mathcal{V}} \sim \mathcal{V}^{\frac{\gamma-1}{D}}. \tag{3.38}$$

A consequence of the absence for closed SAM, for $0 < D < 2$, of relevant geometrical operators other than the point insertion one, is the general hyperscaling law (3.17) relating $\gamma$ to $\nu$ : $\gamma - 1 = -\nu d/D$. Indeed, from (3.38), $\mathcal{Z}_1$ is simply multiplicatively renormalized as $\mathcal{Z}_1(b, \mathcal{V}) = Z^{-d/2} \mathcal{Z}_1^R(b_R, \mathcal{V}\mu^D)$. This validates the hyperscaling hypothesis that $\mathcal{Z}_1 \sim \langle |\vec{r}|^{-d} \rangle \sim \mathcal{V}^{-\nu d/D}$. Equation (3.17) can been checked explicitly at order $\varepsilon$ for the sphere $\mathcal{S}_D$ and the torus $\mathcal{T}_D$.

For an *open* SAM with *free* boundaries, and when $1 \le D < 2$, the boundary operator $\int_{boundary} d^{D-1}x \mathbf{1}$ is *relevant*, requiring a boundary free energy shift $(-f_{D-1} X^{D-1})$. Since this does not enter into the bulk MOPE, it does not modify the renormalizations of $\vec{r}$ and $b$. Furthermore, only for integer $D = 1$ is it *marginally relevant*,[13] as explained in Part I; thus for $D \ne 1$ the hyperscaling relation (3.17) *remains valid*. Only for open polymers at $D = 1$, do the corresponding (zero-dimensional) end-point divergences enter the multiplicative renormalization of $\mathcal{Z}_1$, and $\gamma$ becomes an independent exponent. In polymer theory, an independent exponent actually appears for each *star vertex*.[32]

Previous calculations[11-13] did not involve the massless propagator $\tilde{G}$ on a finite manifold with Neumann boundary conditions, but the simpler propagator $G$ (3.28), corresponding to a finite SA patch immersed in an infinite Gaussian manifold. The same non-renormalization argument, as explained in Ref. 13 and in Part I, yields $\gamma = 1$ for non-integer $D$.

When $D = 2$, operators involving curvature and boundaries become relevant, and (3.17) is not expected to hold, either for closed or open manifolds, as also seen in Part I.

### 3.10. Θ-*Point and Long-Range Interactions*

The above formalism is actually directly applicable to a large class of manifold models where the interaction can be expressed in terms of free field exponentials with

---

[f]Here we consider $\mathcal{Z}_1(\mathcal{V}) \equiv \exp(-f_D X^D)\mathcal{Z}/\mathcal{Z}_0$, i.e., the *dimensionally regularized* partition function, which includes the free energy shift (see Part I, Subsec. 2.2.2).

suitable neutrality constraints $C_a\{\vec{k}_i\}$. Examples of such interactions are the $n$-body contact potentials, or the two-body long-range Coulomb potential $1/|\vec{r} - \vec{r}'|^{d-2}$, represented by modified dipolar constraints $C\{\vec{k}_i\} = |\vec{k}|^{-2}\delta^d(\vec{k}+\vec{k}')$. In these models the MOPE involves the same multi-local operators as in (3.34), with coefficients (3.35) built with the corresponding constraints $C_a$.

As an application of the MOPE, one finds that for a polymerized membrane at the $\Theta$-point where the two-body term $b$ in (3.1) vanishes, the most relevant short-range interaction is either the usual tricritical *three*-body contact potential, with u.c.d. $d_3^* = 3D/(2 - D)$, as for ordinary polymers,[33] or the two-body singular potential $\Delta_{\vec{r}}\delta^d(\vec{r} - \vec{r}')$ with u.c.d. $\tilde{d}_2^* = 2(3D - 2)/(2 - D)$. The latter is the most relevant one when $D > 4/3$ (see Ref. 34).

The very absence of *long-range* interactions in the MOPE shows that those interactions are not renormalized. When considering charged polymerized membranes with a two-body Coulomb potential for instance, the only (marginally) relevant operator at the upper critical dimension is the local elastic energy density : $(\nabla \vec{r})^2$ :, which indicates that only $\vec{r}$ is renormalized. As a consequence, one can show that $\nu = 2D/(d - 2)$ exactly, generalizing a well-known result for polymers.[35]

We did not address here other interesting issues: the approach to the physical $D = 2$ case from the $D < 2$ manifold theory,[27,36] numerical simulations of 2D polymerized membranes,[10] or the question of the actual physical phase (crumpled or flat) of a two-dimensional polymerized membrane in $d$-space.[37] We have concentrated instead on those more fundamental aspects of renormalization theory, that have been driven by the fascinating properties of these fluctuating polymerized membranes.

## Acknowledgments

These notes rely heavily on Ref. 18, and on the articles in Refs. 21 and 22: *Renormalization Theory for Interacting Crumpled Manifolds*, by François David, B.D., and Emmanuel Guitter. For the self-avoiding manifold Edwards model, I followed Refs. 20, 25 and 26: *Renormalization Theory for Self-Avoiding Polymerized Membranes*, by the same authors. I also wish to thank Emmanuel Guitter for his valuable help with the figures, and Thomas C. Halsey for a careful reading of the manuscript.

## References

1. *Statistical Mechanics of Membranes and Surfaces*, Proceedings of the Fifth Jerusalem Winter School for Theoretical Physics (1987), D.R. Nelson, T. Piran, and S. Weinberg eds. (World Scientific, Singapore, 1989).
2. S.F. Edwards, *Proc. Phys. Soc. Lond.* **85**, 613 (1965).
3. J. des Cloizeaux and G. Jannink, *Polymers in Solution, their Modelling and Structure*, (Clarendon Press, Oxford, 1990).
4. X. Wen, C.W. Garland, T. Hwa, M. Kardar, E. Kokufuta, Y. Li, M. Orkisz, and T. Tanaka, *Nature* **355**, 426 (1992).

5. M.S. Spector, E. Naranjo, S. Chiruvolu, and J.A. Zasadzinski, *Phys. Rev. Lett.* **73**, 2867 (1994).
6. C.F. Schmidt, K. Svoboda, N. Lei, I.B. Petsche, L.E. Berman, C. Safinya, and G.S. Grest, *Science* **259**, 952 (1993).
7. S.I. Stupp, S. Son, H.C. Lin, and L.S. Li, *Science* **259**, 59 (1993).
8. H. Rehage, B. Achenbach, and A. Kaplan, Ber. Bunsenges. *Phys. Chem.* **101**, 1683 (1997).
9. Y. Kantor, M. Kardar, and D.R. Nelson, *Phys. Rev. Lett.* **57**, 791 (1986); *Phys. Rev.* **A35**, 3056 (1987).
10. See, *e.g.*, the lectures by G. Gompper and D.M. Kroll in this volume, and the references therein.
11. M. Kardar and D.R. Nelson, *Phys. Rev. Lett.* **58**, 1289 (1987), 2280 (E); *Phys. Rev.* **A38**, 966 (1988).
12. J.A. Aronowitz and T.C. Lubensky, *Europhys. Lett.* **4**, 395 (1987).
13. B. Duplantier, *Phys. Rev. Lett.* **58**, 2733 (1987).
14. J. des Cloizeaux, *J. Phys. France* **42**, 635 (1981).
15. P.G. de Gennes, *Phys. Lett.* **A38**, 339 (1972).
16. M. Benhamou and G. Mahoux, *J. Phys. France* **47**, 559 (1986).
17. N.N. Bogoliubov and O.S. Parasiuk, *Acta Math.* **97**, 227 (1957); K. Hepp, *Commun. Math. Phys.* **2**, 301 (1966); W. Zimmermann, *Commun. Math. Phys.* **15**, 208 (1969).
18. B. Duplantier, *Phys. Rev. Lett.* **62**, 2337 (1989).
19. See also M. Lässig and R. Lipowsky, *Phys. Rev. Lett.* **70**, 1131 (1993).
20. B. Duplantier, T. Hwa, and M. Kardar, *Phys. Rev. Lett.* **64**, 2022 (1990).
21. F. David, B. Duplantier, and E. Guitter, *Phys. Rev. Lett.* **70**, 2205 (1993), hepth/9212102.
22. F. David, B. Duplantier, and E. Guitter, *Nucl. Phys.* **B394**, 555 (1993), cond-mat/9211038.
23. M. Cassandro and P.K. Mitter, *Nucl. Phys.* **B422**, 634 (1994).
24. P.K. Mitter and B. Scoppola, *Commun. Math. Phys.* **209**, 207 (2000).
25. F. David, B. Duplantier, and E. Guitter, *Phys. Rev. Lett.* **72**, 311 (1994), cond-mat/9307059.
26. F. David, B. Duplantier, and E. Guitter, *Renormalization Theory for Self-Avoiding Polymerized Membranes*, cond-mat/9701136.
27. F. David and K. Wiese, *Phys. Rev. Lett.* **76**, 4564 (1996); *Nucl. Phys.* **B487**, 529 (1997), cond-mat/9608022.
28. F. David and K. Wiese, *Nucl. Phys.* **B535**, 555 (1998), cond-mat/9807160.
29. L.M. Blumenthal, *Theory and Applications of Distance Geometry*, (Clarendon Press, Oxford, 1953).
30. M.C. Bergère and Y.-M.P. Lam, *J. Math. Phys.* **17**, 1546 (1976).
31. B. Duplantier, *Nucl. Phys.* **B430**, 489 (1995).
32. B. Duplantier, *J. Stat. Phys.* **54**, 581 (1989).
33. B. Duplantier, *J. Physique* **43**, 991 (1982).
34. K. Wiese and F. David, *Nucl. Phys.* **B450**, 495 (1995), cond-mat/9503126.
35. P. Pfeuty, R.M. Velasco, P.-G. de Gennes, *J. Physique* **38**, (1977) L5.
36. T. Hwa, *Phys. Rev.* **A41**, 1751 (1990).
37. One may consult, *e.g.*, K. Wiese, in *Phase Transitions and Critical Phenomena*, C. Domb and J.L. Lebowitz eds., Vol. 19 (2001), and the new chapters of the present volume by M. Bowick, G. Gompper and D.M. Kroll, and L. Radzihovsky.

# CHAPTER 10

# ANISOTROPIC AND HETEROGENEOUS POLYMERIZED MEMBRANES

Leo Radzihovsky

*Department of Physics, University of Colorado, Boulder, CO 80309*
*E-mail: radzihov@colorado.edu*

In these lectures I describe long scale properties of fluctuating polymerized membranes in the presence of network anisotropy and random heterogeneities. Amazingly, even infinitesimal amount of these seemingly innocuous but physically important ingredients in the membrane's internal structure leads to a wealth of striking qualitatively new phenomena. Anisotropy leads to a "tubule" phase that is intermediate in its properties and location on the phase diagram between previously discussed "crumpled" and "flat" phases. At low temperature, network heterogeneity generates conformationally glassy phases, with membrane normals exhibiting glass order analogous to spin-glasses. The common thread to these distinct membrane phases is that they exhibit universal anomalous elasticity, (singularly length-scale dependent elastic moduli, universal Poisson ratio, etc.), driven by thermal fluctuations and/or disorder and controlled by a nontrivial low-temperature fixed point.

## 1. Preamble

The nature of a membrane's in-plane order, with three well-studied universality classes, the isotropic liquid,[1] hexatic liquid[2] and solid,[3] crucially affects its conformational properties. The most striking illustration of this (discussed in lectures by Yacov Kantor and by David Nelson) is the stabilization in polymerized membranes (but not fluid ones, that are always crumpled[1] beyond a persistence length[4]) of a "flat" phase,[3] with long-range orientational order in the local membrane normals,[5,6] that is favored at low temperature over the entropically preferred high-temperature crumpled state. In a beautiful "order from disorder" phenomenon, a subtle interplay of wild thermal fluctuations with nonlinear membrane elasticity (made possible by a finite shear modulus of a solid) infinitely enhances membrane's bending rigidity, thereby stabilizing the flat phase against these very fluctuations.[3] A universal fluctuation-driven "anomalous elasticity" characterizes the resulting flat phase, with length-scale dependent elastic moduli, non-Hookean stress-strain relation, and a universal negative Poisson ratio.[7–9]

Given such a qualitative distinction between liquid and solid membranes, it is perhaps not too surprising that other in-plane orders can have important qualitative effects on membrane's long-scale properties. In these lectures I will discuss two such ingredients, namely, membrane network in-plane *anisotropy* (Sec. 2) and *heterogeneity* (Sec. 3) in fluctuating *polymerized* membranes, and will show that these seemingly innocuous generalizations, indeed lead to a wealth of new phenomena that are the subject of these lectures.

## 2. Anisotropic Polymerized Membranes

### 2.1. *Motivation and Introduction*

In addition to the basic theoretical motivation, the interest in anisotropic polymerized membranes is naturally driven by a number of possible experimental realizations, some of which are illustrated in Fig. 1. One example is a tethered membrane made through photo-polymerization of a fluid phospholipid membrane exhibiting lipid tilt order. On average, the lipids are tilted relative to the membrane normal, inducing a vector in-plane order and an intrinsic elastic anisotropy that can in principle be aligned with an electric or a magnetic field and polymerized in.

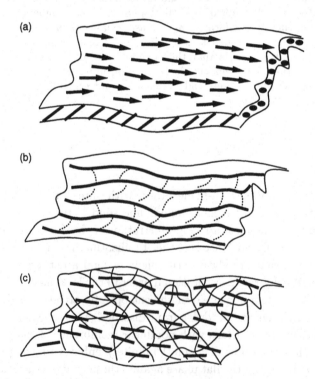

Fig. 1.   Examples of anisotropic polymerized membranes with (a) a polymerized-in lipid tilt order in a bilayer membrane, (b) a weakly crosslinked array of linear polymers, and (c) a spontaneous in-plane nematic order in a nematic elastomer membrane.

Another possible method[10] of fabricating polymerized sheets is by cross-linking a stretched out, aligned array of linear polymers, that would clearly lead to an intrinsically anisotropic tethered membrane. One other promising candidate is a two-dimensional sheet of a nematic elastomer,[11] a material that received considerable attention recently because of its novel elastic and electro-optic properties.[12] These cross-linked liquid-crystal polymer gels exhibit all standard liquid-crystal phases and therefore in their nematic state are highly anisotropic.

Almost 10 years ago, it was discovered[13] that in-plane anisotropy has a dramatic qualitative effect on the global phase diagram of polymerized membranes. As illustrated in Fig. 2, it leads to an entirely new "tubule" phase of a polymerized membrane (see Fig. 3), that is crumpled along one and extended along the other of the two membrane axes, with wild undulations about its average cylindrical geometry.[14]

This is possible because in an anisotropic membrane, where symmetry between $x_\perp$ and $y$ axes is broken, (e.g., curvature moduli $\kappa_{x_\perp} \neq \kappa_y$) the $x_\perp$- and $y$-directed

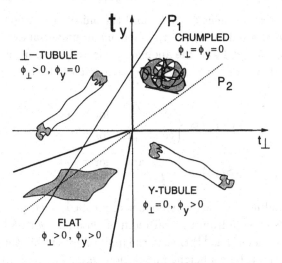

Fig. 2.   Phase diagram for anisotropic tethered membranes showing crumpled, tubule and flat phases as a function of reduced temperatures $t_\perp \propto T - T_c^\perp$, $t_y \propto T - T_c^y$.

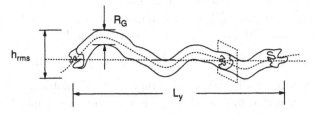

Fig. 3.   Schematic of the y-tubule phase of anisotropic polymerized membrane, with thickness $R_G$ and roughness $h_{\rm rms}$.

tangents $\vec{\zeta}_\perp = \partial_\perp \vec{r}$ and $\vec{\zeta}_y = \partial_y \vec{r}$ will generically order at two distinct temperatures $T_c^\perp \sim \kappa_{x_\perp}$ and $T_c^y \sim \kappa_y$, respectively, thereby allowing two intermediate tubule states: (1) $x_\perp$-extended tubule with $\langle \vec{\zeta}_\perp \rangle \neq 0$, $\langle \vec{\zeta}_y \rangle = 0$, for $T_c^y < T < T_c^\perp$ and (2) $y$-extended tubule with $\langle \vec{\zeta}_\perp \rangle = 0$, $\langle \vec{\zeta}_y \rangle \neq 0$, for $T_c^\perp < T < T_c^y$.

The direct crumpling transition occurs in such more generic anisotropic membrane only for that special set of cuts through the phase diagram (like P$_2$) that pass through the origin and is in fact tetra-critical. Generic paths (like P$_1$) will experience *two* phase transitions, crumpled-to-tubule, and tubule-to-flat, that are in distinct universality classes. The tubule phase is therefore not only generically possible, but actually unavoidable, in membranes with any type or amount of *intrinsic anisotropy*.[18]

As illustrated in Fig. 3 the tubule is characterized by its thickness $R_G$ (the radius of gyration of its crumpled boundary), and its undulations $h_{\mathrm{rms}}$ transverse to its average axis of orientation. Quite generally (for a $y$-extended tubule of an $L_\perp \times L_y$ membrane), they obey the scaling laws:

$$R_G(L_\perp, L_y) = L_\perp^\nu S_R(L_y/L_\perp^z), \quad h_{\mathrm{rms}}(L_\perp, L_y) = L_y^\zeta S_h(L_y/L_\perp^z), \tag{1}$$

where the roughness exponent $\zeta = \nu/z$, and the anisotropy exponent $z = (1+2\nu)/(3-\eta_\kappa) < 1$ are expressed in terms two independent universal exponents: the radius of gyration exponent $\nu < 1$ and the anomalous elasticity exponent $\eta_\kappa$ for the tubule bending rigidity defined by $\kappa \sim L_y^{\eta_\kappa}$. The *universal* scaling functions $S_{R,h}(x)$ have the limiting forms:

$$S_R(x) \propto \begin{cases} x^{\zeta - \nu_p/z}, & \text{for } x \to 0 \\ \text{constant}, & \text{for } x \to \infty \end{cases}, \tag{2}$$

$$S_h(x) \propto \begin{cases} \text{constant}, & \text{for } x \to 0 \\ x^{\frac{3}{2} - \zeta}, & \text{for } x \to \infty \end{cases}, \tag{3}$$

where $\nu_p$ is the radius of gyration exponent of a coiled linear polymer $\approx 3/5$. These asymptotic forms emerge from general requirements (supported by detailed renormalization group calculations[13]) that in the limit of a very thin tubule (in the curled up $\perp$ direction) the tubule's height undulations $h_{\mathrm{rms}}(L_y)$ reproduce statistics of a linear polymer of length $L_y$ and width $L_\perp$. And, that in the opposite limit of vanishing $L_y$, scaling functions must recover the radius of gyration $R_G(L_\perp) \sim L_\perp^{\nu_p}$ of a linear polymer of length $L_\perp$ and thickness $L_y$.

The scaling forms, Eqs. (2) and (3) imply that for a "roughly square" membrane — that is, one with $L_\perp \sim L_y \equiv L$ — in the limit $L \to \infty$

$$R_G(L_\perp \sim L_y \equiv L) \propto L^\nu, \quad h_{\mathrm{rms}}(L_\perp \sim L_y \equiv L) \propto L^{1 - \eta_\kappa z/2}. \tag{4}$$

Combining this with detailed renormalization group calculations, that find a strictly positive $\eta_\kappa$,[13] shows that $h_{\mathrm{rms}} \ll L$ for a large roughly square membrane. Thus, the end-to-end orientational fluctuations $\theta \sim h_{\mathrm{rms}}/L \propto L^{-\eta_\kappa/2} \to 0$ as $L \to \infty$ for such a roughly square membrane, proving that tubule order (which requires orientational persistence in the extended direction) *is* stable against undulations of the tubule embedded in $d = 3$ dimensions.

The rest of Sec. 2 is devoted to developing a model of an anisotropic polymerized membrane and using it to study phase transitions into and out of the tubule phase and the anomalous elastic properties of a fluctuating tubule summarized above.

## 2.2. *Model*

A model for anisotropic membranes is a generalization of the isotropic model discussed in David Nelson's lectures.[6] As there, we characterize the configuration of the membrane by the position $\vec{r}(\mathbf{x})$ in the $d$-dimensional embedding space of the point in the membrane labeled by a $D$-dimensional internal co-ordinate $\mathbf{x}$, with $d = 3$ and $D = 2$ corresponding to the physical case of interest.

Rotational and translational symmetries require that the Landau–Ginzburg free energy $F$ is an expansion in powers of tangent vectors $\vec{\xi}_{\perp,y} = \partial_{\perp,y}\vec{r}$ and their gradients

$$
\begin{aligned}
F[\vec{r}(\mathbf{x})] = \frac{1}{2} \int d^{D-1}x_\perp dy \Big[ & \kappa_\perp (\partial_\perp^2 \vec{r})^2 + \kappa_y (\partial_y^2 \vec{r})^2 + \kappa_{\perp y} \partial_y^2 \vec{r} \cdot \partial_\perp^2 \vec{r} + t_\perp (\partial_\alpha^\perp \vec{r})^2 \\
& + t_y (\partial_y \vec{r})^2 + \frac{u_{\perp\perp}}{2} (\partial_\alpha^\perp \vec{r} \cdot \partial_\beta^\perp \vec{r})^2 + \frac{u_{yy}}{2} (\partial_y \vec{r} \cdot \partial_y \vec{r})^2 + u_{\perp y} (\partial_\alpha^\perp \vec{r} \cdot \partial_y \vec{r})^2 \\
& + \frac{v_{\perp\perp}}{2} (\partial_\alpha^\perp \vec{r} \cdot \partial_\alpha^\perp \vec{r})^2 + v_{\perp y} (\partial_\alpha^\perp \vec{r})^2 (\partial_y \vec{r})^2 \Big] \\
& + \frac{b}{2} \int d^D x \int d^D x' \delta^{(d)} (\vec{r}(\mathbf{x}) - \vec{r}(\mathbf{x}')),
\end{aligned}
\tag{5}
$$

where we have taken the membrane to be isotropic in the $D-1$ membrane directions $\mathbf{x}_\perp$, orthogonal to one special direction, $y$. While at first glance $F$ might appear quite formidable, in fact (aside from the last term), in terms of the tangent order parameter $\vec{\xi}_{\perp,y}$ it has a standard form of the Landau's $\phi^4$ theory. The first three terms in $F$ (the $\kappa$ terms) represent the anisotropic bending energy of the membrane. The elastic constants $t_\perp$ and $t_y$ are the most strongly temperature dependent parameters in the model, changing sign from large, positive values at high temperatures to negative values at low temperatures. The $u$ and $v$ quartic elastic terms are needed to stabilize the membrane when one or both of the elastic constants $t_\perp$, $t_y$ become negative. The final, $b$ term in Eq. (5) represents the self-avoidance of the membranes, i.e., its steric or excluded volume interaction.

## 2.3. *Mean-field theory*

In mean-field theory, we seek a configuration $\vec{r}(\mathbf{x})$ that minimizes the free energy, Eq. (5). The curvature energies $\kappa_\perp (\partial_\perp^2 \vec{r})^2$ and $\kappa_y (\partial_y^2 \vec{r})^2$ are clearly minimized when

$\vec{r}(\mathbf{x})$ is linear in $\mathbf{x}$

$$\vec{r}(\mathbf{x}) = (\zeta_\perp \mathbf{x}_\perp, \zeta_y y, 0, 0, \ldots, 0). \tag{6}$$

Inserting Eq. (6) into Eq. (5), and for now neglecting the self-avoidance, we obtain the mean-field free energy density for anisotropic membranes

$$f_{\text{mft}} = \frac{1}{2}\left[ t_y \zeta_y^2 + t_\perp(D-1)\zeta_\perp^2 + \frac{1}{2}u'_{\perp\perp}(D-1)^2\zeta_\perp^4 + \frac{1}{2}u_{yy}\zeta_y^4 + v_{\perp y}(D-1)\zeta_\perp^2\zeta_y^2 \right]. \tag{7}$$

Minimizing the free energy over order parameters $\zeta_\perp$ and $\zeta_y$ yields two possible phase diagram topologies, depending on whether $u'_{\perp\perp}u_{yy} > v_{\perp y}^2$ or $u'_{\perp\perp}u_{yy} < v_{\perp y}^2$.[19]

For $u'_{\perp\perp}u_{yy} > v_{\perp y}^2$, the phase diagram is given in Fig. 2. For $t_\perp > 0$ and $t_y > 0$, average tangent vectors $\zeta_\perp$ and $\zeta_y$ both vanish, describing a crumpled (collapsed to the origin) membrane.

In the regime between the positive $t_\perp$-axis and the $t_y < 0$ part of the $t_y = (u_{yy}/v_{\perp y})t_\perp$ line, lies the y-tubule phase, characterized by $\zeta_\perp = 0$ and $\zeta_y = \sqrt{|t_y|/u_{yy}} > 0$: the membrane is extended in the y-direction but crumpled in all $D-1$ $\perp$-directions.

The $\perp$-tubule phase is the analogous phase with the $y$ and $\perp$ directions reversed, $\zeta_y = 0$ and $\zeta_\perp = \sqrt{|t_\perp|/u_{\perp\perp}} > 0$, and lies between the $t_\perp < 0$ segment of the line $t_y = (v_{\perp y}/u'_{\perp\perp})t_\perp$ and the positive $t_y$ axis.

Finally, the flat phase, characterized by both

$$\zeta_\perp = [(|t_\perp|u_{yy} - |t_y|v_{\perp y})/(u'_{\perp\perp}u_{yy} - v_{\perp y}^2)]^{1/2} > 0, \tag{8}$$

$$\zeta_y = [(|t_y|u_{\perp\perp} - |t_\perp|v_{\perp y})/(u'_{\perp\perp}u_{yy} - v_{\perp y}^2)]^{1/2} > 0, \tag{9}$$

lies between the $t_\perp < 0$ segment of the line $t_y = (u_{yy}/v_{\perp y})t_\perp$ and the $t_y < 0$ segment of the line $t_y = (v_{\perp y}/u'_{\perp\perp})t_\perp$.

For $u'_{\perp\perp}u_{yy} < v_{\perp y}^2$, the flat phase disappears, and is replaced by a direct first-order transition from $\perp$-tubule to y-tubule along the locus $t_y = (v_{\perp y}/u'_{\perp\perp})t_\perp$ (see Fig. 4). The boundaries between the tubule and the crumpled phases remain the positive $t_y$ and $t_\perp$ axes, as for $u'_{\perp\perp}u_{yy} > v_{\perp y}^2$ case.

## 2.4. Fluctuations and Self-avoidance in the Crumpled and Flat Phases

Although anisotropy has dramatic effects on the phase diagram of polymerized membranes, it does not modify the nature of the crumpled and flat phases.[20] To see this for the crumpled phase, note that when both $t_\perp$ and $t_y$ are positive, all of the other local terms in Eq. (5), i.e., the $\kappa$, $u$, and $v$-terms, are subdominant at long wavelengths. When these irrelevant terms are neglected, a simple change of variables $\mathbf{x}_\perp = \mathbf{x}'\sqrt{t_\perp/t_y}$ makes the remaining energy isotropic. Thus, the entire crumpled phase is identical in its scaling properties to that of isotropic membranes. In particular, the membrane in this phase has a radius of gyration $R_G(L)$ which

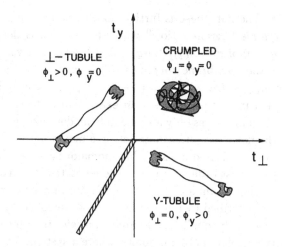

Fig. 4. Phase diagram for anisotropic tethered membranes showing a new tubule phase, for the range of elastic parameters when the intermediate flat phase disappears. A first-order phase transition separates $y$- and $\perp$-tubule phases.

scales with membrane linear dimension $L$ like $L^\nu$, with $\nu = (D+2)/(d+2)$ in Flory theory, and very similar values predicted by $\epsilon$-expansion techniques.[22-24]

Fluctuations in the flat phase can be incorporated by considering small deviations from the mean-field conformation in Eq. 6

$$\vec{r}(\mathbf{x}) = (\zeta_\perp \mathbf{x}_\perp + u_\perp(\mathbf{x}), \zeta_y y + u_y(\mathbf{x}), \vec{h}(\mathbf{x})), \tag{10}$$

where $u_\alpha(\mathbf{x})$ are $D$ in-plane phonon fields and $h_i(\mathbf{x})$ are $d_c = d - D$ out-of-plane height undulation fields. Inserting this into free energy, Eq. (5), with $t_\perp$ and $t_y$ both in the range in which the flat phase is stable, we obtain the uniaxial elastic energy studied by Toner.[25] As he showed, amazingly, the fluctuation renormalized this anisotropic elastic energy into the *isotropic* membrane elastic energy studied previously.[3,7-9] Therefore, in the flat phase, and at sufficiently long scales, the anisotropic membranes behave exactly like isotropic ones.

### 2.4.1. *Anomalous Elasticity of the Flat Phase*

This in particular implies that the flat phase of anisotropic membranes is stable against thermal fluctuations even though it breaks continuous (rotational) symmetry and is two-dimensional.[26] As in isotropic membranes, this is due to the fact that at long wavelengths these very thermal fluctuations drive the effective (renormalized) bend modulus $\kappa$ to infinity,[3,7-9] thereby suppressing effects of these same fluctuations that seek to destabilize the flat phase, resulting in an Esher-like "order-from-disorder" phenomenon.

Specifically, $\kappa(q)$ becomes wavevector dependent, and diverges like $q^{-\eta_\kappa}$ as $q \to 0$. In the flat phase the standard Lamé coefficients $\mu$ and $\lambda^{27}$ are also infinitely renormalized and become wavevector dependent, vanishing in the $q \to 0$ limit as

$\mu(q) \sim \lambda(q) \sim q^{\eta_u}$. The flat phase is furthermore novel in that it is characterized by a universal *negative* Poisson ratio[7,9] which for $D = 2$ is defined as the long wavelength limit $q \to 0$ of $\sigma = \lambda(q)/(2\mu(q) + \lambda(q))$. The transverse undulations in the flat phase, i.e. the membrane roughness $h_{rms}$ scales with the internal size of the membrane as $h_{rms} \sim L^\zeta$, with $\zeta = (4 - D - \eta_\kappa)/2$, exactly. Furthermore, an underlying rotational invariance imposes an exact Ward identity between $\eta_\kappa$ and $\eta_u$, $\eta_u + 2\eta_\kappa = 4 - D$. This leaves only a single independent exponent, characterizing the properties of the flat phase of even anisotropic membranes.

To appreciate how exotic and unusual this anomalous elasticity really is one only needs to observe that most ordered and disordered states of matter (systems with quenched disorder being prominent exceptions[28]), are in a sense trivial, with fluctuations about them described by a harmonic theory controlled by a Gaussian fixed point.[29] That is, generically, qualitatively important effects of fluctuations are confined to the vicinity of isolated critical points, where a system is tuned to be "soft", and characterized by low energy excitations. However, there exists a novel class of systems, with flat phase of polymerized membranes as a prominent member (that also includes smectic[30,31] and columnar liquid crystals,[32] vortex lattices in putative magnetic superconductors,[33] and nematic elastomers[12,34–38]) whose ordered states are a striking exception to this rule. A unifying feature of these phases is their underlying, spontaneously broken rotational invariance, that strictly enforces a particular "softness" of the corresponding Goldstone mode Hamiltonian. As a consequence, the usually small nonlinear elastic terms are in fact comparable to harmonic ones, and therefore must be taken into account. Similarly to their effects near continuous phase transitions (where they induce universal power-law anomalies), but extending throughout an ordered phase, fluctuations drive nonlinearities to qualitatively modify such soft states. The resulting strongly interacting ordered states at long scales exhibit rich phenomenology such as a universal nonlocal elasticity, a strictly nonlinear response to an arbitrarily weak perturbation and a universal ratio of wavevector-dependent singular elastic moduli, all controlled by a nontrivial infrared stable fixed point illustrated in Fig. 5.

### 2.4.2. *SCSA of the Flat Phase*

The study of such anomalous elasticity in polymerized membranes was initiated by Nelson and Peliti using a simple one-loop self-consistent theory, that assumed

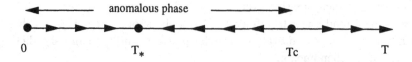

Fig. 5.   Renormalization group flow for anomalously elastic solids, with $T_c$ the transition temperature to the ordered state and $T_*$ a nontrivial infrared stable fixed point controlling properties of the strongly interacting ordered, critical-like state.

a non-renormalization of the in-plane elastic Lame' constants.[3] They found that phonon-mediated interaction between capillary waves leads to a divergent bending rigidity with $\eta_\kappa = 1$ and membrane roughness exponent $\zeta = 1/2$. Subsequent controlled $\epsilon = 4-D$ and $1/d_c$ expansions[7,8] confirmed existence of anomalous elasticity, and discovered an additional effect of softening (screening) of in-plane elasticity by out-of-plane undulations, that lead to a vanishing of long-scale Lame' coefficients $\lambda(q) \sim \mu(q) \sim q^{\eta_u}$.

Here we treat anomalous elasticity within the so-called self-consistent screening approximation (SCSA), first applied to the study of the flat phase of polymerized membranes by Le Doussal and Radzihovsky.[9] The attractive feature of SCSA is that it becomes exact in three complementary limits. By construction, it is exact in the large embedding dimension $d_c \to \infty$ limit and agrees with the systematic $1/d$ analysis of Guitter *et al.*[8] Because of Ward identities associated with rotational invariance it is exact (at any $d$) to lowest order $\epsilon = 4 - D$, i.e., agrees with one-loop results of Aronovitz and Lubensky.[7] (It would be very interesting to check predictions of SCSA to order $\epsilon^2$, however technical difficulties with two-loop calculations have so far precluded this).[39] And finally it gives exact value of $\eta_\kappa(D)$ for $d_c = 0$. Given these exact constraints, it is not surprising that for physical dimensions ($D = 2$, $d = 3$) SCSA exponents and the universal negative Poisson ratio[9] compare so well with latest, largest scale numerical simulations, discussed in lectures by Mark Bowick.[40]

As discussed above, at sufficiently long scales, flat phase of an anisotropic polymerized membrane is identical to that of an isotropic one,[25] and is described by an isotropic effective free energy that is a sum of a bending and in-plane elastic energies:

$$F[\vec{h}, \mathbf{u}] = \int d^D x \left[ \frac{\kappa}{2} (\nabla^2 \vec{h})^2 + \mu u_{\alpha\beta}^2 + \frac{\lambda}{2} u_{\alpha\alpha}^2 \right], \tag{11}$$

where the strain tensor is $u_{\alpha\beta} = \frac{1}{2}(\partial_\alpha u_\beta + \partial_\beta u_\alpha + \partial_\alpha \vec{h} \cdot \partial_\beta \vec{h})$. To implement the SCSA in the flat phase it is convenient to first integrate out the noncritical phonons fields $u_\alpha$ obtaining a quartic theory for interacting Goldstone tangent vector modes $\partial_\alpha \vec{h}$, described by free energy

$$F[\vec{h}] = \int d^D x \left[ \frac{\kappa}{2} (\nabla^2 \vec{h})^2 + \frac{1}{4d_c} (\partial_\alpha \vec{h} \cdot \partial_\beta \vec{h}) R_{\alpha\beta,\gamma\delta} (\partial_\gamma \vec{h} \cdot \partial_\delta \vec{h}) \right], \tag{12}$$

where for convenience, we rescaled Lamé coefficients so that the quartic coupling is of order $1/d_c$. The four-point coupling fourth-rank tensor is given by

$$R_{\alpha\beta,\gamma\delta} = \frac{K - 2\mu}{2(D-1)} P_{\alpha\beta}^T P_{\gamma\delta}^T + \frac{\mu}{2} (P_{\alpha\gamma}^T P_{\beta\delta}^T + P_{\alpha\delta}^T P_{\beta\gamma}^T), \tag{13}$$

where $P_{\alpha\beta}^T = \delta_{\alpha\beta} - q_\alpha q_\beta/q^2$ is a transverse (to **q**) projection operator. The convenience of this decomposition is that $K = 2\mu(2\mu + D\lambda)/(2\mu + \lambda)$ and $\mu$ moduli renormalize independently and multiplicatively.

To determine the renormalized elasticity, we set up a $1/d_c$-expansion[41-43] for 2-point and 4-point correlation functions of $\vec{h}$, and turn them into a closed self-consistent set of two coupled integral equations for self-energy $\Sigma(\mathbf{k})$ that define SCSA

$$\Sigma(\mathbf{k}) = \frac{2}{d_c} k_\alpha k_\beta k_\gamma k_\delta \int_q \tilde{R}_{\alpha\beta,\gamma\delta}(\mathbf{q}) G(\mathbf{k} - \mathbf{q}), \tag{14}$$

$$\tilde{R}_{\alpha\beta,\gamma\delta}(\mathbf{q}) = R_{\alpha\beta,\gamma\delta}(\mathbf{q}) - R_{\alpha\beta,\mu\nu}(\mathbf{q})\Pi_{\mu\nu,\mu'\nu'}(\mathbf{q})\tilde{R}_{\mu'\nu',\gamma\delta}(\mathbf{q}), \tag{15}$$

where $G(\mathbf{k}) \equiv 1/(\kappa k^4 + \Sigma(\mathbf{k})) \equiv 1/(\kappa(k)k^4)$ is the renormalized propagator, $R_{\alpha\beta,\gamma\delta}(\mathbf{q})$ the bare quartic interaction vertex and $\tilde{R}_{\alpha\beta,\gamma\delta}(\mathbf{q})$ the "screened" interaction vertex dressed by the vacuum polarization bubbles $\Pi_{\alpha\beta,\gamma\delta}(\mathbf{q}) = \int_p p_\alpha p_\beta p_\gamma p_\delta G(\mathbf{p})G(\mathbf{q} - \mathbf{p})$, and tensor multiplication is defined above.

In the long wavelength limit these integral equations are solved exactly by a renormalized propagator $G(\mathbf{k}) \approx 1/\Sigma(\mathbf{k}) \approx Z/k^{4-\eta_\kappa}$, with $Z$ a non-universal amplitude. Substituting this ansatz into Eqs. (14), (15) and solving for the renormalized elastic Lamé moduli, we find that indeed they must vanish as a universal power, with $\mu(\mathbf{q}) \propto \lambda(\mathbf{q}) \sim q^{\eta_u}$, and the phonon anomalous exponent $\eta_u = 4 - D - 2\eta_\kappa$ related to $\eta_\kappa$ by dimensional analysis (power-counting on $q$) first obtained by Nelson and Peliti.[3] This recovers the celebrated exponent relation enforced by the underlying rotational invariance of the membrane in the embedding space. The information about remaining independent exponent $\eta_\kappa$ resides in the $\eta_\kappa$-dependent *amplitudes* of the above equations. Cancelling out the nonuniversal scale $Z$, we obtain

$$d_c = \frac{2}{\eta_\kappa} D(D-1) \frac{\Gamma\left[1 + \frac{1}{2}\eta_\kappa\right]\Gamma\left[2 - \eta_\kappa\right]\Gamma\left[\eta_\kappa + D\right]\Gamma\left[2 - \frac{1}{2}\eta_\kappa\right]}{\Gamma\left[\frac{1}{2}D + \frac{1}{2}\eta_\kappa\right]\Gamma\left[2 - \eta_\kappa - \frac{1}{2}D\right]\Gamma\left[\eta_\kappa + \frac{1}{2}D\right]\Gamma\left[\frac{1}{2}D + 2 - \frac{1}{2}\eta_\kappa\right]}, \tag{16}$$

that determines $\eta_\kappa(D, d)$ within SCSA. For $D = 2$ this equation can be simplified, and one finds

$$\eta_\kappa(D = 2, d_c) = \frac{4}{d_c + \sqrt{16 - 2d_c + d_c^2}}. \tag{17}$$

Thus for physical membranes ($d_c = 1$) we obtain: $\eta_\kappa = 0.821, \eta_u = 0.358$ and:

$$\zeta = 1 - \frac{\eta_\kappa}{2} = \frac{\sqrt{15} - 1}{\sqrt{15} + 1} = 0.590. \tag{18}$$

At long length scales SCSA also gives a universal ratio between renormalized in-plane elastic moduli determined by $\tilde{R}_{\alpha\beta,\gamma\delta}(\mathbf{q})$, Eq. (15), and therefore predicts a universal and *negative* Poisson ratio

$$\lim_{q\to 0} \sigma \equiv \frac{\lambda(q)}{2\mu(q) + (D-1)\lambda(q)} = -\frac{1}{3} \tag{19}$$

that compares extremely well with the most recent and largest simulations.[40]

Expanding the result Eq. (16) in $1/d_c$ we obtain:

$$\eta_\kappa = \frac{8}{d_c}\frac{D-1}{D+2}\frac{\Gamma[D]}{\Gamma[\frac{D}{2}]^3\Gamma[2-\frac{D}{2}]} + O\left(\frac{1}{d_c^2}\right), \tag{20}$$

$$= \frac{2}{d_c} + O\left(\frac{1}{d_c^2}\right), \quad \text{for } D = 2, \tag{21}$$

which coincides with the exact result,[7,8] as expected by construction of the SCSA. Similarly, expanding Eq. (16) to first order in $\epsilon = 4 - D$ we find:

$$\eta_\kappa = \frac{\epsilon}{2 + d_c/12}, \tag{22}$$

also in agreement with the exact result.[7] This is not, however, a general property of SCSA, but is special to membranes, and can be traced to the convergence of the vertex and box diagrams.

Because in the flat phase, widely intrinsically separated parts of the membranes (i.e., points $\mathbf{x}$ and $\mathbf{x}'$, with $|\mathbf{x} - \mathbf{x}'|$ large) do not bump into each other (i.e., never have $\vec{r}(\mathbf{x}) = \vec{r}(\mathbf{x}')$), the self-avoidance interaction in Eq. (5) is irrelevant in the flat phase. Hence we expect above predictions to accurately describe conformational properties of physical isotropic and anisotropic polymerized membranes.

## 2.5. *Fluctuations in "Phantom" Tubules*

Thermal fluctuations of the (y-)tubule about the mean-field state $\vec{r}_o(\mathbf{x}) = \zeta_y(y, \vec{0})$ are described by conformation,

$$\vec{r}(\mathbf{x}) = (\zeta_y y + u(\mathbf{x}), \vec{h}(\mathbf{x})), \tag{23}$$

where $\vec{h}(\mathbf{x})$ is a $d - 1$-component vector field orthogonal to the tubule and $u(\mathbf{x})$ is a scalar phonon field along the tubule (taken along $y$-axis). The order parameter $\zeta_y$ is a tubule extension scale that is slightly modified by thermal fluctuations from the mean-field value given in Sec. 2.3 and is determined by the condition that all linear terms in $\vec{h}(\mathbf{x})$ and $u(\mathbf{x})$ in the renormalized elastic free energy exactly vanish. This criterion guarantees that $\vec{h}(\mathbf{x})$ and $u(\mathbf{x})$ represent fluctuations around the true tubule ground state.

Inserting the decomposition Eq. (23) into the free energy Eq. (5), neglecting irrelevant terms (e.g., phonon nonlinearities), and, for the moment ignoring the self-avoidance interaction, gives $F = F_{mft} + F_{el}$, where $F_{mft}$ is given by Eq. (7) (with $\zeta_\perp = 0$) and $F_{el}[u(\mathbf{x}), \vec{h}(\mathbf{x})]$ is the fluctuating elastic free energy part

$$F_{el} = \frac{1}{2}\int d^{D-1}x_\perp dy \left[\kappa(\partial_y^2\vec{h})^2 + t(\partial_\alpha^\perp\vec{h})^2 + g_\perp(\partial_\alpha^\perp u)^2 + g_y\left(\partial_y u + \frac{1}{2}(\partial_y\vec{h})^2\right)^2\right], \tag{24}$$

where $\kappa \equiv \kappa_y$, $t \equiv t_\perp + v_{\perp y}\zeta_y^2$, $g_y \equiv u_{yy}\zeta_y^2/2$, $g_\perp \equiv t + u_{\perp y}\zeta_y^2$, and $\gamma = t_y + u_{yy}\zeta_y^2$ are elastic constants.

The underlying rotational invariance of the tubule in the embedding space is responsible for two essential features of $F_{el}$. Firstly, it enforces a strict vanishing of the $(\partial_y \vec{h})^2$ tension-like term, with curvature $(\partial_y^2 \vec{h})$ as the lowest order harmonic term in the Goldstone tangent mode $\partial_y \vec{h}$. The result is highly anisotropic bulk elastic energy. Rotational invariance also ensures that nonlinear elasticity can only come in through powers of nonlinear strain tensor $E(u, \vec{h}) \equiv \partial_y u + \frac{1}{2}(\partial_y \vec{h})^2$, and that this property must be preserved upon renormalization.

### 2.5.1. *Anomalous Elasticity of the Tubule Phase*

Within harmonic approximation tubule, *bulk* rms transverse height undulations are given by

$$\langle |\vec{h}(\mathbf{x})|^2 \rangle \approx \int_{q_\perp > L_\perp^{-1}} \frac{d^{D-1}q_\perp \, dq_y}{(2\pi)^D} \frac{1}{t q_\perp^2 + \kappa q_y^4} \propto L_\perp^{5/2 - D}. \tag{25}$$

This suggests that for "phantom" tubules, the upper critical dimension $D_{uc} = 5/2$, contrasting with $D_{uc} = 4$ for the flat phase.[7,8] This also implies that rms fluctuations of the tubule normals are given by

$$\langle |\delta n_y(\mathbf{x})|^2 \rangle \propto L_\perp^{3/2 - D}. \tag{26}$$

Since this diverges in the infra-red $L_\perp \to \infty$ for $D \leq D_{lc} = 3/2$, this harmonic bulk mode analysis (ignoring anomalous elasticity and zero modes) suggests that the lower critical dimension $D_{lc}$ below which the tubule is necessarily crumpled is given by $D_{lc} = 3/2$.

As for the flat phase, one needs to assess the role of elastic nonlinearities that appear in free energy $F_{el}$, Eq. (24). One way to do this is to integrate out the phonon $u$ (which, at long scales can be done exactly). This leads to the only remaining nonlinearity in $\vec{h}$

$$F_{\text{anh}}[\vec{h}] = \frac{1}{4} \int d^{D-1}x \, dy \, (\partial_y \vec{h} \cdot \partial_y \vec{h}) \, V_h \, (\partial_y \vec{h} \cdot \partial_y \vec{h})], \tag{27}$$

with the Fourier transform of the kernel given by

$$V_h(\mathbf{q}) = \frac{g_y g_\perp q_\perp^2}{g_y q_y^2 + g_\perp q_\perp^2}. \tag{28}$$

Because of the $|\mathbf{q}|_\perp \approx q_y^2$ ($z = 1/2$) anisotropy of the bulk $\vec{h}$ modes, $V_h(\mathbf{q})$ scales as $g_\perp q_\perp$ at long scales, and therefore is strongly irrelevant near the Gaussian fixed point as long as $g_\perp$ is not renormalized (but see below). It is straightforward to verify to *all* orders in perturbation theory, that in a phantom tubule, there is no such renormalization of $g_\perp$.[13]

However, as asserted earlier, the *full* elasticity Eq. (24), *before u* is integrated out, *is* anomalous, because $g_y(\mathbf{q})$ is driven to zero as $q_y \to 0$, according to

$$g_y(\mathbf{q}) = q_y^{\eta_u} S_g(q_y/q_\perp^z), \tag{29}$$

with $S_g(x)$ universal scaling function:

$$S_g(x) \propto \begin{cases} \text{constant,} & x \to \infty \\ x^{-\eta_u}, & x \to 0 \end{cases}, \tag{30}$$

and exact exponents:

$$z = \frac{1}{2}, \quad \eta_u = 5 - 2D. \tag{31}$$

One simple way to see this is to note that rotational invariance enforces graphical corrections to preserve the form of the nonlinear strain tensor $E(u, \vec{h}) \equiv \partial_y u + \frac{1}{2}(\partial_y \vec{h})^2$. This leads to a relation between $\eta_\kappa$, $\eta_u$ and the anisotropy exponent $z$

$$2\eta_\kappa + \eta_u = 4 - (D-1)/z \tag{32}$$

which, together with the defining relation $z = 2/(4 - \eta_\kappa)$ reduces to

$$2\eta_u - (D-5)\eta_\kappa = 10 - 4D, \tag{33}$$

and for $\eta_\kappa = 0$ gives $z$ and $\eta_u$ in Eq. 31. This is supported by a detailed self-consistent one-loop perturbative calculation of $g_y(\mathbf{q})$ and by direct RG analysis.[13]

For *phantom* membranes with $D = 2$, $\eta_u = 1$ and $z = 1/2$, we find:

$$g_y(\mathbf{q}) \propto \begin{cases} q_y, & q_y \gg \sqrt{q_\perp} \\ \sqrt{q_\perp}, & q_y \ll \sqrt{q_\perp} \end{cases}. \tag{34}$$

This leads to phonon rms fluctuations given by

$$\langle u(\mathbf{x})^2 \rangle = L_\perp^{1/4} S_u\left(L_y/L_\perp^{3/4}\right), \tag{35}$$

with the universal scaling function having the limiting form:

$$S_u(x) \propto \begin{cases} \text{constant,} & x \to \infty \\ x^{1/3}, & x \to 0 \end{cases}. \tag{36}$$

For roughly square membranes, $L_y \sim L_\perp = L$, so, as $L \to \infty$, $L_y/L_\perp^{3/4} \to \infty$ this gives

$$\langle u(\mathbf{x})^2 \rangle = L_\perp^{1/4}, \tag{37}$$

a result that is consistent with simulations by Bowick *et al.*[44]

## 2.5.2. *Zero-modes and Tubule Shape Correlation*

The tubule's shape is characterized by

$$R_G^2 \equiv \langle |\vec{h}(\mathbf{L}_\perp, y) - \vec{h}(0_\perp, y)|^2 \rangle, \quad h_{rms}^2 \equiv \langle |\vec{h}(\mathbf{x}_\perp, L_y) - \vec{h}(\mathbf{x}_\perp, 0)|^2 \rangle. \qquad (38)$$

As illustrated in Fig. 3 $R_G$ measures the radius of a typical cross-section of the tubule perpendicular to its extended (y-) axis, and $h_{rms}$ measures tubule end-to-end transverse fluctuations.

For a phantom tubule these are easily computed *exactly* using tubule free energy, Eq. (24). One important subtlety is that one needs to take into account "zero modes" (Fourier modes with $\mathbf{q}_\perp$ or $q_y = 0$), that, because of anisotropic scaling $q_\perp \sim q_y^2$ of the bulk modes can dominate tubule shape fluctuations.

For $R_G$ one finds

$$R_G^2 = 2(d - D) \left[ \frac{k_B T}{L_y} \int_{L_\perp^{-1}} \frac{d^{D-1} q_\perp}{(2\pi)^{D-1}} \frac{1}{tq_\perp^2} (1 - e^{i\mathbf{q}_\perp \cdot \mathbf{L}_\perp}) \right.$$

$$\left. + k_B T \int_{L_\perp^{-1}, L_y^{-1}} \frac{d^{D-1} q_\perp dq_y}{(2\pi)^D} \frac{(1 - e^{i\mathbf{q}_\perp \cdot \mathbf{L}_\perp})}{tq_\perp^2 + \kappa(\mathbf{q})q_y^4} \right], \qquad (39)$$

with the first and second terms coming from the $q_y = 0$ "zero mode" and the standard bulk contributions, respectively. From its definition, it is clear that the $\mathbf{q}_\perp = 0$ "zero mode" does not contribute to $R_G$. Standard asymptotic analysis of the above integrals gives

$$R_G(L_\perp, L_y) = L_\perp^\nu S_R(L_y/L_\perp^z) \qquad (40)$$

with, for phantom membranes,

$$\nu = \frac{5 - 2D}{4}, \quad z = \frac{1}{2}, \qquad (41)$$

and the limiting form of the universal scaling function $S_R(x)$ given by

$$S_R(x) \propto \begin{cases} 1/\sqrt{x}, & \text{for } x \to 0 \\ \text{constant}, & \text{for } x \to \infty \end{cases}. \qquad (42)$$

This gives

$$R_G \propto L_\perp^\nu \propto L^{1/4}, \quad \text{for } D = 2, \qquad (43)$$

for the physically relevant case of a square membrane $L_\perp \sim L_y \sim L \to \infty$, for which $L_y \gg L_\perp^z$, with bulk mode contribution dominating over the $q_y = 0$ zero mode. This result is in excellent quantitative agreement with simulations of Bowick *et al.*[44] who found $\nu = 0.24 \pm 0.02$, in $D = 2$. It would be interesting to test the full anisotropic scaling prediction of Eqs. 40, 42 by varying the aspect ratio of the membrane in such simulations. For instance, for fixed $L_\perp$ and increasing $L_y$ these predict *no* change in $R_G$. The same should hold if $L_y$ is *decreased* at fixed $L_\perp$: $R_G$ should remain unchanged until $L_y \sim \sqrt{L_\perp}$, at which point the tubule should begin to get thinner (i.e. $R_G$ should decrease).

Equations (40) and (42) also correctly recover the limit of $L_y = $ constant $\ll L_\perp^z \to \infty$, where the tubule simply becomes a phantom, coiled up, $D-1$-dimensional polymeric network of size $L_\perp$ embedded in $d-1$ dimensions, with the radius of gyration $R_G(L_\perp) \sim L_\perp^{(3-D)/2}$. In the physical dimensions ($D = 2$ and $d = 3$) this in particular gives a coiled up ideal polymer of length $L_\perp$ with $R_G \sim L_\perp^{1/2}$, as expected.

Similar analysis for the tubule roughness $h_{\text{rms}}$ gives

$$
h_{\text{rms}}^2 = 2(d - D)\left[\frac{k_B T}{L_\perp^{D-1}} \int_{L_y^{-1}} \frac{dq_y}{(2\pi)} \frac{1}{\kappa(q_y)q_y^4}(1 - e^{iq_y L_y})\right.
$$

$$
\left. + k_B T \int_{L_\perp^{-1}, L_y^{-1}} \frac{d^{D-1}q_\perp dq_y}{(2\pi)^D} \frac{(1 - e^{iq_y L_y})}{tq_\perp^2 + \kappa(\mathbf{q})q_y^4}\right]. \tag{44}
$$

In contrast to $R_G$, only the $\mathbf{q}_\perp = 0$ zero mode (first term) and bulk modes contribute to $h_{\text{rms}}$, giving

$$
h_{\text{rms}}(L_\perp, L_y) = L_y^\zeta S_h(L_y/L_\perp^z) \tag{45}
$$

with, for phantom membranes,

$$
\zeta = \frac{5 - 2D}{2}, \quad z = \frac{1}{2} \tag{46}
$$

and the asymptotics of $S_h(x)$ given by

$$
S_h(x) \propto \begin{cases} \text{constant}, & \text{for } x \to 0 \\ x^{3/2 - \zeta}, & \text{for } x \to \infty \end{cases}. \tag{47}
$$

Equations (40) and (45) give information about the tubule roughness for arbitrarily large size $L_\perp$ and $L_y$, and arbitrary aspect ratio. For the physically relevant case of a square membrane $L_\perp \sim L_y \sim L \to \infty$, for which $L_y \gg L_\perp^z$, in contrast to $R_G$ (where bulk modes dominates), $\mathbf{q}_\perp = 0$ zero mode dominates, leading to

$$
h_{\text{rms}} \propto \frac{L_y^{\zeta + (D-1)/2z}}{L_\perp^{(D-1)/2}} \propto L^{\zeta + (D-1)(1-z)/2z}. \tag{48}
$$

Equations (18), (46) then give, for a $D = 2$ phantom tubule, $\zeta = 1/2$, $z = 1/2$

$$
h_{\text{rms}} \sim \frac{L_y^{3/2}}{L_\perp^{1/2}}, \tag{49}
$$

and therefore predicts for a square phantom membrane wild transverse tubule undulations

$$
h_{\text{rms}} \sim L, \tag{50}
$$

consistent with simulations[44] that find $h_{\text{rms}} \sim L^\gamma$, with $\gamma = 0.895 \pm 0.06$. As with $R_G$, it would be interesting to test the full scaling law Eq. (45) by simulating non-square membranes, and testing for the independent scaling of $h_{\text{rms}}$ with $L_y$ and

$L_\perp$. Note that, unlike $R_G$, according to Eq. (49), $h_{\text{rms}}$ will show immediate growth (reduction) when one increases (decreases) $L_y$ at fixed $L_\perp$.

The above discussion also reveals that our earlier conclusions about the lower critical dimension $D_{lc}$ for the existence of the tubule are strongly dependent on how $L_\perp$ and $L_y$ go to infinity relative to each other; i.e., on the membrane aspect ratio. The earlier conclusion that $D_{lc} = 3/2$ only strictly applies when the bulk modes dominate the physics, which is the case for a very squat membrane, with $L_y \approx L_\perp^z$, in which case $L_y \ll L_\perp$. For the physically more relevant case of a square *phantom* membrane, from the discussion above, we find that tubule phase is just marginally stable with $D_{lc} = 2^-$.

Equations (40) and (45) also correctly recover the limit of $L_\perp^z = \text{constant} \ll L_y \to \infty$, where the tubule simply becomes a polymer (ribbon) of thickness $R_G(L_\perp)$ of length $L_y$ embedded in $d-1$ dimensions. These equations then correctly recover the polymer limit giving

$$h_{\text{rms}} \approx L_P(L_y/L_P)^{3/2}, \qquad (51)$$

with $L_\perp$-dependent persistent length $L_P(L_\perp) \propto L_\perp^{D-1}$. So, as expected for a phantom tubule, if $L_\perp$ does not grow fast enough (e.g. remains constant), while $L_y \to \infty$, the tubule behaves as a linear polymer and crumples along its axis and the distinction between the crumpled and tubule phases disappears.

## 2.6. *Self-avoidance in the Tubule Phase*

Self-avoidance is an important ingredient that must be included inside the tubule phase. Detailed analysis of self-avoidance overturns arguments in Sec. 2.5.1, and leads to anomalous elasticity in the bending rigidity modulus $\kappa(\mathbf{q})$. Self-avoidance therefore also modifies the values of other shape exponents, while leaving the scaling form of correlation functions unchanged.

In the y-tubule phase the self-avoidance interaction $F_{\text{SA}}$ from Eq. 5 reduces to

$$F_{\text{SA}} = v \int dy \, d^{D-1}x_\perp \, d^{D-1}x'_\perp \, \delta^{(d-1)}(\vec{h}(\mathbf{x}_\perp, y) - \vec{h}(\mathbf{x}'_\perp, y)), \qquad (52)$$

with $v = b/2\zeta_y$.

### 2.6.1. *Flory Theory*

The effects of self-avoidance can be estimated by generalizing standard Flory arguments[45] from polymer physics[46] to the extended tubule geometry. The total self-avoidance energy scales as

$$E_{\text{SA}} \propto V\rho^2, \qquad (53)$$

where $V \approx R_G^{d-1}L_y$ is the volume in the embedding space occupied by the tubule and $\rho = M/V$ is the embedding space density of the tubule. Using the fact that the tubule mass $M$ scales like $L_\perp^{D-1} L_y$, we see that

$$E_{\text{SA}} \propto \frac{L_y L_\perp^{2(D-1)}}{R_G^{d-1}}. \tag{54}$$

Using the radius of gyration $R_G \propto L_\perp^\nu$, and considering, as required by the anisotropic scaling, a membrane with $L_\perp \propto L_y^2$, we find that $E_{\text{SA}} \propto L_y^{\lambda_{\text{SA}}}$ around the phantom fixed point, with

$$\lambda_{\text{SA}} = 1 + 4(D-1) - (d-1)\nu. \tag{55}$$

Self-avoidance is relevant when $\lambda_{\text{SA}} > 0$, which, from the above equation, happens for $\nu = \nu_{\text{ph}} = (5 - 2D)/4$ when the embedding dimension

$$d < d_{uc}^{\text{SA}} = \frac{6D - 1}{5 - 2D}. \tag{56}$$

For $D = 2$-dimensional membranes, $d_{uc}^{\text{SA}} = 11$. Thus, self-avoidance is strongly relevant for the tubule phase in $d = 3$, in contrast to the flat phase.

We can estimate the effect of the self-avoidance interactions on $R_G$ in Flory theory, by balancing the estimate Eq. (54) for the self-avoidance energy with a similar estimate for the elastic energy:

$$E_{\text{elastic}} = t \left( \frac{R_G}{L_\perp} \right)^2 L_\perp^{D-1} L_y. \tag{57}$$

Equating $E_{\text{elastic}}$ with $E_{\text{SA}}$, we obtain a Flory estimate for the radius of gyration $R_G$:

$$R_G(L_\perp) \propto L_\perp^{\nu_F}, \quad \nu_F = \frac{D+1}{d+1}, \tag{58}$$

which should be contrasted with the Flory estimate of $\nu_F^c = (D+2)/(d+2)$ for the *crumpled* phase. For the physical case $D = 2$, Eq. (58) gives

$$R_G \propto L_\perp^{3/4}, \tag{59}$$

a result that is known to be *exact* for the radius of gyration of a $D = 1$-polymer embedded in $d = 2$-dimensions.[47] Since the cross-section of the $D = 2$-tubule, crudely speaking, traces out a crumpled polymer embedded in two dimensions (see Fig. 3), it is intriguing to conjecture that $\nu = 3/4$ is also the *exact* result for the scaling of the thickness of the tubule.

### 2.6.2. *Renormalization Group and Scaling Relations*

A new significant complexity that arises and is special to the tubule phase (as compared to the crumpled phase) is the simultaneous presence of local elastic and nonlocal (in the intrinsic space) self-avoidance nonlinearities. The above Flory mean-field analysis (that ignores elastic nonlinearities) is nicely complemented by a renormalization group approach that can handle this complexity of the full model

$F = F_{el} + F_{SA}$, Eqs. (24), (52). Although, as we argued above, elastic nonlinearities are irrelevant in a phantom tubule, in a physical self-avoiding tubule, they are indeed important and lead to an anomalous bending rigidity elasticity.

The correct model, that incorporates the effects of both the self-avoiding interaction and the anharmonic elasticity, is defined by the full tubule free energy

$$F = \frac{1}{2} \int d^{D-1}x_\perp dy \left[ \kappa(\partial_y^2 \vec{h})^2 + t(\partial_\alpha^\perp \vec{h})^2 + g_\perp(\partial_\alpha^\perp u)^2 + g_y \left( \partial_y u + \frac{1}{2}(\partial_y \vec{h})^2 \right)^2 \right]$$
$$+ v \int dy d^{D-1}x_\perp d^{D-1}x'_\perp \delta^{(d-1)}(\vec{h}(\mathbf{x}_\perp, y) - \vec{h}(\mathbf{x}'_\perp, y)). \tag{60}$$

To assess the role of elastic ($g_y$) and self-avoiding ($v$) nonlinearities we implement standard momentum-shell RG.[48] We integrate out perturbatively in $g_y$ and $v$ short-scale fluctuations of modes $u(\mathbf{q})$ and $\vec{h}(\mathbf{q})$ within a cylindrical shell $\Lambda e^{-l} < q_\perp < \Lambda$, $-\infty < q_y < \infty$, and anisotropically rescale lengths ($\mathbf{x}_\perp, y$) and fields ($\vec{h}(\mathbf{x}), u(\mathbf{x})$), so as to restore the ultraviolet cutoff to $\Lambda$:

$$\mathbf{x}_\perp = e^l \mathbf{x}'_\perp, \quad y = e^{zl} y', \tag{61}$$
$$\vec{h}(\mathbf{x}) = e^{\nu l} \vec{h}'(\mathbf{x}'), \quad u(\mathbf{x}) = e^{(2\nu-z)l} u'(\mathbf{x}'), \tag{62}$$

where we have chosen the convenient (but not necessary) rescaling of the phonon field $u$ so as to preserve the form of the rotation-invariant strain operator $\partial_y u + \frac{1}{2}(\partial_y \vec{h})^2$. Under this transformation the free energy returns back to its form, Eq. 60, but with effective length-scale ($l = \log L_\perp$) dependent coupling constants determined by

$$\frac{dt}{dl} = [2\nu + z + D - 3 - f_t(v)]t, \tag{63}$$

$$\frac{d\kappa}{dl} = [2\nu - 3z + D - 1 + f_\kappa(g_y, g_\perp)]\kappa, \tag{64}$$

$$\frac{dg_y}{dl} = [4\nu - 3z + D - 1 - f_g(g_y)]g_y, \tag{65}$$

$$\frac{dg_\perp}{dl} = [4\nu - z + D - 3]g_\perp, \tag{66}$$

$$\frac{dv}{dl} = [2D - 2 + z - (d-1)\nu - f_v(v)]v, \tag{67}$$

where the various $f$-functions represent the graphical (i.e., perturbative) corrections. Since the self-avoiding interaction only involves $\vec{h}$, and (for convenience) the parameters in the $\vec{h}$ propagator ($t$ and $\kappa$) are going to be held fixed at 1, the graphical corrections coming from self-avoiding interaction alone depend only on the strength $v$ of the self-avoiding interaction. Therefore, to *all* orders in $v$, and leading order in $g_y$, $f_t(v)$ and $f_v(v)$ are only functions of $v$ and $f_\kappa(g_y, g_\perp)$ and $f_g(g_y)$ are only functions of $g_y$ and $g_\perp$.

It is important to note that $g_\perp$ suffers no graphical corrections, i.e., Eq. 66 is *exact*. This is enforced by an exact symmetry

$$u(\mathbf{x}_\perp, y) \rightarrow u(\mathbf{x}_\perp, y) + \chi(\mathbf{x}_\perp), \tag{68}$$

where $\chi(\mathbf{x}_\perp)$ is an arbitrary function of $\mathbf{x}_\perp$, under which the nonlinearities in $F$ are invariant.

We further note that there is an additional tubule "gauge"-like symmetry for $g_y = 0$

$$\vec{h}(\mathbf{x}_\perp, y) \rightarrow \vec{h}(\mathbf{x}_\perp, y) + \vec{\phi}(y), \tag{69}$$

under which the only remaining nonlinearity, the self-avoiding interaction, being local in $y$, is invariant. This "tubule gauge" symmetry demands that $f_\kappa(g_y = 0, g_\perp) = 0$, which implies that if $g_y = 0$, there is no divergent renormalization of $\kappa$, *exactly*, i.e., the self-avoiding interaction *alone* cannot renormalize $\kappa$. This *non*-renormalization of $\kappa$ by the self-avoiding interaction, in a truncated (unphysical) membrane model with $g_y = 0$, has been verified to all orders in a perturbative renormalization group calculation.[49]

To see that $\nu$ and $z$ obtained as fixed point solutions of Eqs. (63)–(67) have the same physical significance as the $\nu$ and $z$ defined in the scaling expressions Eqs. (1) for the radius of gyration $R_G$ and tubule wigglyness $h_{\mathrm{rms}}$, we use RG transformation to relate these quantities in the unrenormalized system to those in the renormalized one. This gives, for instance, for the radius of gyration

$$R_G(L_\perp, L_y; t, \kappa, \ldots) = e^{\nu l} R_G(e^{-l} L_\perp, e^{-zl} L_y; t(l), \kappa(l), \ldots). \tag{70}$$

Choosing $l = l_* = \log L_\perp$ this becomes:

$$R_G(L_\perp, L_y; t, \kappa, \ldots) = L_\perp^\nu R_G(1, L_y/L_\perp^z; t(l_*), \kappa(l_*), \ldots). \tag{71}$$

This relation holds for *any* choice of the (after all, arbitrary) rescaling exponents $\nu$ and $z$. However, *if* we make the special choice such that Eqs. (63)–(67) lead to fixed points, $t(l_*)$, $\kappa(l_*), \ldots$ in Eq. (71) go to *constants*, independent of $l_*$ (and hence $L_\perp$), as $L_\perp$ and hence $l_*$, go to infinity. Thus, in this limit, we obtain from Eq. (71)

$$R_G(L_\perp, L_y; t, \kappa, \ldots) = L_\perp^\nu R_G(1, L_y/L_\perp^z; t_*, \kappa_*, \ldots), \tag{72}$$

where $t_*, \kappa_*, \ldots$ are the fixed point values of coupling constants. This result clearly agrees with the scaling forms for $R_G$, Eq. (1) (with analogous derivation for $h_{\mathrm{rms}}$) with the scaling function given by $S_R(x) \equiv R_G(1, x; t_*, \kappa_*, g_y^*, v^*)$.

The recursion relations Eqs. (63)–(67) reproduce all of our earlier phantom membrane results (when $v = 0$, leading to $f_\kappa = 0$), as well as the upper-critical embedding dimension $d_{uc}^{\mathrm{SA}} = (6D - 1)/(5 - 2D)$ for self-avoidance predicted by Flory theory, Eq. (55), *and* the upper critical *intrinsic* dimension $D_{uc} = 5/2$ for anomalous elasticity for phantom membranes (determined by eigenvalues of dimensionless couplings corresponding to $v$ and $g_y$). They also reproduce all of the Flory theory exponents under the approximation that graphical corrections to $t$ and $v$ are the same, i.e., that $f_t(v_*) = f_v(v_*)$.

To analyze the renormalization of $\kappa$ in a self-avoiding tubule, we focus once again on the non-local interaction $F_{anh}$, Eq. (27), mediated by integrated out phonons, with a kernel

$$V_h(\mathbf{q}) = \frac{g_y g_\perp q_\perp^2}{g_y q_y^2 + g_\perp q_\perp^2}, \tag{73}$$

whose long-scale scaling determines renormalization of $\kappa$. If $g_y(\mathbf{q}) q_y^2 \gg g_\perp(\mathbf{q}) q_\perp^2$ (as we saw for a phantom tubule), then at long scales $V_h(\mathbf{q}) \approx g_\perp q_\perp^2 / q_y^2 \sim q_y^2$ in the relevant limit of $q_\perp \sim q_y^2$. Simple power counting around the Gaussian fixed point then shows that this elastic nonlinearity is irrelevant for $D > D_{uc} = 3/2$, and therefore $f_\kappa^* = \eta_\kappa = 0$ for a physical $D = 2$-dimensional tubule, as we argued in Sec. 2.5.

On the other hand, if the scaling is such that $g_\perp(\mathbf{q}) q_\perp^2$ dominates over $g_y(\mathbf{q}) q_y^2$, then $V_h(\mathbf{q}) \approx g_y$, i.e. a constant at long length scales. Simple power-counting then shows that this coupling is relevant for $D < D_{uc} = 5/2$ and the bending rigidity modulus of a $D = 2$-dimensional tubule *is* anomalous in this case.

In the RG language, the relevance of $V_h$ is decided by the sign of the renormalization group flow eigenvalue of $g_\perp(l)$ in Eq. (66)

$$\lambda_{g_\perp} = 4\nu - z + D - 3, \tag{74}$$

which is exactly determined by the fixed-point values of $\nu$ and $z$, since $g_\perp$ suffers no graphical renormalization.

As discussed in previous sections, for a phantom tubule (or for $d > d_{uc}$) $\nu = (5 - 2D)/4$ and $z = 1/2$. For $d$ below but close to $d_{uc}^{SA} = (6D - 1)/(5 - 2D)$ ($= 11$ for $D = 2$), these values are modified by the self-avoiding interaction, but only by order $\epsilon \equiv d - d_{uc}^{SA}$. Hence a $D = 2$-dimensional tubule, embedded in $d$ dimensions close to $d_{uc}^{SA} = 11$, $\lambda_{g_\perp} \approx -1/2$ and $g_\perp(l)$ flows according to

$$\frac{dg_\perp}{dl} = \left[ -\frac{1}{2} + O(\epsilon) \right] g_\perp, \tag{75}$$

i.e. $g_\perp$ is *irrelevant* near $d = 11$ (for $\epsilon \ll 1$), $V_h(\mathbf{q}) \sim g_\perp q_\perp^2 / q_y^2 \sim q_y^{2 - O(\epsilon)}$ is irrelevant to the *bend* elasticity for a $D = 2$-dimensional tubule embedded in high dimensions near $d_{uc}^{SA} = 11$, and, hence, $f_\kappa = \eta_\kappa$ in Eq. (64) vanishes as the fixed point. Therefore, in these high embedding dimensions the full model of a self-avoiding tubule reduces to the *linear* elastic truncated model with self-avoiding interaction as the only nonlinearity, that can be nicely studied by standard expansion in $\epsilon = d_{uc}^{SA} - d$.[49] As we discussed above, the "tubule gauge" symmetry guarantees that in this case the self-avoiding interaction alone cannot renormalize $\kappa$, i.e., $f_\kappa = \eta_\kappa = 0$ for $d$ near $d_{uc}^{SA}$.

This together with Eq. (64) at its fixed point leads to an *exact* exponent relation

$$z = \frac{1}{3}(2\nu + D - 1),\qquad(76)$$

valid for a finite *range* $d_* < d < d_{uc}^{SA}$ of embedding dimensions, and for phantom tubules in any embedding dimension. Where valid, it therefore reduces the tubule problem to a single independent shape exponent.

However, this simple scenario, and, in particular, the scaling relation Eq. (76), is *guaranteed* to break down as $d$ is reduced. The reason for this is that, as $d$ decreases, $\nu$ increases, and eventually becomes so large that the eigenvalue $\lambda_{g_\perp}$ of $g_\perp$ changes sign and becomes positive. As discussed earlier, once this happens, the nonlinear vertex Eq. (73) becomes relevant, and $\kappa$ acquires a divergent renormalization, i.e., $f_\kappa \neq 0$, and bend tubule elasticity becomes anomalous.

Now, it is easy to show, using Eq. (76) and a rigorous lower bound $\nu > (D-1)/(d-1)$ inside $\lambda_{g_\perp}$, that it *must* become positive for $d > d_*^{lb}(D)$ with

$$d_*^{lb}(D) = \frac{4D-1}{4-D} = 7/2, \quad \text{for } D = 2.\qquad(77)$$

Hence, for the case of interest $D = 2$ critical dimension $d_*$ is bounded by $7/2$ from below. In fact, Flory and $\epsilon = 11 - d$ estimates indicate that $d_*(D = 2) \approx 6$.[13]

We therefore conclude that in a physical $D = 2$-dimensional self-avoiding tubule, embedded in $d = 3 < d_* \approx 6$, anharmonic elasticity $F_{anh}$ *is* important at long scales and leads to anomalous and divergent bending rigidity $\kappa(\mathbf{q})$ and $g_y(\mathbf{q})$

$$g_y(\mathbf{q}) = q_y^{\eta_u} S_g(q_y/q_\perp^z), \quad \kappa(\mathbf{q}) = q_y^{-\eta_\kappa} S_\kappa(q_y/q_\perp^z),\qquad(78)$$

with

$$z\eta_\kappa = f_\kappa(g_y^*, g_\perp^*), \quad z\eta_u = f_g(g_y^*).\qquad(79)$$

and asymptotic forms of scaling functions

$$S_g(x \to 0) \to x^{-\eta_u}, \quad S_\kappa(x \to 0) \to x^{\eta_\kappa}.\qquad(80)$$

Another consequence is the breakdown of the exponent relation, Eq. (76), that is replaced by *two* exact relations holding between *four* independent exponents $z$, $\nu$, $\eta_\kappa$, and $\eta_u$

$$z = \frac{1}{3 - \eta_\kappa}(2\nu + D - 1), \quad z = \frac{1}{3 + \eta_u}(4\nu + D - 1),\qquad(81)$$

which automatically contain the rotational symmetry Ward identity

$$2\eta_\kappa + \eta_u = 3 - (D - 1)/z,\qquad(82)$$

formally arising from the requirement that graphical corrections do not change the form of the rotationally invariant strain operator $\partial_y u + \frac{1}{2}(\partial_y \vec{h})^2$. The existence of a nontrivial $d_* > 3$ and its consequences are summarized by Figs. 6 and 7.

The physics behind above somewhat formal derivation of exponent relations (Ward identities) can be further exposed through a simple physical shell argument.

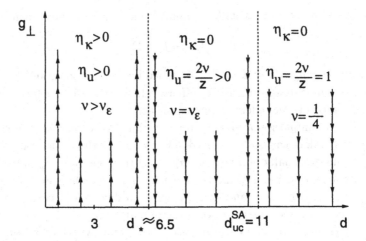

Fig. 6.   Schematic illustration of change in relevance of $g_\perp(l)$ at $d_*$. For embedding dimensions below $d_*$ (which includes the physical case of $d = 3$), $g_\perp(l)$ becomes relevant and (among other phenomena) leads to anomalous bending elasticity with $\kappa(\mathbf{q}) \sim q_y^{-\eta_\kappa}$, that diverges at long length scales.

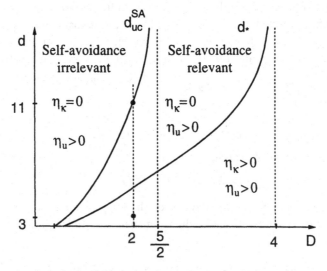

Fig. 7.   Schematic of the tubule "phase" diagram in the embedding $d$ vs intrinsic $D$ dimensions. Self-avoiding interaction becomes relevant for $d < d_{uc}^{SA}(D) = (6D-1)/(5-2D)$, $(= 11$, for $D = 2)$. Below the $d_*(D)$ curve (for which the lower bound is $d_*^{lb}(D) = (4D-1)/(4-D)$) the anharmonic elasticity becomes relevant, leading to anomalous elasticity with a divergent bending rigidity.

As can be seen from Fig. 8, bending of a tubule of radius $R_G$ into an arc of radius $R_c$ induces an in-plane strain energy density $g_y(L_y, L_\perp)(R_G(L_y)/R_c)^2$. Interpreting this additional energy as an effective bending energy density $\kappa_y(L_\perp, L_y)/R_c^2$, leads to the *effective* bending modulus $\kappa_y(L_\perp, L_y)$,

$$\kappa_y(L_\perp, L_y) \sim g_y(L_\perp, L_y)R_G(L_\perp, L_y)^2. \tag{83}$$

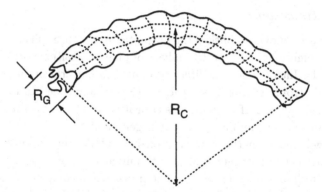

Fig. 8. Illustration of the physical mechanism for the enhancement of the bending rigidity $\kappa$ by the shear $g_y$ elasticity. To bend a polymerized tubule of thickness $R_G$ into an arc of radius $R_c$ requires $R_G/R_c$ fraction of bond stretching and therefore costs elastic shear energy, which when interpreted as bending energy leads to a length-scale dependent renormalization of the bending rigidity $\kappa$ and to the Ward identity Eq. (84).

This leads to a relation between the scaling exponents

$$2\nu = z(\eta_\kappa + \eta_u).  \tag{84}$$

that is contained in Eqs. (81) obtained through RG analysis above.

We note, finally, that all of the exponents must show a jump discontinuity at $d_*$. Therefore, unfortunately, an extrapolation from $\epsilon = 11 - d$-expansion in a truncated model with linear elasticity[49] down to the physical dimension of $d = 3$ (which is rigorously below $d_*$) gives little information about the properties of a real tubule.

The computations for a physical tubule must be performed for $d < d_*$, where both the self-avoidance and the elastic nonlinearities are both relevant and must be handled simultaneously. As we discussed above, for $d < d_*$, the eigenvalue $\lambda_{g_\perp} > 0$, leading to the flow of $g_\perp(l)$ to infinity, which in turn leads to $V_h(\mathbf{q}) = g_y$. Physically this regime of $g_\perp \to \infty$ corresponds to freezing out the phonons $u$, i.e. setting $u = 0$ in the free energy $F[\vec{h}, u]$, Eq. (60), with the resulting effective free energy functional for a physical self-avoiding tubule given by

$$F = \frac{1}{2} \int d^{D-1}x_\perp dy \left[ \kappa(\partial_y^2 \vec{h})^2 + t(\partial_\alpha^\perp \vec{h})^2 + \frac{1}{4} g_y (\partial_y \vec{h})^4 \right]$$
$$+ v \int dy d^{D-1}x_\perp d^{D-1}x'_\perp \delta^{(d-1)}(\vec{h}(\mathbf{x}_\perp, y) - \vec{h}(\mathbf{x}'_\perp, y)). \tag{85}$$

Since one must perturb in $g_y$ around a nontrivial, *strong* coupling fixed point with $v^* = O(1)$, unfortunately, no controlled analysis of above model exists todate and remains a challenging open problem. Nevertheless, above RG analysis combined with Flory estimates and exact exponent relations provides considerable information about shape fluctuations and elasticity of a polymerized tubule.

## 2.7. Phase Transitions

Now that we have solidly established the properties of the three phases of anisotropic polymerized membranes, we turn to analysis of phase transitions between them. As discussed in the Introduction, a direct crumpled-to-flat transition is highly non-generic for anisotropic membranes, as it has to be finely tuned to the tetracritical point illustrated in Fig. 2. If so tuned this transition will be in the same universality class as for isotropic membranes (where it is generic).[6,9]

Here we will focus on the new crumpled-to-tubule and tubule-to-flat transitions, which are generic for membranes with any amount of anisotropy. As for the tubule phase itself and the crumpled-to-flat transition, a complete analysis of these transitions must include both elastic and self-avoiding nonlinearities, a highly non-trivial open problem. Below we will instead present an incomplete solution. First we will present an RG analysis of a phantom (non-selfavoiding) membrane, focusing on the much simpler crumpled-to-tubule transition. We will then complement this study with a scaling theory and Flory approximation of the crumpled-to-tubule and tubule-to-flat transitions in a more realistic self-avoiding membrane.

### 2.7.1. Renormalization Group Analysis of Crumpled-To-Tubule Transition

The crumpled-to-(y-)tubule (CT) transition takes place when $t_y \to 0$, while $t_\perp$ remains positive. This implies that the CT critical point is characterized by highly anisotropic scaling $q_\perp \propto q_y^2$. It leads to a considerable simplification of the full free energy defined in Eq. (5) down to

$$F[\vec{r}(\mathbf{x})] = \frac{1}{2} \int d^{D-1}x_\perp dy \left[ \kappa_y (\partial_y^2 \vec{r})^2 + t_\perp (\partial_\alpha^\perp \vec{r})^2 + t_y (\partial_y \vec{r})^2 + \frac{u_{yy}}{2} (\partial_y \vec{r} \cdot \partial_y \vec{r})^2 \right].$$

(86)

The standard $O(d)$ $\phi^4$ model form of $F$ facilitates systematic analysis using conventional RG methods.[48] Because of the strong *scaling* anisotropy of the quadratic pieces of the free energy, it is convenient to rescale lengths $\mathbf{x}_\perp$ and $y$ anisotropically:

$$\mathbf{x}_\perp = \mathbf{x}_\perp' e^l, \quad y = y' e^{zl},$$

(87)

and rescale the "fields" $\vec{r}(\mathbf{x})$ according to

$$\vec{r}(\mathbf{x}) = e^{\chi l} \vec{r}'(\mathbf{x}').$$

(88)

Under this transformation

$$\kappa_y(l) = \kappa_y e^{(D-1-3z+2\chi)l}, \quad t_\perp(l) = t_\perp e^{(D-3+z+2\chi)l}.$$

(89)

Requiring that both $\kappa_y$ and $t_\perp$ remain fixed under this rescaling (zeroth order RG transformation) fixes the "anisotropy" exponent $z$ and the "roughness" exponent $\chi$:

$$z = \frac{1}{2}, \quad \chi = (5/2 - D)/2.$$

(90)

Although this choice keeps the quadratic in $\vec{r}$ part of $F$, Eq. (86), unchanged, it *does* change the quartic piece:

$$u_{yy}(l) = u_{yy}e^{(D-1-3z+4\chi)l} = u_{yy}e^{(5/2-D)l}, \tag{91}$$

and shows that below the upper critical dimension $D_{uc} = 5/2$, the Gaussian critical point is unstable to elastic nonlinearities, that become comparable to harmonic elastic energies on scales longer than the characteristic length scale $L_\perp^{nl}$

$$L_\perp^{nl} = \left(\frac{\kappa_y}{u_{yy}}\right)^{1/(5/2-D)}. \tag{92}$$

To describe the new behavior that prevails on even *longer* length scales requires a full-blown RG treatment.

Such an analysis[13] leads to corrections to the simple rescaling of $\kappa_y$, $t_\perp$, and $t_y$ (due to $u_{yy}$ non-linearity), characterized by "anomalous" exponents $\eta_\kappa$, $\eta_t$, and $\delta\theta$:

$$\kappa_y(l) = \kappa_y e^{(D-1-3z+2\eta_\kappa+2\chi)l}, \tag{93}$$

$$t_\perp(l) = t_\perp e^{(D-3+z+\eta_t+2\chi)l}, \tag{94}$$

$$t_y(l) = t_y e^{(D-1-z-\delta\theta+2\chi)l} \equiv t_y e^{\lambda_t l}, \tag{95}$$

that give

$$z = \frac{2-\eta_t}{4-\eta_\kappa}, \quad \chi = \frac{10-4D+\eta_\kappa(D-3+\eta_t)-3\eta_t}{8-2\eta_\kappa},$$

valid at the new nontrivial CT critical point.

Once the values of $\eta_t$, $\eta_\kappa$ and $\chi$ at the critical point are determined, the renormalization group gives a relation between correlation functions at or near criticality (small $t_y$) and at small wavevectors (functions that are difficult to compute, because direct perturbation theory is divergent) to the same correlation functions away from criticality and at large wavevectors (functions that can be accurately computed using perturbation theory). For example, the behavior of the correlation lengths near the transition can be deduced in this way:

$$\xi_\perp(t_y) = e^l \xi_\perp(t_y e^{\lambda_t l}), \quad \xi_y(t_y) = e^{zl} \xi_y(t_y e^{\lambda_t l}), \tag{96}$$

assuming that a critical fixed point exists and all other coupling constants have well-defined values at this point. Taking $t_y e^{\lambda_t l} \approx 1$ then gives

$$\xi_\perp(t_y) \approx a t_y^{-\nu_\perp}, \quad \xi_y(t_y) \approx a t_y^{-\nu_y}, \tag{97}$$

where $a \approx \xi(1)$ is the microscopic cutoff and,

$$\nu_\perp = \frac{1}{\lambda_t} = \frac{4-\eta_\kappa}{2(2-\eta_t-2\delta\theta)-\eta_\kappa(2-\eta_t-\delta\theta)}, \quad \nu_y = z\nu_\perp \tag{98}$$

are correlation length exponents perpendicular and along the tubule axis.

The anomalous exponents can be computed by integrating out short-scale degrees of freedom perturbatively in $u_{yy}$. Standard analysis shows that indeed there is a nontrivial critical point (at a finite value of $u_{yy}^* = O(\epsilon)$, $\epsilon \equiv 5/2 - D$)), at which, to all orders $\eta_t = 0$, and to one-loop order, for a physical membrane ($D = 2$, $d = 3$, i.e., $\epsilon = 1/2$)

$$\eta_\kappa = 0, \quad z = 1/2, \quad \chi = 1/4, \quad \nu_\perp \approx 1.227, \quad \nu_y \approx 0.614. \tag{99}$$

It is interesting to note that, in contrast to the treatment of crumpled-to-flat transition in isotropic membranes[6,9], where the critical point was only stable for an unphysically large value of the embedding dimension $d > 219$, the critical point characterizing the crumpled-to-tubule transition discussed here is stable for all $d$. Furthermore, the relatively small value of $\epsilon = 1/2$ (in contrast to for example $\epsilon = 2$ for flat phase and crumpling transition) gives some confidence that above critical exponents for the CT transition in a phantom membrane might even be quantitatively trustworthy.

### 2.7.2. Scaling Theory of Crumpled-To-Tubule and Tubule-To-Flat Transitions

The above RG analysis for the CT transition in phantom membranes, can be nicely complemented by a scaling theory combined with Flory estimates, that can incorporate both the elastic and self-avoiding nonlinearities, as we now describe.

Standard scaling arguments, supported by RG analysis and matching calculations (see e.g., Subsec. 2.5) suggest that near the crumpled-to-tubule transition, for a square membrane of internal size $L$, membrane extensions $R_y$ and $R_G$ along and orthogonal to the developing tubule axis should exhibit scaling form:

$$R_{G,y} = L^{\nu_{ct}^{G,y}} f_{G,y}(t_y L^\phi),$$

$$\propto \begin{cases} t_y^{\gamma_+^{G,y}} L^{\nu_c}, & t_y > 0, L \gg \xi_{ct} \\ L^{\nu_{ct}^{G,y}}, & L \ll \xi_{ct} \\ |t_y|^{\gamma_-^{G,y}} L^{\nu_t^{G,y}}, & t_y < 0, L \gg \xi_{ct} \end{cases} \tag{100}$$

where subscripts $t$, $c$ and $ct$ refer to tubule, crumpled and tubule-to-crumpled transition, respectively, and $\xi_{ct} \propto |t_y|^{-1/\phi}$ is a correlation length for the crumpled-to-tubule transition, $t_y = (T - T_{ct})/T_{ct}$, $T_{ct}$ is the crumpled-to-tubule transition temperature, with $t_y > 0$ corresponding to the crumpled phase. Consistency demands that exponents $\gamma_{+/-}^{G,y}$ are given by

$$\gamma_+^{G,y} = \frac{\nu_c - \nu_{ct}^{G,y}}{\phi}, \quad \gamma_-^{G,y} = \frac{\nu_t^{G,y} - \nu_{ct}^{G,y}}{\phi}. \tag{101}$$

The asymptotic forms in Eq. (100) are dictated by general defining properties of the phases and the CT transition. For example, scaling of both $R_y$ and $R_G$ like $L^{\nu_c}$, with the same exponent $\nu_c$ in the crumpled phase is rooted in the isotropy of that phase. In contrast, extended and highly anisotropic nature of the tubule phase dictates that $\nu_t^G \neq \nu_t^y = 1$. The anisotropy is, however, manifested even in the crumpled phase through the different temperature-dependent amplitudes of $R_G$

and $R_y$, with the aspect ratio $R_y/R_G$ actually *diverging* as $T \to T_{ct}^+$, and membrane begins to extend into a tubule configuration. The former of these vanishes as $t_y \to 0^+$ (since the radius of gyration in the tubule phase is much less than that in the crumpled phase, since $\nu_t < \nu_c$), which implies $\gamma_+^G > 0$, while the latter diverges as $t_y \to 0^+$, since the tubule ultimately extends in that direction, which implies $\gamma_+^y < 0$.

We will now show how these general expectations are born out by the Flory theory. Following analysis very similar to that done in Sec. 2.6.1, but keeping track of temperature-dependent order parameter $\zeta_y$, we find that the Flory approximation to the free energy density in Eq. (86), supplemented with self-avoidance is given by

$$f_{\rm Fl}[R_G, \zeta_y] = t_y \zeta_y^2 + u_{yy} \zeta_y^4 + t_\perp \left(\frac{R_G}{L_\perp}\right)^2 + v \frac{L_\perp^{D-1}}{\zeta_y R_G^{d-1}}. \tag{102}$$

Minimizing this over $R_G$, gives

$$R_G \approx L_\perp^{\nu_t} \left(\frac{v}{t_\perp \zeta_y}\right)^{1/(d+1)}, \tag{103}$$

with the tubule exponent $\nu_t = \frac{D+1}{d+1}$ found earlier, but now with additional temperature and $L_\perp$ dependence of $R_G$ through $\zeta_y(t_y, L_\perp)$ that interpolates between tubule, crumpled and critical behavior. Inserting this expression for $R_G$ into Eq. (102), gives

$$f_{\rm Fl}[\zeta_y] = t_y \zeta_y^2 + u_{yy} \zeta_y^4 + t_\perp^{\frac{d-1}{d+1}} \left(\frac{v}{\zeta_y}\right)^{\frac{2}{d+1}} L_\perp^{-\frac{2(d-D)}{d+1}}. \tag{104}$$

Minimizing $f_{\rm Fl}[\zeta_y]$ in the crumpled phase ($t_y > 0$) gives

$$\zeta_y \approx \left(\frac{v^2 t_\perp^{d-1}}{t_y^{d+1}}\right)^{\frac{1}{2(d+2)}} L_\perp^{-\frac{d-D}{d+2}}, \tag{105}$$

that, as expected vanishes in the thermodynamic limit. For a square ($L \times L$) membrane this then gives for $R_y = \zeta_y L_y$ and $R_G$ (using Eq. (103))

$$R_y \propto t_y^{-\frac{d+1}{2(d+2)}} L_\perp^{\frac{D+2}{d+2}}, \quad R_G \propto t_y^{\frac{1}{2(d+2)}} L_\perp^{\frac{D+2}{d+2}}, \tag{106}$$

which, after comparing with the general form, Eq. (100), gives

$$\nu_c = \frac{D+2}{d+2}, \quad \gamma_+^y = -\frac{d+1}{2(d+2)}, \quad \gamma_+^G = \frac{1}{2(d+2)}, \tag{107}$$

$\nu_c$ reassuringly agrees with the well-known Flory result for the radius of gyration exponent $\nu_c$ for a $D$-dimensional manifold, embedded in $d$ dimensions,[5,22-24] and $\gamma_+^{y,G}$ special to crumpled *anisotropic* membranes.

As anticipated earlier, $\gamma_+^y \neq \gamma_+^G$ implies that intrinsically anisotropic membrane are qualitatively distinct from isotropic ones even in the crumpled phase, as they exhibit a ratio of major to minor moment of inertia eigenvalues (related to $R_G/R_y$) that diverges as CT transition is approached.

Now for the tubule phase, characterized by $t_y < 0$ and a finite order parameter $\zeta_y > 0$, last term in $f_{\mathrm{Fl}}$, Eq. (104) is clearly negligible for $L_\perp > \xi_{cr}$ and simple minimization gives $\zeta_y \approx \sqrt{|t_y|/u_{yy}}$, which, when then inserting into $R_{y,G}$ and comparing with the general scaling forms gives for a square membrane

$$\nu_t^y = 1, \quad \gamma_-^y = \frac{1}{2}, \quad \nu_t^G = \frac{D+1}{d+1}, \quad \gamma_-^G = -\frac{1}{2(d+1)}. \tag{108}$$

Finally, right at the crumpled-to-tubule transition, $t_y = 0$, minimization of $f_{\mathrm{Fl}}[\zeta_y]$ gives

$$\zeta_y \propto L_\perp^{-\frac{(d-D)}{3+2d}} \tag{109}$$

which, when inserted in $R_{y,G}$ gives the advertised critical scaling forms with exponents

$$\nu_{ct}^y = \frac{D+d+3}{2d+3}, \quad \nu_{ct}^G = \frac{2D+3}{2d+3}, \quad \phi = \frac{2(d-D)}{2d+3}. \tag{110}$$

that are reassuringly consistent with our independent calculations of exponents $\gamma_{+,-}^{G,y}$, $\nu_c$, $\nu_t^{G,y}$, and $\nu_{ct}^{G,y}$ using exact exponent relations above. For the physical case of a two dimensional membrane embedded in a three dimensions, $(D = 2, d = 3)$

$$\nu_c = 4/5, \quad \nu_{ct}^G = 7/9, \quad \nu_{ct}^y = 8/9, \quad \nu_t = 3/4, \tag{111}$$
$$\gamma_+^G = 1/10, \quad \gamma_+^y = -2/5, \quad \gamma_-^G = -1/8, \quad \gamma_-^y = 1/2, \quad \phi = 2/9,$$

The singular parts of other thermodynamic variables obey scaling laws similar to that for $R_{G,y}$, Eq. (100). For example, the singular part of the specific heat per particle $C_v \sim \frac{1}{L^D} \frac{\partial^2}{\partial t_y^2} \left( \frac{1}{2} t_y R_y^2 L^{D-2} \right)$, using Eq. (100) exhibits the scaling form

$$C_v = L^\beta g(t_y L^\phi),$$
$$\propto \begin{cases} t_y^{-\alpha_+} L^{\beta-\alpha_+\phi}, & t_y > 0, L \gg \xi_{ct} \\ L^\beta, & L \ll \xi_{ct} \\ |t_y|^{-\alpha_-} L^{\beta-\alpha_-\phi}, & t_y < 0, L \gg \xi_{ct} \end{cases} \tag{112}$$

with $g(x) \approx \frac{d^2}{dx^2} [f_y^2(x)]$, and

$$\beta = 2\nu_{ct}^y - 2 + \phi, \quad \alpha_\pm = -2\gamma_\pm^y + 1, \tag{113}$$
$$\beta = 0, \quad \alpha_+ = \frac{2d+3}{d+2}, \quad \alpha_- = 0, \quad \text{Flory theory.} \tag{114}$$

This leads to the unusual feature that outside the critical regime (i.e. for $L \gg \xi_{ct}$), the singular part of the specific heat above the crumpled-to-tubule transition vanishes in the thermodynamic limit like $L^{-\alpha_++\phi} \sim L^{-2(d-D)/(d+2)} \sim L^{-2/5}$. Similar behavior was also found for the direct crumpled-to-flat transition by Paczuski *et al.*[6]

We now turn to the tubule-to-flat (TF) transition. On both sides of this transition, $R_y = L_y \times O(1)$. Therefore only $R_G$ exhibits critical behavior, which can be summarized by the scaling law:

$$R_G = L^{\nu_y^{\perp,\nu}} f^{\perp}(t_{\perp} L^{\phi_y}),$$

$$\propto \begin{cases} t_{\perp}^{\gamma_+^{tf}} L^{\nu_t}, & t_{\perp} > 0, L \gg \xi_{tf} \\ L^{\nu_y}, & L \ll \xi_{tf} \\ |t_{\perp}|^{\gamma_-^{tf}} L, & t_{\perp} < 0, L \gg \xi_{tf} \end{cases} \tag{115}$$

where $t_{\perp} = (T - T_{tf})/T_{tf}$, $t_{\perp} > 0$ corresponds to the tubule phase, $\xi_{tf} \propto |t_{\perp}|^{-1/\phi_y}$ is the correlation length for this transition, and the exponents obey the scaling relations:

$$\gamma_+^{tf} = (\nu_t - \nu_{tf})/\phi_{tf}, \quad \gamma_-^{tf} = (1 - \nu_{tf})/\phi_{tf}. \tag{116}$$

In Flory theory we find:

$$\phi_{tf} = 1/3, \quad \nu_{tf} = 5/6, \quad \gamma_+^{tf} = -1/4, \quad \gamma_-^{tf} = -1/2. \tag{117}$$

The singular part of the specific heat again obeys a scaling law:

$$C_v = L^{2\nu_y + \phi_y - 2} g(t_{\perp} L^{\phi_y}),$$

$$\propto \begin{cases} t_{\perp}^{-\alpha_+^{tf}} L^{-\kappa_+}, & t_{\perp} > 0, L \gg \xi_{tf} \\ L^{2\nu_y + \phi_y - 2}, & L \ll \xi_{tf} \\ |t_{\perp}|^{-\alpha_-^{tf}} L^{-\kappa_-}, & t_y < 0, L \gg \xi_{tf} \end{cases} \tag{118}$$

where, in Flory theory,

$$\alpha_+^{tf} = 3/2, \quad \alpha_-^{tf} = 0, \quad \kappa_+ = 1/2, \quad 2\nu_{tf} + \phi_{tf} - 2 = \kappa_- = 0. \tag{119}$$

Thus, again, the singular part of the specific heat vanishes (now like $L^{-1/2}$) in the thermodynamic limit above (i.e., on the tubule side of) the transition, while it is $O(1)$ and smooth as a function of temperature in both the critical regime and in the flat phase.

## 3. Random Heterogeneity in Polymerized Membranes

### 3.1. *Motivation*

Soon after a general picture of idealized homogeneous membranes was established, much of the attention turned to effects of random heterogeneity on conformational properties of polymerized membranes.[50] As with many other condensed matter systems (e.g., random magnets, pinned charged density waves, vortex lattices in superconductors)[51] a main motivation is that random inhomogeneity is an inevitable feature in most physical membrane realizations. As illustrated a cartoon of cellular membrane wall (Fig. 1.4 in lectures by Stan Leibler), functional proteins, nanopores (controlling ionic flow through membrane), and other heterogeneities (with sincere apologies to biologists for such crude physicist's terminology) are incorporated into a cellular lipid bilayer. Holes or tears, random variation in the local coordination number (disclinations), dislocations, grain-boundaries, and impurities incorporated into fishnet-like biopolymer spectrin network attached to cellular wall (as e.g., in red blood cells) are also substantial sources of inhomogeneity in membranes. Such defects in the two-dimensional polymer network are also inadvertently introduced during photo-polymerization of synthetic polymerized membranes. If one is only interested in statistical conformational properties of such membranes, these frozen-in heterogeneities can be treated as random quenched disorder, similar to treatment of impurities in other condensed matter contexts.[51]

Consistent with theoretical predictions,[50] the fact that quenched internal disorder can drastically modify the conformational thermodynamics of polymerized membranes was first demonstrated experimentally by Mutz, Bensimon and Brienne.[52] They observed that partially (heterogeneous) polymerized vesicles undergo upon cooling a "wrinkling" transition to a folded rigid glassy stucture that resembles a raisin. Natural interpretation of this important experiment as an evidence for a transition towards a crumpled spin-glass-like state provided strong motivation for further theoretical studies of quenched disorder in polymerized membranes.

Below we will describe a generalized model that includes effects of quenched disorder in a polymerized membrane and will show that (as in-plane anisotropy, discussed in previous section) it has drastic qualitative effects on long-length conformational properties of a polymerized membrane.

## 3.2. *Model of a Heterogeneous Polymerized Membrane*

It is clear that above sources of heterogeneity lead to local random in-plane dilations and compressions and can therefore be modelled by local random stresses $-\sigma_{\alpha\beta}(\mathbf{x})u_{\alpha\beta}$. Geometrically this can be understood as a random preferred background metric $g^0_{\alpha\beta}(\mathbf{x}) = \delta_{\alpha\beta} + \eta^0_{\alpha\beta}(\mathbf{x})$ with strain $\tilde{u}_{\alpha\beta} = \frac{1}{2}(g_{\alpha\beta} - g^0_{\alpha\beta}(\mathbf{x}))$ measured relative to this deformed state, and metric $g_{\alpha\beta} = \partial_\alpha \vec{r} \cdot \partial_\beta \vec{r}$ seeking to relax to $g^0_{\alpha\beta}(\mathbf{x})$. These local in-plane stresses can and will be relaxed by buckling of the membrane into the third dimension that will tend to screen the elastic interaction, thereby lowering the elastic energy by partial trade-off between in-plane elastic energy and membrane bending (curvature $\kappa$) energy. However, as illustrated in Fig. 9(a), because such randomness respects the reflection symmetry relating two sides of the membrane, the induced puckering will locally break this Ising symmetry *spontaneously*, and in a way specific to each configuration of disorder.

A qualitatively distinct form of quenched disorder that *explicitly* breaks reflection symmetry arises from asymmetric inclusions of the type illustrated in Fig. 9(b). These lead to a local preferred extrinsic curvature, that, in the flat phase is described by $-\nabla^2 \vec{h} \cdot \vec{c}(\mathbf{x})$.

Membrane defects will also of course lead to heterogeneous elastic moduli, $\kappa(\mathbf{x})$, $\mu(\mathbf{x})$, and $\lambda(\mathbf{x})$. However, it can be shown that such weak heterogeneity (i.e., as long as the average value of these elastic constants remains larger than their variance, that is they are predominantly positive) has no qualitative effects on membrane

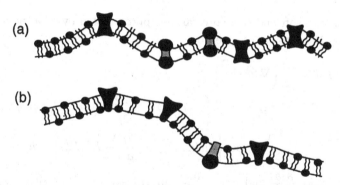

Fig. 9. A cartoon of a bilayer membrane with a reflection- (a) symmetric and (b) asymmetric inclusions that can be modelled by two qualitatively distinct types of disorder, the random stress and random mean curvature disorder, respectively.

long-scale conformations. Consequently, all effects of membrane heterogeneity can be modelled by just two types of quenched disorder, random stress and random curvature. This is perhaps not surprising, given the aforementioned analogy of a membrane with a ferromagnet (with membrane normal $\hat{n}$ playing the role of a spin $\vec{S}$), where too, random bond (that respects $\vec{S} \rightarrow -\vec{S}$) and random field (that is odd under $\vec{S} \rightarrow -\vec{S}$) are the only two qualitatively important types of quenched disorder. One qualitatively important distinction from random magnets that we can already anticipate at this point is that the curvature disorder, that couples to the gradient of the order parameter $\hat{n}$ is far weaker perturbation than its ferromagnetic analog, the random field disorder that couples directly to magnetization. This distinction will lead to importance of the curvature disorder below $D_{uc} = 4$, (same as the random stress disorder and therefore competing with it) contrasting with the upper-critical dimension of $D_{uc} = 6$ of the random-field in a ferromagnet.[51]

The general model of heterogeneous membrane is therefore described by an effective Hamiltonian $F[u_\alpha, \vec{h}]$:

$$F[u_\alpha, \vec{h}] = \int d^D x \left[ \frac{\kappa}{2} \left( \nabla^2 \vec{h} - \frac{\vec{c}(\mathbf{x})}{\kappa} \right)^2 + \mu (u_{\alpha\beta})^2 + \frac{\lambda}{2} (u_{\alpha\alpha})^2 \right.$$
$$\left. - \mu u_{\alpha\beta} \eta_{\alpha\beta}(\mathbf{x}) - \frac{\lambda}{2} u_{\beta\beta} \eta_{\alpha\alpha}(\mathbf{x}) \right], \tag{120}$$

where quenched disorder fields $\vec{c}(\mathbf{x})$ and $\eta_{\alpha\beta}(\mathbf{x})$ can be characterized by zero-mean, Gaussian statistics with second moment given by:

$$\overline{c_i(\mathbf{x}) c_j(0)} = \delta_{ij} \Delta_c(\mathbf{x}), \tag{121}$$

$$\overline{\eta_{\alpha\beta}(\mathbf{x}) \eta_{\gamma\delta}(0)} = \left( \Delta_1(\mathbf{x}) - \frac{1}{D} \Delta_2(\mathbf{x}) \right) \delta_{\alpha\beta} \delta_{\gamma\delta} + \frac{1}{2} \Delta_2(\mathbf{x}) (\delta_{\alpha\gamma} \delta_{\beta\delta} + \delta_{\alpha\delta} \delta_{\beta\gamma}). \tag{122}$$

Another useful form of this model is obtained after the phonons $u_\alpha$ are integrated out, which, at long length scales, can be done exactly since the phonons (unlike $\vec{h}$) are not soft and therefore can be approximated by a harmonic elasticity. The resulting Hamiltonian is given by

$$F[\vec{h}] = \int d^D x \left[ \frac{\kappa}{2} \left( \nabla^2 \vec{h} - \frac{\vec{c}(\mathbf{x})}{\kappa} \right)^2 \right.$$

$$+ \frac{1}{8} \partial_\alpha \vec{h} \cdot \partial_\beta \vec{h} \left\{ 2\mu P^T_{\alpha\gamma} P^T_{\beta\delta} + \frac{2\mu\lambda}{2\mu+\lambda} P^T_{\alpha\beta} P^T_{\gamma\delta} \right\} \partial_\gamma \vec{h} \cdot \partial_\delta \vec{h}$$

$$\left. - \frac{1}{4} \eta_{\alpha\beta}(\mathbf{x}) \left\{ 2\mu P^T_{\alpha\gamma} P^T_{\beta\delta} + \frac{2\mu\lambda}{2\mu+\lambda} P^T_{\alpha\beta} P^T_{\gamma\delta} \right\} \partial_\gamma \vec{h} \cdot \partial_\delta \vec{h} \right] \qquad (123)$$

where $P^T_{\alpha\beta} = \delta_{\alpha\beta} - \frac{\partial_\alpha \partial_\beta}{\nabla^2}$ and $P^L_{\alpha\beta} = \frac{\partial_\alpha \partial_\beta}{\nabla^2}$ are transverse and longitudinal projection operators. For $D = 2$, $d = 3$ membrane $F[\vec{h}]$ simplifies considerably to:

$$F[\vec{h}] = \int d^2 x \left[ \frac{\kappa}{2} \left( \nabla^2 h - \frac{c(\mathbf{x})}{\kappa} \right)^2 + \frac{K}{8} \left( P^T_{\alpha\beta} \partial_\alpha h \partial_\beta h - P^T_{\alpha\beta} \eta_{\alpha\beta}(\mathbf{x}) \right)^2 \right], \qquad (124)$$

For a generic configuration of impurity disorder, the ground state is highly nontrivial as it is determined by simultaneous, but generically conflicting, minimization of the extrinsic and Gaussian curvature $\left( R = \frac{1}{2}((\nabla^2 h)^2 - (\partial_\alpha \partial_\beta h)^2) \right)$ terms

$$\nabla^2 h = \frac{1}{\kappa} c(\mathbf{x}), \quad P^T_{\alpha\beta} \partial_\alpha h \partial_\beta h = P^T_{\alpha\beta} \eta_{\alpha\beta}(\mathbf{x}). \qquad (125)$$

Long-scale properties of such ground state are amenable to statistical treatment, utilizing standard field theoretic machinery.

### 3.3. Weak Quenched Disorder: "Flat-glass"

#### 3.3.1. Short-range Disorder

For many (but not all; see below) realizations of heterogeneity discussed above, such as for example random membrane inclusions, the disorder is short-ranged, and therefore can be characterized by $\delta$-function correlated disorder with variances $\Delta_c(\mathbf{x}) = \Delta_c \delta^D(\mathbf{x})$, $\Delta_{1,2}(\mathbf{x}) = \Delta_{1,2} \delta^D(\mathbf{x})$.

To understand the effects of quenched disorder it is helpful to first study the stability of the flat phase (described by the anomalous elastic fixed point, studied in Sec. 2.4) by performing a simple perturbative calculation in disorder and elastic nonlinearities directly for a physical membrane ($D = 2, d = 3$). Standard analysis[50] then leads to disorder and thermally renormalized bending rigidity $\kappa^D_R$:

$$\kappa^D_R(q) = \kappa + (k_B T \kappa + \Delta_c) \int \frac{d^2 p}{(2\pi)^2} \frac{K \left[ \hat{q}_\alpha P^T_{\alpha\beta}(\mathbf{p}) \hat{q}_\beta \right]^2}{\kappa^2 |\mathbf{q}+\mathbf{p}|^4}$$

$$- (\Delta_1 + \Delta_2) \int \frac{d^2 p}{(2\pi)^2} \frac{K^2 \left[ \hat{q}_\alpha P^T_{\alpha\beta}(\mathbf{p}) \hat{q}_\beta \right]^2}{4\kappa |\mathbf{q}+\mathbf{p}|^4}. \qquad (126)$$

The first, temperature-dependent correction that enhances $\kappa$ is identical to that of a homogeneous membrane and is responsible for the stability of the flat phase of polymerized disorder-free membranes.[3] At low temperature the temperature-independent contributions dominate. The last, random stress contribution leads to a divergent softening of the bending rigidity,[50] while the random curvature term works to increase the bending rigidity,[50,53,54] and thereby works to stabilize the flat phase through the "order-from-disorder" mechanism that is the zero-temperature analog of the thermal one discussed in Sec. 2.4.

Weak disorder should *not* affect the asymptotic behavior of membranes in the flat phase at sufficiently high temperatures, despite its importance at $T = 0$. To see this, assume the disorder is so weak that we can replace the elastic constants on the right hand side of Eq. (126) by wave-vector-dependent quantities $\kappa_R(p)$ and $K_R(p)$ renormalized only by thermal fluctuations in the way controlled by the disorder-free flat phase fixed point. As discussed in Sec. 2.4 these are expected to be singular at long scales, with $\kappa_R(p) \sim p^{-\eta_\kappa}$ and $K_R(p) \sim p^{\eta_u}$.[3,7-9] The expression for $\kappa_R^D(q)$, Eq. (126) becomes

$$\kappa_R^D(q) = \kappa_R(q) + \Delta_c \int \frac{d^2p}{(2\pi)^2} \frac{K_R(\mathbf{p}) \left[ \hat{q}_\alpha P_{\alpha\beta}^T(\mathbf{p}) \hat{q}_\beta \right]^2}{\kappa_R^2(\mathbf{q}+\mathbf{p})|\mathbf{q}+\mathbf{p}|^4}$$

$$- (\Delta_1 + \Delta_2) \int \frac{d^2p}{(2\pi)^2} \frac{K_R^2(\mathbf{p}) \left[ \hat{q}_\alpha P_{\alpha\beta}^T(\mathbf{p}) \hat{q}_\beta \right]^2}{4\kappa_R(\mathbf{q}|\mathbf{q}+\mathbf{p}|^4)}, \tag{127}$$

$$= \kappa_R(q)[1 + \text{const}.\Delta_c q^{\eta_\kappa} - \text{const}.(\Delta_1 + \Delta_2)q^{\eta_u}], \tag{128}$$

where we made use of the exact 2D exponent relation[3,7-9] $2\eta_f + \eta_u = 2$ (a consequence of rotational invariance). Since $\eta_\kappa$ and $\eta_u$ are positive, at finite temperature weak quenched disorder just gives a *subdominant* nonanalytic correction to disorder-free result for $\kappa_R(q)$. Physically this finite temperature irrelevance of disorder comes from singularly soft in-plane elastic moduli and divergent bending rigidity, that, respectively facilitate screening of the in-plane stress induced by impurities and suppress wrinkling effects of curvature disorder.

These perturbative arguments are supported by detailed renormalization group calculations controlled by an $\epsilon = 4 - D$-expansion that show that at finite temperature the Aronovitz-Lubensky fixed point is stable to weak quenched disorder.[50] This is summarized by the RG flow equation of the coupling constants illustrated in Fig. 10.

However, as is clear from the flow diagram, for sufficiently low-temperatures, even weak bare strain disorder becomes strong ($\Delta$'s flow to large values), invalidating above perturbative argument. In this case a full low temperature RG analysis is necessary. As first discovered by Morse, Lubensky and Grest, it shows[9,50,53,54] that interplay of random stress and curvature disorder leads to a new stable zero-temperature fixed point that controls long-scale properties of the disorder-roughened polymerized membrane. Similar to the thermally rough flat phase described by the AL fixed point, the resulting $T = 0$ phase is characterized by a power-law roughness (with $\zeta < 1$) about on average flat configuration. It therefore has all the ingredients of the "flat-glass" phase anticipated by Nelson and Radzihovsky.[50]

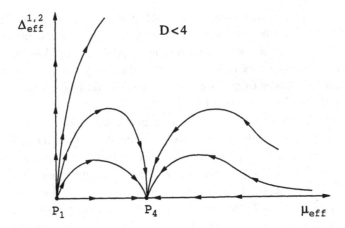

Fig. 10.  RG flow diagram showing the irrelevance of random strain disorder (disorder variance scales to zero at long scales) near the disorder-free finite temperature fixed point P4 controlling properties of the flat phase, and disorder induced instability of the flat phase at vanishing temperature ($\mu_{eff} \propto \mu T$).

These RG results can be nicely complemented by the SCSA analysis.[9,54] This can be done most effectively by first applying the replica formalism,[55] that allows one to work with a translationally invariant effective Hamiltonian. To do this, one introduces $n$ copies of fields $\vec{h}^a$ and $u^a_\alpha$ labeled by the replica index $a$, and integrates out the quenched disorder, thereby obtaining a replicated Hamiltonian. Assuming commutability of the thermodynamic and the $n \to 0$ limits, the relation to the disorder-averaged free energy is established through the identity

$$\ln Z = \lim_{n \to 0} \frac{Z^n - 1}{n}, \tag{129}$$

where $Z$ the partition function.

The membrane roughness is characterized by the full disorder-averaged height correlation function (that can also be related to replicated ones):

$$\overline{\langle(\vec{h}(\mathbf{x}) - \vec{h}(\mathbf{0}))^2\rangle} = \overline{\langle(\vec{h}(\mathbf{x}) - \vec{h}(\mathbf{0}))^2\rangle}_{\text{conn}} + \overline{\langle\vec{h}(\mathbf{x}) - \vec{h}(\mathbf{0})\rangle^2}, \tag{130}$$

$$\sim A_c|\mathbf{x}|^{2\zeta} + A|\mathbf{x}|^{2\zeta'}, \tag{131}$$

where, respectively, the first (connected) and second contributions characterize thermal- and disorder-generated roughness, with corresponding roughness exponents $\zeta$ and $\zeta'$, and the overbar denotes configurational (disorder) averages. The related exponents characterizing the Fourier transform of these parts of the height correlation functions are given by $\zeta = (4 - D - \eta_\kappa)/2$ and $\zeta' = (4 - D - \eta'_\kappa)/2$. Analysis very similar to that done for homogeneous membranes in Subsec. 2.4.2, but with additional replica matrix structure leads to a zero-temperature fixed point, that is marginally unstable to finite temperature.[53,54] It is characterized by $\eta_\kappa = \eta'_\kappa$, with

$$\eta_\kappa(d_c, D) = \eta_\kappa^{\text{pure}}(4d_c, D), \tag{132}$$

where $\eta_\kappa^{\text{pure}}(d_c, D)$ is the SCSA exponent for $\eta_\kappa$ found in Subsec. 2.4.2 characterizing a homogeneous polymerized membrane at a finite temperature.[9,54] The underlying reasons for this amazing connection between roughness exponents at the disorder- and thermally-dominated fixed points is unclear. However, this SCSA prediction agrees with the $1/d_c$- and $\epsilon$-expansions to lowest order in respective small parameters. For a physical membrane, $D = 2$, $d_c = 1$, SCSA predicts:[54]

$$\eta = 2/(2 + \sqrt{6}) = 0.449, \quad \zeta = 0.775, \tag{133}$$

that compares quite well (and much better than the lowest order $\epsilon$- or $1/d_c$-expansions) with the numerical simulation[53] result $\zeta = 0.81 \pm 0.03$ for a heterogeneous polymerized membrane.

### 3.3.2. *Long-range Disorder*

Above analysis of short-range impurity disorder can be easily extended to treat disorder with long-range correlations, that can arise from weakly correlated distribution of frozen-in dislocations and disclinations, random grain boundaries,[56] and from polymerized-in quasi-long-range correlated lipid tilt (or other membrane vector) order. At long scales, such disorder can be characterized by variances with power-law Fourier transforms:

$$\Delta_c(\mathbf{q}) = \Delta_c q^{-z_c}, \tag{134}$$

$$\Delta_{1,2}(\mathbf{q}) = \Delta_{1,2} q^{-z_{1,2}}, \tag{135}$$

where $z_c$ and $z_{1,2}$ are curvature and stress disorder correlation exponents. Such long-range disorder considerably enriches the phase diagram of heterogeneous polymerized membranes, introducing a number of new flat-glass phases, that are summarized as function of value of these range exponents in Fig. 11. For sufficiently short-ranged disorder (both $z_c$ and $z_{1,2}$ small), and for finite and zero temperature, we respectively recover the SCSA exponents for the Aronovitz-Lubensky[7-9] and Morse-Lubensky[9,53,54] fixed points.

More generally, the nature of the stable phases strongly depends on the value of the disorder-range $z$ exponents and divides into three classes: (1) $\zeta > \zeta'$ with temperature dominated roughness, (2) $\zeta < \zeta'$ with disorder dominated roughness, and (3) $\zeta = \zeta'$ with equal scaling of the disorder and thermal contributions to the membrane roughness. Each one of these regions then further subdivides into distinct flat-glass phases depending on whether long-range curvature, stress, or both types of disorders are relevant. That is, in the presence of long-range disorder four new flat-glass phases, stable at finite temperature appear: (i) short-range (SR) curvature and long-range (LR) stress disorder (LRSG in Fig. 11), (ii) LR curvature and SR stress disorder (LRCG), (iii) LR disorder in both curvature and stress disorder (LRMG),

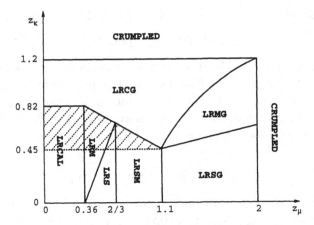

Fig. 11.   Domain of stability of the flat phases as a function of $z_\mu, z_\kappa$. (1) Disorder-dominated phases ($\zeta' > \zeta$): long-range stress glass (LRSG), long-range curvature glass (LRCG), long-range mixed glass (LRMG). (2) Temperature-dominated phases ($\zeta' < \zeta$): LR curvature (LRCAL), LR mixed (LRM) and LR stress (LRS). (3) LRSM: marginal phase with $\zeta = \zeta'$. The shaded area corresponds to a region of thermal phase transitions between several stable phases (LRCG and the others). The region where the membrane crumples is indicated.

and (iv) zero curvature disorder and LR stress disorder (not represented in Fig. 11). In addition to these flat-glass phases three corresponding temperature-dominated flat phases (LRS, LRCAL, LRM), and two phases for which $\zeta = \zeta'$ appear. In the shaded area in Fig. 11 several of these phases are stable. Phase transition controlled by strength of disorder and/or temperature is therefore expected between them.

For sufficiently long range disorder correlations (large $z$'s), the dominant roughness exponent $\zeta'$ reaches 1, presumably indicating disorder-driven crumpling transition, and therefore breakdown of the weak-disorder expansion about (on average) flat phase. The expected energy-driven crumpling instability to a qualitatively distinct *isotropic* "crumpled-glass" state is a priori of entirely different nature than the entropy-driven crumpling transition predicted for phantom membranes.

### 3.4.  *Strong Quenched Disorder: "Crumpled-glass"*

Description of the strongly disordered crumpled-glass phase and the associated transition is significantly more complicated, because in addition to complexities of conventional spin-glasses, nonlocal self-avoiding interaction must be included. Such crumpled phase is characterized by a vanishing average tangent field $\overline{\langle \partial_\alpha r_i \rangle}$ (hence crumpled), but with a nonzero crumpled-glass order parameter $\overline{\langle \partial_\alpha r_i \rangle \langle \partial_\beta r_j \rangle}$ analogous to Edwards-Anderson spin-glass order parameter in disordered magnets.[51] Some progress toward description of such crumpled-glass phase in phantom membranes was made by Radzihovsky and Le Doussal,[57] by utilizing a $1/d$-expansion.[41–43]

In the limit of large embedding dimension, $d \to \infty$, the *homogeneous* membrane model can be solved exactly.[8] In contrast, for a *disordered* membrane even in the $d \to \infty$ limit the exact solution of the crumpled-glass phase appears to be intractable. The difficulty arises from the tensor structure of the crumpled-glass order parameter, that leads to a problem of matrix field theory, a notoriously difficult problem. However, some progress can be made within an additional mean-field like approximation that ignores fluctuations in the tensor crumpled-glass order parameter. To simplify technical aspects of the presentation it is convenient to specialize to a purely scalar stress-only disorder, described by random Gaussian, zero-mean dilations and compressions in the locally preferred metric, $g^0_{\alpha\beta}(\mathbf{x}) = \delta_{\alpha\beta}(1+\delta g^0(\mathbf{x}))$. A much more questionable (but technically necessary) approximation is omission of the self-avoiding interaction, that is undoubtedly important in the crumpled-glass phase.

Because of the isotropic nature of the crumpled-glass phase, the model must be formulated in terms of $d$-dimensional conformation vector $\vec{r}(\mathbf{x})$. The effective Hamiltonian is given by

$$F[\vec{r}] = d \int d^D x \left[ \frac{\kappa}{2} |\nabla^2 \vec{r}|^2 + \frac{\mu}{4} (\partial_\alpha \vec{r} \cdot \partial_\beta \vec{r} - g^0_{\alpha\beta}(\mathbf{x}))^2 + \frac{\lambda}{8} (\partial_\alpha \vec{r} \cdot \partial_\alpha \vec{r} - g^0_{\alpha\alpha}(\mathbf{x}))^2 \right] \tag{136}$$

where the elastic moduli were rescaled by $d$ so as to obtain sensible and nontrivial results in the limit $d \to \infty$. Replicating $F$ allows averaging over quenched disorder.[55] Then, introducing two Hubbard-Stratanovich fields $\chi_{\alpha\beta}$ and $Q^{ij}_{ab\alpha\beta}$ to decouple replica diagonal and off-diagonal nonlinearities, respectively, leads to an effective Hamiltonian that is quadratic in $\vec{r}$. This allows formal integration over $\vec{r}$, that is conveniently done around background configuration $\vec{r}_0$. Now, ignoring fluctuations in the Hubbard-Stratanovich fields, the values of the order parameters $\partial_\alpha \vec{r}_0$ and $Q^o_{ab\alpha\beta ij}$ are determined by minimizing the resulting replicated free energy, together with the equation of constraint relating $\chi_{a\alpha\beta}$ to these order parameters.

Assuming that the replica symmetry breaking does not occur until higher order in $1/d$, as it happens in the random anisotropy axis model[58,59] we look for the saddle point replica-symmetric solution of the following form,

$$\vec{r}^o_a = \zeta x_\alpha \hat{e}_\alpha, \quad \chi^o_{a\alpha\beta} = \chi \delta_{\alpha\beta}, \quad Q^o_{ab\alpha\beta ij} = Q \delta_{\alpha\beta} \delta_{ij} (1 - \delta_{ab}). \tag{137}$$

The corresponding saddle-point equations for $\zeta$, $\chi$ and $Q$ are given by:

$$(1 - \zeta^2) + 2\chi(\alpha + \beta D) = \frac{T}{2D} \int_0^\Lambda \frac{d^D k}{(2\pi)^D} \left[ \frac{2}{\kappa k^2 + \chi + \frac{\hat{A}}{2T}Q} + \frac{\hat{A}Q/T}{\left(\kappa k^2 + \chi + \frac{\hat{A}}{2T}Q\right)^2} \right], \tag{138}$$

$$\zeta^2 = Q \left( 1 - \frac{\hat{A}}{2D} \int_0^\Lambda \frac{d^D k}{(2\pi)^D} \frac{1}{\left(\kappa k^2 + \chi + \frac{\hat{A}}{2T}Q\right)^2} \right), \tag{139}$$

$$\left( \chi + \frac{Q\hat{A}}{2T} \right) \zeta = 0, \tag{140}$$

where $\Delta$ is scalar stress disorder variance and $\hat{A} = (2\mu + D\lambda)\Delta$. For the special case of homogeneous membranes, $\Delta = 0$, these equations reassuringly reduce to

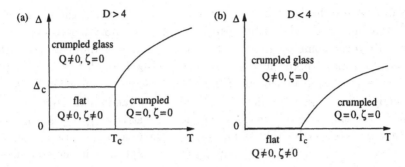

Fig. 12. A phase diagram for a disordered polymerized membrane showing (a) crumpled, flat- and crumpled-glass phases for $D > 4$. (b) In the physical case $D = 2 < 4$ the flat phase does not appear in this $d \to \infty$ limit, but is expected to when $1/d$ corrections are taken into account.

those found by David *et al.*[8] and describe the crumpled-to-flat transition exactly to leading order in $1/d$.

For a heterogeneous membrane ($\hat{\Delta} > 0$) there are three distinct solutions to these saddle-point equations, corresponding to three different possibilities for the values of the pair of order parameters $\zeta$ and $Q$.

$$\zeta = 0, \quad Q = 0, \tag{141}$$

$$\zeta \neq 0, \quad Q \neq 0, \tag{142}$$

$$\zeta = 0, \quad Q \neq 0, \tag{143}$$

that correspond to the crumpled phase, flat phase and crumpled-glass phase of the membrane, respectively.

Critical properties of these three phases and phase boundaries between them can be obtained from a straightforward analysis of Eqs. (138)–(140).[57] The phase behavior is summarized in Fig. 12, illustrating that within this approximation the lower-critical dimension for the flat phase in the presence of quenched disorder is $D^{\Delta}_{lc} = 4$, and therefore for this model (in $d \to \infty$ limit) only crumpled-glass and thermal crumpled phases survive in $D = 2$ membranes.

For the flat phase saddle point equations give

$$\zeta^2 = A\left(1 - \frac{T}{T_c}\right)\left(1 - \frac{\Delta}{\Delta_c}\right), \quad Q = A\left(1 - \frac{T}{T_c}\right), \tag{144}$$

where critical crumpling transition temperature $T_c$ and critical value of disorder $\Delta_c$ (not to be confused with the variance of the curvature disorder from Secs. 3.2–3.3.2), defining the rectangular boundaries of the flat phase, are given by

$$T_c^{-1} = \frac{1}{D} \int_0^{\Lambda} \frac{d^D k}{(2\pi)^D} \frac{1}{\kappa k^2}, \quad \Delta_c^{-1} = \frac{1}{2D} \int_0^{\Lambda} \frac{d^D k}{(2\pi)^D} \frac{1}{\kappa^2 k^4}, \tag{145}$$

and $A^{-1} = 1 + (\alpha + D\beta)\hat{\Delta}/T$. The power-law vanishing of these order parameters according $\zeta \sim (T_c - T)^{\beta^\zeta_T}$, $\zeta \sim (\hat{\Delta}_c - \hat{\Delta})^{\beta^\zeta_\Delta}$ and $Q \sim (T_c - T)^{\beta^Q_T}$, defines the corresponding $\beta$ exponents: $\beta^\zeta_T = 1/2$, $\beta^\zeta_\Delta = 1/2$, $\beta^Q_T = 1$.

Outside this rectangular region $\zeta$ vanishes and the membrane undergoes a crumpling transition out of the flat phase. For $\hat{\Delta} \leq \hat{\Delta}_c$ and as $T \to T_c$ the transition is to the crumpled phase, while for $T \leq T_c$ and as $\hat{\Delta} \to \hat{\Delta}_c$ flat phase is unstable to the crumpled-glass phase.

The glass susceptibility near the transition from the flat phase to the crumpled-glass phase can also be easily calculated by introducing an external field $h_{ij\alpha\beta} = h\delta_{ij}\delta_{\alpha\beta}$ conjugate to the crumpled-glass order parameter $Q_{ab\alpha\beta ij}$. The resulting saddle-point equations then lead to the crumpled-glass susceptibility, $\chi_{sg} = \partial Q / \partial h \sim (\hat{\Delta}_c - \hat{\Delta})^{-\gamma_{sg2}}$, with $\gamma_{sg2} = 1$.

Similarly, upon approach to the flat phase from the crumpled-glass (characterized by $Q \neq 0$ and $\zeta = 0$) the tangent susceptibility $\chi_\zeta$ diverges as tangent order $\zeta$ spontaneously develops. Turning on an external field $f$ that couples to the tangent order parameter, leads to $\chi_\zeta = \partial \zeta / \partial f \sim (\Delta - \Delta_c)^{-\gamma_{\zeta 2}}$ giving $\gamma_{\zeta 2} = 2/|4 - D|$, that, as expected diverges at the lower-critical dimension $D_{lc}^\Delta = 4$ of the flat phase.

Finally, we look at the transition between the crumpled and crumpled-glass phases. The crumpled-glass susceptibility $\chi_{sg}$ near this transition is given by $\chi_{sg} \sim (\Delta_c(T) - \Delta)^{-\gamma_{sg1}}$, with $\gamma_{sg1} = 1$, as at the flat-to-crumpled-glass transition, except for the modified phase boundary that is nonzero for any $D$:

$$\Delta_c^{-1}(T) = \frac{1}{2D} \int_0^\Lambda \frac{d^D k}{(2\pi)^D} \frac{1}{(\kappa k^2 + \chi(T))^2}, \tag{146}$$

and together with saddle-point equations and $Q = 0$, also defines the phase boundary between the crumpled and crumpled-glass phases for $D > 2$, $\Delta_c(T) - \Delta_c \sim (T - T_c)^\phi$, with $\phi$ the crossover exponent $\phi = |D - 4|/(D - 2)$.

As noted above, $d \to \infty$ analysis predicts an instability of the flat phase of $D = 2$ membranes to any amount of disorder ($D_{lc}^\Delta = 4$). A computation of $1/d$ corrections for the disordered membrane is technically quite challenging and remains an open problem. However, quite generally, anomalous elasticity generated by $1/d$ corrections (e.g., finite $\eta_\kappa = O(1/d)$ exponent[8]) strongly suggests the lowering of $D_{lc}^\Delta$ down to $D_{lc}^\Delta = 4 - O(1/d)$. This is supported by the Harris criterion applied to the buckling transition, controlled by the Aronovitz-Lubensky fixed point, that leads to stability of the flat phase (and the AL fixed point) as long as $\eta_u$ remains positive. This is also consistent with the $\epsilon = 4 - D$-expansion analysis of Radzihovsky and Nelson (performed for arbitrary $d$),[50] discussed in Subsec. 3.3.1, that the lower-critical dimension is reduced down to $D_{lc}^\Delta = 4 - 4/d$. A phase diagram for $D \leq 4$ consistent with the nature of the $1/d$ corrections is illustrated in Fig. 13.

Finally we observe that the crumpled-glass phase can be destroyed by applying an external tension to the membrane's boundaries. The metastable degenerate ground states would disappear and the average of the local tangents would no longer vanish. In this respect an external stress would be analogous to an external magnetic field in spin systems. As the stress is reduced the membrane would slowly return to the glassy phase but with some hysteresis. The line separating the regions of stable

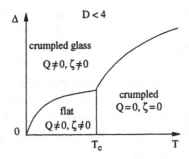

Fig. 13. Conjectured phase diagram for a disordered membrane $D \leq 4$ when $1/d$ corrections are taken into account, that allow a region of flat phase of size $1/d$ to appear.

and metastable degenerate states is then the analogue of the d'Almeida-Thouless line[60] studied in great detail for the real spin-glasses.[51]

## 4. Interplay of Anisotropy and Heterogeneity: Nematic Elastomer Membranes

We would like to conclude these lectures with a discussion of a new exciting realization of polymerized membranes, nematic elastomer membranes.[11] The motivation for their study is driven by recent experimental progress in the synthesis of nematic liquid-crystal elastomers,[12] statistically isotropic and homogeneous gels of crosslinked polymers (rubber), with main- or side-chain mesogens, that can *spontaneously* develop nematic orientational order, accompanied by a spontaneous uniaxial distortion illustrated in Fig. 14.

Even in the absence of fluctuations, *bulk* nematic elastomers were predicted[34] and later observed to display an array of fascinating phenomena,[12,35] the most striking of which is the vanishing of stress for a range of strain, applied transversely to the spontaneous nematic direction. This striking softness is generic, stemming

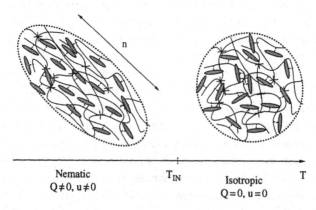

Fig. 14. Spontaneous uniaxial distortion of nematic elastomer driven by isotropic-nematic transition.

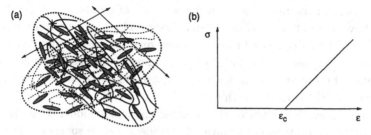

**Fig. 15.** (a) Simultaneous reorientation of the nematic director and of the uniaxial distortion is a low-energy nemato-elastic Goldstone mode of an ideal elastomer, that is responsible for its softness and (b) its flat (vanishing stress) stress-strain curve for a range of strains.

from the *spontaneous* orientational symmetry breaking by the nematic state,[34,35] accompanied by a Goldstone mode,[61] that leads to the observed soft distortion and strain-induced director reorientation,[62] illustrated in Fig. 15. The hidden rotational symmetry also guarantees a vanishing of one of the five elastic constants,[35] that usually characterize harmonic deformations of a three-dimensional uniaxial solid.[27] Given the discussion in Sec. 2.4.1, not surprisingly, the resulting elastic softness leads to qualitative importance of thermal fluctuations and local heterogeneity. Similar to their effects in smectic and columnar liquid crystals,[30-32] thermal fluctuations lead to anomalous elasticity (universally length-scale dependent elastic moduli) in bulk homogeneous elastomers with dimensions below 3,[36,37] and below 5, when effects of the random network heterogeneity are taken into account.[38]

This rich behavior of bulk elastomers provided strong motivation to study nematic elastomer membranes ($D$-dimensional sheets of nematic elastomer fluctuating in $d$ dimensions).[11] A model of such a membrane must incorporate both network anisotropy and heterogeneity discussed in previous sections. However, an important distinction from *explicitly* anisotropic membranes discussed in Sec. 2 is that the nematic anisotropy is *spontaneously* chosen in the amorphous (initially statistically isotropic) elastomer matrix. At harmonic level this in-plane rotational symmetry can be captured by a two-dimensional harmonic effective Hamiltonian

$$\mathcal{H}_{NE}^0 = \frac{1}{2} \int d^2x \big[ \kappa_{xx}(\partial_x^2 h)^2 + \kappa_{yy}(\partial_y^2 h)^2 + 2\kappa_{xy}(\partial_x^2 h)(\partial_y^2 h) + K_y(\partial_y^2 u_x)^2$$
$$+ K_x(\partial_x^2 u_y)^2 + \lambda_x(\varepsilon_{xx})^2 + \lambda_y(\varepsilon_{yy})^2 + 2\lambda_{xy}\varepsilon_{xx}\varepsilon_{yy} \big], \qquad (147)$$

with $\varepsilon_{\alpha\alpha} = \partial_\alpha u_\alpha$ (no sum over $\alpha$ implied here), and characterized by a uniaxial phonon elasticity with a vanishing transverse shear modulus, $\mu_{xy}$. This latter feature is what distinguishes a nematic elastomer membrane from an *explicitly* anisotropic membrane discussed in Sec. 2. The vanishing in-plane shear modulus captures at the harmonic level the invariance of the free energy under infinitesimal rotation of the nematic axis and the accompanying uniaxial distortion of the elastomer matrix. To ensure an in-plane stability curvature phonon elastic energies are included in $\mathcal{H}_{NE}^0$.

As a result, in the putative flat nematic phase of an elastomer membrane, the phonons have qualitatively "softer" harmonic elasticity than in a conventional polymerized membrane discussed in Sec. 2. Consequently, as in other "soft" systems (e.g., smectic and columnar liquid crystals phases; see discussion in Sec. 2.4.1), in the presence of thermal fluctuations and/or heterogeneities, *nonlinear* elastic terms are essential for the correct description.

First principles derivation of the nematic elastomer model, that incorporates (hidden) in-plane rotational invariance at nonlinear level is somewhat involved and we refer an interested reader to recent detailed work on this subject.[36,37] In short, one starts out with a model of a statistically homogeneous and isotropic elastic membrane coupled to a nematic in-plane order parameter $Q_{\alpha\beta}$. The rotational symmetry is then spontaneously broken by the nematic ordering at the isotropic-nematic transition, that also induces a spontaneous uniaxial in-plane distortion of the elastomer matrix, characterized by a strain tensor $u_{\alpha\beta}^0$. Expansion about this flat uniaxial state, ensuring underlying in-plane and embedding space rotational invariance leads to a nonlinear elastic Hamiltonian of a nematic elastomer membrane. Its form is that of the $\mathcal{H}_{NE}^0$, Eq. (147), but with the harmonic strain $\varepsilon_{\alpha\beta}$ replaced by a nonlinear strain tensor $w_{\alpha\beta}$, that incorporates both in-plane and height nonlinearities of the form $(\partial_x u_y)^2$ and $(\partial_x h)^2$, respectively.

In $D = 2$ both phonon and height anharmonic terms are strongly relevant (their perturbative corrections grow with length scale) and therefore must be both taken into account. One approach is to generalize the above model to a $D$-dimensional nematic elastomer membrane and perform an RG calculation controlled by an $\epsilon$-expansion. However, it is easy to show that upper-critical dimensions for these nonlinearities are different, with height undulations becoming relevant below $D_{uc}^h = 4$, and smectic-like and columnar-like in-plane nonlinearities with upper-critical dimensions of $D_{uc}^{sm} = 3$ and $D_{uc}^{col} = 5/2$, respectively. Hence for $D > 3$, in-plane phonon nonlinearities can be neglected, with height undulation nonlinearities (of the type studied in Secs. 2.4.2, 2.5.1 the only remaining relevant ones for $D < 4$, and controllable close to $D = 4$ with an $\epsilon = 4 - D$-expansion.

Generically, one would expect these undulation nonlinearities to renormalize bending rigidities $\kappa_{\alpha\beta}$ as well as in-plane elastic moduli $\lambda_{\alpha\beta}$, leading to anomalous elasticity with $\kappa_{\alpha\beta}(\mathbf{q}) \sim q^{-\eta_\kappa}$, $\lambda_{\alpha\beta}(\mathbf{q}) \sim q^{\eta_\lambda}$. As discussed in Subsec. 2.4.2, here too, rotational invariance imposes an exact Ward identity between exponents:

$$2\eta_\kappa + \eta_\lambda = \epsilon = 4 - D. \tag{148}$$

However, it is not difficult to show,[11] that once in-plane nonlinearities are neglected (legitimate for $D > 3$), the harmonic phonons can be integrated out exactly, and lead to a purely harmonic effective Hamiltonian $\mathcal{H}[h]$. Therefore there is a strict *non*renormalization of the bending rigidity tensor $\kappa_{\alpha\beta}$ for $D > 3$. This together with the Ward identity, Eq. (148) gives

$$\eta_\kappa = 0, \quad \eta_\lambda = 4 - D, \tag{149}$$

a result that is supported by a detailed renormalization group calculation.[11] This analysis also makes contact and recovers some of the results previously obtained in the studies of isotropic polymerized membranes. In particular, the previously seemingly unphysical, the so-called "connected fluid"[7] is realized as a fixed point of a nematically-ordered elastomer membrane that is unstable to the globally stable nematic-elastomer fixed point.[11]

Despite of some success, there are obvious limitations of above description, most notably in its application to the physical case of $D = 2$ elastomer membranes and inclusion of the (usually more dominant) local network heterogeneity. The first shortcoming primarily has to do with the neglect of in-plane elastic nonlinearities, which, near the Gaussian fixed point become relevant for $D < 3$. While it is very likely that the subdominance of these in-plane nonlinearities relative to the undulation ones will persist some amount *below* $D = 3$,[63] we expect that in the physical case of $D = 2$ all three nonlinearities need to be treated on equal footing. Carrying this out in a consistent treatment remains an open and challenging problem.

Secondly, elastomers are only statistically homogeneous and isotropic, exhibiting significant local heterogeneity in the polymer network. As we saw in Sec. 3, such internal quenched disorder has rich qualitative effects in ordinary polymerized membranes. Furthermore, recent work by Xing and Radzihovsky has demonstrated,[38] that interplay between nonlinear elasticity and random strains and torques (due to network heterogeneity) leads to disorder controlled anomalous elasticity even in three-dimensional bulk nematic elastomers. Because nematic elastomer membranes are far softer than ordinary polymerized membranes and their bulk analogs, we expect network heterogeneity to have strong and rich effects in these systems. Considerable research remains to work out the resulting phenomenology.

It is interesting to conclude with a general discussion of the global conformational phase behavior of nematic elastomer membranes. As with ordinary polymerized membranes, upon cooling, isotropic elastomer membranes should undergo a crumpling (flattening) transition from the crumpled to flat-isotropic phase. Upon further cooling, an in-plane (flat) isotropic-to-(flat) nematic transition can take place, leading to a flat membrane with a spontaneous in-plane nematic order. As for explicitly anisotropic membranes discussed in Sec. 2, such nematically-ordered elastomer membranes should undergo further transition to a nematically-ordered tubule phase. Because of the in-plane rotation symmetry that is *spontaneously* (as opposed to explicitly) broken, such nematic tubule will exhibit in-plane elasticity ("soft" phonons) that is qualitatively distinct from tubules discussed in Sec. 2.5.1, and will constitute a distinct phase of elastic membranes.[64] One interesting scenario of phase progression is the nematic-flat to nematic-tubule to nematic-flat reentrant phase transitions, driven by competition between growth of nematic order (anisotropy) and suppression of membrane's out-of-plane undulations upon cooling, as schematically illustrated in Fig. 16. Considerable research to elucidate the nature of the resulting fascinating phases and transitions remains.[64]

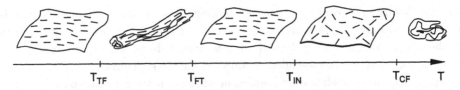

Fig. 16. A possible phase diagram for ideal nematic elastomer membranes. As the temperature is lowered a crumpled membrane undergoes a transition to isotropic-flat phase at $T_{CF}$, followed by a 2D in-plane isotropic-nematic like transition to an anisotropic (nematic) flat phase. As $T$ is lowered further, this anisotropic flat phase becomes unstable to a nematic tubule phase, where it continuously crumples in one direction but remains extended in the other. At even lower temperature, a tubule-flat transition takes place at $T_{TF}$.

## 5. Summary

In these notes, I have presented a small cross-section of the beauty and richness of fluctuating polymerized membranes. I have demonstrated the importance of the in-plane order in determining the long-scale conformations of these elastic sheets, by discussing in-plane anisotropy and local random heterogeneity and showed that these lead to a rich and highly nontrivial phenomenology.

## 6. Acknowledgments

The material presented in these lectures is an outcome of research done with a number of wonderful colleagues. The physics of anisotropic membranes presented in Sec. 1 is primarily based on extensive work done with John Toner. Section 2 on heterogeneities in polymerized membranes is based on many years of fruitful collaboration with David Nelson and Pierre Le Doussal. And the final section is based on a collaboration with Xiangjun Xing, Tom Lubensky and Ranjan Mukhopadhyay. I am indebted to these colleagues for much of my insight into the material presented here. This work was supported by the National Science Foundation through grants DMR-0321848 and MRSEC DMR-0213918, the Lucile and David Packard Foundation and the A. P. Sloan Foundation.

## References

1. L. Peliti and S. Leibler, *Phys. Rev. Lett.* **54**, 1690 (1985); D. Foster, *Phys. Lett.* **A114**, 115 (1986); A. M. Polyakov, *Nucl. Phys.* **B268**, 406 (1986); F. David, *Europhys. Lett.* **2**, 577 (1986).
2. F. David, E. Guitter and L. Peliti, *J. Phys.* **49**, 2059 (1987); E. Guitter and M.Kardar, *Europhys. Lett.* **13**, 441 (1990). J. Park and T. C. Lubensky *Phys. Rev. E* **53**, 2648 (1996).
3. D. R. Nelson and L. Peliti, *J. Phys.* (Paris) **48**, 1085 (1987).
4. Crumpling, however, only takes place beyond an exponentially long persistence length $\xi_p = ae^{4\pi\kappa/3k_BT}$, that, in a typical liquid membrane at room temperature far exceeds its size, and therefore even a liquid membrane appears not crumpled.
5. Y. Kantor, M. Kardar and D. R. Nelson, *Phys. Rev. Lett.* **57**, 791 (1986).

6. M. Paczuski, M. Kardar and D. R. Nelson, *Phys. Rev. Lett.* **60**, 2638 (1988).
7. J. A. Aronovitz and T. C. Lubensky, *Phys. Rev. Lett.* **60**, 2634 (1988); J. A. Aronovitz, L. Golubovic and T. C. Lubensky, *J. Phys.* (Paris) **50**, 609 (1989).
8. F. David and E. Guitter, *Europhys. Lett.* **5**, 709 (1988); E. Guitter, F. David, S. Leibler and L. Peliti, *J. Phys.* (Paris) **50**, 1789 (1989).
9. P. Le Doussal and L. Radzihovsky, *Phys. Rev. Lett.* **69**, 1209 (1992). SCSA is incredibly successful in that for the flat phase of polymerized membranes it predicts exponents that are *exact* in $d \to \infty$, $d = D$ and correct to a leading order in $\epsilon = 4 - D$, thereby showing agreement with all known exact results on polymerized membranes. This approximation was introduced by Bray to treat a ferromagnetic critical point in A. J. Bray, *Phys. Rev. Lett.* **32**, 1413 (1974).
10. D. Bensimon, private communication.
11. X. Xing, R. Mukhopadhyay, T. Lubensky and L. Radzihovsky, *Phys. Rev. E* **68**, 021108 (2003).
12. M. Warner and E. M. Terentjev, *Prog. Polym. Sci.* **21**, 853 (1996), and references therein; E. M. Terentjev, *J. Phys. Cond. Mat.* **11**, R239 (1999).
13. L. Radzihovsky and J. Toner, *Phys. Rev. Lett.* **75**, 4752 (1995); *Phys. Rev. E* **57**, 1832 (1998).
14. We remind the reader here that this "tubule" phase is an anisotropic phase of *polymerized* membranes with a finite in-plane shear rigidity. This should not be confused with a similar, but distinct "tubule phase" observed in *liquid* lipid membranes studied in Refs. 15–17.
15. B. N. Thomas *et al.*, *Science* **267**, 1635 (1995).
16. C.-M. Chen, *Phys. Rev. E* **59**, 6192 (1999).
17. A. Rudolph, J. Calvert, P. Schoen and J. Schnur, "Technological Developments of Lipid Based Tubule Microstructures", in *Biotechnological Applications of Lipid Microstructures*, ed. B. Gaber *et al.* (Plenum, 1988); R. Lipkin, *Science* **246**, 44 (December 1989).
18. Actually, in most current experimental realizations, the polymerization is random and leads to an isotropic membrane, making the isotropic case quite generic.
19. M. E. Fisher and D. R. Nelson, *Phys. Rev. Lett.* **32**, 1350 (1974); D. R. Nelson, J. M. Kosterlitz, and M. E. Fisher, *Phys. Rev. Lett.* **33**, 813 (1974), and *Phys. Rev.* **B13**, 412 (1976); A. Aharony, in *Phase Transitions and Critical Phenomena* , edited by C. Domb and M. S. Green (Academic, New York, 1976), Vol. 6, and *Bull. Am. Phys. Soc.* **20**, 16 (1975).
20. In these lectures we focus on the anisotropy in the membrane polymer network. This contrasts strongly with the anisotropy of the embedding space that membrane is fluctuating in. In this latter very interesting case considered in the work of Tokuyasu and Toner[21] (with anisotropy introduced by application of an electric field), anisotropy has important qualitative effects even in the flat phase, where it lowers the upper-critical dimension for elastic nonlinearities down to $D_{uc} = 5/2$ and leads to a new anomalous elasticity fixed point with new universal exponents.
21. T. A. Tokuyasu and J. Toner, *Phys. Rev. Lett.* **68**, 3721 (1992).
22. M. Kardar and D. R. Nelson, *Phys. Rev. Lett.* **58**, 1289 (1987); **58**, 2280(E); *Phys. Rev. A* **38**, 966 (1988).
23. Y. Kantor and D. R. Nelson, *Phys. Rev. Lett.* **58**, 1289 (1987); Y. Kantor, M. Kardar and D. R. Nelson, *Phys. Rev. A* **35**, 3056 (1987).
24. J. A. Aronovitz and T. C. Lubensky, *Europhys. Lett.* **4**, 395 (1987).
25. J. Toner, *Phys. Rev. Lett.* **62**, 905 (1988).

26. P. Hohenberg, *Phys. Rev.* **158**, 383 (1967); N. D. Mermin and H. Wagner, *Phys. Rev. Lett.* **17**, 1133 (1966); S. Coleman, *Commun. Math. Phys.* **31**, 259 (1973).

27. L. D. Landau and E. M. Lifshitz, *Theory of Elasticity*, Pergamon Press (1975).

28. J. L. Cardy and S. Ostlund, *Phys. Rev. B* **25**, 6899 (1982); D. S. Fisher, *Phys. Rev. B* **31**, 7233 (1985); L. Radzihovsky and J. Toner, *Phys. Rev. B* **60**, 206 (1999).

29. For example, spin waves in a ferromagnetic state or antiferromagnetic state and phonons in an ordinary crystal (even in 2d), at long scales are faithfully described by a harmonic Hamiltonian, controlled by a Gaussian fixed point.

30. G. Grinstein and R. Pelcovits, *Phys. Rev. Lett.* **47**, 856 (1981); *Phys. Rev. A* **26**, 915 (1982).

31. L. Radzihovsky and J. Toner, *Phys. Rev. Lett.* **78**, 4414 (1997); *Phys. Rev. B* **60**, 206 (1999); B. Jacobsen, K. Saunders, L. Radzihovsky and J. Toner, *Phys. Rev. Lett* **83**, 1363 (1999).

32. K. Saunders, L. Radzihovsky and J. Toner, *Phys. Rev. Lett.* **85**, 4309 (2000).

33. L. Radzihovsky, A. M. Ettouhami, K. Saunders and J. Toner, *Phys. Rev. Lett.* **87**, 27001 (2001).

34. L. Golubovic and T. C. Lubensky, *Phys. Rev. Lett.* **63**, 1082 (1989).

35. T. C. Lubensky, R. Mukhopadhyay, L. Radzihovsky and X. Xing, *Phys. Rev. E* **66**, 011702 (2002).

36. X. Xing and L. Radzihovsky, *Europhys. Lett.* **61**, 769 (2003), X. Xing, Ph.D. Thesis, University of Colorado (2003).

37. O. Stenull and T. C. Lubensky, *Europhys. Lett.* **61**, 779 (2003).

38. X. Xing and L. Radzihovsky, *Phys. Rev. Lett.* **90**, 168301 (2003).

39. L. Radzihovsky and B. Jacobsen, unpublished.

40. M. Falcioni, M. Bowick, E. Guitter and G. Thorleifsson, *Europhys. Lett.* **38**, 67 (1997).

41. S. Coleman, R. Jackiw and H. Politzer, *Phys. Rev. D* **10**, 2491 (1974); R. G. Root, *Phys. Rev. D* **10**, 3322 (1974).

42. This is analogous to the familiar $1/n$ expansion for critical phenomena, in which one expands about the number of spin components $n \to \infty$ limit.

43. Z. Justin, *Field Theory and Critical Phenomena*, Oxford Press (1994).

44. M. Bowick, M. Falcioni and G. Thorleifsson, *Phys. Rev. Lett.* **79**, 885 (1997).

45. In the case of polymers, Flory theory agrees with the *exact* predictions for the radius of gyration exponent $\nu$ in *all* dimensions $d$ where such exact predictions exist; in $d = 4, d = 2$, and $d = 1$, Flory theory recovers the exact results of $\nu = 1/2$, $\nu = 3/4$, and $\nu = 1$, respectively. And in $d = 3$ dimensions (where an exact result is not available) it agrees with the $\epsilon$-expansion to better than 1%.

46. M. Doi and S. F. Edwards, *The Theory of Polymer Dynamics*, Oxford University, New York, 1986; P. G. de Gennes, *Scaling Concepts in Polymer Physics*, Cornell University, Ithaca, 1979.

47. B. Nienhuis, *Phys. Rev. Lett.* **49**, 1062 (1982); B. Duplantier in *Fields, Strings and Critical Phenomena*, Les Houches Lectures, ed. E. Brezin and J. Zinn-Justin (1990).

48. K. G. Wilson and J. Kogut, *Phys. Rep. C* **12**, 77 (1977).

49. M. Bowick and E. Guitter, *Phys. Rev. E* **56** 7023 (1997).

50. L. Radzihovsky and D. R. Nelson, *Phys. Rev. A* **44**, 3525 (1991); *Europhys. Lett.* **16**, 79 (1991).

51. K. Binder and A. P. Young, *Rev. Mod. Phys.* **58**, 801 (1986); M. Chan, N. Mulders and J. Reppy, *Physics Today* **30**, August (1996); D. S. Fisher, G. Grinstein and A. Khurana, *Physics Today* **41**, December (1988); *Spin Glasses and Random Fields*, ed. A. P. Young, World Scientific, Singapore; T. Nattermann and S. Scheidl, *Advances in Physics* **49**, 607 (2000); *Liquid Crystals in Complex Geometries*, ed. G. P. Crawford and S. Žumer Taylor & Francis, London (1996).

52. M. Mutz, D. Bensimon and M. J. Breinne, *Phys. Rev. Lett.* **67**, 923 (1991).
53. D. Morse, T. C. Lubensky and G. S. Grest, *Phys. Rev. A* **45** R2151 (1992); D. Morse and T. C. Lubensky, *Phys. Rev. A* **46**, 1751 (1992).
54. P. LeDoussal and L. Radzihovsky, *Phys. Rev. B* **R48**, 3548 (1992).
55. S. F. Edwards and P. W. Anderson, *J. Phys. (Paris) F* **5**, 965 (1975).
56. D. R. Nelson and L. Radzihovsky, *Phys. Rev. A* **46**, 7474 (1992).
57. L. Radzihovsky and P. Le Doussal, *Journal de Physique I* **2**, 599 (1991).
58. Y. Y. Goldschmidt, *Phys. Rev. B* **30**, 1632 (1984).
59. A. Khurana, A. Jagannathan and J. M. Kosterlitz, *Nucl. Phys. B* **240**, [FS12] 1 (1984).
60. J. R. d'Almeida and D. J. Thouless, *J. Phys. A* **64**, L743 (1989).
61. We discuss here only "ideal" gels that exhibit a statistically isotropic and homogeneous high-temperature phase, in contrast to the so-called "semisoft" gels that are intrinsically anisotropic and therefore do not exhibit true nemato-elastic Goldstone mode.
62. H. Finkelmann, I. Kundler, E. M. Terentjev and M. Warner, *J. Phys. II* **7**, 1059 (1997); G. C. Verwey, M. Warner and E. M. Terentjev, *J. Phys. II (France)* **6**, 1273–1290 (1996); M. Warner, *J. Mech. Phys. solids* **47**, 1355 (1999).
63. The situation is qualitatively similar to that of the $\phi^4$ theory describing Ising transition. There too, the dominant nonlinearity is $\phi^4$ that becomes relevant for $D < 4$, with the subdominant $\phi^6$ nonlinearity near the Gaussian fixed point turning on only for $D \leq 3$. Despite of this one expects and indeed finds that $\phi^6$ operator can be neglected even for $D = 3$ and $D = 2$, as the relevance analysis needs to be done near the interacting (Wilson-Fisher) and not near the Gaussian critical point. In the same way, we expect that subdominance of in-plane nonlinearities will survive even below $D = 3$, but probably not all the way down to the physical case of $D = 2$.
64. X. Xing and L. Radzihovsky, University of Colorado preprint (2003).

# CHAPTER 11

# FIXED-CONNECTIVITY MEMBRANES

Mark J. Bowick

*Physics Department, Syracuse University , Syracuse, NY 13244-1130*
*E-mail: bowick@physics.syr.edu*

The statistical mechanics of flexible surfaces with internal elasticity and shape fluctuations is summarized. Phantom and self-avoiding isotropic and anisotropic membranes are discussed, with emphasis on the universal negative Poisson ratio common to the low-temperature phase of phantom membranes and all strictly self-avoiding membranes in the absence of attractive interactions. The study of crystalline order on the frozen surface of spherical membranes is also treated.

## 1. Introduction

The statistical mechanics of polymers, which are *one-dimensional* chains or loops to a first approximation, has proven to be a rich and fascinating field.[1-3] The success of physical methods applied to polymers relies on universality — many of the macroscopic length scale properties of polymers are independent of microscopic details such as the chemical identity of the monomers and their interaction potential.[4]

Membranes are two-dimensional (2D) generalizations of polymers. The generalization of polymer statistical mechanics to membranes, surfaces fluctuating in three dimensions, has proven to be very rich because of the richer spectrum of shape and elastic deformations available. In contrast to polymers, there are *distinct* universality classes of membranes distinguished by the nature of their short-range order. There are crystalline, fluid and hexatic membrane analogues of the corresponding phases of strictly two-dimensional systems (monolayers) where shape fluctuations are frozen.[5-7]

The closest membrane analogue to a polymer is a 2D fishnet-like mesh of nodes with a fixed coordination number for each node. A fixed-connectivity membrane with spherical topology from the world of art[8] is shown in Fig. 1. Bonds are assumed to be unbreakable while the nodes themselves live in flat $d$-dimensional Euclidean space $\mathbf{R}^d$, with a physical membrane corresponding to the case $d = 3$. The intrinsic crystalline order of fixed-connectivity membranes with, say, typical coordination number 6, leads to the alternative terminology crystalline membranes. They are also referred to as polymerized or tethered membranes. In general the Hamiltonian for a fixed-connectivity membrane will include both intrinsic elastic contributions

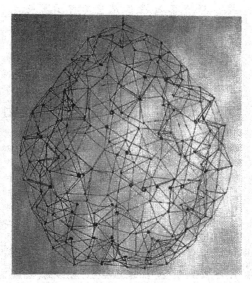

Fig. 1.  Esfera (Sphere) 1976: Gertrude Goldschmidt (Gego). Stainless steel wire — 97 × 88 cm (Patricia Phelps de Cisneros Collection, Caracas, Venezuela).

(compression and shear) and shape (bending) contributions, since the membrane undergoes both types of deformation.[9,10]

Flexible membranes are an important member of the enormous class of *soft* condensed matter systems.[3,11-13] Soft matter responds easily to external forces and has physical properties that are often dominated by the entropy of thermal or other statistical fluctuations.

This chapter will describe the properties of fixed-connectivity membranes with focus on the universal negative Poisson ratio that illustrates the novel elastic behavior of the extended (*flat*) phase of physical membranes, the tubular phase of anisotropic membranes and ordering on frozen curved membrane topographies.

## 2. Physical Examples of Membranes

One can polymerize suitable chiral oligomeric precursors to form molecular sheets.[14] This approach is based directly on the idea of creating an intrinsically two-dimensional polymer. Alternatively one can permanently cross-link fluid-like Langmuir–Blodgett films or amphiphilic layers by adding certain functional groups to the hydrocarbon tails and/or the polar heads,[15,16] as shown schematically in Fig. 2.

The 2D-cytoskeletons of certain cell membranes are beautiful and naturally occurring fixed-connectivity membranes that are essential to the function and stability of the cell as a whole.[17-21] The simplest and most thoroughly studied example is the cytoskeleton of mammalian erythrocytes (red blood cells). The human body has roughly $5 \times 10^{13}$ red blood cells. The red blood cell cytoskeleton is a protein network

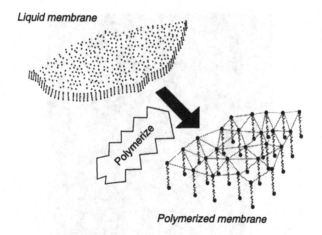

Fig. 2. Making a fixed-connectivity membrane by polymerizing a fluid membrane.

Fig. 3. An electron micrograph of a 0.5 micron square region of a red blood cell cytoskeleton at magnification 365,000:1. The skeleton is negatively stained and has been artificially spread to a surface area nine to ten times as great as in the native membrane. Image courtesy of Daniel Branton (Dept. of Biology, Harvard University).

whose links are spectrin tetramers (of length approximately 200 nm) meeting at junctions composed of short actin filaments (of length 37 nm and typically 13 actin monomers long)[22–24] (see Figs. 3, 4). There are roughly 70,000 triangular faces in the entire mesh which is bound as a whole by ankyrin and other proteins to the cytoplasmic side of the other key component of the cell membrane, the fluid phospholipid bilayer. Without the cytoskeleton the lipid bilayer would disintegrate into

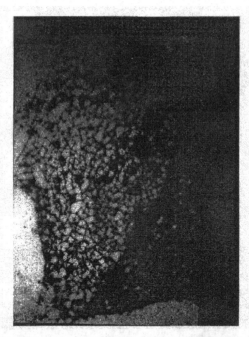

Fig. 4.   An extended view of the spectrin/actin network which forms the cytoskeleton of the red blood cell membrane. Image courtesy of Daniel Branton.

a thousand little vesicles and certainly the red blood cell would not be capable of the shape deformations required to squeeze through narrow capillaries.

There are also inorganic realizations of fixed-connectivity membranes. Graphitic oxide (GO) membranes are micron size sheets of solid carbon, with thicknesses on the order of 10 Å, formed by exfoliating carbon with a strong oxidizing agent. Their structure in an aqueous suspension has been examined by several groups.[25-27] Metal dichalcogenides such as $MoS_2$ have also been observed to form rag-like sheets.[28] Finally similar structures occur in the large sheet molecules, shown in Fig. 5, and believed to be an ingredient in glassy $B_2O_3$.

## 3. Phase Diagrams

Let us consider the general class of $D$-dimensional elastic and flexible manifolds fluctuating in $d$-dimensional Euclidean space. Such manifolds are described by a $d$-dimensional vector $\vec{r}(\mathbf{x})$, where $\mathbf{x}$ labels the $D$-dimensional internal coordinates, as illustrated in Fig. 6. A physical membrane, of course, corresponds to the case $(D = 2, d = 3)$.

The Landau free energy of a membrane must be invariant under global translations, so the order parameter is given by derivatives of the embedding $\vec{r}$, viz. the tangent vectors $\vec{t}_\alpha = \frac{\partial \vec{r}}{\partial x^\alpha}$, with $\alpha = 1, \ldots, D$. Invariance under rotations in both

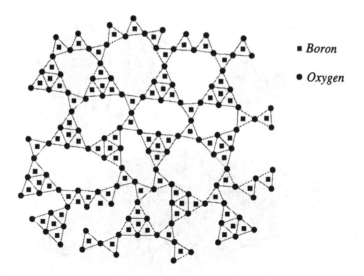

Fig. 5. The sheet molecule $B_2O_3$.

- ■ *Boron*
- ● *Oxygen*

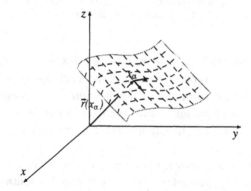

Fig. 6. The parametrization of a *membrane* with internal coordinates $\mathbf{x}$ and bulk coordinates $\vec{r}(\mathbf{x})$.

the internal and bulk space limits the Landau free energy to the form[5,29,30]

$$F(\vec{r}) = \int d^D\mathbf{x} \left[ \frac{t}{2}(\partial_\alpha\vec{r})^2 + u(\partial_\alpha\vec{r}\partial_\beta\vec{r})^2 + v(\partial_\alpha\vec{r}\partial^\alpha\vec{r})^2 + \frac{\kappa}{2}(\partial^2\vec{r})^2 \right]$$
$$+ \frac{b}{2} \int d^D\mathbf{x}\, d^D\mathbf{y}\, \delta^d\big(\vec{r}(\mathbf{x}) - \vec{r}(\mathbf{y})\big), \tag{1}$$

where higher order terms are irrelevant in the long wavelength limit. The physics of Eq. (1) depends on the elastic moduli $t$, $u$ and $v$, the bending rigidity $\kappa$ and the strength of self-avoidance $b$. The limit $b = 0$ describes a *phantom* membrane that may self-intersect with no energy cost.

For small deformations from a reference ground state one may write $\vec{r}(\mathbf{x})$ as

$$\vec{r}(\mathbf{x}) = \big(\zeta\mathbf{x} + \mathbf{u}(\mathbf{x}), \vec{h}(\mathbf{x})\big), \tag{2}$$

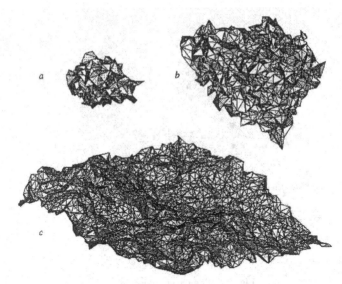

Fig. 7. Typical configurations of phantom membranes: (a) the crumpled phase, (b) the critical *crumpling* phase and (c) the *flat* or bulk-orientationally-ordered phase. Images are from the simulations of Ref. 31.

where $\mathbf{u}(\mathbf{x})$ are $D$ "internal" phonon modes and $\vec{h}(\mathbf{x})$ $d - D$ "out-of-plane" height fluctuations. The case $\zeta = 0$ corresponds to a mean field isotropic crumpled phase for which typical equilibrium membrane configurations have fractal Hausdorff dimension $d_H$ ($d_H = \infty$ for phantom membranes) and there is no distinction between the internal phonons and the height modes. The crumpled phase is illustrated in Fig. 7(a).

The regime $\zeta \neq 0$ describes a membrane which is "flat" up to small fluctuations. The full rotational symmetry of the free energy is spontaneously broken. The fields $\vec{h}$ are the Goldstone modes and scale differently than the phonon fields $\mathbf{u}$. Figure 7(c) is a visualization of a typical configuration in the "flat" phase.

Phantom membranes are by far the easiest to treat analytically and numerically. They may even be physically realizable by synthesizing membranes from strands that cut and repair themselves on a sufficiently short time scale that they access self-intersecting configurations. One can also view the analysis of the phantom membrane as the first step in understanding the more physical self-avoiding membrane. Combined analytical and numerical studies have yielded a thorough understanding of the phase diagram of phantom fixed-connectivity membranes.

### 3.1. *Phantom Membranes*

The phantom membrane free energy is

$$F(\vec{r}) = \int d^D \mathbf{x} \left[ \frac{t}{2}(\partial_\alpha \vec{r})^2 + u(\partial_\alpha \vec{r} \partial_\beta \vec{r})^2 + v(\partial_\alpha \vec{r} \partial^\alpha \vec{r})^2 + \frac{\kappa}{2}(\partial^2 \vec{r})^2 \right]. \qquad (3)$$

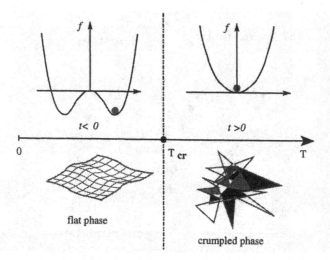

Fig. 8. The mean field free energy density $f$ of fixed-connectivity membranes as a function of the order parameter $t$, together with a schematic of the low temperature flat ordered phase and the high temperature crumpled disordered phase.

The mean field effective potential, using the expansion of Eq. (2), is

$$V(\zeta) = D\zeta^2 \left( \frac{t}{2} + (u + vD)\zeta^2 \right),$$ (4)

with minima

$$\zeta^2 = \begin{cases} 0 & t \geq 0 \\ -\frac{t}{4(u+vD)} & t < 0 \end{cases}.$$ (5)

This implies a "flat" (extended) phase for $t < 0$ and a crumpled phase for $t > 0$, separated by a continuous crumpling transition at $t = 0$, as sketched in Fig. 8. Of course anything is possible in mean field theory but a variety of analytic and numerical calculations indicates the true phase diagram of phantom membranes is qualitatively like Fig. 9.[9] The crumpled phase is described by a line of equivalent Gaussian fixed points (GFPs). There is a crumpling transition line in the $\kappa - t$ plane containing an infrared stable fixed point (CTFP) which describes the long wavelength properties of the crumpling transition. Finally, for large enough values of $\kappa$ and negative values of $t$, the system is in a "flat" phase whose properties are dictated by an infrared stable flat phase fixed point (FLFP).

### 3.1.1. *The Crumpled Phase*

In the crumpled phase, the free energy Eq. (3), for $D \geq 2$, simplifies to

$$F(\vec{r}) = \frac{t}{2} \int d^D \mathbf{x} \, (\partial_\alpha \vec{r})^2 + \text{irrelevant terms},$$ (6)

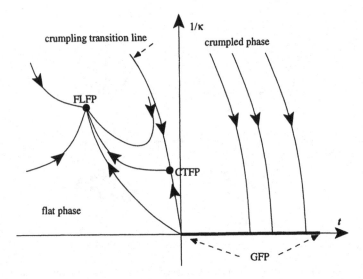

Fig. 9.   Schematic plot of the phase diagram for phantom membranes.

since the model is completely equivalent to a linear sigma model with $O(d)$ internal symmetry in $D \geq 2$ dimensions and therefore all derivative operators in $\vec{r}$ are irrelevant by power counting.[32,33] The parameter $t$ labels equivalent Gaussian fixed points, as depicted in Fig. 9. In renormalization group language there is a marginal direction for positive $t$. The large distance properties of this phase are described by simple gaussian fixed points with the exact connected 2-point function:

$$G(\mathbf{x}) \sim \begin{cases} |\mathbf{x}|^{2-D} & D \neq 2 \\ \log |\mathbf{x}| & D = 2 \end{cases}.$$ (7)

The associated critical exponents may also be computed exactly. The Hausdorff (fractal) dimension $d_H$, or equivalently the size exponent $\nu = D/d_H$, is given (for the physical case $D = 2$) by

$$d_H = \infty \ (\nu = 0) \Rightarrow R_G^2 \sim \log L,$$ (8)

where $R_G^2$ is the radius of gyration and $L$ is the linear membrane size. This result is confirmed by numerical simulations of fixed-connectivity membranes in the crumpled phase where the logarithmic behavior of the radius of gyration is accurately checked.[31,34−45]

### 3.1.2. The Crumpling Transition

Near the crumpling transition the membrane free energy is given by

$$F(\vec{r}) = \int d^D\mathbf{x} \left[ \frac{1}{2}(\partial^2\vec{r})^2 + u(\partial_\alpha\vec{r}\partial_\beta\vec{r})^2 + \hat{v}(\partial_\alpha\vec{r}\partial^\alpha\vec{r})^2 \right],$$ (9)

where the bending rigidity has been scaled out and $\hat{v} = v - \frac{u}{D}$. By naive power counting the directions defined by the couplings $u$ and $\hat{v}$ are relevant for $D \leq 4$ and the model is amenable to an $\varepsilon = 4 - D$-expansion. The $\beta$ functions are given by[9]

$$\beta_u(u_R, v_R) = -\varepsilon u_R + \frac{1}{8\pi^2}\left\{\left(\frac{d}{3} + \frac{65}{12}\right)u_R^2 + 6u_Rv_R + \frac{4}{3}v_R^2\right\} \tag{10}$$

and

$$\beta_v(u_R, v_R) = -\varepsilon v_R + \frac{1}{8\pi^2}\left\{\frac{21}{16}u_R^2 + \frac{21}{2}u_Rv_R + (4d+5)v_R^2\right\}. \tag{11}$$

These two coupled beta functions have a fixed point only for $d > 219$.[29] This suggests that the crumpling transition is first order for $d = 3$. Other analyses, however, indicate a continuous crumpling transition. A revealing extreme limit of membranes was studied by David and Guitter.[46] This is the limit of infinite elastic constants in the flat phase. Since the elastic terms in the Hamiltonian scale like $q^2$ in momentum space, as compared to $q^4$ for the bending energy, this limit exposes the dominant infrared behavior of the membrane. In this "stretchless" limit the elastic strain tensor $u_{\alpha\beta}$ must vanish and the Hamiltonian is constrained, very much in analogy to a nonlinear sigma model. The Hamiltonian becomes

$$H_{NL} = \int d^D\sigma \frac{\kappa}{2}(\Delta\vec{r})^2, \tag{12}$$

together with the constraint $\partial_\alpha \vec{r} \partial_\beta \vec{r} = \delta_{\alpha\beta}$. Remarkably, the $\beta$-function for the inverse bending rigidity, $\alpha = 1/\kappa$, may be computed within a large-$d$ expansion, giving

$$\beta(\alpha) = q\frac{\partial\alpha}{\partial q} = \frac{2}{d}\alpha - \left(\frac{1}{4\pi} + \frac{\text{const.}}{d}\right)\alpha^2. \tag{13}$$

For $d = \infty$ there is no stable fixed point and the membrane is always crumpled. To next order in $1/d$, however, Eq. (13) reveals an ultraviolet stable fixed point at $\alpha = 8\pi/d$, corresponding to a continuous crumpling transition. The size exponent at the transition is found to be[47]

$$d_H = \frac{2d}{d-1} \Rightarrow \nu = 1 - \frac{1}{d} = \frac{2}{3} \text{ (for } d = 3). \tag{14}$$

Le Doussal and Radzihovsky[48] analyzed the Schwinger-Dyson equations for the model of Eq. (9) keeping up to four point vertices. The result for the Hausdorff dimension and size exponent is

$$d_H = 2.73 \Rightarrow \nu = 0.73. \tag{15}$$

Finally Monte Carlo Renormalization Group simulations[49] of the crumpling transition find a continuous transition with exponents

$$d_H = 2.77(10) \Rightarrow \nu = 0.71(3). \tag{16}$$

Thus three independent analyses find a continuous crumpling transition with a consistent size exponent.

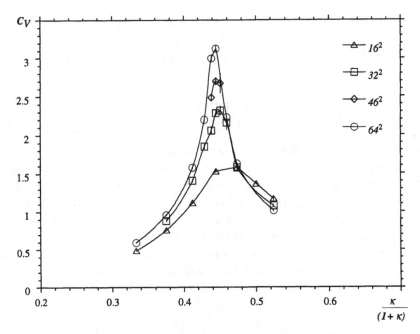

Fig. 10.   Plot of the specific heat observable from the simulations of Ref. 31. The growth of the specific heat peak with system size indicates a continuous transition.

Further evidence for the crumpling transition being continuous is provided by numerical simulations[9,44,45] where the analysis of observables like the specific heat (see Fig. 10) or the radius of gyration radius give textbook continuous phase transitions, although the value of the exponents at the transition are difficult to determine precisely.

### 3.1.3. *The Flat Phase*

In a flat membrane (see Fig. 11), it is natural to introduce the strain tensor

$$u_{\alpha\beta} = \partial_\alpha u_\beta + \partial_\beta u_\alpha + \partial_\alpha h \partial_\beta h. \tag{17}$$

The free energy Eq. (3) in these variables becomes

$$F(\mathbf{u}, h) = \int d^D \mathbf{x} \left[ \frac{\hat{\kappa}}{2} (\partial_\alpha \partial_\beta h)^2 + \mu u_{\alpha\beta} u^{\alpha\beta} + \frac{\lambda}{2} (u_\alpha^\alpha)^2 \right], \tag{18}$$

where irrelevant higher derivative terms have been dropped. One recognizes the standard Landau free energy of elasticity theory,[50] with Lamé coefficients $\mu$ and $\lambda$, plus an extrinsic curvature term, with bending rigidity $\hat{\kappa}$. These couplings are related to the original ones in Eq. (3) by $\mu = u\zeta^{4-D}$, $\lambda = 2v\zeta^{4-D}$, $\hat{\kappa} = \kappa\zeta^{4-D}$ and $t = -4 \left( \mu + \frac{D}{2}\lambda \right) \zeta^{D-2}$.

   The large distance properties of the flat phase for fixed-connectivity membranes are completely described by the free energy of Eq. (18). Since the bending rigidity

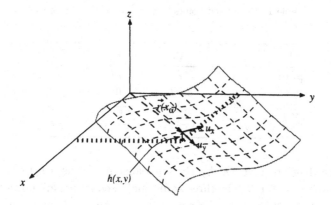

Fig. 11.  Membrane coordinates appropriate for analyzing fluctuations in the flat phase.

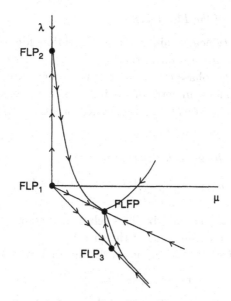

Fig. 12.  Phase diagram for the phantom flat phase. There are three infra-red unstable fixed points, labelled by FLP1, FLP2 and FLP3, but the physics of the flat phase is governed by the infra-red stable fixed point (FLFP).

may be scaled out at the crumpling transition, the free energy becomes a function of $\frac{\mu}{\kappa^2}$ and $\frac{\lambda}{\kappa^2}$. The $\beta$-functions for the couplings $\mu$ and $\lambda$ in the $\varepsilon$-expansion are[51,52]

$$\beta_\mu(\mu_R, \lambda_R) = -\varepsilon\mu_R + \frac{\mu_R^2}{8\pi^2}\left(\frac{d_c}{3} + 20A\right);$$

$$\beta_\lambda(\mu_R, \lambda_R) = -\varepsilon\lambda_R + \frac{1}{8\pi^2}\left(\frac{d_c}{3}\mu_R^2 + 2(d_c + 10A)\lambda_R\mu_R + 2d_c\lambda_R^2\right),$$

(19)

where $d_c$ is the codimension $d - D$ and $A = \frac{\mu_R + \lambda_R}{2\mu_R + \lambda_R}$. These coupled $\beta$-functions possess four fixed points (see Fig. 12), whose values are shown in Table 1.

Table 1.  The fixed points and critical exponents of the flat phase.

| FP | $\mu_R^*$ | $\lambda_R^*$ | $\eta$ | $\eta_u$ |
|---|---|---|---|---|
| FLP1 | 0 | 0 | 0 | 0 |
| FLP2 | 0 | $2\varepsilon/d_c$ | 0 | 0 |
| FLP3 | $\dfrac{12\varepsilon}{20+d_c}$ | $\dfrac{-6\varepsilon}{20+d_c}$ | $\dfrac{\varepsilon}{2+d_c/10}$ | $\dfrac{\varepsilon}{1+20/d_c}$ |
| FLFP | $\dfrac{12\varepsilon}{24+d_c}$ | $\dfrac{-4\varepsilon}{24+d_c}$ | $\dfrac{\varepsilon}{2+d_c/12}$ | $\dfrac{\varepsilon}{1+24/d_c}$ |

The phase diagram revealed by the $\varepsilon$-expansion is thus a little more complex than that sketched in Fig. 9. The three additional fixed points are infra-red unstable, however, and can only be reached for very specific values of the Lamé coefficients.

### 3.1.4. *The Properties of the Flat Phase*

Figure 7(c) shows a typical equilibrium configuration for a membrane that has developed a preferred orientation in the bulk — the surface normals clearly have long-range order. In this phase the membrane is a rough extended two-dimensional structure. The rotational symmetry of the full free energy is spontaneously broken from $O(d)$ to $O(d-D) \times O(D)$. The remnant rotational symmetry is realized in Eq. (18) as

$$h_i(\mathbf{x}) \to h_i(\mathbf{x}) + A^{i\alpha}\mathbf{x}_\alpha;$$
$$u_\alpha(\mathbf{x}) \to u_\alpha - A^{i\alpha}h_i - \frac{1}{2}\delta^{ij}A^{i\alpha}A^{\beta j}\mathbf{x}_\beta, \tag{20}$$

where $A^{i\alpha}$ is a $D \times (d-D)$ matrix. This relation provides Ward identities which greatly simplify the renormalization of the theory.

The phonon and height propagators in the infrared limit are given by

$$\Gamma_{uu}(\vec{p}) \sim |\vec{p}|^{2+\eta_u};$$
$$\Gamma_{hh}(\vec{p}) \equiv |\vec{p}|^4 \kappa(\vec{p}) \sim |\vec{p}|^{4-\eta}, \tag{21}$$

where the last equation defines the anomalous bending rigidity $\kappa(\vec{p}) \sim |\vec{p}|^{-\eta}$. The two scaling exponents $\eta_u$ and $\eta$ are related by the scaling relation[51]

$$\eta_u = 4 - D - 2\eta, \tag{22}$$

which follows from the Ward identities (Eq. (20)) associated with the remnant rotational symmetry . The roughness exponent $\zeta$, which measures the growth with system size of the rms height fluctuations transverse to the flat directions, is determined from $\eta$ by the further scaling relation

$$\zeta = \frac{4-D-\eta}{2}. \tag{23}$$

The long wavelength properties of the flat phase are described by the FLFP (see Fig. 12). Since the FLFP occurs at non-zero renormalized values of the Lamé coefficients, the associated critical exponents are clearly non-Gaussian. These key critical exponents have also been determined by independent methods.

Large scale simulation of membranes in the flat phase model were performed in Ref. 31. The results obtained for the critical exponents are very accurate:

$$\eta_u = 0.50(1); \quad \eta = 0.750(5); \quad \zeta = 0.64(2). \tag{24}$$

A review of numerical results may be found in Refs. 44, 45.

The SCSA approximation[48] gives a beautiful result for general $d$:

$$\eta(d) = \frac{4}{d_c + \sqrt{16 - 2d_c + d_c^2}}, \tag{25}$$

which for $d = 3$ gives

$$\eta_u = 0.358; \quad \eta = 0.821; \quad \zeta = 0.59. \tag{26}$$

Finally the large-d expansion[46] gives

$$\eta = \frac{2}{d} \Rightarrow \eta(3) = \frac{2}{3}. \tag{27}$$

The numerical simulations are in qualitative agreement with both the SCSA and large-d analytical estimates.

On the experimental side we are fortunate to have two measurements of the key critical exponents for the flat phase of fixed-connectivity membranes. The static structure factor of the red blood cell cytoskeleton has been measured by small-angle x-ray and light scattering, yielding a roughness exponent of $\zeta = 0.65(10)$.[24] Freeze-fracture electron microscopy and static light scattering of the conformations of graphitic oxide sheets reveal flat sheets with a fractal dimension $d_H = 2.15(6)$. Both these measured values are in good agreement with the best analytic and numerical predictions, but the errors are still too large to discriminate between different analytic calculations and to accurately substantiate the numerical simulations.

The Poisson ratio[50] of a phantom fixed-connectivity membrane (which measures the transverse elongation due to a longitudinal stress) is universal and within the SCSA approximation is given by

$$\sigma(D) = -\frac{1}{D+1} \Rightarrow \sigma(2) = -\frac{1}{3}. \tag{28}$$

This result has also been checked in numerical simulations.[53,54] Rather remarkably, it turns out to be negative. While Ref. 53 finds $\sigma \approx -0.15$ the latter simulation[54] finds $\sigma \approx -0.32$. Materials with a negative Poisson ratio have been dubbed *auxetics*.[55] The wide variety of potential applications of auxetic materials suggests a fascinating role for flexible fixed-connectivity membranes in materials science (see Sec. 4).

A final critical regime of a flat membrane is achieved by subjecting the membrane to external tension.[47] This gives rise to a low temperature phase in which the membrane has a domain structure, with distinct domains corresponding to flat phases with different bulk orientations. This describes, physically, a *buckled* membrane whose equilibrium shape is no longer planar.

## 3.2. Self-avoiding Membranes

Physically, realistic fixed-connectivity membranes will have large energy barriers to self-intersection. That is they will generally be self-avoiding. Self-avoidance is familiar in the physics of polymer chains and may be treated by including the Edwards-type delta-function repulsion of the Hamiltonian in Eq. (1). A detailed summary of our current understanding is given in Refs. 9, 10. The essential finding is that self-avoidance eliminates all but the flat phase.

### 3.2.1. Numerical Simulations

Numerical simulations are currently essential in understanding the statistical mechanics of self-avoiding membranes because the treatment of nonlinear elasticity together with non-local self-avoidance is currently beyond the realm of analytic techniques.

Two discretizations of membranes have been adopted to incorporate self-avoidance. The *balls and springs* class of models begins with a network of $N$ particles in a intrinsically triangular array and interacting via a nearest-neighbor elastic potential

$$V_{NN}(\vec{r}) = \begin{cases} 0 & \text{for } |\vec{r}| \leq b \\ \infty & \text{for } |\vec{r}| > b \end{cases},$$ (29)

where the free parameter $b$ plays the role of a tethering length. An additional hard sphere steric repulsion forbids *any* node to be closer than a distance $\sigma$ from any other node:

$$V_{\text{steric}}(\vec{r}) = \begin{cases} \infty & \text{for } |\vec{r}| \leq \sigma \\ 0 & \text{for } |\vec{r}| > \sigma \end{cases}.$$ (30)

Early simulations[34,35] of this class of model gave a first estimate of the fractal dimension for physical membranes compatible with the Flory estimate $d_H = 2(d + D)/(2 + D) = 2.5$.[36] The system sizes simulated, however, were quite small and subsequent simulations for larger systems found that the membrane is flat.[56,57] This result is remarkable when one recalls that there is no explicit bending rigidity.

A plausible explanation[58] for the loss of the crumpled phase is that next-to-nearest neighbor excluded volume effects induce a positive bending rigidity, driving the model to the FLFP. The structure function of the self-avoiding model has been computed numerically[59] and found to compare well with the analytical structure function for the flat phase of phantom fixed-connectivity membranes. In particular the roughness exponents are comparable.

The induced bending rigidity may be lowered by taking a smaller excluded volume.[60] The flat phase persists to very small values of $\sigma$ with eventual signs of a crumpled phase, probably due to effective loss of self-avoidance. A more comprehensive study,[61] in which the hard sphere radius is taken to zero with an excluded volume potential which is a function of the internal distance along the lattice, concluded that self-avoidance implies flatness in the thermodynamic limit of large membranes.

Self-avoidance may also be implemented by modelling impenetrable triangular meshes. This has the advantage that there is no restriction on the bending angle between adjacent cells and therefore no induced bending rigidity.[62]

The first simulations of the plaquette model[63] found a Hausdorff dimension in rough agreement in agreement with the Flory estimate 2.5 but this has not held up in subsequent work. A subsequent simulation[64] found $d_H \approx 2.3$ and extensive recent work employing more sophisticated algorithms and extending to much larger membranes confirm the loss of the crumpled phase.[62]

Some insight into the lack of a crumpled phase for self-avoiding fixed-connectivity membranes is offered by the study of folding.[65–71] Folding corresponds to the limit of infinite elastic constants[46] with the further approximation that the space of bending angles is discretized. One quickly discovers that the reflection symmetries of the allowed folding vertices forbid local folding (crumpling) of surfaces. There is therefore essentially no entropy for crumpling. There is, however, local unfolding and the resulting statistical mechanical models are non-trivial. The lack of local folding is the discrete equivalent of the long-range curvature-curvature interactions that stabilize the flat phase. The dual effect of the integrity of the surface (time-independent connectivity) and self-avoidance is so powerful that crumpling seems to be impossible in low embedding dimensions.

### 3.2.2. *The Properties of the Self-avoiding Fixed Point*

For the physically relevant case $d = 3$ numerical simulations thus find that there is no crumpled phase. Furthermore, the flat phase is *identical* to the flat phase of the phantom membrane.[62] The roughness exponent $\zeta_{SA}$ from numerical simulations of self-avoidance at $d = 3$ using ball-and-spring models[72] and impenetrable plaquette models[62] and the roughness exponent at the FLFP, Eq. (24), compare extremely well

$$\zeta_{SA} = 0.64(4), \quad \zeta = 0.64(2). \tag{31}$$

The numerical evidence thus strongly indicates that the SAFP is exactly the same as the FLFP and that the crumpled self-avoiding phase is absent in the presence of purely repulsive potentials (see Fig. 13). This conjecture is strengthened by the finding that the Poisson ratio of self-avoiding membranes is the same as that of flat phantom membranes[73] (see Sec. 4). This identification of fixed points enhances the significance of the FLFP treated earlier.

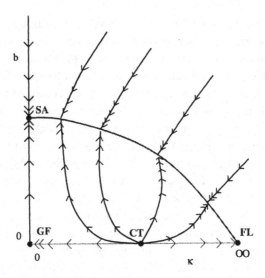

Fig. 13. The conjectured phase diagram for self-avoiding fixed-connectivity membranes in 3 dimensions. With any degree of self-avoidance the renormalization group flows are to the flat phase fixed point of the phantom model (FL).

## 4. Poisson Ratio and Auxetics

In the classical theory of elasticity[50] an arbitrary deformation of a $D$-dimensional elastic body may be decomposed into a pure shear and a pure compression:

$$u_{ij} = \left[u_{ij} - \frac{1}{D}(\text{Tr}\,u)\delta_{ij}\right] + \frac{1}{D}(\text{Tr}\,u)\delta_{ij}, \qquad (32)$$

where Tr denotes the trace and the term in square brackets is a pure shear (volume-preserving but shape changing) while the second term is a pure compression (shape-preserving but volume-changing). The elastic free energy is then given by

$$F_{el} = \mu \left[u_{ij} - \frac{1}{D}(\text{Tr}\,u)\delta_{ij}\right]^2 + \frac{1}{2}K\,(\text{Tr}\,u)^2, \qquad (33)$$

where $\mu$ is the shear modulus and $K$ is the bulk modulus. This free energy may be written equivalently as

$$F_{el} = \mu\,u_{ij}u_{ij} + \frac{1}{2}\lambda\,(\text{Tr}\,u)^2, \qquad (34)$$

with the elastic Lamé coefficient $\lambda$ related to the bulk and shear moduli by

$$K = \lambda + \frac{2\mu}{D}. \qquad (35)$$

For the physical membrane, $D = 2$, this reads $K = \lambda + \mu$. Thermodynamic stability requires that both $K$ and $\mu$ be positive, otherwise the free energy could be spontaneously lowered by pure compressional or pure shear deformations, respectively.

The Poisson ratio $\sigma$ is defined as the ratio of transverse contractile strain to longitudinal tensile strain for an elastic body subject to a uniform applied tension T. For tension applied uniformly in, say, the $x$-direction

$$\sigma = -\frac{\delta y/y}{\delta x/x}, \tag{36}$$

the Poisson ratio is easily found to be

$$\sigma = \frac{K - \mu}{K + \mu}, \tag{37}$$

for $D = 2$, and

$$\sigma = \frac{1}{2}\left(\frac{3K - 2\mu}{3K + \mu}\right) \tag{38}$$

for $D = 3$. Thermodynamic stability is only possible for $-1 \leq \sigma \leq 1$ for $D = 2$ and $-1 \leq \sigma \leq \frac{1}{2}$ for $D = 3$. The upper bounds (1 and $\frac{1}{2}$ respectively) are approached for materials that have vanishing shear modulus compared to their bulk modulus (rubber-like) and the lower bounds ($-1$) for materials with negligible bulk modulus in comparison to their shear modulus ("anti-rubber").[74] Clearly, the Poisson ratio may be negative (auxetic) for $K < \mu$ ($D = 2$) and $K < \frac{2}{3}\mu$ ($D = 3$). Most materials get thinner when stretched and fatter when squashed – auxetic materials are uncommon. The earliest known example, dating from more than a century ago, is that of a pyrite (FeS$_2$) crystal,[75] which has a Poisson ratio, in certain crystallographic directions, of $\sigma \approx -0.14$. More recently, some isotropic polyester foams have been created with Poisson ratios as large as $\sigma \sim -0.7$.[77,78] The potential of auxetic materials in materials science is nicely reviewed in Ref. 79. One of the rare naturally occurring auxetics is SiO$_2$ in its $\alpha$-crystobalite phase.[80,81] Cristobalite is one of the three distinct crystalline forms of SiO$_2$, together with quartz and tridymite. Its Poisson ratio reaches a maximum negative value of $-0.5$ in some directions, with orientationally-averaged values for single-phased aggregates of $-0.16$.

The underlying mechanism driving fixed-connectivity membranes auxetic ($\sigma = -\frac{1}{3}$) has schematic similarities to that illustrated in Fig. 14. Submitting a membrane to tension will suppress its out-of-plane fluctuations, forcing it entropically to expand in both in-plane directions. More physically, the out-of-plane undulations renormalize the elastic constants (the Lamé coefficients), in such a way that the long-wavelength bulk modulus is less than the shear modulus, which is the signature of a two-dimensional auxetic material. The soft matter origin of the universal negative Poisson ratio of fixed-connectivity membranes provides a fundamentally new paradigm for the design of novel materials. The best current experimental measurements of the Poisson ratio of the red blood cell cytoskeleton[82] find $\sigma \approx +\frac{1}{3}$ from separate determinations of the bulk and shear modulus. The cytoskeleton still has the fluid lipid bilayer attached, however, and this may influence the pure cytoskeletal

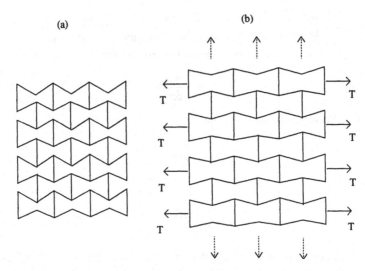

Fig. 14. Mechanical model of an auxetic material: (a) in the absence of applied stress and (b) under applied lateral stress $T$. The lateral stretching accompanying the applied stress forces the material out in the transverse dimension.

elasticity. A direct measurement of the Poisson ratio for a flexible fixed-connectivity membrane remains an important and challenging task.

Auxetic materials have desirable mechanical properties such as higher in-plane indentation resistance, transverse shear modulus and bending stiffness. They have clear applications as sealants, gaskets and fasteners. They may also be promising materials for artificial arteries, since they can expand to accommodate sudden increases in blood flow.

We can model a realistic fixed-connectivity membrane with an elastic free energy and either large bending rigidity or self-avoidance. This is of practical importance in modelling since, for example, we may replace the more complicated non-local self-avoidance term with a large bending rigidity.

It would be very interesting to know if nature utilizes the auxetic character of the red-blood cell spectrin cytoskeleton in the elastic deformations of red blood cells as they pass through fine blood capillaries. As such cells deform, the membrane skeleton can unfold, which might help to transport large molecules or expose reactive chemical groups.[83]

## 5. Anisotropic Membranes

An anisotropic membrane is a fixed-connectivity membrane in which the elastic moduli or the bending rigidity in one distinguished direction are different from those in the remaining $D - 1$ directions. Such a membrane may be described by a $d$-dimensional vector $\vec{r}(\mathbf{x}_\perp, y)$, where now the $D$ dimensional internal coordinates are split into $D - 1\mathbf{x}_\perp$ coordinates and the orthogonal distinguished direction $y$.

Requiring invariance under translations, $O(d)$ rotations in the embedding space and $O(D-1)$ rotations in the internal space, the equivalent of Eq. (1) becomes

$$F(\vec{r}(\mathbf{x})) = \frac{1}{2} \int d^{D-1}\mathbf{x}_\perp \, dy \Big[ \kappa_\perp (\partial_\perp^2 \vec{r})^2 + \kappa_y (\partial_y^2 \vec{r})^2 + \kappa_{\perp y} \partial_y^2 \vec{r} \cdot \partial_\perp^2 \vec{r} + t_\perp (\partial_\alpha^\perp \vec{r})^2$$

$$+ t_y (\partial_y \vec{r})^2 + \frac{u_{\perp\perp}}{2} (\partial_\alpha^\perp \vec{r} \cdot \partial_\beta^\perp \vec{r})^2 + \frac{u_{yy}}{2} (\partial_y \vec{r} \cdot \partial_y \vec{r})^2 + u_{\perp y} (\partial_\alpha^\perp \vec{r} \cdot \partial_y \vec{r})^2$$

$$+ \frac{v_{\perp\perp}}{2} (\partial_\alpha^\perp \vec{r} \cdot \partial_\alpha^\perp \vec{r})^2 + v_{\perp y} (\partial_\alpha^\perp \vec{r})^2 (\partial_y \vec{r})^2 \Big]$$

$$+ \frac{b}{2} \int d^D \mathbf{x} \int d^D \mathbf{x}' \delta^d (\vec{r}(\mathbf{x}) - \vec{r}(\mathbf{x}')). \tag{39}$$

This model has eleven free parameters – three distinct bending rigidities, $\kappa_\perp, \kappa_y$ and $\kappa_{\perp y}$, seven elastic moduli, $t_\perp, t_y, u_{\perp\perp}, u_{yy}, u_{\perp y}, v_{\perp\perp}$ and $v_{\perp y}$, and the strength of self-avoidance coupling $b$.

As before we decompose displacements as

$$\vec{r}(\mathbf{x}) = \Big( \zeta_\perp \mathbf{x}_\perp + \mathbf{u}_\perp(\mathbf{x}), \ \zeta_y y + u_y(\mathbf{x}), \ \vec{h}(\mathbf{x}) \Big), \tag{40}$$

with $\mathbf{u}_\perp$ being the $D-1$-dimensional intrinsic phonon modes, $u_y$ the intrinsic phonon mode in the distinguished direction $y$ and $\vec{h}$ the $d-D$-dimensional out-of-plane fluctuation modes. If $\zeta_\perp = \zeta_y = 0$, the membrane is crumpled and if both $\zeta_\perp$ and $\zeta_y$ do not vanish, the membrane is flat. There is, however, the possibility that $\zeta_\perp = 0$ and $\zeta_y \neq 0$ or $\zeta_\perp \neq 0$ and $\zeta_y = 0$. This describes a *tubular* phase, in which the membrane is crumpled in some internal directions but flat in the remaining ones.[84,85] Figure 15 displays a typical equilibrium configuration from the tubular phase, along with the low and high-temperature flat and crumpled phases for a phantom anisotropic membrane.

Let us deal with the phantom anisotropic membrane first. Both analytical[86] and numerical work[87] has established that the phase diagram contains a crumpled, tubular and flat phase. The crumpled and flat phases are equivalent to the isotropic ones, so anisotropy turns out to be an irrelevant interaction in those phases. The new physics is contained in the tubular phase.

## 5.1. *Phantom Tubular Phase*

### 5.1.1. *The Phase Diagram*

We first describe the mean field theory phase diagram and then the effect of fluctuations. There are two situations, depending on the value of a certain function $\Delta$, which depends on the elastic constants $u_{\perp\perp}, v_{\perp y}, u_{yy}$ and $v_{\perp\perp}$.[84-86]

For $\Delta > 0$ the mean field solution exhibits crumpled, flat and tubular phases. When $t_y > 0$ and $t_\perp > 0$ the model is crumpled. Lowering the temperature so that one of the $t$ couplings becomes negative drives the membrane to the tubular

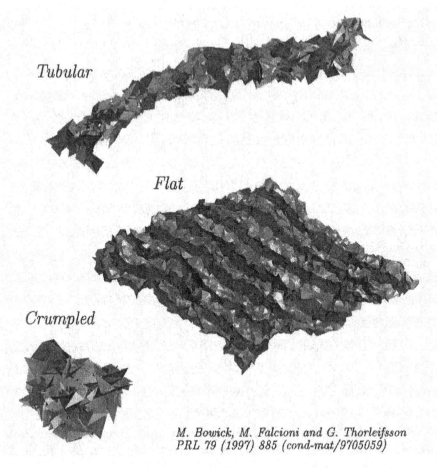

*Tubular*

*Flat*

*Crumpled*

M. Bowick, M. Falcioni and G. Thorleifsson
PRL 79 (1997) 885 (cond-mat/9705059)

Fig. 15.   The three phases of anisotropic crystalline membranes.

phase (either a $\perp$ or $y$-tubule). Lowering the temperature still further flattens the membrane. For $\Delta < 0$ the flat phase disappears from the mean field solution, leaving only the crumpled and tubular phases separated by a continuous transition. Tubular phases are the stable low temperature stable phases in this regime. This mean field result is summarized in Fig. 16.

Beyond mean field theory, the Ginzburg criterion applied to this particular model suggests that the phase diagram is stable for physical membranes $D = 2$ at any embedding dimension $d$. The mean field description should be qualitatively correct even for the full model.

Numerical simulations have spectacularly confirmed this beautiful analytic prediction.[87] Changing the temperature generates a sequence of continuous phase transitions crumpled-to-tubular and tubular-to-flat, in total agreement with the $\Delta > 0$ case above (see Fig. 16).

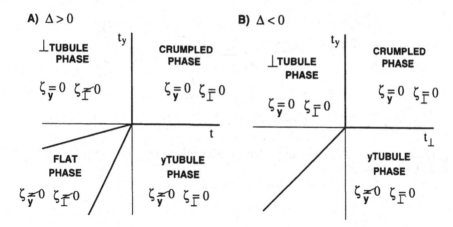

Fig. 16. The phase diagram for anisotropic phantom membranes.

### 5.1.2. *The Crumpled Anisotropic Phase*

In this phase $t_y > 0$ and $t_\perp > 0$, and the free energy Eq. (39) reduces , for $D \geq 2$, to

$$F(\vec{r}(\mathbf{x})) = \frac{1}{2} \int d^{D-1}\mathbf{x}_\perp dy \left[ t_\perp (\partial_\alpha^\perp \vec{r})^2 + t_y (\partial_y \vec{r})^2 \right] + \text{irrelevant Terms}. \qquad (41)$$

By redefining the $y$ coordinate to be $y' = \frac{t_\perp}{t_y} y$ this reduces to Eq. (6), with $t \equiv t_\perp$. Anisotropy is clearly irrelevant in the crumpled phase.

### 5.1.3. *The Flat Phase*

In the flat phase intrinsic anisotropies are only apparent at short-distances and therefore should be irrelevant in the infrared limit. This argument may be made more precise.[88] The flat phase is thus equivalent to the flat phase of isotropic membranes.

## 5.2. *The Tubular Phase*

We now turn to the study of the novel tubular phase, both in the phantom case and with self-avoidance. Since the physically relevant case for membranes is $D = 2$ the $y$-tubular and $\perp$-tubular phase are the same.

The key critical exponents characterizing the tubular phase are the size (or Flory) exponent $\nu$, giving the scaling of the tubular diameter $R_g$ with the extended ($L_y$) and transverse ($L_\perp$) sizes of the membrane, and the roughness exponent $\zeta$ associated with the growth of height fluctuations $h_{\text{rms}}$ (see Fig. 17):

$$\begin{aligned} R_g(L_\perp, L_y) &\propto L_\perp^\nu S_R(L_y/L_\perp^z); \\ h_{\text{rms}}(L_\perp, L_y) &\propto L_y^\zeta S_h(L_y/L_\perp^z), \end{aligned} \qquad (42)$$

where $S_R$ and $S_h$ are scaling functions[84,85] and $z$ is the anisotropy exponent.

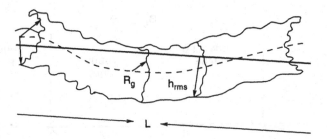

Fig. 17.   A schematic illustration of a tubular configuration indicating the radius of gyration $R_g$ and the height fluctuations $h_{rms}$.

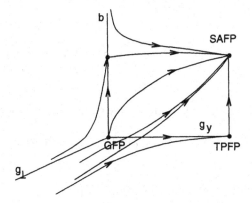

Fig. 18.   The phase diagram for self-avoiding anisotropic membranes with the Gaussian fixed point (GFP), the tubular phase fixed point (TPFP) and the self-avoidance fixed point (SAFP).

The general free energy described in Eq. (39) may be simplified considerably in a $y$-tubular phase:[89,90]

$$F(u, \vec{h}) = \frac{1}{2} \int d^{D-1}\mathbf{x}_\perp dy \left[ \kappa(\partial_y^2 \vec{h})^2 + t(\partial_\alpha \vec{h})^2 \right.$$

$$\left. + g_\perp(\partial_\alpha u + \partial_\alpha \vec{h} \partial_y \vec{h})^2 + g_y \left( \partial_y u + \frac{1}{2}(\partial_y \vec{h})^2 \right)^2 \right]$$

$$+ \frac{b}{2} \int dy d^{D-1}\mathbf{x}_\perp d^{D-1}\mathbf{x}'_\perp \delta^{d-1}(\vec{h}(\mathbf{x}_\perp, y) - \vec{h}(\mathbf{x}'_\perp, y)), \qquad (43)$$

reducing the number of free couplings to five. The coupling $g_\perp$, furthermore, is irrelevant by standard power counting. The most natural assumption is to set it to zero. In that case the phase diagram one obtains is shown in Fig. 18. Without self-avoidance, i.e. $b = 0$, the Gaussian Fixed Point (GFP) is unstable and the long-wavelength behavior of the membrane is controlled by the tubular phase fixed point (TPFP). Any amount of self-avoidance, however, leads to a new fixed point, the Self-avoiding Tubular fixed point (SAFP), which describes the large distance properties of self-avoiding tubules.

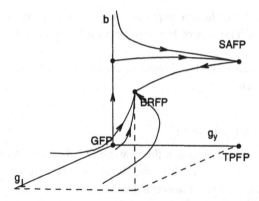

Fig. 19. The phase diagram for self-avoiding anisotropic membranes with the Gaussian fixed point (GFP), the tubular phase fixed point (TPFP), the self-avoidance fixed point (SAFP) and the bending rigidity fixed point (BRFP).

Radzihovsky and Toner advocate a different scenario.[85] For sufficiently small embedding dimensions $d$, including the physical $d = 3$ case, these authors suggest the existence of a new bending rigidity renormalized fixed point (BRFP), which is the infra-red fixed point describing the actual properties of self-avoiding tubules (see Fig. 19).

Here we follow the arguments presented in Refs. 89, 90 and consider the model defined by Eq. (43) with the $g_\perp$-term set to zero. One can prove then than there are some general scaling relations among the critical exponents. All three exponents may be expressed in terms of a single exponent

$$\zeta = \frac{3}{2} + \frac{1-D}{2z};$$
$$\nu = \zeta z. \qquad (44)$$

Remarkably, the phantom case, as described by Eq. (43), can be solved exactly. The result for the size exponent is

$$\nu_{ph}(D) = \frac{5 - 2D}{4}, \qquad \nu_{ph}(2) = \frac{1}{4} \qquad (45)$$

with the remaining exponents following from the scaling relations Eq. (44).

The self-avoiding case may be treated with techniques similar to those in isotropic case. The size exponent may be estimated within the Flory approximation, yielding

$$\nu_{Fl} = \frac{2}{d_H} = \frac{D+1}{d+1}. \qquad (46)$$

The Flory estimate is an uncontrolled approximation. Fortunately, a $\varepsilon$-expansion, adapting the multi-local operator product expansion technique[91-93] to the case of tubules, is also possible.[89,90] The resulting renormalization group $\beta$-functions provide evidence for the phase diagram shown in Fig. 18. Extrapolation techniques also provide estimates for the size exponent, the most accurate value being $\nu = 0.62$ for the physical case. The rest of the exponents may be computed from the scaling relations.

Numerical simulations so far, however, do *not* find a tubular phase in the case of strict self-avoidance.

## 6. Order on Curved Surfaces

Imagine we instantaneously freeze a fluctuating membrane so that it has some fixed but curved shape. We can then ask about the nature of the ground state of particles distributed on this surface and interacting with some microscopic pair-wise repulsive potential. The relevant physics is clearly related to the infinite bending rigidity limit (flat phase) of elastic membranes. In such a membrane the topology and topography are fixed.

Spherical particles on a flat surface pack most efficiently in a simple triangular lattice, as illustrated in Fig. 20. In the dense limit each particle "kisses" six of its neighbors.[94] Such six-coordinated triangular lattices cannot, however, be perfectly wrapped on the curved surface of a sphere; topology alone requires there be defects in coordination number. The panels on a soccer ball and the spherical carbon molecule $C_{60}$ (buckyball)[97,98] are good illustrations of the necessity of defects for a spherical triangulation — they have 12 pentagonal faces (each the dual of a 5-coordinated defect) in addition to 20 hexagonal faces (each dual to a regular 6-coordinated node). The necessary packing defects can be characterized by their topological charge, $q$, which is the departure of their coordination number $c$ from the preferred flat space value of 6 ($q = 6 - c$). These coordination number defects

Fig. 20.    A 2.5 micron scan of 0.269 micron diameter polystyrene spheres crystallized into a regular triangular lattice — taken from http://invsee.asu.edu/nmodules/spheresmod/.

are point-like topological defects called *disclinations*[99] and they detect intrinsic Gaussian curvature located at the defect. A profound theorem of Euler[100,101] states that the total disclination charge of *any* triangulation whatsoever of the sphere must be 12![102] A total disclination charge of 12 can be achieved in many ways, however, which makes the determination of the minimum energy configuration of repulsive particles, essential for crystallography on a sphere, an extremely difficult problem. This was recognized nearly 100 years ago by J.J. Thomson,[103] who attempted, unsuccessfully, to explain the periodic table in terms of rigid electron shells. Similar problems arise in fields as diverse as multi-electron bubbles in superfluid helium,[104] virus morphology,[105–108] protein s-layers,[109,110] giant molecules[111,112] and information processing.[95,96] Indeed, both the classic Thomson problem, which deals with particles interacting through the Coulomb potential, and its generalization to other interaction potentials, are still open problems after almost 100 years of attention.[113–115]

The spatial curvature encountered in curved geometries adds a fundamentally new ingredient to crystallography not found in the study of order in spatially flat systems. As the number of particles on the sphere grows, isolated charge 1 defects (5s) will induce too much strain. This strain can be relieved by introducing additional dislocations, consisting of pairs of tightly bound 5-7 defects,[116,117] which don't spoil the topological constraints because their net disclination charge is zero. Dislocations, which are themselves point-like topological defects in two dimensions, disrupt the translational order of the crystalline phase but are less disruptive of orientational order.[117]

Recent work on an experimental realization of the generalized Thomson problem has allowed us to explore the lowest energy configuration of the dense packing of repulsive particles on a spherical surface and to confront a previously developed theory with experiment.[118] We create two-dimensional packings of colloidal particles on the surface of spherical water droplets and view the structures with optical microscopy. Above a critical system size, the thermally equilibrated colloidal crystals display distinctive high-angle grain boundaries, which we call "scars". These grain boundaries are found to end entirely within the crystal, which is never observed on flat surfaces because the energy penalty is too high.

The experimental system is based on the self-assembly of one micron diameter cross-linked polystyrene beads adsorbed on the surface of spherical water droplets (of radius $R$), themselves suspended in a density-matched oil mixture.[119] The polystyrene beads are almost equally happy to be in oil or water (the bead/oil surface tension is close to the bead/water surface tension) and therefore diffuse freely *until* they find the oil-water interface and stick there. Particle assembly on the interface of two distinct liquids dates to the pioneering work of Pickering[120] and was beautifully exploited by Pieranski[121] some time ago. The particles are imaged with phase contrast using an inverted microscope. After determining the center of mass of each bead, the lattice geometry is analyzed by original Delaunay triangulation algorithms[122] appropriate to spherical surfaces.

We analyze the lattice configurations of a collection of 40 droplets. A typical small spherical droplet with system size, $R/a = 4.2$, where $a$ is the mean particle spacing, is shown in Fig. 21a. The associated Delaunay triangulation is shown in Fig. 21b. The only defect is one isolated charge +1 disclination. Extrapolation to the entire surface of the sphere is statistically consistent with the required 12 total disclinations.

Qualitatively different results are observed for larger droplet sizes as defect configurations with excess dislocations appear. Although some of these excess dislocations are isolated, most occur in the form of distinctive $(5-7-5-7-\cdots-5)$ chains, each of net charge +1, as shown in Fig. 21d. These chains form high-angle (30°) grain boundaries, or scars, which terminate freely within the crystal. Such a feature is energetically prohibitive in equilibrium crystals in flat space. Thus, although grain boundaries are a common feature of 2D and 3D crystalline materials, arising from a mismatch of crystallographic orientations across a boundary, they usually terminate

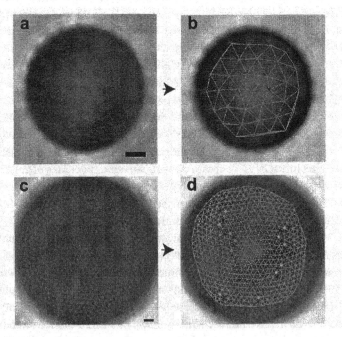

Fig. 21. Light microscope images of particle-coated droplets. Two droplets (a) and (c) are shown, together with their associated defect structures (b) and (d). Panel (a) shows an ≈13% portion of a small spherical droplet with radius $R = 12.0$ microns and mean particle spacing $a = 2.9$ microns ($R/a = 4.2$), along with the associated triangulation (b). Charge +1(−1) disclinations are shown in red and yellow respectively. Only one +1 disclination is seen. Panel (c) shows a cap of spherical colloidal crystal on a water droplet of radius $R = 43.9$ microns with mean particle spacing $a = 3.1$ microns ($R/a = 14.3$), along with the associated triangulation (d). In this case the imaged crystal covers about 17% of the surface area of the sphere. The scale bars in (a) and (c) are 5 microns.

at the boundary of the sample in flat space because of the excessive strain energy associated with isolated terminal disclinations. Termination within the crystal is a feature unique to curved space.

Of key interest is the number of excess dislocations per chain as a function of the dimensionless system size $R/a$. This is plotted in Fig. 22. Scars only appear for droplets with $R/a \geq 5$. These results provide a critical confirmation of a theoretical prediction that $R/a$ must exceed a threshold value $(R/a)_c \approx 5$, corresponding to $M \approx 360$ particles, for excess defects to proliferate in the ground state of a spherical crystal.[123] The precise value of $(R/a)_c$ depends on details of the microscopic potential, but its origin is easily understood by considering just one of the 12 charge $+1$ disclinations required by the topology of the sphere. In flat space such a topological defect has an associated energy that grows quadratically with the size of the system,[117] since it is created by excising a $2\pi/6$ wedge of material and gluing the boundaries together.[11,117] The elastic strain energy associated with this defect grows as the area. In the case of the sphere the radius plays the role of the system size. As the radius increases, isolated disclinations become much more energetically costly. This elastic strain energy may be reduced by the formation of linear dislocation arrays, i.e. grain boundaries. The energy needed to create these additional dislocation arrays is proportional to a dislocation core energy $E_c$ and scales linearly with the system size.[123] Such screening is inevitable in flat space (the plane) if one forces an extra disclination into the defect-free ground state. Unlike the situation in flat space, grain boundaries on the sphere can freely terminate,[123–126] consistent with the scars seen on colloidal droplets.

A powerful analytic approach to determining the ground state of particles distributed on a curved surface has been developed.[123,126,127] The original particle problem is mapped to a system of interacting disclination defects in a continuum

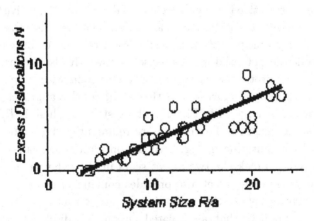

Fig. 22. Excess dislocations as a function of system size. The number of excess dislocations per minimal disclination $N$ as a function of system size $R/a$, with the linear prediction given by theory shown as a solid red line.

elastic curved background. The defect-defect interaction is universal with the particle microscopic potential determining two free parameters — the Young modulus $K_0$ of the elastic background and the core energy $E_c$ of an elementary disclination. A rigorous geometrical derivation of the effective free energy for the defects is given in Ref. 128. An equivalent derivation may also be given by integrating out the phonon degrees of freedom from the elastic Hamiltonian,[29] with the appropriate modifications for a general distribution of defects. The energy of a two-dimensional crystal embedded in an arbitrary frozen geometry described by a metric $g_{ij}(\mathbf{x})$ is given by

$$H = E_0 + \frac{Y}{2} \iint d\sigma(x)d\sigma(y) \left\{ [s(x) - K(x)] \frac{1}{\Delta^2} [s(y) - K(y)] \right\}, \qquad (47)$$

where the integration is over a fixed surface with area element $d\sigma(x)$ and metric $g_{ij}$, $K$ is the Gaussian curvature, $Y$ is the Young modulus in flat space and $s(x) = \sum_{i=1}^{N} \frac{\pi}{3} q_i \delta(x, x_i)$ is the disclination density $[\delta(x, x_i) = \delta(x - x_i)/\sqrt{\det(g_{ij})}]$. Here 5- and 7-fold defects correspond to $q_i = +1$ and $-1$, respectively. Defects like dislocations or grain boundaries can be built from these $N$ elementary disclinations. $E_0$ is the energy corresponding to a perfect defect-free crystal with no Gaussian curvature; $E_0$ would be the ground state energy for a 2D Wigner crystal of electrons in the plane.[129] Equation (47), restricted to a sphere, gives[123]

$$H = E_0 + \frac{\pi Y}{36} R^2 \sum_{i=1}^{N} \sum_{j=1}^{N} q_i q_j \chi(\theta^i, \psi^i; \theta^j, \psi^j) + N E_c, \qquad (48)$$

where $E_c$ is a defect core energy, $R$ is the radius of the sphere and $\chi$ is a function of the geodesic distance $\beta_{ij}$ between defects with polar coordinates $(\theta^i, \psi^i \,; \theta^j, \psi^j)$:

$$\chi(\beta) = 1 + \int_0^{\frac{1-\cos\beta}{2}} dz \, \frac{\ln z}{1 - z}. \qquad (49)$$

The potential is attractive for opposite charged defects and repulsive for like-charged defects. Many predictions of this model are universal in the sense that they are insensitive to the microscopic potential. This enables us to make definite predictions even though the colloidal potential is not precisely known. It also means that our model system serves as a prototype for any analogous system with repulsive interactions and spherical geometry. To further test the validity of this approach, we show a typical ground state for large M in Fig. 23. The system size here is $R/a = 12$, similar to the droplet in Fig. 21d. The results are remarkably similar to the experimentally observed configuration in Fig. 21d; the only difference is a result of thermal fluctuations, which break the two defect scars in the experiment. This agreement between theory and experiment also provides convincing evidence that these scars are essential components of the equilibrium crystal structure on a sphere.

The theory predicts that an isolated charge +1 disclination on a sphere is screened by a string of dislocations of length $\cos^{-1}(5/6)R \approx 0.59R$.[123] One can use the variable linear density of dislocations to compute the total number of

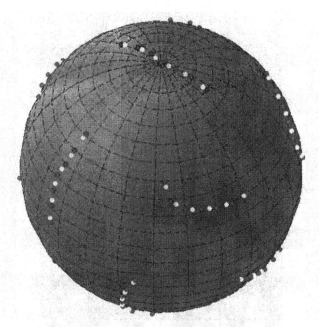

Fig. 23.  Model grain boundaries. This image is obtained from a numerical minimization for a system size comparable to the large droplet in Fig. 22(c,d).

excess dislocations $N$ in a scar. One finds that $N$ grows for large $(R/a)$ as $\frac{\pi}{3}\left[\sqrt{11} - 5\cos^{-1}(5/6)\right]\frac{R}{a} \approx 0.41\frac{R}{a}$, independently of the microscopic potential. This prediction is universal, and is in remarkable agreement with the experiment, as shown by the solid line in Fig. 22.

We expect these scars to be widespread in nature. They should occur, and hence may be exploited, in sufficiently large stiff viral protein capsids, giant spherical fullerenes, spherical bacterial surface layers (s-layers), provided that the spherical geometry is not too distorted. Terminating strings of heptagons and pentagons might serve as sites for chemical reactions or even initiation points for bacterial cell division[109] and will surely influence the mechanical properties of spherical crystalline shells.

The polyhedral siliceous cytoskeleton of the unicellular non-motile ocean organism *Aulosphaera* (a member of the species *Phaeodaria*[130]) is shown in Fig. 24. A triangulation revealing three scars, two of which are branched, is shown in Fig. 25. The skeleton itself is such a perfect triangular lattice that it coincides with the Delaunay triangulation determined by its vertices. The case of viral capsids has been analyzed in Ref. 108, where it is shown that, rather than scarring, icosahedral packings become unstable to faceting for sufficiently large virus size, in analogy with the buckling instability of disclinations in two-dimensional crystals.[5,131]

Scarred spherical crystals may provide the building blocks (atoms) of micron-scale molecules[132,133] and materials. While topology dictates the overall number of scars (12), the details of the geometry and defect energetics determine the length

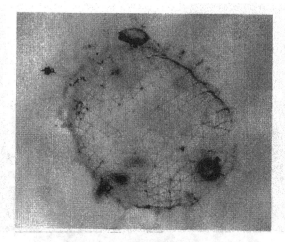

Fig. 24.

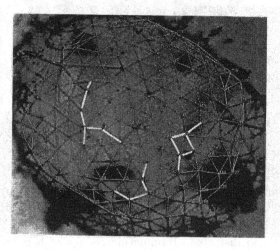

Fig. 25.

and structure of the scars themselves. It is possible that scarred colloidosomes will ultimately yield complex self-assembled materials with novel mechanical or opto-electronic properties.[132]

New structures arise if one changes the structure of the colloid of the topology of the surface they coat. Nelson has analyzed the case of nematic colloids coating a sphere.[133] In this case the preferred number of elementary disclination defects is 4, allowing for the possibility of colloidal atoms with tetrahedral functionality and $sp^3$-type bonding. The case of toroidal templates with 6-fold bond-orientational (hexatic) order has been analyzed recently.[127] It is found that defects are energetically favored in the ground state for fat torii or moderate vesicle size. A schematic of a "typical" ground state is shown in Fig. 26.

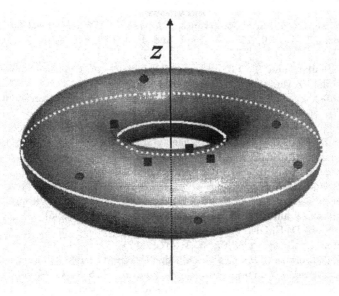

Fig. 26. A typical ground state for a toroidal hexatic. Five-fold disclinations are shown as solid circles (red) and 7-fold disclinations as solid squares (blue).

## Acknowledgments

The work described here has been carried with a large number of talented collaborators without whom the work would never have been completed. My thanks go to Alex Travesset, Angelo Cacciuto, Marco Falcioni, Emmanuel Guitter, Gudmar Thorleifsson, David Nelson, Kostas Anagnostopoulos and Andreas Bausch.

## References

1. P. G. de Gennes, *Scaling Concepts in Polymer Physics* (Cornell University Press, Ithaca, NY, 1979).
2. G. des Cloiseaux and J. F. Jannink, *Polymers in Solution: Their Modelling and Structure* (Clarendon Press, Oxford, 1990).
3. P. G. de Gennes and J. Badoz, *Fragile Objects* (Copernicus, Springer-Verlag, New York, 1996).
4. Y. Kantor, "Properties of Tethered Surfaces" (this volume).
5. D. R. Nelson and L. Peliti, *J. Phys. (France)* **48**, 1085 (1987).
6. See F. David, *Geometry and Field Theory of Random Surfaces and Membranes*, this volume.
7. D. R. Nelson, "Defects in superfluids, superconductors and membranes," in *Fluctuating Geometries in Statistical Mechanics and Field Theory*, edited by F. David, P. Ginsparg and J. Zinn-Justin (North Holland, Amsterdam, 1966), pp. 423–477.
8. "Gego's Galaxies: Setting Free the Line," R. Storr, *Art in America*, 108–113 (June, 2003).
9. M. J. Bowick and A. Travesset, "The Statistical Mechanics of Membranes," *Phys. Rep.* **344**, 255–310 (2001); in "Renormalization Group Theory at the Turn of the Millennium," eds. D. O'Connor and C. R. Stephens [arXiv:cond-mat/0002038].

10. K. J. Wiese, *Polymerized Membranes, a Review,* in "Phase Transitions and Critical Phenomena," Vol. **19**, 253: C. Domb and J. L. Lebowitz (Eds.), Academic Press (2001).

11. P. M. Chaikin and T. C. Lubensky, *Principles of Condensed Matter Physics* (Cambridge University Press, Cambridge, 1995).

12. T. C. Lubensky, *Solid State Commun.* **102**, 187 (1997) [arXiv:cond-mat/9609215].

13. T. A. Witten, *Rev. Mod. Phys.* **71**, S367 (1999).

14. S. I. Stupp, S. Son, H. C. Lin and L. S. Li, *Science* **259**, 59 (1993).

15. J. H. Fendler and P. Tundo, *Acc. Chem. Res.* **17**, 3 (1984).

16. J. H. Fendler, *Chem. Rev.* **87**, 877 (1987).

17. See Chapter 10 of B. Alberts, D. Bray, J. Lewis, M. Raff, K. Roberts and J. D. Watson, *Molecular Biology of the Cell* (Garland Publishing, New York, 1994).

18. *Structure and Dymanics of Membranes,* Handbook of Biological Physics Vol. 1, edited by R. Lipowsky and E. Sackmann (Elsevier, Amsterdam, 1995).

19. C. Picart and D. E. Discher, *Biophys. J.* **77**, 865 (1999).

20. E. Sackmann, *Chem. Phys. Chem.* **3**, 237 (2002).

21. D. Boal, *Mechanics of the Cell* (Cambridge University Press, Cambridge, UK, 2002).

22. T. J. Byers and D. Branton, *Proc. Nat. Acad. Sci. U.S.A.* **82**, 6153 (1985).

23. A. Elgsaeter, B. T. Stokke, A. Mikkelsen and D. Branton, *Science* **234**, 1217 (1986).

24. C. F. Schmidt, K. Svoboda, N. Lei, I. B. Petsche, L. E. Berman, C. R. Safinya and G. S. Grest, *Science* **259**, 952 (1993).

25. T. Hwa, E. Kokufuta and T. Tanaka, *Phys. Rev.* **A44**, R2235 (1991).

26. X. Wen *et al.*, *Nature* **355**, 426 (1992).

27. M. S. Spector, E. Naranjo, S. Chiruvolu and J. A. Zasadzinski, *Phys. Rev. Lett.* **73**, 2867 (1994).

28. R. R. Chianelli, E. B. Prestridge, T. A. Pecoraro and J. P. DeNeufville, *Science* **203**, 1105 (1999).

29. See D. R. Nelson, *The Theory of the Crumpling Transition,* this volume.

30. M. Paczuski, M. Kardar and D. R. Nelson, *Phys. Rev. Lett.* **60**, 2638 (1988).

31. M. J. Bowick, S. M. Catterall, M. Falcioni, G. Thorleifsson and K. N. Anagnostopoulos, *J. Phys. I (France)* **6**, 1321 (1996) [arXiv:cond-mat/9603157].

32. K. Wilson and J. Kogut, *Phys. Rep.* **12**, 75 (1974).

33. N. Goldenfeld, *Lectures on Phase Transitions and the Renormalization Group* (Westview Press, 1992).

34. Y. Kantor, M. Kardar and D. R. Nelson, *Phys. Rev. Lett.* **57**, 791 (1986).

35. Y. Kantor, M. Kardar and D. R. Nelson, *Phys. Rev.* **A35**, 3056 (1987).

36. See B. Duplantier, this volume.

37. M. Baig, D. Espriu and J. Wheater, *Nucl. Phys.* **B314**, 587 (1989).

38. J. Ambjørn, B. Durhuus and T. Jonsson, *Nucl. Phys.* **B316**, 526 (1989).

39. R. Renken and J. Kogut, *Nucl. Phys.* **B342**, 753 (1990).

40. R. G. Harnish and J. Wheater, *Nucl. Phys.* **B350**, 861 (1991).

41. J. Wheater and T. Stephenson, *Phys. Lett.* **B302**, 447 (1993).

42. M. Baig, D. Espriu and A. Travesset, *Nucl. Phys.* **B426**, 575 (1994).

43. J. Wheater, *Nucl. Phys.* **B458**, 671 (1996).

44. G. Gompper and D. M. Kroll, *J. Phys.: Condens. Matter* **9**, 8795 (1997).

45. G. Gompper and D. M. Kroll, *Curr. Opin. Colloid Interface Sci.* **2**, 373 (1997).

46. F. David and E. Guitter, *Europhys. Lett.* **5**, 709 (1988).

47. E. Guitter, F. David, S. Leibler and L. Peilit, *Phys. Rev. Lett.* **61**, 2949 (1988); *J. Phys. France* **50**, 1787 (1989).

48. P. Le Doussal and L. Radzihovsky, *Phys. Rev. Lett.* **69**, 1209 (1992).

49. D. Espriu and A. Travesset, *Nucl. Phys.* **B468**, 514 (1996).
50. L. D. Landau and E. M. Lifshitz, *Theory of Elasticity*, Vol. **7** of *Course of Theoretical Physics* (Pergamon Press, Oxford, UK, 1986).
51. J. A. Aronovitz and T. C. Lubensky, *Phys. Rev. Lett.* **60**, 2634 (1988).
52. J. A. Aronovitz, L. Golubović and T. C. Lubensky, *J. Phys. France* **50**, 609 (1989).
53. Z. Zhang, H. T. Davis and D. M. Kroll, *Phys. Rev.* **E53**, 1422 (1996).
54. M. Falcioni, M. J. Bowick, E. Guitter and G. Thorleifsson, *Europhys. Lett.* **38**, 67 (1997) [arXiv:cond-mat/9610007].
55. K. E. Evans, M. A. Nkansah, I. J. Hutchinson and S. C. Rogers, Nature **353**, 124 (1991).
56. M. Plischke and D. Boal, *Phys. Rev.* **A38**, 4943 (1988).
57. F. F. Abraham, W. Rudge and M. Plischke, *Phys. Rev. Lett.* **92**, 1757 (1989).
58. F. F. Abraham and D. R. Nelson, *J. Phys. France* **51**, 2653 (1990).
59. F. F. Abraham and D. R. Nelson, *Science* **249**, 393 (1990).
60. B. Boal, E. Levinson, D. Liu and M. Plischke, *Phys. Rev.* **A40**, 3292 (1989).
61. Y. Kantor and K. Kremer, *Phys. Rev.* **E48**, 2490 (1993).
62. M. Bowick, A. Cacciuto, G. Thorleifsson and A. Travesset, *Eur. Phys. J.* **E5**, 149 (2001) [arXiv:cond-mat/0006383].
63. A. Baumgartner, *J. Phys. I France* **1**, 1549 (1991).
64. D. M. Kroll and G. Gompper, *J. Phys. France* **3**, 1131 (1993).
65. P. Di Francesco and E. Guitter, *Europhys. Lett.* **26**, 455 (1994) [arXiv:cond-mat/9402058].
66. P. Di Francesco and E. Guitter, *Phys. Rev.* **E50**, 4418 (1994) [arXiv:cond-mat/9406041].
67. M. Bowick, P. DiFrancesco, O. Golinelli and E. Guitter, *Nucl. Phys.* **B450**, 463 (1995) [arXiv:cond-mat/9502063].
68. M. Bowick, P. DiFrancesco, O. Golinelli and E. Guitter, p. 21 in *Common Trends in Condensed Matter and High Energy Physics*; eds. A. Barone and A. Devoto (Istituto Italiano per gli Studi Filosofici, Naples, 2000) [arXiv:cond-mat/9610215].
69. M. Bowick. O. Golinelli, E. Guitter and S. Mori, *Nucl. Phys.* **B495**, 583 (1997) [arXiv:cond-mat/99611105].
70. E. Cirillo, G. Gonnella and A. Pelizzola, *Phys. Rev.* **E53**, 1479 (1996) [arXiv:hep-th/9507160].
71. E. Cirillo, G. Gonnella and A. Pelizzola, *Phys. Rev.* **E53**, 3253 (1996) [arXiv:hep-th/9512069].
72. G. Grest, *J. Phys. I (France)* **1**, 1695 (1991).
73. M. Bowick, A. Cacciuto, G. Thorleifsson and A. Travesset, *Phys. Rev. Lett* **87**, 148103 (2001) [arXiv:cond-mat/0103173].
74. Poisson himself "proved" that the Poisson ratio must be $\frac{1}{4}$ for an elastic cylinder under tension, based on a faulty molecular-force model due to Laplace and the neglect of shear forces.[76]
75. A. E. H. Love, *A Treatise on the Mathematical Theory of Elasticity* (Dover, New York, 1944), 4th edition, p. 163.
76. *Siméon-Denis Poisson et La Science de son Temps*, M. Métivier, P. Costabel and P. Dugac (L'Ecole Polytechnique Press, Palaiseau, France, 1981).
77. R. Lakes, *Science* **235**, 1038 (1987).
78. For more details, including a recipe for making auxetic foam, see http://silver.neep.wisc.edu/ lakes/Poisson.html.
79. K. E. Evans and A. Alderson, *Adv. Mater.* **12**, 617 (2000).

80. A. Yeganeh-Haeri, D. J. Weidner and J. B. Parise, *Science* **257**, 650 (1992).
81. N. Keskar and J. R. Chelikowsky, *Nature (London)* **358**, 222 (1992).
82. D. E. Discher, N. Mohandas and E. A. Evans, *Science* **266**, 1032 (1994).
83. R. Lakes, *Nature (London)* **414**, 503 (2001).
84. L. Radzihovsky and J. Toner, *Phys. Rev. Lett.* **75**, 4752 (1995) [arXiv:cond-mat/9510172].
85. L. Radzihovsky and J. Toner, *Phys. Rev.* **E57**, 1832 (1998) [arXiv:cond-mat/9708046].
86. See L. Radzihovsky, this volume.
87. M. Bowick, M. Falcioni and G. Thorleifsson, *Phys. Rev. Lett.* **79**, 885 (1997) [arXiv:cond-mat/9708046].
88. J. Toner, *Phys. Rev. Lett.* **62**, 905 (1988).
89. M. Bowick and E. Guitter, *Phys. Rev.* **E56**, 7023 (1997) [arXiv:cond-mat/9705045].
90. M. Bowick and A. Travesset, *Phys. Rev.* **E59**, 5659 (1999) [arXiv:cond-mat/9808214].
91. B. Duplantier, T. Hwa and M. Kardar, *Phys. Rev. Lett.* **70**, 2205 (1993) [arXiv:hep-th/9212102].
92. F. David, B. Duplantier and E. Guitter, *Nucl. Phys.* **B394**, 555 (1993) [arXiv:hep-th/9211038].
93. F. David, B. Duplantier and E. Guitter, arXiv:cond-mat/9702136.
94. The kissing number for sphere packings in 3 dimensions is 12, in 8 dimensions 240 and in 24 dimensions 196,650.[95,96]
95. N. J. A. Sloane, *Sci. Am.* **250**, 116 (1984).
96. J. H. Conway and N. J. A. Sloane, *Sphere Packings, Lattices and Groups* (Third Edition, Springer-Verlag, New York, 1998).
97. H. W. Kroto *et al.*, *Nature* **318**, 162 (1985).
98. M. F. Jarrold, *Nature* **407**, 26 (2000).
99. See F. David, this volume.
100. L. Euler, *Opera Omnia*, series i, Vol. **26**, Orell Füssli Verlag, 1953 (see also http://www.ics.uci.edu/~eppstein/junkyard/euler/).
101. P. Hilton and J. Pedersen, *Amer. Math. Monthly* **103**, 121 (1996).
102. 12 here is obtained from $6\chi$, where the Euler characteristic $\chi$ of a space is a topological invariant equal to 2 for the sphere.
103. J. J. Thomson, *Phil. Mag.* **7**, 237 (1904).
104. P. Leiderer, *Z. Phys. B* **98**, 303 (1995).
105. D. L. D. Caspar and A. Klug, *Cold Spring Harbor Symposia on Quantitative Biology* Vol. **XXVII** (Basic Mechanisms in Animal Virus Biology), 1 (1962).
106. C. J. Marzec and L. A. Day, *Biophys. Jour.* **65**, 2559 (1993).
107. V. J. Reddy *et al.*, *J. Virol.* **75**, 11943 (2001) (see also http://mmtsb.scripps.edu/viper/viper.html).
108. J. Lidmar, L. Mirny and D. R. Nelson, arXiv:cond-mat/0306741.
109. U. B. Sleytr, M. Sára, D. Pum and B. Schuster, *Prog. Surf. Sci.* **68**, 231 (2001).
110. D. Pum, P. Messner and U. B. Sleytr, *J. Bacteriology* **173**, 6865 (1991).
111. T. Liu, *J. Am. Chem. Soc.* **124**, 10942 (2002).
112. A. Müller, P. Kögerler and A. W. M. Dress, *Coord. Chem. Rev.* **222**, 193 (2001).
113. The determination of the aymptotics of the ground state energy of spherical crystals with a logarithmic potential appeared as Problem **7** of Smale's list of key mathematical problems for the 21st century: S. Smale, *Math. Intelligencer* **20**, No. 2, 7 (1998).
114. E. L. Altschuler *et al.*, *Phys. Rev. Lett.* **78**, 2681 (1997).
115. T. Erber and G. M. Hockney, *Advances in Chemical Physics* (eds I. Prigogine and S. A. Rice) Vol. **XCVIII**, 495 (1997).

116. R. Cotterill, *The Cambridge Guide to the Material World* (Cambridge University Press, Cambridge, England, 1985).

117. D. R. Nelson, *Defects and Geometry in Condensed Matter Physics* (Cambridge University Press, Cambridge, 2002).

118. A. R. Bausch, M. J. Bowick, A. Cacciuto, A. D. Dinsmore, M. F. Hsu, D. R. Nelson, M. G. Nikolaides, A. Travesset and D. A. Weitz, *Science* **299**, 1716 (2003) [arXiv:cond-mat/0303289].

119. A. D. Dinsmore *et al.*, *Science* **298**, 1006 (2002); the particular system we explored is obtained by confining cross-linked polystyrene beads (IDC, Portland, USA) to the surface of water droplets.

120. S. U. Pickering, *J. Chem. Soc.* **97**, 2001 (1907).

121. P. Pieranski, *Phys. Rev. Lett.* **45**, 569 (1980).

122. B. Delaunay, *Bull. Acad. Sci. USSR: Class. Sci. Math. Nat.* **7**, 793 (1934).

123. M. J. Bowick, D. R. Nelson and A. Travesset, *Phys. Rev.* B **62**, 8738 (2000) [arXiv:cond-mat/9911379].

124. M. J. W. Dodgson and M. A. Moore, *Phys. Rev.* B **60**, 3816 (1997).

125. A. Perez-Garrido and M. A. Moore, *Phys. Rev.* B **60**, 15628 (1999) and references therein.

126. M. Bowick, A. Cacciuto, D. R. Nelson and A. Travesset, *Phys. Rev. Lett.* **89**, 185502 (2002) [arXiV:cond-mat/0206144].

127. M. Bowick, D. R. Nelson and A. Travesset, arXiv:cond-mat/0309709 [Phys. Rev. E (2004)].

128. M. Bowick and A. Travesset, *J. Phys.* **A34**, 1535 (2001) [arXiv:cond-mat/0005356].

129. For the case of a Coulomb potential, see L. Bonsall and A. A. Maradudin, *Phys. Rev.* B **15**, 1959 (1977).

130. O. R. Anderson, *Radiolaria* (Springer-Verlag, New York, 1983).

131. S. Seung and D. R. Nelson, *Phys. Rev.* **A38**, 1005 (1988).

132. R. Kamien, *Science* **299**, 1671 (2003).

133. D. R. Nelson, *Nano Lett.* **2**, 1125 (2002).

# CHAPTER 12

# TRIANGULATED-SURFACE MODELS OF FLUCTUATING MEMBRANES

G. Gompper

*Institut für Festkörperforschung, Forschungszentrum Jülich,*
*D-52425 Jülich, Germany*
*E-mail: g.gompper@fz-juelich.de*


D.M. Kroll

*Supercomputing Institute, University of Minnesota, 599 Walter Library,*
*117 Pleasant Street S.E., Minneapolis, MN 55455, USA*
*E-mail: krolldm@msi.umn.edu*

The thermal behavior of membranes — surfaces of nearly vanishing tension — depends strongly on their internal state, which can be either polymerized, fluid, crystalline, or hexatic. Thermal fluctuations have a dramatic effect on the conformation and elastic properties of membranes. In this review we describe both the continuum models of membranes which are used for theoretical analyses as well as the triangulated-network models employed in simulations. The fruitful interaction between these two approaches, which has lead to recent progress in this field, is emphasized. The essential results of recent research on fluctuating membranes are summarized, and the effects of bending rigidity, self-avoidance, topological defects in the crystalline phase, external compressive forces, internal degrees of freedom such as the composition in multi-component membranes, and topology fluctuations are discussed in detail.

## 1. Introduction

Membranes are two-dimensional sheets of atoms or molecules which are different from the three-dimensional medium in which they are embedded. Because their thickness is on the order of nanometers, but lateral extension is on the order of microns, membranes are very flexible and can be easily deformed. This separation of length scales also implies that membranes can, in many cases, be modeled as idealized surfaces (of vanishing thickness). In this approach, the shape and fluctuation spectrum of the membrane is determined by an elastic energy; the molecular properties, architecture, and interactions of the membrane constituents enter only through the functional form of the elastic energy and the values of the elastic moduli.

The most important example of membranes are bilayers composed of amphiphilic molecules such as surfactants or lipids. Amphiphilic molecules self-assemble in aqueous solution into a large variety of phases. The driving force for this aggregation process is the hydrophobic effect, which strongly disfavors contact between water and non-polar solutes such as hydrocarbons. Amphiphilic molecules therefore aggregate in such a way that the hydrocarbon tails are shielded from contact with water by a layer of polar head groups. This leads to the formation of spherical and cylindrical micelles, and of bilayer membranes. Micelles typically form when the size of the head groups is larger than that of the tails, while bilayer membranes are stable when head and tail are of roughly equal size.

Lipid bilayers are the basic building block of the plasma membrane as well as the membranes of all cell organelles in living organisms. Cell membranes are typically composed of several species of lipids, have different lipid composition in the inner and outer leaflet of the bilayer, and contain many embedded, adsorbed, or attached proteins.

In an amphiphilic membrane, the molecules are densely packed because the head groups have to be sufficiently close together to prevent water from penetrating into the hydrocarbon region, the membrane is free to adjust its surface area so as to minimize the free energy of the amphiphiles, and the molecular solubility of amphiphiles in water is typically extremely small so that only very few amphiphiles can be found in the embedding fluid. The area of a membrane is therefore essentially constant and the interfacial tension of membranes is either very small or vanishes completely. The shapes and fluctuation spectra of fluid membranes are therefore determined by their curvature elasticity.

As in any two-dimensional condensed-matter system, membranes can exist in several different phases. For single-component membranes, fluid or crystalline phases are possible. In multi-component membranes, phase separation into domains of different composition can occur. Moreover, due to the embedding in a higher-dimensional space, membranes can also change their topology, e.g. from vesicles to lamellar or bicontinuous phases. The coupling of shear elasticity, crystalline defects and composition to the shape and topology of the membranes leads to many new and unexpected phenomena which have been extensively studied over the last several years.

Due to their extreme flexibility, membranes typically exhibit large thermal fluctuations. Models for computer simulations of membrane behavior can be grouped into two classes. *Atomistic models* start from the individual molecules and their interactions. Such models can be used to predict membrane properties for specific surfactants or lipids, for example, as a function of amphiphile chain length. However, due to the large number of degrees of freedom, atomistic models are restricted to ensembles of a few thousand amphiphilic molecules. *Coarse-grained models* start from the continuum description of a membrane in terms of an elastic energy. These models can therefore be used to study membrane behavior on the scale of micrometers. The two classes of models are largely complementary, and should be considered as two parts of a *multi-scale modeling approach* to predict the behavior of fluctuating membranes.

In order to make coarse-grained models amenable to computer simulations, the membrane surface has to be discretized, usually in the form of a triangulated network. In this chapter, we review the essential results of recent research involving continuum models and simulations of network models of fluctuating membranes. A nice review of random-surface models from the viewpoint of a high energy physicist has appeared some time ago (Wheater 1994). Various related and complementary aspects of the physics of membranes are discussed in recent books and reviews of Nelson *et al.* (1989), Helfrich (1990), Lipowsky (1991), Gompper and Schick (1994), Safran (1994), Gelbart *et al.* (1995), Grest and Murat (1995), Lipowsky and Sackmann (1995), David *et al.* (1996), Seifert (1997), Gompper and Kroll (1997c), Safran (1999), Bowick and Travesset (2001), Wiese (2001), and Boal (2002).

## 2. Polymerized Membranes

### 2.1. *Elastic Free Energy and Flory Theory*

Tethered or polymerized membranes are fixed connectivity surfaces. A simple example of a surface of this type is a collection of hard spheres tethered by strings of finite extension into a triangular network embedded in 3-dimensional space. A coarse-grained description of the network is obtained by replacing the positions $r_i$ of the individual particles by a coarse-grained average coordinate vector $r(x)$ which is a function of the continuous *internal* coordinate $x$ of particles in the network. The *external* coordinate $r$ is the coarse-grained average of the positions of particles in the vicinity of point $x$.

The form of the Landau–Ginzburg free-energy functional, $\mathcal{F}$, which describes the energy of an arbitrary configuration of a tethered network, is dictated by symmetry considerations. The basic symmetries are invariance with respect to translations and rotations. The first implies that $\mathcal{F}$ can only depend on derivatives such as $\partial_\alpha r \equiv \partial r/\partial x_\alpha$ and $\partial_\alpha \partial_\beta r$, while the second requires that all terms consist of scalar products of the form $\partial_\alpha r \cdot \partial_\beta r$. For an isotropic network, an expansion in powers of $\partial_\alpha r$ yields the free-energy functional (Paczuski *et al.* 1988)

$$\mathcal{F}_1 = \int d^D x \left[ \frac{t}{2}(\partial_\alpha r)^2 + u(\partial_\alpha r \cdot \partial_\beta r)^2 + v(\partial_\alpha r \cdot \partial_\alpha r)^2 + \frac{\kappa}{2}(\partial_\alpha^2 r)^2 + \cdots \right], \quad (1)$$

where $D = 2$, $\alpha = 1, 2$ for membranes, and summation over repeated indices is implied. The terms in Eq. (1) have simple physical interpretations. The first, with coefficient $t$, represents a Hookean elasticity, while the terms with coefficients $u$ and $v$ are anharmonic elastic terms. The last term is a bending energy that arises from bond-bending forces. The coefficient $\kappa$ is the bending rigidity. Both in-plane elasticity and bending rigidity are macroscopic manifestations of the internal connectivity.

In the subsequent discussion, it is often useful to consider the generalization of Eq. (1) to $D$-dimensional networks embedded $d$-dimensional space. In this case,

**r** is a *d*-dimensional vector, and $\alpha = 1, 2, \ldots, D$. For example, $D = 1$ corresponds to a linear molecule such as a polymer, and for $D = 3$, the model describes a gel. Non-integer $D$ might approximate fractal structures.

Because the entropy-generated elastic energy of the network is minimized when it has a small size $R$ in space, surface elements, or particles, that are distant neighbors along the network backbone, can be in close physical contact. Their interaction can then no longer be ignored. At high temperatures, the particles only feel the hardcore interactions, which can be described by an interaction potential of the form $\frac{1}{2}b(T) \int d^D x \int d^D y \, \delta^{(d)} [\mathbf{r}(\mathbf{x}) - \mathbf{r}(\mathbf{y})]$. As in polymers, the coefficient $b(T)$ is related to the second virial coefficient in a solution of such networks, and can change sign as a function of the thermodynamic fields, such as temperature. Higher order terms are then necessary to stabilize the system. Their contribution, $\mathcal{F}_2$, to the free energy functional is (Paczuski *et al.* 1988)

$$\mathcal{F}_2 = \frac{1}{2}b(T) \int d^D x_1 \int d^D x_2 \, \delta^{(d)} [\mathbf{r}(\mathbf{x}_1) - \mathbf{r}(\mathbf{x}_2)]$$
$$+ c(T) \int d^D x_1 \int d^D x_2 \int d^D x_3 \, \delta^{(d)}[\mathbf{r}(\mathbf{x}_1) - \mathbf{r}(\mathbf{x}_2)]\delta^{(d)} [\mathbf{r}(\mathbf{x}_2) - \mathbf{r}(\mathbf{x}_3)]. \tag{2}$$

The probability of a particular configuration of the network is determined by the free energy $\mathcal{F} = \mathcal{F}_1 + \mathcal{F}_2$.

The scaling behavior can be studied using a Flory-type (mean field) approximation. Consider a network of linear dimension $L$ and let $R$ be its size in space. In $\mathcal{F}_1$, we approximate terms of the type $\partial_\alpha r_i$ by $R/L$, and $\int d^D x$ by $L^D$. In $\mathcal{F}_2$, the fact that $\int d^d r \delta^{(d)}(\mathbf{r}) = 1$ suggests approximating $\delta^{(d)}(\mathbf{r})$ by $R^{-d}$. These estimates lead to the free-energy estimate (Paczuski *et al.* 1988)

$$\mathcal{F} \sim tR^2 L^{D-2} + wR^4 L^{D-4} + \kappa R^2 L^{D-4} + bL^{2D}/R^d + cL^{3D}/R^{2d}, \tag{3}$$

where $w = u + Dv$. Since $B$ and $t$ can change sign with temperature, the terms with $c > 0$ and $w > 0$ are necessary to ensure stability.

A detailed discussion of the predictions of this approach is provided by Paczuski *et al.* (1988) and Nelson (1989). Here we only consider the most important cases.

### 2.1.1. *The Crumpled Phase*

Both $t$ and $b$ are positive at sufficiently high temperatures; in this case, these two terms asymptotically dominate the rest. Minimizing Eq. (3) with respect to $R$, one finds

$$R \sim (b/t)^{1/(d+2)} L^{(D+2)/(d+2)}. \tag{4}$$

This corresponds to a *crumpled* network with a non-trivial fractal (or Haussdorf) dimension. Generally, a scaling exponent $\nu$ defined by the relation

$$R \sim L^\nu \tag{5}$$

is used to characterize the extent to which a $D$-dimensional network is crumpled in the $d$-dimensional embedding space. Alternatively, one can use the fractal dimension $d_f$ which relates the mass ($\sim L^D$) of the network to its spatial extent,

$$L^D \sim R^{d_f}, \tag{6}$$

to characterize the conformation. The relationship between these two exponents is $\nu = D/d_f$. The Flory estimate (4) for $\nu$ in this case is, therefore,

$$\nu_f = (D+2)/(d+2), \tag{7}$$

and

$$d_f = \frac{(d+2)D}{(D+2)}. \tag{8}$$

### 2.1.2. The Flat Phase

Result (4) indicates that $R$ diverges for $t \to 0$. For $t < 0$, the anharmonic term $wR^4 L^{d-4}$ is needed for stability. The competition between these two terms leads to $R \sim |t|^{1/2} L$, which clearly describes an expanded *flat* phase.

### 2.1.3. The Crumpling Transition

The transition at $t = 0$ between the flat and crumpled phases is called the *crumpling transition*. Assuming that both $b$ and $w$ are positive, one finds that

$$R \sim (b/w)^{1/(d+4)} L^{\nu_c}, \tag{9}$$

with $\nu_c = (D+4)/(d+4)$, at the transition. Paczuski *et al.* (1988) have shown that the scaling behavior in the vicinity of the crumpling transition is described by the scaling function

$$R \sim L^{\nu_c} \Psi(t L^y), \tag{10}$$

where $y = 2(d-D)/(d+4)$, and $\Psi(0) = $ const. $\Psi(x) \to |x|^{\phi\pm}$ for $x \to \pm\infty$, with $\phi_- = 1/2$, and $\phi_+ = -1/(d+2)$. While different crossover exponents $y_+$ and $y_-$ cannot be ruled out in general, they are the same in the present mean-field analysis.

### 2.1.4. Phantom Networks and the Influence of Self-Avoidance

Finally, if we ignore the interaction terms in $\mathcal{F}_2$ and assume $t > 0$, we have what is commonly called a *phantom* membrane in which the configurational free energy is determined by a network of Hookean springs which is allowed to self-intersect. This model is exactly soluble, and it is easy to see that

$$R \sim \begin{cases} L^{(2-D)/2} & \text{for } D < 2 \\ \{\ln(L)\}^{1/2} & \text{for } D = 2. \end{cases} \tag{11}$$

For $D > 2$, fluctuations are too weak to prevent the complete collapse of the network. Result (11) can also be obtained by either requiring that the Hookean term in Eq. (3), $R^2 L^{D-2}$, is of order one, or by dimensional analysis of the Gaussian free-energy functional. Assume now that $R$ scales as in Eq. (11), and consider the scaling behavior of the interaction term proportional to $b$ in $\mathcal{F}_2$. One finds that this term scales as $L^{2D}/R^d \sim L^{2D-d(2-D)/2}$. The exponent on the right hand side of this equation vanishes at the *upper critical dimension* (Kardar and Nelson 1987; Duplantier 1987; Aronowitz and Lubensky 1987),

$$d_{uc} = 4D/(2 - D). \qquad (12)$$

This term is *irrelevant*, i.e. it scales to zero in the limit of large system sizes if $d > d_{uc}$; the scaling behavior of the network is not changed by self-avoidance in this case. For polymers $(D = 1), d_{uc} = 4$. In dimensions greater than four, the conformation of a polymer is therefore described by the random walk exponent $\nu = 1/2$. In $d = 3$, the Flory result (7) is an excellent approximation to the true scaling behavior. For two-dimensional tethered networks, however, self-avoidance can only be neglected when $d = \infty$, in agreement with the fact that the fractal dimension of the non-interacting surface is infinite (Gross 1984). In view of this result, and the quality of the Flory approximation for polymers, one might expect that expression (4), with $D = 2$, would provide a reasonably accurate description of the scaling behavior of self-avoiding tethered networks. As we shall see later, this is not the case. In view of this, it is worth noting that if we assume the Flory scaling relation (4), $n$-body interactions are *relevant* for

$$d < \frac{2nD}{2(n-1) - D}. \qquad (13)$$

For polymers, it follows that three-body interactions are relevant only in dimensions $d < 2$. In contrast, for $D = 2$, three-body terms are relevant below six dimensions, and four-body terms are relevant below four dimensions. In fact, in $d = 3$, 3-, 4-, and 5-body interactions are all relevant, and 6-body terms are marginal. All these interaction terms can be expected to influence the scaling behavior of the surface in $d = 3$ and need to be treated self-consistently (Hwa 1990; Grest 1991).

## 2.2. Tethered Networks

Polymerized membranes can be modeled by a network of particles that are connected together to form a regular two-dimensional array embedded in $d = 3$ spatial dimensions. While the network is generally taken to be a triangular array, the type of lattice is, to a large extent, unimportant (Baig *et al.* 1994). Similarly, the exact form of the "binding" potential between neighboring particles is also irrelevant. However, it is essential that the bonds between adjacent particles cannot be broken so that the connectivity is fixed. This guarantees that the network has a finite shear modulus. Each particle or vertex is labeled by a two-dimensional *internal* coordinate vector $\mathbf{x} \equiv (x_1, x_2)$, with discrete $x_1$ and $x_2$, denoting its place in the network. For a triangular mesh, two primitive vectors $\{\mathbf{a}^{(1)}, \mathbf{a}^{(2)}\}$ of equal length $\ell_0$ making an angle of 60° define the lattice. The location of the vertices in the lattice are given by

$$\mathbf{x} = m\,\mathbf{a}^{(1)} + n\,\mathbf{a}^{(2)}, \tag{14}$$

where $m$ and $n$ are integers.

### 2.2.1. *Tether-and-Bead and Gaussian Models*

Two general classes of nearest-neighbor interaction potentials have been used in simulation studies of polymerized membranes. In the first, the vertices are point particles with harmonic nearest-neighbor interactions (Ambjørn *et al.* 1989; Baig *et al.* 1989). In the second, hard spheres of radius $\sigma_0$ are placed at each vertex, and the spheres interact via tethers of maximum extension $\ell_0$ (Kantor and Nelson 1987). A variant of the tether-and-bead model, in which the hard sphere and tethering potentials are replaced by anharmonic interaction potentials, is used in molecular dynamics simulations (Abraham *et al.* 1989).

The simplest tethering potential, $V(r)$, is one which causes the particles to behave as if tethered by a string,

$$V(r) = \begin{cases} 0 & \text{if } r < \ell_0 \\ \infty & \text{otherwise.} \end{cases} \tag{15}$$

The potential $V(r)$ acts only between tethered nearest neighbors; it ensures that the distance between nearest neighbors is less than $\ell_0$. If this is the only interaction between particles, we have a "phantom" tethered surface which can self-intersect when large fluctuations bring distant segments of the network into close spatial proximity. Phantom membranes are very flexible surfaces, and there is no resistance to bending. They can roll up or collapse at no energy cost.

The total interaction energy of the network is

$$E_{nn} = \sum_{\langle \mathbf{x},\mathbf{x}' \rangle} V(|\mathbf{r}(\mathbf{x}) - \mathbf{r}(\mathbf{x}')|). \tag{16}$$

The thermal behavior of the network can be determined exactly for the Gaussian potential $V(r) = (K/2)(r/a)^2$. In this case, the average value of the mean-squared separation $|\mathbf{r}(\mathbf{x}) - \mathbf{r}(\mathbf{x}')|^2$ is (Kantor *et al.* 1987)

$$\langle |\mathbf{r}(\mathbf{x}) - \mathbf{r}(\mathbf{x}')|^2 \rangle \simeq \frac{k_B T d a^2}{\pi \sqrt{3} K} \ln(|\mathbf{x} - \mathbf{x}'|/a) \tag{17}$$

for $|\mathbf{r}(\mathbf{x}) - \mathbf{r}(\mathbf{x}')| \gg a$. The radius of gyration squared,

$$R_G^2 = \frac{1}{2A^2} \int d^2x \int d^2x' \langle |\mathbf{r}(\mathbf{x}) - \mathbf{r}(\mathbf{x}')|^2 \rangle, \tag{18}$$

where $A$ is the area of the network, therefore scales as $R_G^2 \sim \ln(L)$ with the linear size $L$ of the membrane [see Eq. (11)] — independent of the spatial dimension.

Simulations have shown that the same scaling behavior is obtained for the tethering potential (15). Furthermore, a simple Migdal–Kadanoff bond-moving renormalization-group approximation has been used (Kantor *et al.* 1987) to show that potentials of this form are mapped into a Gaussian under iteration, supporting the view that the logarithmic scaling behavior of the radius of gyration squared is universal for networks with central force nearest-neighbor interactions. A related class of models with an elastic energy proportional to the sum of the areas of the elementary triangles have also been studied (Gross 1984; Billoire *et al.* 1984). The results indicate that this model also belongs to the same universality class. However, as pointed out by Ambjørn *et al.* (1985), this model has the pathology that the partition function is dominated by surfaces with infinite spikes in the thermodynamic limit. This happens because essentially all the surface area is taken up by a small number of elementary surface triangles, with all others becoming vanishingly small. It is clear that models with the interaction energy (16) do not share this problem.

### 2.2.2. *Self-Avoidance and Bending Energy*

In simulations, self-avoidance can be guaranteed by placing a particle at each vertex which is large enough so that it cannot pass through the network mesh. In Monte Carlo simulations, the particles are taken to be hard spheres of diameter $\sigma_0$. In this case, $V(r)$ given in Eq. (15) is augmented by the potential

$$V_{HS}(r) = \begin{cases} \infty & \text{if } r < \sigma_0 \\ 0 & \text{otherwise} \end{cases} \tag{19}$$

between *all* beads. Self-avoidance requires $\ell_0/\sigma_0 < \sqrt{3}$. Since these potentials do not introduce an energy scale into the problem, the results are independent of temperature, and the free energy is solely due to entropic effects. Such potentials may be expected to generate small persistence lengths, and thus reduce the crossover effects (Kantor *et al.* 1986). In molecular dynamics simulations, it is convenient to use a softer interaction potential, and a purely repulsive Lennard–Jones potential is generally used. For any of these choices, the long-wavelength elastic properties of the network should be similar to real polymerized membranes.

In Monte Carlo simulations, an explicit bending rigidity can be added. For tethered networks, the commonly used discretization (Kantor and Nelson 1987) is

$$E_b^{\text{norm}} = \frac{1}{2}\lambda_b \sum_{\langle ij \rangle} |\mathbf{n}_i - \mathbf{n}_j|^2 = \lambda_b \sum_{\langle ij \rangle} (1 - \mathbf{n}_i \cdot \mathbf{n}_j), \tag{20}$$

where the sum runs over all pairs of neighboring triangles, and $\mathbf{n}_i$ is the surface normal vector of triangle $i$.

## 2.3. *Simulation Methods*

### 2.3.1. *Basic Algorithm*

The total energy $E$ of a network configuration is given by the sum of the nearest-neighbor, $E_{nn}$, and bending, $E_b$, energies. A number of Monte Carlo, molecular dynamics, and Langevin methods have been used to determine the thermodynamic behavior of tethered networks.

Monte Carlo methods are the simplest. A Monte Carlo step consists of an attempt to update the position of each vertex by a random displacement in the cube $[-s, s]^3$. Updates are accepted with a probability equal to $\min[1, \exp(-\delta H_{n,o}/k_B T)]$, where

$$\delta H_{n,o} = E_{\text{old}} - E_{\text{new}}. \tag{21}$$

$s$ is chosen so that approximately 50% of the attempts are accepted. If there is no explicit bending energy, so that Eqs. (15) and (19) are the only interaction potentials, all moves are accepted which do not violate the hard sphere or maximum-tether-length restrictions.

A wide variety of molecular dynamics procedures have also been used to study polymerized networks. Simulations have been performed in the micro-canonical ensemble (Abraham *et al.* 1989; Abraham and Nelson 1990b), constant temperature ensemble — using both constraint and Nosé thermostating (Zhang *et al.* 1996) as well as heat-bath algorithms (Grest and Murat 1990; Grest 1991; Petsche and Grest 1993; Zhang *et al.* 1993) — , and the $(T, \sigma)$-ensemble (Zhang *et al.* 1996), where $\sigma$ is the lateral tension.

### 2.3.2. *Periodic Boundary Conditions*

The curling observed near the edge of tethered networks in simulations with free-edge boundary conditions makes it quite difficult to determine the true scaling behavior of the "bulk membrane" (Abraham and Nelson 1990b). These finite-size effects can be minimized by performing simulations using periodic boundary conditions. Tensionless networks can then be simulated by allowing the projected area of the membrane to fluctuate. Two schemes are possible. In the first, both the size and shape of the simulation cell is allowed to fluctuate; this procedure is required if one wants to determine, for example, the Poisson ratio, or determine both Lamé constants in the solid phase of a planar two-dimensional network. In the second, the projected area fluctuates, but the cell shape is fixed; this approach is more stable in simulations of crystalline membranes in which the thermal generation of lattice defects is taken into account (compare Sec. 4 below).

The simulation procedure described here is based on the isothermal-isobaric Monte Carlo methods introduced by McDonald (1972) and subsequently generalized by Yashonath and Rao (1985). The simulation cell shape is described by the matrix $\mathbf{h} = (\mathbf{a}, \mathbf{b})$ of the two vectors $\mathbf{a}$ and $\mathbf{b}$ that span the simulation cell. The position of a particle in the cell is

$$\mathbf{r}_i = \mathbf{h}\mathbf{s}_i = \xi_i\mathbf{a} + \eta_i\mathbf{b}, \tag{22}$$

where $\mathbf{s}_i = (\xi_i, \eta_i)^T$, with $0 \le \xi_i, \eta_i \le 1$. The simulation is performed in the scaled coordinate $\{\mathbf{s}_i\}$. For a tethered membrane, a Monte Carlo step consists of a random displacement of all beads in the square $[-\delta, \delta]^3$. The step size $\delta$ is again chosen such that approximately 50% of the attempted bead moves are accepted. Every five sweeps or so the independent elements of the matrix $\mathbf{h}$ are updated using a standard Metropolis algorithm with the weight $\exp(-\delta H_{n,o}/k_B T)$, where (McDonald 1972; Yashonath and Rao 1985; Allen and Tildesley 1992)

$$\delta H_{n,o} = \delta V_{n,o} + \sigma(A_n - A_o) - k_B T \ln(A_n/A_o). \tag{23}$$

The projected area $A$ is given by $A = \det\{\mathbf{h}\}$, and the subscripts $o$ and $n$ refer to the old and new states, respectively. $\delta V_{n,o}$ is the total change in interaction potential on going from state $o$ to $n$. Because of the hard-sphere nature of the beads and the fixed tether lengths, it is either zero or infinite, depending upon whether or not the cell update leads to bead overlap or neighbor distances which exceed the tether length. The last term in Eq. (23) arises from the Jacobi determinant of the transformation (22). Moves are then accepted with a probability equal to $\min(1, \exp[-\delta H_{n,o}/k_B T])$ (Allen and Tildesley 1992). If the shape of the simulation cell is kept fixed, only the projected area fluctuates, and Eq. (23) is used to determine the probability that an area update is accepted. In the case of fluctuating cell shape, the most general parameterization of the cell is $\mathbf{a} = (a_1, a_2), \mathbf{b} = (b_1, c_1)$. We choose the vector $\mathbf{a}$ to point in the $x$-direction, i.e. we set $a_2 = 0$. This choice breaks the rotational symmetry of the whole system, and requires an additional Faddeev–Popov term (David 1989) in the weight $\exp(-\delta H_{n,o}/k_B T)$, which now reads (Gompper and Kroll 1997b).

$$\delta H_{n,o} = \delta V_{n,o} + \sigma(A_n - A_o) - k_B T \ln(A_n/A_o) - k_B T \ln(a_{1n}/a_{1o}). \tag{24}$$

By treating the three parameters $(a_1, b_1, c_1)$ as dynamical variables, the length of both sides as well as the inner angle of the simulation cell fluctuate independently.

### 2.3.3. *Determination of Elastic Constants*

The elastic constants of a planar network can be expressed in terms of correlations of the strain tensor (Parrinello and Rahman 1980; Parrinello and Rahman 1981)

$$\epsilon = \frac{1}{2}[(\mathbf{h}_0^T)^{-1}\mathbf{G}\,\mathbf{h}_0^{-1} - \mathbf{1}], \tag{25}$$

where $\mathbf{1}$ is the unit matrix and the matrix of reference basis vectors $\mathbf{h}_0$ is determined by the requirement $\langle\epsilon\rangle = 0$. The metric tensor is $\mathbf{G} = \mathbf{h}^T\mathbf{h}$. In particular,

$$A\langle\epsilon_{ij}\epsilon_{kl}\rangle = k_B T S_{ijkl}, \tag{26}$$

where $\mathcal{A}$ is the area of the network and

$$S_{ijkl} = -\frac{\lambda}{4\mu(\mu + \lambda)}\delta_{ij}\delta_{kl} + \frac{1}{4\mu}(\delta_{ik}\delta_{jl} + \delta_{il}\delta_{jk}) \tag{27}$$

is the elastic compliance tensor, where $\lambda$ and $\mu$ are the two-dimensional Lamé coefficients. Using Eqs. (26) and (27), one finds that

$$\langle \epsilon_{11}\epsilon_{22} \rangle = -\frac{k_B T}{\langle A \rangle}\frac{\lambda}{4\mu(\lambda + \mu)}, \tag{28}$$

$$\langle \epsilon_{12}\epsilon_{12} \rangle = -\frac{k_B T}{\langle A \rangle}\frac{1}{4\mu}, \tag{29}$$

and

$$\langle \epsilon_{11}\epsilon_{11} \rangle = \langle \epsilon_{22}\epsilon_{22} \rangle = \frac{k_B T}{\langle A \rangle}\frac{\lambda + 2\mu}{4\mu(\lambda + \mu)} \equiv \frac{k_B T}{\langle A \rangle K_0}, \tag{30}$$

where $\langle A \rangle = \det\{\mathbf{h}_0\}$, and $K_0$ is the Young modulus.

These results can be used to determine the area compressibility

$$k_B T K_A = \frac{K_B T}{\lambda + \mu} = \langle A \rangle \langle (\epsilon_{11} + \epsilon_{22})^2 \rangle \tag{31}$$

and the Poisson ratio

$$\sigma_P = \frac{\lambda}{(\lambda + 2\mu)} = -\frac{\langle \epsilon_{11}\epsilon_{22} \rangle}{\langle \epsilon_{11}\epsilon_{11} \rangle} = 1 - \frac{1}{2}K_0 K_A. \tag{32}$$

Equation (31) is the linear-response approximation to the exact area compressibility

$$k_B T K_A = (\langle A^2 \rangle - \langle A \rangle^2)/\langle A \rangle. \tag{33}$$

In the crystalline phase, these two expressions axe equivalent in the thermodynamic limit. However, in a finite system, the last equality in Eq. (32) only holds when the linear-response result (31) is used.

In order to determine the strain correlation functions, it is convenient to express the elements of the strain tensor in terms of the basis vectors $\{\mathbf{a}, \mathbf{b}\}$ of the simulation cell. Choosing the 1-axis to coincide with the direction of $\mathbf{a}$, one finds

$$\epsilon_{11} = (\mathbf{a}^2/c^2 - 1)/2, \tag{34}$$

$$\epsilon_{12} = \epsilon_{21} = \mathbf{a} \cdot \mathbf{g}/(2c^2), \tag{35}$$

and

$$\epsilon_{22} = (\mathbf{g}^2/c^2 - 1)/2, \tag{36}$$

where $\mathbf{g} \equiv (2\mathbf{b} - \mathbf{a})/\sqrt{3}$. The equilibrium length, $c$, of the cell edge is defined by the thermal averages $c^2 \equiv \langle \mathbf{a}^2 \rangle = \langle \mathbf{b}^2 \rangle = \langle \mathbf{g}^2 \rangle$, and the angle between the mean directions of the two basis vectors is $\pi/3$.

## 2.4. Fluctuations About the Flat Phase

### 2.4.1. Free Energy and Renormalization Group Results

Fluctuations in the ordered, flat phase can be studied in an expansion about the flat state by introducing in-plane phonon modes $u_\alpha$ ($\alpha = 1, \ldots, D$) and out-of-plane undulation modes $f_a$ ($a = D+1, \ldots, d$) and setting (Paczuski *et al.* 1988)

$$\mathbf{r}(\mathbf{x}) = m(x_\alpha - u_\alpha)\mathbf{e}_\alpha + f_a\mathbf{e}_a. \tag{37}$$

The $\{\mathbf{e}_\alpha\}$ are a set of orthonormal in-plane basis vectors, and $\{\mathbf{e}_a\}$ are a set of orthonormal basis vectors in the subspace normal to plane of the network. To leading order in gradients of $u_\alpha$ and $f_a$, the free energy (1) reduces to

$$\mathcal{F} = \int d^D x \left[ \frac{1}{2}\kappa(\nabla^2 \mathbf{f})^2 + \mu u_{\alpha\beta}^2 + \frac{1}{2}\lambda u_{\gamma\gamma}^2 \right], \tag{38}$$

where $u_{\alpha\beta} = [\partial_\alpha u_\beta + \partial_\beta u_\alpha + \partial_\alpha \mathbf{f} \cdot \partial_\alpha \mathbf{f}]/2$ is the strain matrix, and the elastic constants are $\mu = 4um^4$ and $\lambda = 8vm^4$.

Nelson and Peliti (1987) have shown that the long-range orientational order which occurs in $D = 2$ networks at large bending rigidities is due to a long-range interaction between local Gaussian curvatures mediated by transverse phonons of the crystalline membrane. Using a simple one-loop self-consistent theory for $D = 2$, *assuming* non-vanishing elastic constants, they showed that phonon-mediated interactions between capillary waves lead to a renormalization, or wave-vector dependence, of the bending rigidity of the form

$$\kappa(q) \sim q^{-\eta}, \tag{39}$$

with $\eta = 1$. The theory of fluctuations about the flat phase was extended by Aronowitz and Lubensky (1988) to general $D$ and $d$. An $\epsilon = 4 - D > 0$ expansion confirmed that the flat phase is described by non-trivial scaling behavior, with $\kappa(q)$ scaling as in Eq. (39), but with *anomalous*, scale-dependent elastic constants

$$\lambda(q) \sim \mu(q) \sim q^\omega, \tag{40}$$

with $\omega > 0$. It was also shown (Aronowitz and Lubensky 1988) that as a consequence of rotational invariance,

$$\omega = 4 - D - 2\eta. \tag{41}$$

An explicit renormalization-group calculation to first order in $\epsilon \equiv 4 - D$ yielded the result $\eta = 12(4 - D)/(24 + d - D)$. For flat, planar networks ($D = 2, d = 3$) this implies $\eta = 24/25 = 0.96$. The Poisson ratio, $\sigma_p = \lim_{q \to 0} \lambda(q)/[\lambda q + 2\mu(q)]$, is predicted to be universal, with $\sigma_p = -1/5$, independent of both $d$ and $D$.

More recently, Le Doussal and Radzihovsky (1992) used the self-consistent screening approximation, and find $\eta = 4/(1 + \sqrt{15}) \approx 0.821$ for $D = 2$ in three dimensions. This is currently the most accurate theoretical estimate for $\eta$. The Poisson ratio is predicted to be $\sigma_P = -1/3$.

The renormalized elastic constants enter an effective, long-wavelength free energy for the Fourier-transformed phonon, $\mathbf{u}$, and undulation, $\mathbf{f}$, modes. For the physically relevant case $D = 2, d = 3$, $\mathbf{u}$ is a two-dimensional vector, and $f$ is a scalar. The effective free-energy functional for these modes is (Abraham and Nelson 1990b).

$$\mathcal{F}_{\text{eff}} = \frac{1}{2} \int \frac{d^2 q}{(2\pi)^2} \{ \kappa(q) q^4 |f_q|^2 + \mu(q) q^2 |u_q|^2 + [\mu(q) + \lambda(q)] |\mathbf{q} \cdot \mathbf{u}_q|^2 \}. \qquad (42)$$

The size of the out-of-plane fluctuations in a network with a characteristic linear dimension $L$ is

$$\langle f^2(\mathbf{x}) \rangle \approx \frac{k_B T}{2\pi} \int_{L^{-1}}^{a^{-1}} \frac{q \, dq}{q^4 \kappa(q)} \sim L^{2\zeta}, \qquad (43)$$

where $\zeta = 1 - \eta/2$ and $a$ is a short-distance cutoff on the order of the lattice spacing of the network. Similarly, the amplitude of in-plane phonon fluctuations are given $\langle |\mathbf{u}(\mathbf{x})|^2 \rangle \sim L^{\omega}$

### 2.4.2. Simulations of the Flat Phase

Monte Carlo and molecular dynamics simulations have shown that both self-avoiding polymerized membranes as well as phantom tethered membranes with a sufficiently large bending rigidity are in the flat phase. Several methods, such as the analysis of the moments-of-inertia tensor (Boal et al. 1989), the anisotropic scattering intensity (Abraham and Nelson 1990a), and the pressure exerted on two confining walls (Gompper and Kroll 1991b; Gompper and Kroll 1991a), have been employed to determine the exponents $\eta$ or $\zeta$ which characterize the out-of-plane fluctuations. A typical configuration of a polymerized membrane between two walls is shown in Fig. 1. Most of these simulations were performed using free-edge boundary conditions. The most precise values of the exponents, however, have been obtained from simulations of vesicles (Zhang et al. 1993; Petsche and Grest 1993) and tension-less membranes with periodic boundary conditions (Zhang et al. 1996). In the first case, one finds $\eta = 0.81 \pm 0.03$, or $\zeta = 0.60$ (Zhang et al. 1993), and $\zeta = 0.58 \pm 0.02$ (Petsche and Grest 1993), while in the second, $\zeta = 0.59 \pm 0.02$ (Zhang et al. 1996). These values are in good agreement with the theoretical result of Le Doussal and Radzihovsky (1992) quoted above. This settles a debate concerning the possible absence of renormalization of the in-plane elastic constants (Lipowsky and Giradet 1990; Abraham 1991).

In $d = 3$, a negative value of the Poisson ratio has been confirmed in simulations (Zhang et al. 1996; Falcioni et al. 1997; Bowick et al. 2001a). Whereas the value obtained using periodic boundary conditions, $\sigma_p = -0.15 \pm 0.01$ (Zhang et al. 1996), is about a factor two smaller than the theoretical expectation quoted above, simulations with free-edge boundary conditions yielded $\sigma_p \approx -0.32$ (Falcioni et al. 1997; Bowick et al. 2001a). The value $\sigma_P = -0.34$ was measured in Monte Carlo simulations of tethered networks with free-edge boundary conditions in $d = 4$ (Barsky and Plischke 1994).

(a)

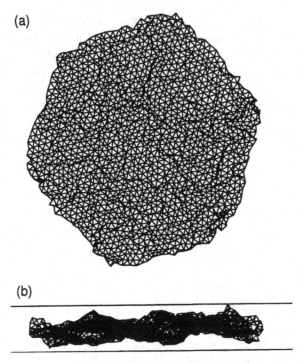

(b)

Fig. 1.   Typical configuration of a polymerized membrane (containing 1801 vertices with tether length $\ell_0 = 1.6\sigma_0$) confined between two parallel walls of separation $9\sigma_0$. Projections (a) onto the wall and (b) parallel to the wall are shown. From Gompper and Kroll (1991a).

## 2.4.3. *Effect of Self-Avoidance*

The simplest model of tethered self-avoiding membranes is a triangular network of purely repulsive spheres of diameter $\sigma_0$. Self-intersection is prohibited if the potentials are chosen so that the spheres cannot penetrate an elementary triangle of the network. It was first shown by Abraham *et al.* (1989), and later confirmed by others (Ho and Baumgärtner 1989; Abraham and Nelson 1990a; Abraham and Nelson 1990b; Grest 1991; Gompper and Kroll 1991b; Gompper and Kroll 1991a) that self-avoidance completely suppresses the crumpling transition, so that self-avoiding membranes are in the flat phase even without an explicit bending rigidity.

It has been argued by Abraham and Nelson (1990b) that a bending rigidity is generated for entropic reasons by excluded volume interactions, even if there is no explicit bending energy. In fact, such a term is generated already upon the introduction of next nearest-neighbor interactions, and for the standard tether-and-bead model discussed in the previous paragraph, this effective rigidity may be larger than that required to produce the flat phase in phantom membranes. This effective rigidity is proportional to the temperature, so that changing the temperature does not modify the strength of this effect. This argument leaves open the possibility that more flexible surfaces might crumple. Subsequently, simulations have been performed on tethered networks in which linear chains of $n$ monomers are connected to form either a hexagonal lattice (Abraham 1992; Abraham and Goulian 1992) or a triangular lattice (Petsche and Grest 1993). The largest simulations have been performed on networks consisting of $N = 29,420$ monomers for open membranes (Abraham 1992; Abraham and Goulian 1992) and $N = 25,002$ for vesicles (Petsche and Grest 1993), both for $n = 8$. Video enhanced differential contrast microscopy images of isolated red blood cell cytoskeletons in high salt (Schmidt *et al.* 1993) show a convincing similarity between the conformations of the red blood cell cytoskeletons and the simulated network.

There have also been several studies of models in which the effective size of the particles at the vertices, $\sigma_0$, is decreased (Abraham *et al.* 1989; Boal *et al.* 1989; Kantor and Kremer 1993; Barsky and Plischke 1994), or the triangular network is site (Grest and Murat 1990) or bond (Plischke and Fourcade 1991) diluted. Self-avoidance can also be guaranteed by requiring that the elementary surface triangles of the network do not self-intersect (Baumgärtner 1991; Baumgärtner and Renz 1992; Kroll and Gompper 1993; Bowick *et al.* 2001a; Bowick *et al.* 2001b). This results in a very flexible surface which can fold on itself without any cost in energy. The resulting surface is therefore much rougher than those constructed with hard-core repulsion for small $n$, and is, in fact, very similar to the tethered-chain models in the limit of large $n$. In all these cases — with the exception of Baumgärtner (1991) and Baumgärtner and Renz (1992), where a crumpled phase has been claimed —, however, the membranes were found to remain flat, even though the local bending rigidity is quite low. This suggests that the interactions inducing the rigidity, whatever they are, are relevant under renormalization.

Polymerized membranes have also been studied in spatial dimensions larger than three. As discussed in Sec. 2.1.4, self-avoidance should become less important in this case, and the network is more likely to be crumpled. Indeed, a crumpled phase is found in dimensions $d \geq 5$ (Grest 1991; Barsky and Plischke 1994). This finding is in agreement with analyses of the generalized Edwards model with two-body interactions using a Gaussian variational approximation (Goulian 1991) and an expansion in large embedding space $d$ (Le Doussal 1992). It was found that the flat phase is stable for $d = 3$ and that tethered membranes are crumpled for $d > 4$, with the scaling exponent $\nu$ approaching one (with logarithmic corrections) for $d \searrow 4$. For $d > 4$, their results for the fractal dimension $d_f$ are much closer to the simulation results than the Flory prediction (8). Guitter and Palmeri (1992) used a variational approach for large $d$, and found that for $d = 3$, tethered membranes lie exactly on the boundary of the stable flat phase. Le Doussal and Radzihovsky (1992) have used the self-consistent screening approximation to obtain the estimate that the upper critical dimension is approximately 4.98.

An alternative interpretation of the simulation results is suggested by the prediction of Flory theory that four-body interactions are relevant below four dimensions and three-body interactions are relevant below six dimensions (see Sec. 2.1.4). Thus, if three-body interactions are sufficient to generate an effective rigidity, Flory theory predicts that tethered networks should crumple above six dimensions. While the critical dimensions suggested by Flory theory should not be taken too seriously, this argument may provide some insight into origin of observation that self-avoiding tethered surfaces are crumpled only for $d \geq 5$.


## 2.5. *Heterogeneous Polymer-Fluid Networks*

Heterogeneous membranes, which are composite fluid-polymerized networks, have unusual elastic properties. A prominent example of this type of network is the cell membrane of mammalian red blood cells (Elgsaeter *et al.* 1986; Steck 1989). This membrane consists of a lipid bilayer with an attached quasi-hexagonal network of spectrin tetramers. The lipid membrane provides a large area compression modulus and a high flexibility to bending deformations, while the polymer network provides the stiffness required to recover the biconcave equilibrium shape of the red blood cell after being squeezed through narrow capillaries of only a third of their size. Other examples of heterogeneous networks are partially polymerized sheets of certain phospholipid molecules (Sackmann *et al.* 1985).

Several models have been used for the simulations of fluid membranes with attached polymer networks. The first study of this problem (Boal *et al.* 1992) employed the model of fluid membranes described in Sec. 3.2.1 below. In addition to the "fluid" tethers, a second set of connections is introduced on a small subset of vertices (one in every 36), which form a hexagonal network of fixed connectivity. These "spectrin" tethers have a maximum length $s_{\max}$, which is the main model parameter. Spectrin tethers are allowed to intersect, as they only represent the in-plane projections of the three-dimensional polymer chains. Monte Carlo simulations of this model in two dimensions show (Boal *et al.* 1992) that the dimensionless area compression modulus $K_A \sigma_0^2$ decreases with increasing $s_{\max}$, but quickly reaches a plateau for $s_{\max} \gtrsim 8\sigma_0$. The area compression modulus for large $s_{\max}$ is therefore mainly determined by the fluid component. The shear modulus, on the other hand, is found to decrease rapidly with increasing $s_{\max}$; it is determined by the polymer network. Furthermore, the Poisson ratio $\sigma_P$ of the spectrin network was shown to be *negative*. Such materials expand transversely when stretched longitudinally. Note that no out-of-plane fluctuations of the membrane are taken account in this model. Thus, the physical mechanism for $\sigma_P < 0$ must be *different* from the case discussed in Secs. 2.4 and 2.4.2 above.

To gain more insight into this unexpected result, a simplified model has been considered (Boal *et al.* 1993). In this model, the fluid component is not taken into account explicitly, but only via a lateral tension on the spectrin network which determines the average area per vertex. The nearest-neighbor interactions were taken to be either square-well potentials or Hookean springs, and both self-avoiding and phantom membranes were examined. In all of these cases, a *negative* Poisson ratio was found over some range of the lateral tension.

Finally, the polymeric nature of the spectrin network has been explicitly modeled (Boal 1994). In this case, the bonds in the hexagonal network are replaced by short polymer chains, with chain lengths in the range 4–30 (compare Sec. 2.4.3). The midpoint of each chain is then constrained to move in a plane which represents the bilayer; the plane also acts as a repulsive hard wall to all other segments. This model has been used to calculate the elastic moduli of the red blood cell membrane (Boal 1994; Boal and Boey 1995; Boey *et al.* 1998).

Models involving explicit polymer chains contain too many degrees of freedom to be used in studies of the behavior of an entire red blood cell under large deformation, as in micropipette aspiration experiments. Discher *et al.* (1998) have therefore constructed an effective network model of the type discussed in Sec. 2.2, but with an energy

$$E_{\text{net}} = \sum_I C/A_I + \sum_{\langle ij \rangle} V_{\text{bond}}(r_{ij}), \qquad (44)$$

where the first sum runs over all triangles of area $A_I$, the second over all bonds. The first term mimics the low area compressibility of lipid bilayer, the second the non-linear elasticity of the polymer network. Using this model, it was shown that only pre-stressed networks yielded optimal agreement with the fluorescence imaging data of Discher *et al.* (1994).

## 2.6. *Shape of Spherical Shells and Forced Crumpling*

### 2.6.1. *Scaling Theory of Stretching Ridges*

A flat tethered surface cannot be deformed into a spherical shell without the introduction of topological defects. For the generic case of hexagonal lattice symmetry, the flat network consists of a triangular network of 6-fold coordinated vertices. In this case, Euler's theorem requires the introduction of exactly twelve 5-fold vertices (or disclinations) to form a spherical shell. The most symmetric surface which can be formed in this way is an icosahedron; fullerenes, such as the $C_{60}$ molecule, are well-known examples of this type of structure. The 5-fold coordinated vertices can be viewed as disclinations in an otherwise 6-fold coordinated medium. The large strains associated with an isolated 5-fold disclination with "disclinicity" $s = 2\pi/6$ centered in a flat disc of size $R$ imply a very strong dependence of the energy

$$E_5 \approx \frac{s^2}{32\pi} K_0 R^2 \qquad (45)$$

on the disc radius. It was first noticed by Seung and Nelson (1988) that for discs larger than a critical buckling radius $R_b \approx \sqrt{154\kappa/K_0}$ [for $\bar{\kappa}/\kappa = -1$, where $\bar{\kappa}$ is the saddle splay modulus — see Eq. (61) below], the membrane can lower its energy

by buckling out-of-plane and forming a cone. If this occurs, the disclination energy
only grows logarithmically with $R$,

$$E_5 \approx (\pi\kappa/3) \ln(R/R_b) + \frac{s^2}{32\pi} K_0 R_b^2. \tag{46}$$

Buckling might also be expected to occur in spherical shells such as an icosahedron. If the bending rigidity $\kappa$ is much larger than $K_0 R^2$, the closed surface should
be a sphere with an in-plane elastic energy proportional to $K_0 R^2$. In fact, it has
been found that the elastic energy of 12 disclinations on an undeformed sphere of
radius $R$ is (Bowick *et al.* 2001c)

$$E \approx 0.604 \frac{s^2}{4\pi} K_0 R^2. \tag{47}$$

If the value of $K_0 R^2$ is increased for fixed $\kappa$, however, the critical buckling radius
will eventually be reached, and the elastic energy of the surface should be well
approximated by that of the 12 cones created by the buckled disclinations. If this
is the case, the elastic energy of the surface scales as the $\ln(R/R_b)$. What happens
for still larger values of $K_0 R^2$? It turns out that as the disclinations become more
buckled the whole structure becomes more faceted. This leads to stress condensation along ridges which are connecting the vertices of the icosahedron (Witten and
Li 1993).

The structure and energy of boundary layer solutions around the ridges and
isolated vertices have recently attracted a great detail of attention, and the boundary
layer around a ridge has come to be called a "stretching ridge" because it comes
about through the balance of bending and stretching energy on the fold line, where
both energies are of comparable magnitude (Lobkovsky *et al.* 1995).

This scaling behavior at a "stretching ridge" is somewhat subtle, and can be
determined as follows (Lobkovsky 1996; Lobkovsky and Witten 1997). For the purpose of this review, it is sufficient to consider small displacement gradients, so that
the strain tensor is $u_{ij} = \frac{1}{2}(\partial_i u_j + \partial_i u_j + \partial_i f \partial_i f)$. The energy of the membrane,
$\mathcal{F}$, is given by Eq. (38) with $D = 2$ and $d = 3$. Minimizing the total energy with
respect to variations of $\mathbf{u}$ and $f$ yields

$$\kappa \nabla^4 f = \partial_i(\sigma_{ij}\partial_j f), \tag{48}$$

$$\partial_i \sigma_{ij} = 0, \tag{49}$$

where $\sigma_{ij} = 2\mu u_{ij} + \lambda u_{kk}\delta_{ij}$ is the stress tensor. Because $\sigma_{ij}$ is symmetric and has
zero divergence, it can be expressed in terms of a scalar potential, the Airy stress
function $\chi$, as $\sigma_{ij} = \epsilon_{ik}\epsilon_{jl}\partial_k\partial_l\chi$, so that Eqs. (48) and (49) become

$$\kappa \nabla^4 f = [\chi, f], \tag{50}$$

$$\frac{1}{K_0}\nabla^4 \chi = -\frac{1}{2}[f, f], \tag{51}$$

where $[a, b] \equiv \epsilon_{ik}\epsilon_{jl}(\partial_i\partial_j a)(\partial_k\partial_l b)$ and $\epsilon_{ij}$ is the antisymmetric $2 \times 2$ tensor. The
parameter $K_0 \equiv 4\mu(\mu + \lambda)/(2\mu + \lambda)$ in Eq. (51) is the 2d Young modulus of the

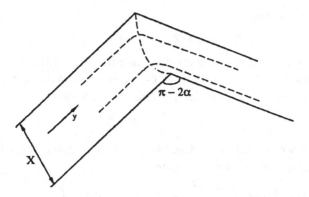

Fig. 2. Long strip of width $X$ bent through a dihedral angle $\pi - 2\alpha$. From Lobkovsky (1996).

membrane. Equations (50) and (51) are the well-known von Kármán equations for thin plates. The first von Kármán equation is the statement of local force equilibrium. The second states that the Gaussian curvature, $-\frac{1}{2}[f, f]$, acts as a source for the stress field. For a thin shell of thickness $h$ constructed from a $3d$ isotropic elastic material, $\kappa = Yh^3/12(1 - \nu_3^2)$ and $K_0 = Yh$, where $Y$ in the Young modulus and $\nu_3$ the Poisson ratio of the $3d$ material.

The simplest way to determine the scaling behavior of stretching ridges is to consider a "minimal" ridge consisting of a defect-free elastic sheet with coordinates in the domain $x \in (-X/2, X/2)$, $y \in (-\infty, \infty)$. Normal forces are applied to the edge in order to bend the strip by an angle $\pi - 2\alpha$ (see Fig. 2). Both the membrane stresses and torques vanish at the boundary (except at $y = 0$), so that

$$\partial_i \partial_j f = \partial_i \partial_j \chi = 0 \quad \text{at} \quad x = \pm X/2. \tag{52}$$

The sharp vertices at the ends of the ridge lead to a singularity in the curvature, $\partial^2 f/\partial y^2 = \alpha\delta(y)$, at $y = 0, x = \pm X/2$, so that $f$ at the boundary is given by

$$f(\pm X/2, y) = \alpha|y|, \tag{53}$$

up to an arbitrary linear function of $x$ and $y$. Note that the coefficient $\alpha$ in Eq. (53) is the bending angle $\alpha$ in Fig. 2 (Lobkovsky 1996).

The von Kármán equations can be put in non-dimensional form by defining $\tilde{\chi} = \chi/\kappa, \tilde{f} = f/X, \tilde{x} = x/X$, and $\tilde{y} = y/X$, so that

$$\nabla^4 \tilde{f} = [\tilde{\chi}, \tilde{f}], \tag{54}$$

$$\bar{\lambda}^2 \nabla^4 \tilde{\chi} = -\frac{1}{2}[\tilde{f}, \tilde{f}], \tag{55}$$

where $\bar{\lambda} \equiv \sqrt{\kappa/K_0}/X$. For $\bar{\lambda} = 0$, there is no cost for bending, and the strip forms a sharp crease, $f(x, y) = \alpha|y|$. The scaling behavior for finite $\bar{\lambda}$ can be found by performing the rescaling

$$\bar{x} = \tilde{x}, \quad \bar{y} = \bar{\lambda}^\beta \tilde{y}, \quad \bar{f} = \bar{\lambda}^\beta \tilde{f}, \quad \bar{\chi} = \bar{\lambda}^\delta \tilde{\chi}, \tag{56}$$

in Eqs. (54) and (55) and equating the dominant terms on either side of the resulting equations in the $\bar{\lambda} \to 0$ limit. Note that the boundary condition (53) requires that $\bar{f}$ and $\tilde{y}$ be rescaled by the same factor. This procedure leads to the result

$$\beta = -\frac{1}{3}, \quad \delta = \frac{2}{3}. \tag{57}$$

For the boundary conditions considered here, the leading contributions to the bending and stretching energies for $\bar{\lambda} \to 0$ are

$$\mathcal{F}_b = \kappa \bar{\lambda}^{-1/3} \int d\bar{x} d\bar{y} \left( \frac{\partial^2 \bar{f}}{\partial \bar{y}^2} \right)^2 \quad \text{and} \quad \mathcal{F}_s = \kappa \bar{\lambda}^{-1/3} \int d\bar{x} d\bar{y} \left( \frac{\partial^2 \bar{\chi}}{\partial \bar{y}^2} \right)^2, \tag{58}$$

where the values for $\beta$ and $\delta$ given in (57) have been used. Since the scaled quantities do not depend on $\bar{\lambda}$ in the small-$\bar{\lambda}$ limit, this means that $\mathcal{F}_b \sim \mathcal{F}_s \sim \kappa \bar{\lambda}^{-1/3} \sim \kappa(X/h)^{1/3}$, where the last relation applies to thin shells of thickness $h$. Similarly, it follow that the transverse curvature at the middle of the ridge scales as $C_{yy} = \partial^2 f/\partial y^2 \sim \bar{\lambda}^{-1/3}/X \sim h^{-1/3} X^{-2/3}$, and the mid-ridge longitudinal strain scales as $u_{xx} = (1/K_0)(\sigma_{xx} - \nu_2 \sigma_{yy}) \simeq (1/K_0)\partial^2 \chi/\partial y^2 \sim \kappa \bar{\lambda}^{1/3}$, where $\nu_2$ is the two-dimensional Poisson ratio. A similar analysis predicts that the sag, which is the vertical deflection of the ridges shape from that of a perfectly sharp crease, also scales as $X^{2/3}$ in the small $\bar{\lambda}$ limit. Witten and Li (1993) first obtained these results using scaling arguments.

### 2.6.2. Simulated Shapes of Spherical Shells

The first verification of the scaling behavior of stretching ridges of length $X$ described in the previous section came from studies of the asymptotic shape of large fullerene balls — flat-sided icosahedra with smooth edges — (Witten and Li 1993; Lobkovsky et al. 1995; Zhang et al. 1995) and other regular polyhedra (Lobkovsky et al. 1995; Lobkovsky 1996).

The total energy of a closed sphere consists of both bending and elastic contributions. Since the disclination energy is independent of the two-dimensional Poisson ratio and the contribution of the Gaussian curvature is a topological invariant, the energy of the vesicle depends only on the dimensionless parameter $\gamma \equiv K_0 R^2/\kappa$. For small values of $\gamma$, the stretching contribution is approximately 12 times the energy of a disclination, and the bending energy is approximately $8\pi\kappa + 4\pi\bar{\kappa}$. For very small values of $\gamma$, the bending energy dominates, and vesicle is spherical. In analogy with Eqs. (45) and (46), one expects

$$E/\kappa \approx \begin{cases} 6B\gamma/\gamma_b + D & \text{for } \gamma < \gamma_b \\ 6B[1 + \ln(\gamma/\gamma_b)] + D & \text{for } \gamma > \gamma_b \end{cases} \tag{59}$$

for larger values of $\gamma$, where $B$ and $D$ are constants and $\gamma_b$ is the value of $\gamma$ at the buckling radius (Lidmar et al. 2003). Finally, for $\gamma \gg 1$, the analysis described in

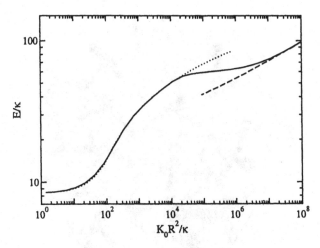

Fig. 3. Total energy of a closed triangulated surface of icosahedral symmetry as a function of $K_0 R^2/\kappa$. The dotted and dashed lines are fits to Eqs. (59) and (60), respectively. From Lidmar *et al.* (2003).

Sec. 2.6.1 predicts a crossover to a vesicle shape close to that of an icosahedron with sharp facets. In this regime, the energy should scale as

$$E/\kappa \approx C\gamma^{1/6} + \text{const.},\qquad(60)$$

with $C \approx 3.8$ (Zhang *et al.* 1995; Lidmar *et al.* 2003). Simulation results for the total energy $E/\kappa$ of a tethered vesicle of icosahedral symmetry are shown in Fig. 3. Fits to the functional forms given in Eqs. (59) and (60) are in good agreement with the estimates quoted above. The transition from bending-energy to stretching-energy dominated regimes occurs at $\gamma \approx 100$. Note that the crossover to "stretching ridge" scaling behavior does not occur until $\gamma \approx 10^7$.

### 2.6.3. *Forced Crumpling of Elastic Sheets*

The crumpling of a sheet of paper or of a car fender in an accident are but two examples of the forced crumpling of thin elastic sheets. In both of these cases, as well as for a large class of compressive boundary conditions, the energetically preferred configurations of crumpled thin sheets consist of flat regions bounded by straight folds connecting conical peaks. The scaling behavior at stretching ridges is expected to apply to a much broader class of ridge configurations than just the minimal ridge discussed in Sec. 2.6.1, and there is growing evidence that a generic compression of an elastic sheet will yield a similar spontaneous condensation of energy into a network of narrow ridges.

Recently, Kramer and Witten (1997) studied this question more carefully. They use a triangular network of springs with bending rigidity to model the elastic sheet. This network is then compressed by slowly decreasing the radius of a confining sphere. They found that after compression, most of the elastic energy is contained

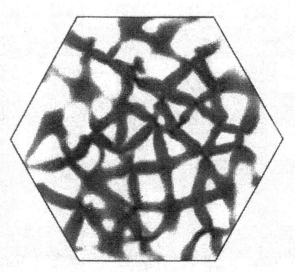

Fig. 4.   Curvature energy distribution in a hexagonal sheet of diameter $L = 160a_0$ (where $a_0$ is the equilibrium spring length) which has been crushed into a sphere of radius $R_s \simeq L/6$. Darker regions have higher energy density. From Kramer and Witten (1997).

in point-like vertices and in a network of "stretching ridges" (see Fig. 4), consistent with the scenario described above. In their simulations of phantom networks, the length of a ridge in the confining sphere is approximately proportional to the sphere radius, $R_s$, so that the number of ridges in a sheet of internal dimension $L$ is approximately $(L/R_s)^2$. Since the energy of a single ridge of length $X$ scales as $\kappa(X/h)^{1/3}$, if it is assumed that the ridges can be treated as independent, the total energy of the (phantom) crushed sheet should scale as $\kappa(R_s/h)^{1/3}(L/R_s)^2 \sim R_s^{-5/3}$; this result is supported by the simulation data (Kramer and Witten 1997).

The buckling of (isolated) ridges under compression by an external force which is applied parallel to the ridge axis has also been investigated (Lobkovsky and Witten 1997; DiDonna and Witten 2001; DiDonna et al. 2001; DiDonna 2002). By rescaling the compressed configuration with the same thickness scaling laws as discussed above, it was demonstrated that the response of the ridge to this kind of forcing is completely described by the same 1/3 power ridge scaling law (DiDonna and Witten 2001; DiDonna et al. 2001; DiDonna 2002). It was also shown that the onset of the buckling instability depends only on the ratio of the strain along the ridge to the curvature across it, and that for a given thickness to ridge length ratio, it always occurs at the same maximum ratio of stress to curvature. It was found numerically for a wide range of boundary conditions that ridges buckle when the forcing increases their elastic energy by 20% over their resting state value. While we are still a long way from understanding all aspects of crumpling, these results provide a promising start for developing a statistical mechanics of crumpled sheets.

In a highly crumpled sheet, resistance to further deformation results almost entirely from the work required to deform and break the ridges which span the volume occupied by the sheet. The behavior of the effective stiffness of a crumpled Mylar sheet has recently been studied experimentally (Matan *et al.* 2002). It was found that the size of the crumpled material decreased logarithmically over seven decades in time when placed under a fixed compressive force, indicating that there is dissipation, either due to friction or plastic flow in regions of high curvature. It is therefore unclear to what extent an analysis based only on conservative forces can fully describe the energetics of crumpling in real materials. Nevertheless, after the sheet has been precrumpled several times to avoid hysteresis effects, the spatial extent of the crumpled material was found to decrease as a power of the applied force, with an exponent 0.53(4), in reasonable agreement with the exponent 0.375 predicted by the simple scaling picture of crumpling described above (Matan *et al.* 2002).

## 3. Fluid Membranes and Vesicles

### 3.1. *Spontaneous Curvature Model and Area-Difference-Elasticity Model for Bilayer Vesicles*

The Hamiltonian of fluid membranes is not only invariant under rotations and translations, but also under reparametrizations. This additional invariance is due to the fluid structure, which does not allow a preferred coordinate system, and therefore cannot support shear stress. Fluid membranes are compressible, but the compressibility modulus is usually rather large, so that they are often studied in the incompressible limit, where the membrane area is fixed. In this case, the only contribution to the configurational energy is the bending energy (Canham 1970; Helfrich 1973; Evans 1974)

$$\mathcal{H}_b = \int dS \left[ \frac{1}{2}\kappa(H - C_0)^2 + \bar{\kappa}K \right], \tag{61}$$

where $\kappa$ is the bending rigidity, $\bar{\kappa}$ the saddle-splay modulus, $C_0$ the spontaneous curvature, and $H = C_1 + C_2$ and $K = C_1 C_2$ are the trace and determinant of the curvature tensor, respectively. For fixed topology, the second term in Eq. (61) is a constant, by the Gauss–Bonnet theorem. Morse and Milner (1995) have shown that a finite compressibility does not change the scaling behavior; we therefore ignore compressibility effects in the following.

In bilayer vesicles, not only the total number of lipid molecules, but also the number of lipid molecules of each monolayer is often conserved on typical experimental time scales, i.e. on time scales smaller than the "flip-flop" time for lipid exchange. However, while the total number of lipid molecules fixes the total membrane area $A$, it is not sufficient to consider a fixed area difference $\Delta A$ between inner and outer monolayer. Instead, it is important to incorporate the fact that the areas of the individual monolayers can respond elastically to a tensile stress. This implies the Hamiltonian (Miao *et al.* 1994)

$$\mathcal{H}_{ADE} = \frac{\kappa}{2} \int dS[C_1 + C_2 - C_0]^2 + \frac{\kappa_A}{2} \frac{\pi}{A\delta^2}(\Delta A - \Delta A_0)^2 \tag{62}$$

of the area-difference-elasticity (ADE) model, where $\Delta A_0$ is the relaxed, initial area difference, $\delta$ is the separation between the neutral surfaces of the two mono-layers, and

$$\Delta A = \delta \int dS(C_1 + C_2). \tag{63}$$

When all lengths are measured in units of $R_0 = [A/(4\pi)]^{1/2}$, the radius of a sphere with the same area, it is easily seen that in addition to $\kappa$, the Hamiltonian depends on two parameters, the dimensionless ratio $\tilde{\alpha} = \kappa_A/\kappa$ of the area stretching modulus and the bending rigidity, and

$$\bar{C}_0 = C_0 + \tilde{\alpha}\pi\frac{\Delta A_0}{A\delta}, \tag{64}$$

where terms independent of the vesicle shape have been omitted. Equation (64) demonstrates that the average shape *and* the fluctuation spectrum depend only on $\tilde{\alpha}$ and on an effective spontaneous curvature $\bar{C}_0$, so that effects of $C_0$ and of $\Delta A_0$ cannot be determined separately.

Note that the ADE model reduces to the spontaneous curvature model (61) in the limit $\tilde{\alpha} \to 0$. For typical phospholipids, it has been estimated that $\tilde{\alpha}$ is of order unity (Miao *et al.* 1994).

## 3.2. *Randomly-Triangulated-Surface Models for Fluid Membranes*

### 3.2.1. *Dynamic Triangulation*

For simulation studies of fluid membranes, the network model introduced in Sec. 2.2 has to be modified to allow for the diffusion of vertices in the membrane. This is done by making the connectivity of the network a dynamic variable. The simplest way to do so is to cut and reattached tethers between the four beads of two neighboring triangles (Kazakov *et al.* 1985; Boulatov et al. 1986; Billoire and David 1986; Ho and Baumgärtner 1990; Kroll and Gompper 1992a; Boal and Rao 1992a), as illustrated in Fig. 5. This bond-flip is only possible if the initially connected vertices have at least four neighbors each, so that the triangular nature of the network is conserved. Also, the bond-flip is only possible if the distance of the two initially disconnected vertices falls within the tether length. The bond-flip algorithm has the advantage that it preserves both the two-dimensional connectivity and the topology of the network.

### 3.2.2. *Bending Energy*

More care has to be taken in the discretization of the bending energy of fluid membranes than for polymerized membranes. The commonly used discretization

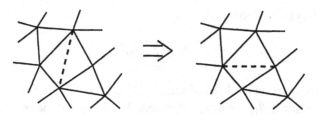

Fig. 5. Fluidity of the membrane is obtained by dynamic triangulation, in which a bond (dashed line) is removed from the network and replaced by a new one, which connects the two previously unconnected vertices of two neighboring triangles.

(Kantor and Nelson 1987) of the bending energy for polymerized membranes is

$$E_b^{\text{norm}} = \frac{1}{2}\lambda_b \sum_{\langle ij \rangle} |\mathbf{n}_i - \mathbf{n}_j|^2 = \lambda_b \sum_{\langle ij \rangle}(1 - \mathbf{n}_i \cdot \mathbf{n}_j), \tag{65}$$

where the sum runs over all pairs of neighboring triangles, and $\mathbf{n}_i$ is the surface normal vector of triangle $i$, compare Eq. (20). In the continuum limit, the difference $\mathbf{n}_i - \mathbf{n}_j$ becomes the gradient of the unit-normal-vector field, and

$$E_b^{\text{norm}} \rightarrow \mathcal{H}_{\text{norm}} = \frac{1}{2}\kappa \int dS \, g^{ij} \partial_i \mathbf{n} \cdot \partial_j \mathbf{n}_j, \tag{66}$$

where $g^{ij}$ is the contravariant metric tensor. Equation (66) is equivalent to the bending energy $\mathcal{H}_b$, Eq. (61), with $\bar{\kappa} = -\kappa$ (Seung and Nelson 1988). The relationship between the bending rigidity $\lambda_b$ in Eq. (20) and $\kappa$ can be determined by either discretizing Eq. (66) on a random surface, as described by Gompper and Kroll (1996) and Itzykson (1986), or by covering a sphere or cylinder with a number $N_\Delta$ of equilateral triangles and taking the limit $N_\Delta \rightarrow \infty$. Surprisingly, the result of the latter procedure depends on the shape of the surface! While, both explicit discretization and coverings of a sphere yield $\lambda_b = \sqrt{3}\kappa$ (Kroll and Gompper 1992a; Gompper and Kroll 1996), coverings of a cylinder yield $\lambda_b = 2\kappa/\sqrt{3}$ (Seung and Nelson 1988; Gompper and Kroll 1996). This problem is discussed rather extensively by Gompper and Kroll (1996).

Discretizations of the squared Laplacian form of the bending energy

$$\mathcal{H}_{\text{Lap}} = \frac{\kappa}{2} \int dS (\Delta \mathbf{R})^2 \equiv \frac{\kappa}{2} \int dS \, H^2 \tag{67}$$

do not share this pathology (Gompper and Kroll 1996). A general introduction to methods for discretizing operators on triangulated random surfaces is given by Itzykson (1986). On a triangulated surface, the mean curvature at node $i$ is

$$H = \mathbf{n} \cdot \Delta \mathbf{R} \rightarrow H_i = \frac{1}{\sigma_i}\mathbf{n}_i \cdot \sum_{j(i)} \frac{\sigma_{ij}}{l_{ij}}(\mathbf{R}_i - \mathbf{R}_j), \tag{68}$$

where $\mathbf{n}_i$ is the surface normal at node $i$ and the sum is over the neighbors of site $i$. $l_{ij}$ is the distance between the two nodes $i$ and $j$, $\sigma_{ij}$ is the length of a bond in the dual lattice (Itzykson 1986), and

$$\sigma_i = \frac{1}{4} \sum_{j(i)} \sigma_{ij} l_{ij} \tag{69}$$

is the area of the virtual dual cell of vertex $i$. The length $\sigma_{ij}$ in Eqs. (68) and (69) is given by $\sigma_{ij} = l_{ij}[\cot(\theta_1) + \cot(\theta_2)]/2$, where $\theta_1$ and $\theta_2$ are the two angles opposite link $ij$ in the triangles $(ijk)$ and $(ijk')$, respectively. Note that since $\cot(\theta) < 0$ if $\theta$ is obtuse, $\sigma_{ij}/l_{ij}$ can be negative if the sides of the two triangles are significantly different. Although there are some sum rules, such as $\sum_i \sigma_i = A$, where $A$ is the area of the surface, there is no guarantee, in general, that the $\sigma_{ij}$, or even the $\sigma_i$, are positive (Itzykson 1986). While this causes no problems in simulation studies of the self-avoiding tether-and-bead models, it can in certain related models, where the vertices are point particles, and the nearest-neighbor interaction potential is harmonic (Espriu 1987; Baillie et al. 1990).

Since $\mathbf{n} \parallel \Delta \mathbf{R}$ for surfaces embedded in three dimensions, Eq. (68) implies that the Laplacian squared bending energy can be written as (Itzykson 1986; Espriu 1987)

$$E_b^{\text{Lap}} = \frac{\tau}{2} \sum_i \sigma_i (\Delta \tilde{\mathbf{R}})_i^2 = \frac{\tau}{2} \sum_i \frac{1}{\sigma_i} \left[ \sum_{j(i)} \frac{\sigma_{ij}}{l_{ij}} (\mathbf{R}_i - \mathbf{R}_j) \right]^2, \tag{70}$$

with $\tau = \kappa$. Other discretizations of the bending energy which involve similar local averages of the mean curvature have been used by Gompper and Goos (1994) and Jülicher (1994).

In simulations performed using Gaussian spring models, the discretization (70) cannot be used, because there are large fluctuations in both the size and shape of the elementary triangles; some of the $\sigma_i$ can then be very small or negative, resulting in unphysical contributions to the bending energy. In this case, a simpler version of the bending energy (Espriu 1987; Baillie et al. 1990),

$$\tilde{E}_b^{\text{Lap}} = \frac{\tau}{2} \sum_i \frac{1}{\Omega_i} \left[ \sum_{j(i)} (\mathbf{R}_i - \mathbf{R}_j) \right]^2, \tag{71}$$

has been employed, where $\Omega_i$ is the sum of the areas of the surface triangles adjacent to site $i$. The form of $\tilde{E}_b^{\text{Lap}}$ follows from Eq. (70) by substituting $\sigma_{ij}/l_{ij} = 1/\sqrt{3}$, the result for equilateral triangles, and noting that $\sum_i \sigma_i = \sum_i \Omega_i/3$.

### 3.3. Phase Diagram of Fluid Vesicles at Low Bending Rigidities

The phase diagram of self-avoiding fluid vesicles as a function of the bending rigidity $\kappa$ and a pressure increment $\Delta p$ between the vesicle's interior and exterior has been determined from Monte Carlo simulations (Gompper and Kroll 1994; Gompper and Kroll 1995b). For very small bending rigidities and small or negative $\Delta p$, vesicles are found to collapse into a branched-polymer-like phase (Kroll and Gompper 1992a;

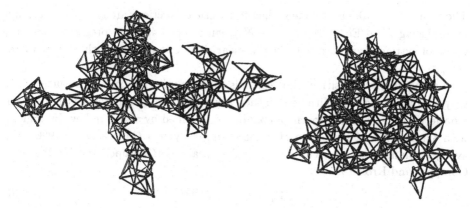

Fig. 6. Typical configurations of fluid vesicles (containing 247 vertices with tether length $\ell_0 = 1.67\sigma_0$) in the entropy-dominated regime, with bending rigidity $\kappa/\kappa_B T \ll 1$. (a) Branched-polymer-like configuration at small pressure increment $\Delta p$. (b) Inflated configuration at sufficiently large and positive $\Delta p$. From Gompper and Kroll (1995b).

Kroll and Gompper 1992b; Boal and Rao 1992a; Baillie and Johnston 1992), which is characterized by the scaling laws

$$\langle V \rangle \sim N \tag{72}$$

$$\langle R_g^2 \rangle \sim N^{\nu_{bp}} \tag{73}$$

for the average volume and radius of gyration, with $\nu_{bp} = 1$. A typical configuration is shown in Fig. 6(a). A more detailed characterization can be obtained by studying random walks on these surfaces; their behavior determines the spectral dimension $d_s$. The mean-square displacement after $t$ steps of a random walk on a surface of $N$ monomers is expected to scale as (Gefen *et al.* 1983; Komura and Baumgärtner 1990)

$$\langle [\mathbf{r}(t) - \mathbf{r}(0)]^2 \rangle = N^\nu f(t/N^{2/d_s}) \tag{74}$$

where the scaling function $f(x) \sim x^{d_s/d_f}$ for $x \ll 1$ and $f(x) = \text{const.}$ for $x \gg 1$. Simulation data indeed collapse on a universal scaling function for $d_s = 1.25 \pm 0.04$ (Kroll and Gompper 1992b), which is in excellent agreement with the best estimates for branched polymers (Havlin *et al.* 1984). Earlier claims of a crumpled phase of fluid membranes characterized by $\nu_{bp} \simeq 0.8$ (Baumgärtner and Ho 1990; Ho and Baumgärtner 1990) and $d_s = 2$ (Komura and Baumgärtner 1990) have not been confirmed.

With increasing pressure increment $\Delta p$, a first-order transition to an "inflated" phase occurs (Gompper and Kroll 1992a; Gompper and Kroll 1992b; Dammann *et al.* 1994). The transition pressure $p^*$ scales with the membrane size $N$ as

$$p^* \sim N^{-\zeta_+} \tag{75}$$

with an exponent in the range $\zeta_+ = 0.5$ (Gompper and Kroll 1992a) to $\zeta_+ = 0.65 \pm 0.05$ (Gompper and Kroll 1994) and $\zeta_+ = 0.69 \pm 0.01$ (Dammann *et al.* 1994). The differences in the values of $\zeta_+$ are mainly due to the different data analysis.

The value of $\zeta_+ = 0.5$ is obtained when finite-size corrections are taken into account by replacing $N$ in Eq. (75) by $(N - N_0)$, with $N_0 = 40$. The presently available range of system sizes does not allow a clear distinction between these two values of $\zeta_+$.

Just above the transition, the vesicle is roughly spherical, but its surface is still very rough, see Fig. 6b. With further increasing pressure, it approaches the shape of a perfect sphere. This approach can again be described by a scaling law. In analogy with the analysis of the shape of inflated ring polymers (Maggs *et al.* 1990), the average volume of the vesicle is predicted to scale as (Gompper and Kroll 1992a; Gompper and Kroll 1992b)

$$\langle V \rangle = V_0 p^{3\omega_+} N^{3\nu_+} \tag{76}$$

where

$$\omega_+ = \frac{1-\nu}{3\nu - 1}, \quad \nu_+ = \frac{\nu}{3\nu - 1}. \tag{77}$$

Monte Carlo simulations confirm this scaling form with an exponent $\nu = 0.787 \pm 0.020$ (Gompper and Kroll 1992a; Gompper and Kroll 1992b). A similar scaling analysis applies to the moments of inertia $\lambda_i$, which determine the asphericity of vesicle shapes. The same arguments which lead to Eq. (76) indicate that (Gompper and Kroll 1992b; Baumgärtner 1993)

$$\frac{3\langle \lambda_i \rangle}{\langle \lambda_1 + \lambda_2 + \lambda_3 \rangle} - 1 = \Gamma_i(\langle V \rangle N^{-3\nu/2}). \tag{78}$$

The scaling functions $\Gamma_i(x)$ are expected to decay asymptotically as $\Gamma_i(x) \sim x^{-1/(3-3\nu)}$ as the vesicle shape becomes spherical. Monte Carlo data for the smallest and largest eigenvalues nicely follow this power-law behavior, with an exponent $\nu$ which is very close to the value quoted above (Gompper and Kroll 1992b).

The entropy-dominated phases are stable for $\lambda_b/k_B T \lesssim 2.5$. Such vesicles shapes have been observed experimentally for DMPC bilayer membranes containing 2 mole % of the bipolar lipid denoted as bola lipid by Duwe *et al.* (1990). For larger bending rigidities, the curvature energy dominates, and prolates, discocytes and stomatocytes are the shapes of minimal free energy (compare Seifert *et al.* (1991), Seifert (1997)). The full phase diagram obtained from simulations of a network of $N = 247$ vertices is shown in Fig. 7.

It is interesting to note that the branched polymer behavior for small bending rigidity remains unchanged when the genus of the vesicle is allowed to fluctuate (Jeppesen and Ipsen 1993).

### 3.4. *Quasi-Spherical Vesicles*

For the fluctuations of an almost spherical vesicle of radius $R_0$, the radial position vector of the vesicle at solid angle $\Omega \equiv (\theta, \phi)$ can be written as

$$r(\Omega) = R_0[1 + u(\Omega)], \tag{79}$$

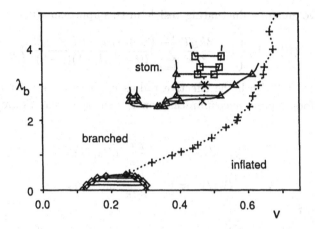

**Fig. 7.** Phase diagram of fluid vesicles in the pressure ensemble as a function of the reduced volume $v = \langle V \rangle / V_0$ and the bending rigidity $\lambda_b$, for $N = 247$. Compressibility maxima are shown by a dotted line, the dumbbell-to-metastable discocyte transition by a dashed line. From Gompper and Kroll (1995b).

where $u(\Omega)$ is the dimensionless amplitude of radial displacement. An expansion of $u$ in spherical harmonics reads

$$u(\Omega) = \sum_{l=0}^{l_M} \sum_{m=-l}^{l} u_{lm} Y_{lm}(\Omega), \tag{80}$$

where $l_M$ is a large wavenumber cutoff determined by the number of degrees of freedom; since only motion normal to the vesicle surface is relevant, $(l_M + 1)^2 = N$ in the present case. The excess bending energy $\Delta E \equiv (\kappa/2)[\int dS \, H^2 - 16\pi]$, area $A$, and volume $V$ of the vesicle can be written (to order $u^2$) as (Milner and Safran 1987; Helfrich 1986)

$$\Delta E = \frac{\kappa}{2} \sum_{l,m} |u_{lm}|^2 l(l+1)(l-1)(l+2), \tag{81}$$

$$A = 4\pi R_0^2 (1 + u_0)^2 + R_0^2 \sum_{l>0} |u_{lm}|^2 [1 + l(l+1)/2], \tag{82}$$

and

$$V = \frac{4\pi}{3} R_0^3 (1 + u_0)^3 + R_0^3 \sum_{l>0} |u_{lm}|^2, \tag{83}$$

where $u_0 = u_{00}/(4\pi)^{1/2}$.

The constant-area constraint is incorporated by choosing $u_0$ to satisfy $A = 4\pi R_0^2$, which implies

$$(1 + u_0)^2 = 1 - \frac{1}{4\pi} \sum_{l>0} |u_{lm}|^2 [1 + l(l+1)/2]. \tag{84}$$

The average volume of a fluctuating vesicle in this approximation is then given by

$$\langle V \rangle = V_0 \left\{ 1 + \frac{3k_BT}{8\pi\kappa} \sum_{l=2}^{l_M} \frac{(2l+1)[1-l(l+1)/2]}{l(l+1)(l-1)(l+2)} \right\}, \tag{85}$$

where $V_0 = 4\pi R_0^3/3$. Both the effective radius $r_{\text{eff}} = R_0(1+u_0)$ and the average volume therefore depend *logarithmically* on the surface area (Gompper and Kroll 1996),

$$\langle r_{\text{eff}} \rangle / R_0 \approx 1 - \frac{k_BT}{4\pi\kappa} \ln(l_M/2) \tag{86}$$

$$\langle V \rangle / V_0 \approx 1 - \frac{3k_BT}{8\pi\kappa} \ln(l_M/2) \tag{87}$$

where $l_M \sim \sqrt{A}$.

A similar approach can be used to describe fluctuations of vesicles at constant volume (Milner and Safran 1987). However, it is very difficult to incorporate constraints of constant area *and* constant volume simultaneously. It is possible, however, to employ a Lagrange multiplier, $\sigma$, to ensure that the average area equals a prescribed value. The energy is then given by $\mathcal{H}_b + \sigma A$, so that $\sigma$ can be interpreted as an effective membrane tension. In this case, the equipartition theorem yields (Milner and Safran 1987)

$$\langle |u_{lm}|^2 \rangle = \frac{k_BT}{\kappa} \frac{1}{(l+2)(l-1)[l(l+1)+Q]} \tag{88}$$

for a membrane with spontaneous curvature $C_0$, with $Q = 2(C_0R_0)^2 - 4C_0R_0 + \sigma R_0^2/\kappa$. The additional term has two important consequences: (i) $C_0R_0$ and $\sigma R_0^2/\kappa$ appear only in the form of a single constant $Q$, so that they cannot be determined independently, and (ii) the spectrum at small mode number $l$ is determined by the value of $Q$.

A systematic theory of vesicle fluctuations with several constraints, such as fixed area *and* fixed volume, has been developed recently by Seifert (1995) and Heinrich *et al.* (1997) (see also Seifert (1997)).

### 3.5. *Renormalization of the Bending Rigidity*

#### 3.5.1. *Renormalization Group Theory*

The interaction between the undulation modes leads to a renormalization of the bending rigidity (Helfrich 1985; Peliti and Leibler 1985; Cai *et al.* 1994),

$$\kappa_R(\ell) = \kappa - \alpha_\kappa \frac{k_BT}{4\pi} \ln(\ell/a) \tag{89}$$

on length scale $\ell$, where $a$ is a microscopic length. The prefactor $\alpha_\kappa$ has been predicted to be universal. However, there has been a long-standing debate about its value; both $\alpha_\kappa = 3$ (Peliti and Leibler 1985; Förster 1986; Kleinert 1986; David and Leibler 1991; Cai *et al.* 1994) and $\alpha_\kappa = 1$ (Helfrich 1985) have been obtained.

In this case, thermal fluctuations on short scales *soften* the membrane on larger length scales. A persistence length $\xi_p$ can then be defined by $\kappa_R(\xi_p) = 0$, and Eq. (89) implies $\xi_p = a \exp[(4\pi\kappa)/(\alpha_\kappa k_B T)]$.

However, a value $\alpha_\kappa = -1$ has also been suggested recently (Helfrich 1998; Pinnow and Helfrich 2000). Negative $\alpha_\kappa$ implies a *stiffening* of the membrane at long length scales.

### 3.5.2. *Scaling of the Vesicle Volume*

The scaling behavior of the average volume $\langle V \rangle$ of fluctuating vesicles of (approximately) constant area $A$ can be used to determine the renormalization of the bending rigidity from Monte Carlo simulations. The analogy with the behavior of the enclosed area of ring polymers in two spatial dimension (Camacho *et al.* 1991) suggests the scaling form

$$\langle V \rangle A^{-3/2} = \Theta_V(\sqrt{A}/\xi_p), \qquad (90)$$

where $\xi_p$ is the persistence length (see Sec. 3.5.1). Data obtained from Monte Carlo simulations are indeed consistent with this scaling ansatz for $\alpha_\kappa = 3.3 \pm 0.5$ (Gompper and Kroll 1995b; Ipsen and Jeppesen 1995). However, the analytic result (87) for $\langle V \rangle$ in the limit of large $\kappa$ is *inconsistent* with the scaling ansatz (90). Corrections to scaling therefore have to be taken into account (Gompper and Kroll 1996). To gain some insight into the behavior for moderate $\kappa$, the effect of the scale dependence of the bending rigidity has been incorporated in Eq. (85) by replacing $\kappa$ by $\kappa_R(\ell)$, compare Eq. (89). This implies (Gompper and Kroll 1996)

$$\langle V \rangle / V_0 - 1 + \frac{3}{2\alpha_\kappa} \ln(4\pi\kappa/\alpha_\kappa) = \Theta_V(\sqrt{A}/\xi_p). \qquad (91)$$

The Monte Carlo data scale about as well with the new scaling form — again with $\alpha_\kappa \simeq 3$ — as they do with Eq. (90).

### 3.5.3. *Undulation Modes of Quasi-Spherical Vesicles*

Another possibility for measuring the renormalization of the bending rigidity is to determine the fluctuation amplitudes of undulation modes from simulations of dynamically triangulated quasi-spherical vesicles. The basic idea is to use Eqs. (79) and (80) as a parameterization of simulated shapes. This is done by performing a least-squares fit of the spherical-harmonics expansion to the radial distances of the vertex coordinates of a sequence of configurations in a simulation; averaging over configurations then yields $\langle |u_{lm}|^2 \rangle$. The number of spherical harmonics is taken to be roughly a factor 2 smaller than the number of vertices in order to smooth out small fluctuations on the scale of single triangles. The mean squared amplitude of

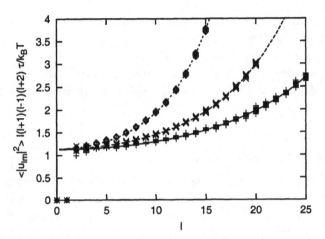

Fig. 8. Scaled mean-squared fluctuation amplitudes, $\langle|u_{lm}|^2\rangle l(l+1)(l-1)(l+2)\tau/k_BT$ as a function of the mode number $l$, for $\tau = 10k_BT$ and vesicles containing $N = 407$ ($\diamond$), $N = 847$ ($\times$) and $N = 1447$ ($+$) vertices.

fluctuations should have the form

$$\langle|u_{lm}|^2\rangle = \frac{\kappa_B T}{\kappa_R(l)} \frac{1}{l(l+1)(l-1)(l+2)}, \qquad (92)$$

compare Eq. (88). A plot of $\langle|u_{lm}|^2\rangle l(l+1)(l-1)(l+2)$ as a function of the mode number $l$ is shown in Fig. 8.

For a triangulated surface, the spectrum at large $l$ must have corrections due to the discretization of the Laplacian on a discrete mesh. The size of these corrections can be estimated for a Laplacian on a flat, perfect hexagonal lattice with lattice constant $a$, where for $\mathbf{q}$ in the principal lattice directions,

$$\langle|u(\mathbf{q})|^2\rangle = \frac{k_B T}{\kappa} \frac{9}{16} [3 - \cos(qa) - 2\cos(qa/2)]^{-2}. \qquad (93)$$

As can be seen from Fig. 8, $q^4\langle|u(\mathbf{q})|^2\rangle$ obtained from Eq. (93) with $qa \simeq \pi l/l_M$ provides a good fit to the data when the lattice constant $a$ is used as a fit parameter. It is reassuring that $a$ is found to be very close to the average nearest-neighbor distance $\langle\ell\rangle$, and to be essentially independent of $N$.

The range of $l$-values is currently too small to see the renormalization of the bending rigidity directly. We therefore interpret the $\kappa$-values extracted from Fig. 8 as effective bending rigidity on the scale of the vesicle size. These effective bending rigidities are plotted in Fig. 9 as a function of $\tau$, the coefficient of the discretized bending energy (70), for three different vesicle sizes. The figure shows that $\kappa_{\text{eff}}$ decreases with increasing membrane size. Furthermore, the magnitude of this effect is in good agreement with the theoretical result (89), with $\alpha_\kappa = 3$. Using the relation $\kappa_{\text{eff}}(N) = \kappa_0(\tau) - (3k_BT/4\pi)\ln(l_M)$ with $l_M = \sqrt{N} - 1$, it is possible to extrapolate back to the scale of the average bond length to obtain the bare bending rigidity $\kappa_0(\tau) = (0.10 + 0.08\tau + \tau^2)/(0.35 + \tau)$, compare Fig. 9.

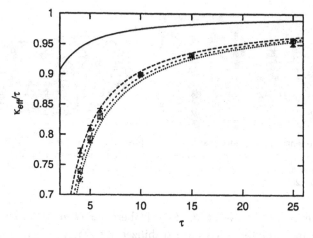

Fig. 9.   Effective bending rigidity $\kappa_{\mathrm{eff}}$ obtained from a fit of the undulation spectrum to Eqs. (92) and (93) for vesicles containing $N = 407$ (▲), $N = 847$ (□) and $N = 1447$ (▼) vertices. The lines show the effect expected from the renormalization (89) with $\alpha_\kappa = 3$, and the extrapolated bare bending rigidity $\kappa_0(\tau)$ (full line).

### 3.6.  *Fluctuations of Non-Spherical Vesicles*

There exist several techniques for measuring the bending modulus (Duwe *et al.* 1990; Evans and Rawicz 1990; Meleard *et al.* 1997). The method most widely used for membranes of higher bending rigidity is the flicker spectroscopy of giant, quasi-spherical vesicles (Duwe *et al.* 1990; Meleard *et al.* 1997). Although this technique can be used to determine precise values of $\kappa$, it has the disadvantage that analytical results required for extracting $\kappa$ from experimental data are only available in the quasi-spherical limit, compare Sec. 3.4. In this limit, the membrane is under a lateral tension which dominates the long-wavelength part of the spectrum (88). To determine $\kappa$, it is therefore necessary to measure the spectrum up to high mode number, which in turn requires very large vesicles. In addition, since the volume-to-area ratio changes with temperature, a vesicle which is quasi-spherical at one temperature is not at another, so that the bending modulus of a single vesicle cannot be determined as a function of temperature. Finally, membrane fluctuations are insensitive to the effective spontaneous curvature in this limit, so that $\bar{C}_0$ cannot be determined.

These shortcomings of traditional undulation analyses can be avoided by utilizing Monte Carlo simulations of dynamically triangulated vesicles to generate data for a wide range of reduced volumes and spontaneous curvatures, which can then be used to extract the elastic parameters of the membrane from flicker spectroscopy data, compare Fig. 10. Since the fluctuation spectra contain information on both the bending modulus and the spontaneous curvature in this case, it is possible to determine *both* elastic constants *simultaneously* in one experiment, and to study their dependence on environmental conditions.

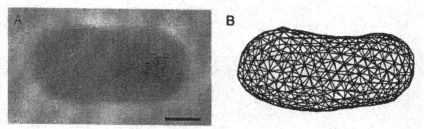

Fig. 10.   Experimental and theoretical vesicle shapes. (A) Phase contrast micrograph (reduced volume $v = 0.828$). The scale bar corresponds to 5 $\mu$m. (B) Simulation snapshot. The parameters in the simulations are $v = 0.825$, $\kappa/k_B T = 25$, $\tilde{\alpha} = 0.90$, $\bar{C}_0 R_0 = -0.28$ and $g = 0.37$. From Döbereiner *et al.* (2003).

In the experiments (Duwe *et al.* 1990; Döbereiner *et al.* 1999; Döbereiner *et al.* 1997), fluctuating prolate vesicles are stabilized by gravity — due to a small density difference of the solvent inside and outside the vesicle — on the bottom of a temperature-controlled micro-chamber. The focal plane of a phase contrast microscope is adjusted to include the long axis of the vesicle, and shape contours are obtained by real time video image analysis. Choosing a coordinate system in which the $x$-coordinate lies along the long axis of the vesicle, the contours are then represented in polar coordinates $(r, \varphi)$ as

$$r(\varphi) = r_0 \left[ 1 + \sum_n a_n \cos(n\varphi) + \sum_n b_n \sin(n\varphi) \right], \qquad (94)$$

where the angle $\varphi$ is measured from the positive $x$-axis. The mean values $\langle a_n \rangle$ describe the mean vesicle shape; for the oriented contours used here, $\langle b_n \rangle = 0$. The mean-square amplitudes $\langle \Delta a_n^2 \rangle \equiv \langle (a_n - \langle a_n \rangle)^2 \rangle$ measure the thermal fluctuations of the vesicles about their mean shape.

This setup can be exactly reproduced in simulation studies (Döbereiner *et al.* 2003). The substrate is modelled by a hard wall at $z = 0$, so that the $z$-coordinates of all vertices are restricted to be in the half-space $z > 0$. The effect of the gravitational field, which acts on the heavier fluid inside the vesicle, is transformed into a surface integral with the help of Gauss's divergence theorem; this adds the term

$$\mathcal{H}_g = \kappa g \int \frac{dS}{R_0^4} \frac{z^2}{2} \mathbf{e}_z \cdot \mathbf{n} \qquad (95)$$

to the ADE-Hamiltonian defined by Eqs. (62) and (63), where $g = (g_0 \Delta \rho R_0^4)/\kappa$ is a dimensionless parameter which measures the strength of the gravitational field, $g_0$ is the gravitational acceleration, $R_0 = (A/(4\pi))^{1/2}$ is the radius of a sphere of the same membrane area $A$, and $\Delta \rho$ the density difference between the fluids inside and outside of the vesicle. The same fitting procedure as described in Sec. 3.5.3 can be used to extract the amplitudes $u_{lm}$ of the spherical harmonics.

An example of the simulation data is shown in Fig. 11. The vesicle is in the oblate phase for $\bar{c}_0 = \bar{C}_0 R_0 \lesssim -1$, i.e. the contour is almost circular and $\langle a_2 \rangle \ll 1$.

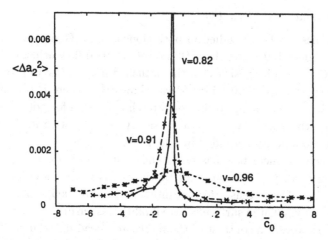

Fig. 11. Mean squared fluctuation amplitude $\langle a_2^2 \rangle - \langle a_2 \rangle^2$ of the elongational mode of non-spherical vesicles as a function of the reduced spontaneous curvature $\bar{c}_0 \equiv \bar{C}_0 R_0$, for three different reduced volumes $v$ as indicated. The peak at $\bar{c}_0 \simeq -1$ signals the oblate-to-prolate transition. From Döbereiner *et al.* (2003).

$\langle a_2 \rangle$ does not vanish identically, because the contours are oriented before averaging. At $\bar{c}_0 \simeq -1$, a phase transition to a prolate phase occurs, in which the vesicle has an elongated average shape. The peak in $\langle a_2^2 \rangle - \langle a_2 \rangle^2$ signals a second-order-like transition. Similarly, a peak in $\langle a_3^2 \rangle - \langle a_3 \rangle^2$ indicates a budding transition. Note that there is a pronounced dependence of the fluctuation amplitudes on the reduced volume. Precise values of $\kappa, \bar{c}_0$ and $v$ can be extracted from the comparison of measured and simulated values of $\langle a_2 \rangle, \langle a_3 \rangle, \langle (\Delta a_2)^2 \rangle, \langle (\Delta a_3)^2 \rangle, \langle (\Delta a_4)^2 \rangle$ and $\langle (\Delta a_5)^2 \rangle$.

This method has been applied to SOPC vesicles in a solution which contains a reactant with a light-sensitive $p$H only on the outside of the vesicle (Döbereiner *et al.* 2003). In this system, the magnitude of the $p$H change depends on the illumination intensity, so that it can be changed quickly and reversibly. The asymmetric $p$H change is expected to induce a spontaneous curvature in the membrane, but to have little effect on the reduce volume or the bending rigidity. An analysis of the flicker spectroscopy data showed that this is indeed the case. Both the reduced volume and the bending rigidity, $\kappa = (32 \pm 1)k_BT$, remained essentially constant, independent of the value of the $p$H, while the spontaneous curvature exhibited a strong, almost linear dependence on $p$H.

### 3.7. *Dynamics of Vesicles in External Fields*

The dynamical behavior of vesicles in external flow fields and the driven transport of vesicles through narrow passages are problems of fundamental interest with potential biological (Lipowsky and Sackmann 1995) and medical applications (Cevc *et al.* 1996).

### 3.7.1. *Elongational and Shear Flow*

Vesicle dynamics has been studied in both elongational (Gompper and Kroll 1993; Kroll and Gompper 1995) and shear (Kraus *et al.* 1996) flow fields. In the first case, very low bending rigidity vesicles (in the branched-polymer phase) were considered in the free draining approximation. It was shown that there is no sharp crumple-stretch transition (in analogy to the coil-stretch transition for polymers (de Gennes 1974)); rather, the vesicles were found to slowly elongate with increasing flow rate until they are completely extended in a long, thin cylinder.

The behavior in shear flow has been studied in the opposite limit, that of very large bending rigidity (no thermal fluctuations). Hydrodynamic interactions were included using an Oseen tensor formalism (Kraus *et al.* 1996). The membrane was modelled by a triangulated surface, where fluidity was ensured by bond flips at regular intervals. However, in contrast to the tether-and-bead model described so far, no beads were involved, the area of each triangle was kept constant, and bond flips were used to prevent the lengths of the triangle edges from differing too strongly. Since thermal fluctuations were not considered, the approach breaks down for dimensionless shear rates less than $k_B T/\kappa$, where typical velocities of rotational diffusion are comparable to the velocity of shear flow. The stationary state of the vesicles was found to be an elongated ellipsoid for all reduced volumes $v = V/V_0 > 0.52$, even for very small shear rates $\dot{\gamma}$. In particular, for $v \lesssim 0.75$, oblate discocytes are locally stable in the absence of an external flow field but still transform to the same shape as prolate vesicles when suspended in shear flow. The stationary state of a vesicle in the flow field was characterized by both a finite inclination angle $\theta$ between the longest axis of the vesicle and the flow direction, and a "tank-treading" tangential motion of the membrane with rotation frequency $\omega$. It was shown (Kraus *et al.* 1996) that the average reduced rotation frequency, $\bar{\omega}/\dot{\gamma}$, and the inclination angle decrease with decreasing reduced volume $v$ — with $\bar{\omega}/\dot{\gamma} = 0.5$ and $\theta = \pi/4$ in the spherical limit, in agreement with results for rigid spheres and fluid drops with infinite surface tension (van de Ven 1989). Both quantities were found to be independent of the shear rate within the numerical accuracy.

The same method has been used to study the dynamics of three-dimensional fluid vesicles in steady shear flow in the vicinity of a wall (Sukumaran and Seifert 2001). Three cases were considered. First, for a neutrally buoyant vesicle placed near the wall, it was found that the lift velocity is linearly proportional to the shear rate and decreases as the vesicle-wall distance increases. Second, a stationary hovering state was found for a vesicle filled with a denser liquid, and the viscous lift force was estimated to be $F_{\text{lift}} \approx 0.5\pi\eta\dot{\gamma}R_0^4/h^2$, where $R_0$ is the effective vesicle radius and $h$ is the mean distance from the center of the vesicle to the substrate. A similar result had been obtained earlier using a lubrication analysis and scaling approach (Seifert 1999). Finally, the dynamic unbinding transition of vesicles bound to a surface by nonspecific interactions was studied. It was found that there is a critical shear rate — proportional to the adhesion strength — above which

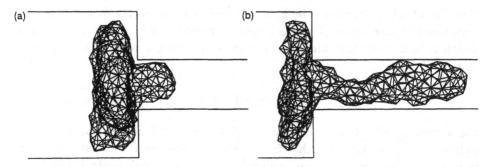

Fig. 12. Typical configurations of a vesicle of size $N = 407$ and bending rigidity $\lambda_b = 2.0k_BT$ moving in a tube at two different stages of squeezing through a cylindrical pore of radius $3\sigma_0$. The driving force is (a) $f\sigma_0^3 = 0.06k_BT$, and (b) $f\sigma_0^3 = 0.09k_BT$. From Gompper and Kroll (1995b).

the vesicle unbinds from the wall. For smaller shear rates, the vesicle tank treads along the substrate. The observed "bound → pinned → unbound or free" unbinding scenario is similar to that determined in two dimensions by Cantat and Misbah (1999).

### 3.7.2. *Vesicles in Micro-Channels*

The driven transport of vesicles through a linear array of narrow pores by an applied (electric or gravitational) field has been studied by Monte Carlo simulations (Gompper and Kroll 1995a). Two typical configurations for different strengths of the driving field are shown in Fig. 12. In the free draining approximation, the mobility of the vesicles was found to increase sharply when the strength $f$ of the driving field exceeded a threshold value $f^*$. For $f > f^*$, the mobility saturates at a value which is essentially independent of the strength of the driving field. The threshold field strength $f^*$ was found to depend on pore radius $r_p$, vesicle area $A$, and bending rigidity $\kappa$ as

$$f^* \sim \kappa^{1+\beta_t} A^{-3/2+\eta_t} r_p^{-2\eta_t}, \tag{96}$$

with $\beta_t \simeq 0.2$ and $\eta_t \simeq 2.4$ (Gompper and Kroll 1995a). The strongly non-linear transport properties of vesicles in this geometry can be understood from the balance of bending and potential energies. The potential energy dominates for small and large protrusions $\Delta z$ of the membrane into the pore, while the bending energy dominates for $\Delta z$ of order $r_p$. This leads to a nucleation barrier, the height of which is determined by the field strength. In the zero-temperature limit, the barrier height vanishes at $f = f^*$. In this limit, $\beta_t = 0$ and $\eta_t \simeq 1.55$ have been obtained from an analysis of the shape equations (Gompper and Kroll 1995a).

### 3.8. *Fluid Membranes with Edges*

The mechanical stability of lipid bilayer membranes against rupture is crucial for the survival of cells. This stability can be explained by the line tension of the free edge of a bilayer, where the lipid molecules have to adjust to the high local curvature of the monolayer. The (free) energy of a single hole of area $A$ and perimeter length $L$ in a membrane with line tension $\lambda_e$ and lateral tension $\sigma$ is given by (Litster 1975)

$$F_{\text{hole}} = \lambda_e L - \sigma A. \qquad (97)$$

For a circular hole of radius $r$, one has $A = \pi r^2$ and $L = 2\pi r$. Small holes of radius $r < r_c = \lambda_e/\sigma$ are therefore stable and disappear spontaneously after formation, while large holes of radius $r > r_c$ grow and rupture the membrane.

The picture presented above ignores the entropic contributions from the fluctuations of the hole perimeter. For sufficiently large holes, the entropy should be similar to the entropy of ring polymers, which scales linearly with the polymer length. This leads to a renormalization of the (effective) line tension, $\lambda_{\text{eff}} = \lambda_e - b$, in Eq. (97), where $b$ is a constant of order $k_B T$. The effective line tension can become very small or even negative, so that holes can form spontaneously even in membranes without tension (Shillcock and Boal 1996).

The dynamic triangulation method described in Sec. 3.2.1 has to be extended in order to model membranes with holes. In addition to bond-flipping, Monte Carlo moves involving both the removal of bonds as well as the insertion of new bonds at the edges of holes have to be introduced. The procedure is described by Shillcock and Boal (1996) and Shillcock and Seifert (1998).

Monte Carlo simulations of a tether-and-bead model for fluid membranes with holes confirm the picture described above (Shillcock and Boal 1996). Furthermore, they show that for tensionless membranes, large holes appear at a reduced line tension $\lambda_e \sigma_0 \simeq 1.24$ for tether length $\ell_0 = \sqrt{3}\sigma_0$. In this case, holes of *fixed* perimeter length show the same scaling behavior as self-avoiding random walks. In contrast, holes in membranes under compression scale as branched polymers. In the case of membranes under tension, no simple scaling behavior has been found.

The simulations have also been used to determine the rupture rate for membranes under tension. The observed dependence of the rupture rate on the line tension follows roughly an exponential Arrhenius law, with a free-energy barrier intermediate between those of circular and self-avoiding holes.

The edge tension is also the reason that membranes form vesicles. It is easy to see that a planar, circular patch of a fluid membrane is energetically unfavorable compared to a spherical vesicle when the patch radius exceeds $r^* = 4(2\kappa + \bar{\kappa})/\lambda_e$ (Helfrich 1974). The effect of thermal fluctuations has been studied by Monte Carlo simulations (Boal and Rao 1992b). The transition from open to closed topology is found to persist even in the limit of vanishing bending rigidity, where it occurs at a reduced tension $\lambda_e \sigma_0$ of order unity (Boal and Rao 1992b). The reason is again the extra entropy of the boundary fluctuations.

## 3.9. *Two-Component Fluid Membranes*

### 3.9.1. *Strong-Segregation Limit and Domain-Induced Budding*

The shape and fluctuations of two-component fluid membranes are again controlled by the curvature energy. In addition, there is a contribution from the line tension of the domain boundary. The total energy of a two-component membrane in the strong segregation limit is given by (Jülicher and Lipowsky 1993; Harden and MacKintosh 1994; Jülicher and Lipowsky 1996; Góźdź and Gompper 1998)

$$\mathcal{H} = \frac{\kappa_A}{2} \int dS (H - C_0^A)^2 + \frac{\kappa_B}{2} \int dS (H - C_0^B)^2 + \lambda \oint ds, \qquad (98)$$

where $\lambda$ is the line tension, and the bending rigidities $\kappa_A$, $\kappa_B$, and spontaneous curvatures $C_0^A$ and $C_0^B$ are in general different for the two components. For simplicity, it is often assumed that the saddle-splay modulus $\bar{\kappa}$ is the same for both components, so that the contribution of the Gaussian curvature is a constant and does not have to be considered.

In two-component membranes, a transition can occur from a nearly planar "cap" phase to a "budded" state, in which the domain has an (almost) spherical shape which is connected by a small neck to the embedding membrane. The location of this transition can be found approximately from the following estimate (Lipowsky 1992; Lipowsky 1993). For a planar membrane, the energy of the domain of radius $R_A$ is given by $E_{\text{planar}} = 2\pi\lambda R_A + (\pi/2)\kappa(C_0^A R_A)^2$. On the other hand, a complete bud has the energy $E_{\text{bud}} = (\pi/2)\kappa(4/R_A - C_0^A)^2$. These two energies are equal at $\lambda R_A/\kappa + 2C_0^A R_A = 4$. Thus, a budding transition occurs with increasing line tension $\lambda$ or with increasing spontaneous curvature $C_0^A$.

More detailed calculations (Jülicher and Lipowsky 1993; Jülicher and Lipowsky 1996; Góźdź and Gompper 2001) confirm this picture, and yield more precise quantitative results.

### 3.9.2. *Triangulated-Surface Models*

The tether-and-bead model can easily be generalized to membranes with two fluid components. In this case, the two components $A$ and $B$ can either be placed on the surface triangles (Kumar *et al.* 2001) or on the vertices (Kumar and Rao 1996; Kumar and Rao 1998). In the first case, the interactions of the two component mixture can be described by an Ising Hamiltonian, where the binary spin variables describe the occupation of the triangles with either of the two components. Since the number of neighboring triangles is always three in this case, the energy of the domain boundary is proportional to the number of bonds at which $A$- and $B$-triangles meet, and is therefore independent of the membrane shape near the domain boundary, as it should (Kumar *et al.* 2001). In contrast, when the Ising model with vertex occupation variables is used, the interaction energy depends on the number of neighbors. It is therefore favorable for the system to minimize the number of bonds which connect $A$- and $B$-vertices. Since the number of neighbors of a site is coupled to the local Gaussian curvature — with few neighbors implying a positive, and many neighbors a negative, Gaussian curvature —, the discretized curvature in combination with an Ising model with vertex variables may lead to artifacts.

However, it is not difficult to cure this problem. All that needs to be done is to use the *length* of the domain boundary instead of the *number* of bonds connecting $A$- and $B$-vertices. This is very natural in the Itzykson discretization (70) of the curvature energy, since variables $\sigma_{ij}$, which are the lengths of the bonds in the dual lattice, are already calculated anyway. The discretized version of the energy of the domain boundary is then

$$\mathcal{H}_l = \lambda \sum_{\langle ij \rangle_{AB}} \sigma_{ij}, \qquad (99)$$

where $\langle ij \rangle_{AB}$ denotes the bonds connecting $A$- and $B$-vertices.

### 3.9.3. *Phase Separation and Budding Dynamics of Two-Component Membranes*

The model in which the components occupy surface triangles has been used to study the dynamics of phase separation coupled to the membrane shape (Kumar *et al.* 2001). An example for the shape changes during the phase separation process is shown in Fig. 13. These simulations suggest the following scenario for the dynamics after a quench from the homogeneously mixed phase of a spherical vesicle into the two-phase region. The budding process exhibits three distinct time regimes: (i) formation and growth of intra-membrane domains; (ii) formation of many small buds; and (iii) coalescence of small buds into larger ones.

The formation of small intra-membrane domains occurs very rapidly. After this initial regime, the growth of domains in regime (i) follows the usual Oswald ripening and Lifshitz–Slyozov–Wagner behavior of phase-separation kinetics for flat,

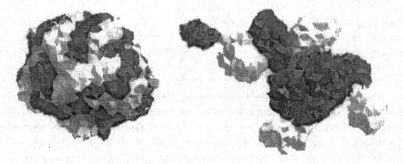

Fig. 13. Two characteristic configurations in the phase separation dynamics of a two-component vesicle consisting of 1200 white $A$ and 1200 gray $B$ triangles. Both components have zero spontaneous curvature and a bending rigidity of $1.0\,k_BT$. The snapshots are taken at times (a) $t = 4 \times 10^3$ and (b) $t = 9 \times 10^4$ (in units of Monte Carlo steps per vertex) after a quench from an initial state with a random distribution of the $A$ and $B$ components. From Kumar *et al.* (2001).

two-dimensional systems, which is characterized by a growth of the (linear) domain size as $t^{1/3}$ with time $t$, and correspondingly by a reduction of the overall length of the domain boundary as $t^{-1/3}$. The crossover time at which buds begin to form depends on the spontaneous curvature $C_0^A$ of the domains, and is the shorter the larger $C_0^A$. At the end of the budding regime (ii), essentially all domains have formed buds. In the final coalescence regime (iii), the buds diffuse on the membrane and sometimes fuse. This regime is again characterized by scaling laws which describe the long-time behavior. The number of buds, $N_{bud}$, was found to decay as

$$N_{bud} \sim t^{-\theta} \tag{100}$$

for large time $t$ with $\theta = 1/2$. For the Rouse dynamics employed in the simulations, this power law can be explained as follows (Kumar *et al.* 2001). The bud has a diffusion constant $D_{bud} \sim k_B T / R_{bud}^2$. For two buds to coalesce, they must move a distance of order $\ell_{bud}$, the average bud distance. The corresponding diffusion time $t_D$ is given by $t_D \sim \ell_{bud}^2 / D_{bud} \sim R_{bud}^4$, since $\ell_{bud} \sim R_{bud}$. Thus, the average bud size and the distance between buds increases as $R_{bud} \sim \ell_{bud} \sim t^{1/4}$, so that $N_{bud} \sim \ell_{bud}^{-2} \sim t^{-1/2}$, as indicated above.

This argument can be generalized to include the effect of hydrodynamic interactions (Kumar *et al.* 2001). Indeed, the only change occurs in the friction coefficient, which is now given by the Stokes expression $\eta R_{bud}$, where $\eta$ is the viscosity of the solvent. This implies $R_{bud} \sim \ell_{bud} \sim t^{1/3}$. Therefore, $N_{bud} \sim \ell_{bud}^{-2} \sim t^{-2/3}$, which implies that $\theta = 2/3$ in Eq. (100) in the presence of hydrodynamic interactions.

## 4. Crystalline and Hexatic Membranes

### 4.1. *Melting in Two Dimensions*

#### 4.1.1. *Theory of Kosterlitz, Thouless, Halperin, Nelson and Young (KTHNY)*

The ideal crystal structure for spherical particles in two dimensions is a close-packed triangular lattice. However, phonons destroy the long-range order of the crystalline phase. In particular, the structure factor, $S(\mathbf{q}) = \langle \rho_q \rho_{-q} \rangle$, where $\rho_q$ is the Fourier transform of the density operator, has cusp-like singularities at reciprocal lattice vectors $\mathbf{G}$. Similarly, $\langle \rho_{\mathbf{G}}(\mathbf{R}) \rangle = \langle \exp[i\mathbf{G} \cdot \mathbf{u}(\mathbf{R})] \rangle$ decays as a power of the linear system size $L$ for $L \to \infty$, so that there is no long-range order in a two-dimensional crystal.

However, as first noted by Mermin (1968), a defect-free two-dimensional crystal still exhibits long-range bond-orientational order. Orientational order in two dimensions can be measured using the complex order parameter

$$\psi_6(\mathbf{r}) = e^{6i\theta(\mathbf{r})}, \tag{101}$$

where $\theta(\mathbf{r})$ is the local bond angle. In a crystalline solid, $\theta(\mathbf{r})$ is given by the anti-symmetric part of the strain tensor,

$$\theta(\mathbf{r}) = \frac{1}{2}[\partial_x u_y(\mathbf{r}) - \partial_y u_x(\mathbf{r})]. \tag{102}$$

$\psi_6$ is the appropriate order parameter for triangular lattices since one is only interested in bond order modulo 60° rotations. Since $\theta(\mathbf{r})$ can be expressed in terms of the displacement field $\mathbf{u}(\mathbf{r})$ and the elastic energy (38) is quadratic in $\mathbf{u}(\mathbf{r})$, it is straightforward to show that the correlation function $\langle \psi_6^*(\mathbf{r})\psi_6(\mathbf{0}) \rangle$ in a defect-free two-dimensional solid approaches a constant for $r \to \infty$, so that there is true long-range orientational order.

The (topological) defects associated with the continuum elastic theory of a solid are dislocations and disclinations. An isolated pair of "plus" (5-fold) and "minus" (7-fold) disclinations is a dislocation. Dislocations are characterized by the amount a path around the defect, which would close on a perfect lattice, fails to close. This mismatch (or Burgers vector) is a lattice vector of the underlying triangular lattice. The Burgers vector is independent of the exact contour which has been chosen, and points at right angles to a line connecting the 5-fold to the 7-fold coordinated vertices in the dislocation core. A dislocation with Burgers vector $\mathbf{b}$ in a two-dimensional crystal of radius $R$ has an energy of order $K_0 b^2 \ln(R/a)$, where $K_0$ is the $2d$ Young modulus and $a$ is the lattice spacing. Because of the energy cost of creating a dislocation, no isolated dislocations are present at low temperatures. However, since a dislocation can be located at $(R/a)^2$ different positions, an entropy $2k_B \ln(R/a)$ is associated with a dislocation. Since both of these contributions to the free energy have the same $R$ dependence, there is a critical melting temperature $k_B T_m \sim K_0 b^2$ above which entropy dominates, dislocations proliferate, and the crystal melts to a hexatic phase (Nelson and Halperin 1979). Disclinations, on the other hand, have an energy of order $K_0 s^2 R^2$ in the solid phase, where $s$ is the disclination charge. The energy associated with free disclinations is therefore so large that they are absent in $2d$ crystals.

Detailed renormalization-group calculations are consistent with the scenario described above. In particular, above the dislocation-unbinding temperature $T_m$, dislocations are separated by a divergent correlation length $\xi_+(T) \sim \exp[b/|T - T_m|^{\bar{\nu}}]$, with $\bar{\nu} \approx 0.37$. Furthermore, the $2d$ Young modulus approaches a *universal* value at $T_m^-$. Above $T_m$, the Lamé constants vanish at long wavelengths. Although disclinations remain very tightly bound at all temperatures up to $T_m$, screening by a density $n_f(T) \approx \xi_+^{-2}$ of free dislocations produces a weaker logarithmic binding of disclinations for $T > T_m$.

Above $T_m$ there is residual bond-orientational order; it is this bond-orientational order that distinguishes this phase — which is called a *hexatic* — from the fluid phase. Because the bond-orientational order parameter (101) has a fixed magnitude and there are no external fields, the leading order contribution of bond-orientational order to the free energy in the hexatic phase arises from gradients of the local bond angle $\theta$ (Halperin and Nelson 1978; Nelson and Halperin 1979),

$$\mathcal{F}_H = \frac{1}{2} K_H(T) \int d^2 r (\nabla \theta)^2, \tag{103}$$

where $K_H$ is the hexatic stiffness. Orientational order-parameter correlations therefore decay as

$$\langle \psi_6^*(\mathbf{r}) \psi_6(\mathbf{0}) \rangle \sim r^{-\eta_6(T)}, \tag{104}$$

if $K_H$ is finite, with $\eta_6 = 18 k_B T / [\pi K_H(T)]$. The hexatic phase is therefore characterized by short-range translational order, as in fluid membranes, and quasi-long-range bond-orientational order (Kosterlitz and Thouless 1973; Halperin and Nelson 1978; Young 1979).

In the hexatic phase, screening by free dislocations leads to logarithmic interaction between pairs of disclinations with equal but opposite disclinicity. It should therefore not be surprising that disclinations also eventually unbind at a higher temperature, $T_i$. At the transition, $\eta_6(T_i) = 1/4$, while for $T > T_i$, $\langle \psi_6^*(\mathbf{r}) \psi_6(\mathbf{0}) \rangle \sim \exp[-r/\xi_\psi(T)]$, with $\xi_\psi(T) \sim \exp[b'/(T - T_i)^{1/2}]$. At both transitions, the free energy has an essential singularity, $\mathcal{F} \sim \xi^{-2}$. Both transition are therefore continuous.

### 4.1.2. Simulation Results for Network Models in Two Dimensions

The models of self-avoiding fluid membranes discussed in this review are generally constructed by placing hard spheres at each vertex of a triangulated surface and connecting neighboring vertices by a tethering potential. In two space dimensions, the behavior of this type of network resembles that of a gas of hard spheres (Strandburg 1986; Zollweg and Chester 1992), the primary difference being that the average density is now determined by the tether length rather than an external pressure. Each bond configuration corresponds to a Delaunay triangulation for a two-dimensional network of particles. The Monte Carlo procedure — which consists of random bead displacements and bond flips as described in Sec. 3.2.1 — can therefore be viewed as a simultaneous dynamical updating of both the Delaunay construction and the particle coordinates of a two-dimensional network of particles. One unusual property of the network model is that in contrast to atomistic models, the average area of the network *decreases* on melting (Gompper and Kroll 2000). In atomistic models, most of the area increase upon melting is due to the creation of "geometrical voids" rather than an increase in the most probable nearest-neighbor spacing (Glaser and Clark 1993). In fact, the most probable nearest-neighbor distance actually decreases slightly upon melting — in accordance with the results for the network model. Simulations of atomistic models have shown (Glaser and Clark 1993) that the bond-length distribution has two peaks in disordered regions of the dense fluid near the melting transition. The primary peak is near the most probable nearest-neighbor separation; however, the second is at a separation approximately a factor of $\sqrt{2}$ larger. Tether lengths corresponding to this second peak are not possible in the tethered fluid model due to the maximum tether length constraint. Another consequence of this suppression of density fluctuations is that the tethered fluid freezes at a significantly lower density than the hard-sphere fluid.

Gompper and Kroll (2000) studied the freezing transition of a network model for tensionless membranes confined to two dimensions. The simulations were performed using periodic boundary conditions, as described in Sec. 2.3.3. For the calculation of the elastic constants of the solid phase, both the size and shape of the simulation cell were allowed to fluctuate. However, in order to avoid extreme shape fluctuations in the fluid phase, the cell shape was fixed when studying the crystalline-to-fluid transition. In this case, the simulation cell was an equilateral parallelogram with an internal angle of $60°$. Translational and bond-orientational order parameters and elastic constants were determined as a function of the tether length, and a finite-size scaling analysis was used to show that the crystal melts via successive dislocation and disclination unbinding transitions, in qualitative agreement with the predictions of the KTHNY theory of melting. The critical tether length at the crystalline-to-hexatic transition, $\ell_0^{ch}(N)$, was found to scale as

$$\ell_0^{ch}(N) = 2a^*/\ln(N) + \ell_0^{ch}(\infty), \tag{105}$$

with $a^* = 0.2175$ and $\ell_0^{ch}(\infty) = 1.4796$. The critical tether length at the hexatic-fluid transition, $\ell_0^{hf}(N)$, scales as

$$\ell_0^{hf}(N) = (2b^*)^2/\ln^2(N) + \ell_0^{hf}(\infty), \tag{106}$$

with $b^* = 0.5840$ and $\ell_0^{hf}(\infty) = 1.5164$, implying that the correlation length at this transition is given by $\xi_\psi \equiv \exp[b^*/(\ell_0 - \ell_0^{hf}(\infty))^{1/2}]$, in agreement with the prediction of the KTHNY theory (Nelson and Halperin 1979). The hexatic phase is therefore stable over only a very small interval of tether lengths.

## 4.2. Freezing of Flexible Membranes

### 4.2.1. Continuum Model and Renormalization Group Results

A flexible membrane can relieve the strain field surrounding a defect by buckling out-of-plane and trading stretching for bending energy. Nelson and Peliti (1987) have argued that the energy of a buckled dislocation remains *finite* as the size of the system tends to infinity. Detailed calculations by Seung and Nelson (1988) have shown that this is indeed the case. Free dislocations are therefore expected to be present at any finite temperature, so that the low temperature phase of a flexible membrane is a hexatic. Since buckled disclinations have a logarithmically divergent energy, compare Eq. (46), the hexatic phase should still be separated from a high-temperature isotropic fluid by a finite temperature disclination unbinding transition.

As discussed above, a hexatic membrane will sustain a local orientational order, which is described by a local orientational order parameter $\mathbf{m}(\mathbf{x})$. The orientational order parameter is a unit vector tangent to the membrane; it can be expressed in terms of its components in the frame of the tangent vectors $\partial_i \mathbf{R}$ as

$$\mathbf{m}(\mathbf{x}) = m^i(\mathbf{x})\partial_i \mathbf{R}(\mathbf{r}) \quad \text{with} \quad m^2 = m^i m^j g_{ij} = 1, \tag{107}$$

where $g_{ij} = (\partial \mathbf{R}/\partial x_i) \cdot (\partial \mathbf{R}/\partial x_j)$ is the metric tensor. Since $\mathbf{m}$ has a fixed magnitude, and there are no external fields aligning $\mathbf{m}$ along a particular direction, the lowest non-trivial contribution to the energy arises from its gradients (Nelson and Peliti 1987; David 1989; Park and Lubensky 1996c), so that

$$\mathcal{F}_A = \frac{1}{2}K_H \int dS g^{ij} D_i \mathbf{m} \cdot D_i \mathbf{m}, \tag{108}$$

where $D_i$ is the covariant derivative. $\mathbf{m}(\mathbf{x})$ can also be expressed in terms of the local bond angle $\theta(\mathbf{x})$ by associating with every point on the surface two tangent orthonormal vectors $\mathbf{e}_1(\mathbf{x})$ and $\mathbf{e}_2(\mathbf{x})$; in this case,

$$\mathbf{m}(\mathbf{x}) = \cos\theta(\mathbf{x})\mathbf{e}_1(\mathbf{x}) + \sin\theta(\mathbf{x})\mathbf{e}_2(\mathbf{x}) = \sum_\alpha m_\alpha \mathbf{e}_\alpha. \tag{109}$$

In order to compare order parameters $\mathbf{m}(\mathbf{x})$ and $\mathbf{m}(\mathbf{x}')$ at two different points $\mathbf{x}$ and $\mathbf{x}'$, one has to parallel transport the order parameter at $\mathbf{x}$ along the geodesic to $\mathbf{x}'$. Under parallel transport in direction $dx^i$, the unit vectors $\mathbf{e}_1$ and $\mathbf{e}_1$ are rotated by an angle $A_i dx^i$; the covariant derivative is then given by

$$D_i m_\alpha = \partial_i m_\alpha + \epsilon_{\alpha\beta} A_i m_\beta, \tag{110}$$

where $\epsilon_{\alpha\beta}$ is the antisymmetric tensor with $\epsilon_{12} = -\epsilon_{21} = 1$. The vector field $A_i$ is the *spin connection*, whose curl is the Gaussian curvature,

$$\gamma^{ij} \partial_i A_j = K, \tag{111}$$

where $\gamma^{ij} = \epsilon_{ij}/\sqrt{g}$. With the use of Eq. (110), the free energy (108) can finally be written as (Nelson and Peliti 1987; David 1989; Park and Lubensky 1996c)

$$\mathcal{F}_A = \frac{1}{2}K_H \int dS g^{ij}(\partial_i\theta - A_i)(\partial_j\theta - A_j). \tag{112}$$

As discussed above, the excitations which destroy the hexatic order and cause a Kosterlitz–Thouless transition to the fluid phase, are disclinations. They give rise to a singular contribution, $\theta_{\text{dclin}}$, to the bond vector field; the disclination density, $s(\mathbf{x})$, is given by

$$s(\mathbf{x}) = \gamma^{ij} \partial_i \partial_j \theta_{\text{dclin}}. \tag{113}$$

The contribution of disclinations to the free energy (112) is (Park and Lubensky 1996c; Park and Lubensky 1996a; Deem and Nelson 1996)

$$\mathcal{F}_C = -\frac{1}{2}K_H \int dS(s - K)\frac{1}{\Delta_g}(s - K), \tag{114}$$

where $\Delta_g = D^i D_i$ is the Laplacian on the surface with metric tensor $g_{ij}$. Note first that the relevant quantity in Eq. (114) is not the disclination density or the Gaussian curvature separately, but rather the difference between them. The disclination

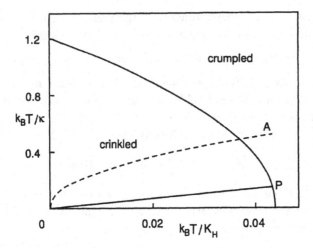

Fig. 14.   Phase diagram for hexatic membranes in the $(k_BT/K_H) - (k_BT/\kappa)$ plane. The solid curved line, $(k_BT/K_H)+(3/32\pi)(k_BT/\kappa)^2 = \pi/72$, is the critical line separating the crinkled from the crumpled phase. The line $0P$ is the crinkled line $4(k_BT/\kappa) = (k_BT/K_H) + (3/32\pi)(k_BT/\kappa)^2$, and $P$ is the crinkled-to-crumpled fixed point. The renormalization-group calculations underlying this diagram are valid below the line $0A$. Redrawn from Park and Lubensky (1996c).

density can therefore be screened by the Gaussian curvature. Second, there is a long-range Coulomb interaction between the densities $[s(\mathbf{x}) - K(\mathbf{x})]$ at different parts of the surface.

The phase behavior of this model has been studied in detail (Nelson and Peliti 1987; David *et al.* 1987; Guitter and Kardar 1990; Park and Lubensky 1996c; Park and Lubensky 1996b; Park and Lubensky 1996a; Deem and Nelson 1996; Nelson 1996). The phase diagram which has now emerged, based on a careful renormalization-group analysis (Park and Lubensky 1996c), is shown in Fig. 14. There is a hexatic, "crinkled" phase, which is characterized by an algebraic decay of the correlation function of surface normal vectors, at large bending rigidity $\kappa$ and hexatic stiffness $K_H$. A Kosterlitz–Thouless transition to a fluid, "crumpled" phase occurs not only with decreasing $K_H$, but *also* with decreasing $\kappa$.

It has been argued by Nelson (1996) that the free energy of isolated five- and seven-fold disclinations in hexatic membranes need not be identical. This asymmetry has been confirmed by explicit calculations of the shapes and energies of these disclinations (Park and Lubensky 1996a; Deem and Nelson 1996). However, this does *not* lead to two distinct Kosterlitz–Thouless defect-proliferation temperatures in the thermodynamic limit — with periodic boundary conditions —, since a "charge neutrality" condition dictates identical numbers of five- and seven-fold disclinations in this case. On the other hand, for membranes of finite size with free edges, the lower energy of five-fold disclinations indicates a tendency towards the formation of spherical vesicles (Nelson 1996; Deem and Nelson 1996); even in this case, however, it has been argued that in the basin of attraction of the hexatic fixed point, thermal fluctuations always drive the system into an "unbuckled" regime, where disclinations proliferate at the same critical temperature (Deem and Nelson 1996).

The coupling of both the local and global geometry of membranes to their internal structure can lead to many interesting and important effects. In the present context, a particularly important example of this has been discussed by Park (1996), where it is shown, in an analysis of the sine-Gordon model of membrane freezing, that shape fluctuations of spherical vesicles cause disclinations to be screened at length scales larger than $R(\kappa/K_H)^{1/2}$, where $R$ is the radius of the vesicle. In the sine-Gordon model, the principal coupling of the sine-Gordon field $\phi$ to geometry occurs via an interaction proportional to $\phi K$, where $K$ is the local Gaussian curvature, with an imaginary coefficient. Shape fluctuations on a sphere, like gauge fluctuations in an infinite two-dimensional superconductor, generate a mass term for $\phi$ that smears out the phase transition. When the screening length is much larger than the system size, screening is unimportant, and a normal Kosterlitz–Thouless transition from a low temperature hexatic to a high temperature fluid phase should occur. However, if the screening length is smaller than the system size, there are unbound disclinations at all non-zero temperatures, and strictly speaking, the hexatic phase does not exist.

The coupling of two-dimensional hexatic order to undulation modes in spherical and cylindrical geometries has also been studied (Lenz and Nelson 2001; Lenz and Nelson 2003). Although hexatic order has a negligible effect on the undulation modes in the planar case, it was shown to lead to significant changes in the amplitude of the low wave-vector modes of the spectrum in curved geometries.

### 4.2.2. *Simulation Results for Triangulated Surfaces*

Simulation studies of the freezing of flexible vesicles of spherical topology have provided strong evidence in support of the suggestion that the low temperature phase of flexible membranes is an hexatic (Gompper and Kroll 1997a; Gompper and Kroll 1997b). In particular, it was found that for short tether lengths (low temperatures), the density, $n_{\mathrm{dloc}}$, of "free" dislocations — defined as pairs of five- and seven-fold vertices which have only 6-fold coordinated vertices as nearest neighbors — with Burgers vector $|\mathbf{b}| = \langle \ell \rangle$, where $\langle \ell \rangle$ is the mean nearest-neighbor separation, scales as

$$\frac{k_B T}{\kappa} \ln(n_{\mathrm{dloc}}) = -\Theta\left(\frac{\kappa}{K_0 \langle \ell \rangle^2}\right), \tag{115}$$

indicating that the free energy of dislocations is indeed finite even for the shortest tether lengths studied. Unfortunately, the spherical topology of the vesicles prevented a detailed characterization of the transition in terms of the orientational order parameter. Instead, the internal order of the membrane was monitored indirectly from the behavior of several quantities such as the system-size dependence of the reduced volume, the density of "free" 7-fold disclinations, $\langle n_{\mathrm{disc}}^7 \rangle$, and the bond flip acceptance rate. The resulting phase diagram in the $(k_B T/[K_0 \langle \ell \rangle^2])$–$(k_B T/\kappa)$

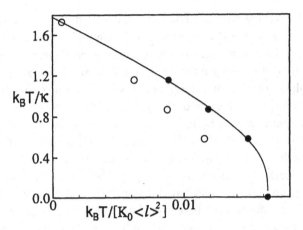

Fig. 15.  Phase diagram in the $(k_B T/[K_0 \langle \ell \rangle^2]) - (k_B T/\kappa)$ plane, where $K_0$ is the Young modulus of a polymerized membrane of the same tether length, and $\langle \ell \rangle \simeq (1 + \ell_0)/2$ is the average nearest-neighbor distance. The solid symbols ($\bullet$) are estimates for the phase boundary obtained from simulations of membranes fluctuating about a planar reference state and from simulations of melting of a network model confined to two dimensions. The open symbols ($\circ$) are estimates obtained from simulations of spherical vesicles. The solid line is a guide to the eye. From Gompper and Kroll (2002).

plane is indicated by open circles in Fig. 15. Note that these estimates of the hexatic-fluid phase boundary are smaller than those obtained from simulations of membranes fluctuating about a planar reference state. This is consistent with the predictions of Park (1996) that shape fluctuations of spherical vesicles cause disclinations to be screened at length scales larger than $R(\kappa/K_H)^{1/2}$. If this screening length is smaller than the system size, there are unbound disclinations at all non-zero temperatures, and strictly speaking, the hexatic phase does not exist, although there may still be a sharp crossover in many quantities, as observed by Gompper and Kroll (1997a) and Gompper and Kroll (1997b).

In order to study the transition in more detail, simulations were subsequently performed for membranes fluctuating about a planar reference state (Gompper and Kroll 2002). Periodic boundary conditions were used, and the simulations were performed for zero spreading pressure. Because of the flat reference state, it was possible to analyze the bond-orientational order parameter susceptibility in detail. Simulations were performed for three values of the bending rigidity. The results for the phase boundary in the $(k_B T/[K_0 \langle \ell \rangle^2])-(k_B T/\kappa)$ plane are plotted as bullets ($\bullet$) in Fig. 15. This phase diagram has a striking similarity with the one obtained from renormalization-group calculations (see Fig. 14). Note, however, that (i) because it was not possible to determine the hexatic stiffness, the data are plotted as a function of the two-dimensional Young modulus of a tethered network of the same tether length, while the field-theoretical model uses the hexatic stiffness $K_H$, and (ii) the stability of the hexatic phase extrapolates to values of the bending rigidity as low as $\kappa/k_B T \simeq 0.56$, while renormalization-group calculations suggest a larger limit of $\kappa/k_B T \simeq 0.83$. It is important to note, however, that the field-theoretical calculations are strictly valid only for sufficiently large $\kappa$.

For $\kappa/k_B T \gtrsim 1.15$, the simulation results clearly showed that the transition is continuous, and that the critical behavior is consistent with the predictions of (Park and Lubensky 1996c; Park and Lubensky 1996b; Park 1996). For smaller values of the bending rigidity, the situation is less clear, although results for $\kappa/k_B T = 0.87$ are consistent with a first-order transition for the system sizes which were simulated. However, it cannot be ruled out that this behavior is a finite-size artifact, and that the transition becomes continuous for larger system sizes.

### 4.3. *Budding of Crystalline Domains in Fluid Membranes*

A famous and biologically very important example of domain formation and budding is the adsorption of clathrin molecules on the plasma membrane. Clathrin molecules are three-armed proteins which assemble to form a regular hexagonal network on the membrane surface. By forming first a coated pit and then a complete bud, see Fig. 16, these clathrin coats control endo- and exocytosis, i.e. the formation and detachment of small transport vesicles from the cell membrane. This process can be driven by a change of $p$H, compare Fig. 16, which induces a change of the spontaneous curvature, as discussed in Sec. 3.6. The formation of clathrin cages is therefore an example for the budding of crystalline patches embedded in a fluid lipid membrane.

The main difference between a fluid and a crystalline membrane domain is the in-plane shear elasticity and quasi-long-range positional order of the crystalline phase. A flat, crystalline membrane therefore cannot be deformed into a spherical bud without the introduction of topological defects. For the generic case of 6-fold

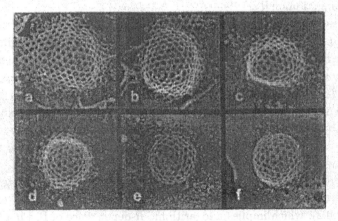

Fig. 16. Rounded clathrin coated pits in normal chick fibroblasts (a–d) and coated pits on membrane fragments derived from cells which have been broken open and left in $p$H7 buffer before fixation and drying (e–f). The width of field of view of the individual pictures is $0.4\,\mu$m. Reproduced from Heuser (1989) by copyright permission of The Rockefeller University Press.

coordinated vertices in the crystalline phase, Euler's theorem requires an excess of exactly 12 5-fold disclinations to form a spherical cage. Disclinations are topological defects which cannot be generated locally. In order to explain bud formation, it is therefore necessary to understand how the disclinations appear inside the crystalline domain. Two principal mechanisms are possible: (i) an edge-acquisition mechanism, in which disclinations form at the edge of the domain and then diffuse into the interior, and (ii) an interior-acquisition mechanism, in which dislocation pairs are generated in the interior; each of these dislocations subsequently dissociates, leaving behind a 5-fold disclination while the 7-fold disclination diffuses to the domain boundary. Another important question is the dependence of the domain size $R_A$ on the line tension and spontaneous curvature at the budding transition. It has been suggested by Mashl and Bruinsma (1998) that $C_0$ at the transition *increases* linearly with $R_A$, in contrast to the fluid case, where $C_0$ decreases as $1/R_A$.

In order to study crystalline domains in fluid membranes, and to address the questions raised above, crystalline order has to be induced in part of the membrane by choosing an appropriately small tether length. This implies that the tether length is not uniform, but depends on the type of the two connected vertices. A natural choice is to use two tether lengths, $\ell_A$ and $\ell_B$, for the $AA$- and $BB$-bonds, respectively, and to set $\ell_{AB} = (\ell_A + \ell_B)/2$.

Let $C_0^B$ and $C_0^A$ denote the spontaneous curvatures in the fluid and the crystalline domains, respectively. Monte Carlo simulations of this model with $C_0^B = 0$ and a range of values of $C_0^A$ and the line tension $\lambda$ have been used to determine the budding phase diagram (Kohyama *et al.* 2003). It was found that the location of the budding transition is well described by

$$\lambda R_A/\kappa + \gamma_0 C_0^A R_A = \Gamma(R_A), \tag{116}$$

with $\gamma_0 = 0.84$ and $\Gamma(R_A) = 3.28 + 0.0004(K_0/\kappa)R_A^2$. $\Gamma$ is therefore a weakly increasing function of $R_A$.

This result can be understood in terms of the scaling behavior of crystalline and hexatic domains (Kohyama *et al.* 2003). The various scaling regimes of $\Gamma(R_A)$ are summarized in Fig. 17. For a spherical, crystalline bud, the bending energy dominates at sufficiently small radii, which implies $\Gamma = 4$, compare Sec. 3.9.1. As the bud-size increases, the stretching energy increases and gives an additional contribution proportional to $K_0 R_A^2$, see Eq. (47). When the stretching energy of a spherical bud becomes equal or larger than its bending energy, the bud can lower its total energy by deforming into an icosahedral shape with rounded edges. The theory of stretching ridges described in Sec. 2.6.1 then implies that $\Gamma(R) \sim (K_0/\kappa)^{1/6} R^{1/3}$. This asymptotic behavior for very large buds is preceded by a regime which is dominated by the bending energy of the cone-shaped corners of the icosahedron (see Sec. 2.6.2), which implies a logarithmic $R$-dependence of $\Gamma$. Finally, if the bud radius becomes much larger than the buckling radius (see Secs. 2.6.1 and 4.2), the membrane enters the hexatic phase, where $\Gamma(R) \sim (K_H/\kappa)\ln(R)$, with $K_H$ the hexatic stiffness.

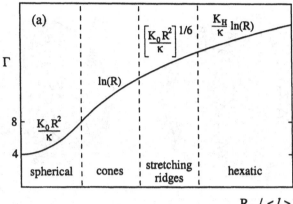

Fig. 17. Schematic plot of the scaling regimes of $\Gamma(R_A)$, compare Eq. (116), which characterizes he dependence of the budding transition on the domain size $R_A$ of a planar crystalline domain. Different scaling regimes are indicated. From Kohyama *et al.* (2003).

The dynamics of defect formation after a quench from a defect-free domain into the budded state can be seen in the snapshots of Fig. 18. The figure clearly demonstrates that the defects first form at the domain boundary and then diffuse into the interior. This result strongly supports the exterior acquisition hypothesis.

Another interesting feature can be seen in Fig. 18c and 18d. The topological disclinations do not appear as isolated defects, but are accompanied by lines of dislocations — which are grain boundaries between two crystal orientations. It has been shown very recently by Bowick *et al.* (2001c) that such defect scars screen the strain field of a disclination and thereby lower the energy of the crystalline membrane on a sphere.

The budding scenario discussed here provides a nice example of how various seemingly disparate aspects of the physics of flexible membranes can be used to provide insight into the formation of clathrin coated pits.

## 5. Membranes of Fluctuating Topology

### 5.1. *Microemulsion and Sponge Phases*

For lipid bilayers, the bending energy is large and membrane fusion or fission is suppressed by a large energy barrier. It is therefore often an excellent approximation to model these systems as membranes of fixed topology. However, when $\kappa/k_B T$, the bending rigidity in thermal units, is sufficiently small, as in many surfactant systems, the membranes can easily change their topology. The essential new control parameters are now the saddle-splay modulus $\bar{\kappa}$, and the membrane volume fraction $\Psi$, i.e. the area $S$ of the membrane times its (constant) thickness $\delta$, divided by the total available volume $V$. Many different phases can be observed as a function of the membrane volume fraction (Gelbart *et al.* 1995).

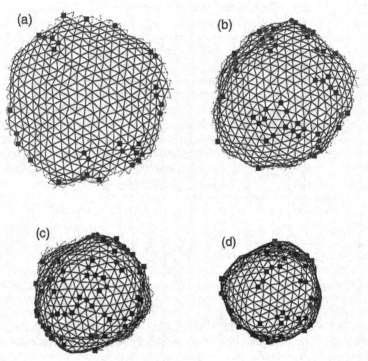

Fig. 18. Snapshots of configurations of a crystalline domain during the budding process after a quench from a defect-free initial state. The pictures show a top view of the crystalline domain; the fluid part of the membrane is not shown. 5-fold and 7-fold coordinated vertices are marked by squares and circles, respectively. The parameters are $N_A = 368$, $N_B = 2442$, $\ell_A/\sigma_0 = 1.50$, $\ell_B/\sigma_0 = 1.68$, $C_0^A\sigma_0 = 0.2$, $C_0^B = 0$, and $\lambda\sigma_0 = 2.0 k_B T$. Snapshots are shown at time (a) $t = 0.1 \times 10^6$, (b) $t = 0.5 \times 10^6$, (c) $t = 2 \times 10^6$, and (d) $t = 3 \times 10^6$ Monte Carlo steps after the quench. From Kohyama et al. (2003).

Examples include several liquid crystalline phases such as the lamellar phase, the hexagonal phase of cylindrical micelles, cubic phases of spherical micelles, and cubic bicontinuous ("plumber's nightmare") phases. Bicontinuous structures, which are characterized by an exponential decay of correlations in the membrane positions, are particularly intriguing. These phases — which are called *microemulsions* — have been observed experimentally in many binary and ternary amphiphilic systems. Microemulsions are macroscopically homogeneous and optically isotropic mixtures of oil, water and amphiphiles. On a mesoscopic scale, they consist of two multiply-connected and intertwined networks of oil- and water-channels which are separated by an amphiphilic monolayer. A similar phase, the *sponge phase*, appears in binary systems of water and amphiphile, where now the two labyrinths are occupied by water, separated by an amphiphilic bilayer. Pictures obtained by freeze-fracture microscopy (Strey *et al.* 1992), where the sample is quickly frozen, cut, and then studied with an electron microscope, are very suggestive in

Fig. 19.   Freeze-fracture microscopy picture of a sponge phase. From Strey *et al.* (1992).

this case, because the sample has a preference to break along the bilayer mid-surface, so that the three-dimensional structure of the membrane becomes visible. An example is shown in Fig. 19, which clearly shows the saddle-like geometry of the amphiphile film. The intuitive picture of microemulsion and sponge phases as fluid versions of bicontinuous cubic phases is strongly supported by these experiments.

A theoretical understanding of the statistical mechanics of such membranes ensembles, however, is only beginning to emerge (Porte 1992; Gompper and Schick 1994; Safran 1994; Morse 1997). This is not surprising, since the statistical mechanics of surfaces which can change not only their shape, but also their topology, is extremely complicated. In principle, a partition function of the form

$$Z = \sum_{\text{topologies}} \int' \mathcal{D}\mathbf{R}(\mathbf{x}) \exp\{-\mathcal{H}[\mathbf{R}(\mathbf{x})]/k_B T\} \qquad (117)$$

has to be calculated, where $\mathcal{D}\mathbf{R}(\mathbf{x})$ denotes the integration over all possible shapes of the surface with parameterization $\mathbf{R}(\mathbf{x})$ at fixed topology, where $\mathbf{x}$ is a two-dimensional coordinate system on the surface. However, this integral is not just over all possible parameterization $\mathbf{R}(\mathbf{x})$ of a surface of fixed topology, but has to be restricted to those parameterizations which lead to physically different shapes in the embedding space; this is indicated by the prime. Finally, the contributions off all different topologies have to be summed over. It is clear that this problem is sufficiently complex that no exact solution will be found anytime soon. Therefore, approximations have to be made in order to get some insight into the behavior of these phases.

## 5.2. Theoretical Predictions

### 5.2.1. Gaussian Random Fields

A very useful approach, which has been worked out by Berk (1987), Berk (1991), Teubner (1991), Pieruschka and Marčelja (1992), Pieruschka and Safran (1993), and Pieruschka and Safran (1995), is to describe membranes as level surfaces of Gaussian Random Fields (GRF). The starting point is a Gaussian free-energy functional of the form

$$\mathcal{H}_0[\phi] = \frac{1}{2} \int d^3q \, \nu(\mathbf{q})^{-1} \Phi(\mathbf{q}) \Phi(-\mathbf{q}), \tag{118}$$

where $\Phi(\mathbf{q})$ is the Fourier transform of a scalar (concentration) field. Using Eq. (118), the average geometry of the $\Phi(\mathbf{r}) = \alpha$ level surfaces can be calculated for arbitrary spectral density $\nu(\mathbf{q})$. In particular, the surface density, $\langle S/V \rangle$, the mean curvature $\langle H \rangle$, the Gaussian curvature $\langle K \rangle$, and the mean curvature squared $\langle H^2 \rangle$ can be calculated exactly (Teubner 1991):

$$\langle S/V \rangle = \frac{2}{\pi} \exp\left[-\frac{\alpha^2}{2}\right] \sqrt{\frac{1}{3}\langle q^2 \rangle_\nu}, \tag{119}$$

$$\langle K \rangle = -\frac{1}{6}\langle q^2 \rangle_\nu (1 - \alpha^2), \tag{120}$$

$$\langle H \rangle = \frac{1}{2}\alpha\sqrt{\frac{\pi}{6}\langle q^2 \rangle_\nu}, \tag{121}$$

$$\langle H^2 \rangle = \langle K \rangle + \frac{1}{5}\frac{\langle q^4 \rangle_\nu}{\langle q^2 \rangle_\nu}, \tag{122}$$

where

$$\langle q^n \rangle_\nu = \int \frac{d^3q}{(2\pi)^3} q^n \nu(\mathbf{q}). \tag{123}$$

Since $\langle H \rangle$ is a linear function of $\alpha$, compare Eq. (121), the level parameter $\alpha$ is proportional to the spontaneous curvature $C_0$. In particular, for $\alpha = 0$ the mean curvature of the surface vanishes; this corresponds to a balanced system, where $C_0 = 0$. The surface density $\langle S/V \rangle$ decreases rapidly with increasing $|\alpha|$.

The GRF-approach is most predictive when the Gaussian model of random interfaces is related to the statistical mechanics of membranes by a variational approximation (Pieruschka and Safran 1993; Pieruschka and Safran 1995). In this case, the spectral density $\nu(\mathbf{q})$ in the functional (118) is determined by the requirement that the $\phi(\mathbf{r}) = 0$ level surfaces mimic the behavior of interfaces controlled by the curvature Hamiltonian (61) as closely as possible. The usual variational approach employs the Feynman–Bogoljubov inequality,

$$F \leq F_0 + \langle \mathcal{H} - \mathcal{H}_0 \rangle_0, \tag{124}$$

where $\mathcal{H}$ and $F$ are the Hamiltonian and the free energy of the system of interest, respectively, and $\mathcal{H}_0$ and $F_0$ the same quantities of the reference system. In this way, an upper bound for the true free energy is obtained. A complication arises in the case of random surfaces because the GRF-Hamiltonian is defined everywhere in space, while the curvature Hamiltonian is only defined on the level surface. The curvature energy therefore does not restrict fluctuations of the field $\Phi(\mathbf{r})$ away from the level surface. In order to suppress such fluctuations, one usually makes the *mean-spherical approximation*, and introduces the constraint $\langle \Phi(\mathbf{r})^2 \rangle = 1$.

In this variational approach, the free energy per unit volume of balanced microemulsions (with $C_0 = 0$) is given by

$$\mathcal{F}/V = \frac{S}{V}[2\kappa\langle H^2 \rangle + \bar{\kappa}\langle K \rangle] - \frac{k_B T}{2} \sum_{\mathbf{q}} \ln \nu(\mathbf{q}), \tag{125}$$

where the first two averages are given by Eqs. (122) and (120), respectively, and the last term is the free energy of the Gaussian random field. The average $\langle \mathcal{H}_0 \rangle_0$ is a constant, independent of $\nu(\mathbf{q})$, and has therefore been omitted from Eq. (125). The functional derivative $\delta\mathcal{F}/\delta\nu(\mathbf{q}) = 0$ of Eq. (125) then gives the functional form (Pieruschka and Safran 1993)

$$\nu(\mathbf{q}) = \frac{a}{q^4 - bq^2 + c} \tag{126}$$

for the spectral density, where the parameters $a$, $b$ and $c$ are now related to the curvature elastic moduli $\kappa$ and $\bar{\kappa}$ and the surface density $S/V$. The first interesting result is that the spectral density is found to be *independent* of $\bar{\kappa}$. Exact expressions for the parameters can be found in Endo *et al.* (2001). To leading order in $\kappa_B T/\kappa$, the parameters are given by

$$a = \frac{15\pi^2}{16} \frac{k_B T}{\kappa} \frac{S}{V}, \quad b = \frac{3}{2}\pi^2 \left(\frac{S}{V}\right)^2, \quad c = \left(\frac{3\pi^2}{4}\right)^2 \left(\frac{S}{V}\right)^4. \tag{127}$$

The spectral density (126) is equivalent to the scattering intensity in bulk contrast.

The free energy of the sponge phase can also be calculated from the GRF approach. To leading order in an expansion in $k_B T/\kappa$, the free-energy density $f = F/V$ is found to be (Pieruschka and Safran 1995)

$$f = \frac{\pi^2}{40}[2\kappa - 5\bar{\kappa}] \left(\delta\frac{S}{V}\right)^3 - \frac{k_B T}{12} \ln\left(\delta\frac{S}{V}\right), \tag{128}$$

where $\delta$ is again the thickness of the amphiphilic interface. This implies that for small membrane volume fractions $\Psi = \delta S/V$, the entropic term dominates over the energy term, and that the sponge phase becomes unstable.

The weak point of this approach is that it is not clear whether the calculated entropy is actually equivalent to the physical conformational entropy of the membranes (Morse 1997). The main problem is that the curvature energy only controls the shape of the $\Phi(\mathbf{r}) = 0$ level surface, while the values of the scalar field $\Phi$ at all other points in space are not affected by it. The fluctuations of $\Phi$ in these oil- and water-regions are mainly determined by the mean-spherical constraint $\langle \Phi^2(\mathbf{r}) \rangle = 1$.

Obviously, an appreciable contribution to the total entropy arises from the fluctuations of $\Phi$ in these 'bulk' regions. This would not affect the predictions of the model if the 'bulk' contributions were independent of the interface positions. Unfortunately, there is currently no argument that this is indeed the case.

### 5.2.2. Small-Scale Membrane Fluctuations, Scale-Dependent Rigidity, and Phase Behavior

Another approach is to start directly from the description of the surfactant membrane as a fluctuating surface with curvature energy (61). Consider again the case of balanced microemulsions and sponge phases, for which the spontaneous curvature vanishes, $C_0 = 0$.

In the absence of thermal fluctuations, the range of stability of the lamellar phase as a function of $\kappa$ and $\bar{\kappa}$ can be deduced from the following argument. The curvature Hamiltonian is written in the form

$$\mathcal{H} = \int dA \left\{ \frac{1}{2}\kappa_+ (C_1 + C_2)^2 + \frac{1}{2}\kappa_- (C_1 - C_2)^2 \right\}, \tag{129}$$

where $\kappa_+ = \kappa + \bar{\kappa}/2$ and $\kappa_- = -\bar{\kappa}/2$. Clearly, both $\kappa_+$ and $\kappa_-$ have to be positive in order for the lamellar phase to be stable, which implies $-2\kappa < \bar{\kappa} < 0$. At $\kappa_+ = 0$ (equivalent to $\bar{\kappa} = -2\kappa$), an instability towards a phase of microscopic vesicles occurs, while at $\kappa_- = 0$ (equivalent to $\bar{\kappa} = 0$), the lamellar phase becomes unstable with respect to a plumber's nightmare phase with a microscopic lattice constant.

It is important to emphasize that for vanishing spontaneous curvature, the phase behavior as a function of the amphiphile volume fraction cannot be understood on the basis of the curvature energy alone — i.e. without considering the effect of thermal fluctuations (Porte 1992). The reason is that the curvature Hamiltonian is conformally invariant in three spatial dimensions; this implies, in particular, that it is invariant under a simultaneous rescaling of all length scales. Since the curvature energy is scale invariant, the energy *density* scales as the third power of an inverse length, i.e. as $(S/V)^3$. Therefore, the curvature energy of any given structure — spherical, cylindrical, lamellar, cubic, or random — scales in exactly the same way with decreasing amphiphile volume fraction, so that there can be no phase transitions. The influence of thermal fluctuations therefore need to be considered in order to understand the phase behavior.

It has been suggested by Safran et al. (1986), Cates et al. (1988), Morse (1994), and Golubović (1994) that the free energy of the sponge phase can be obtained by integrating out the membrane fluctuations on scales less than the typical domain size. This integration over small-scale fluctuations leads to renormalized, scale-dependent curvature moduli

$$\kappa_R(\ell/\delta) = \kappa - \alpha_\kappa \frac{k_B T}{4\pi} \ln(\ell/\delta), \tag{130}$$

$$\bar{\kappa}_R(\ell/\delta) = \bar{\kappa} - \bar{\alpha}_\kappa \frac{k_B T}{4\pi} \ln(\ell/\delta), \tag{131}$$

on length scale $\ell$, with $\alpha_\kappa = 3$ (compare Sec. 3.5.1) and $\bar{\alpha}_\kappa = -10/3$ (David 1989), respectively, and $\ell/\delta = \Psi^{-1}$, where $\Psi$ is the membrane volume fraction. This implies that the curvature Hamiltonian (129) has to be replaced by (Morse 1994; Golubović 1994)

$$F = \int dA \left\{ \frac{1}{2} \kappa_{+,R}(\Psi^{-1})(C_1 + C_2)^2 + \frac{1}{2} \kappa_{-,R}(\Psi^{-1})(C_1 - C_2)^2 \right\} \tag{132}$$

with

$$\kappa_{\pm,R}(\Psi^{-1}) = \kappa_\pm + \alpha_\pm \frac{k_B T}{4\pi} \ln \Psi , \tag{133}$$

where $\alpha_+ = \alpha_\kappa + \bar{\alpha}_\kappa/2 = 4/3$ and $\alpha_- = -\bar{\alpha}_\kappa/2 = 5/3$. The stability argument discussed above now implies that $\kappa_{+,R}$ and $\kappa_{-,R}$ have to be positive for the free energy (132) to be stable against collapse of the structure to molecular scales. There are therefore instabilities at $\kappa_{+,R}(\Psi^{-1}) = 0$ and $\kappa_{-,R}(\Psi^{-1}) = 0$. The latter instability can be identified with the emulsification failure of the sponge phase, so that the phase boundary is predicted to occur at (Morse 1994; Golubović 1994)

$$\ln \Psi = \frac{6\pi}{5} \frac{\bar{\kappa}}{\kappa_B T}. \tag{134}$$

This result can be understood intuitively as follows. For sufficiently large membrane volume fraction, both $\kappa_{+,R}$ and $\kappa_{-,R}$ are positive. The system therefore tries to minimize $(C_1 + C_2)^2$ and $(C_1 - C_2)^2$. This can be achieved by decreasing both $C_1$ and $C_2$, i.e. by swelling a given structure as much as possible — the lamellar phase is stable at this value of $\Psi$. On the other hand, as soon as $\kappa_{+,R}$ or $\kappa_{-,R}$ become negative at some small value of $\Psi$, the free energy can be reduced by collapsing the structure. With decreasing length, however, $\kappa_{+,R}$ and $\kappa_{-,R}$ increase and finally become positive. The collapse therefore stops at the length scale determined by Eq. (134).

The renormalization of $\kappa$ and $\bar{\kappa}$ implies that the free-energy density, $f$, of the sponge phase should behave as (Roux *et al.* 1990; Porte *et al.* 1991; Roux *et al.* 1992)

$$f = (A + B \ln \Psi) \Psi^3. \tag{135}$$

As explained above, the term proportional to $\Psi^3$ comes from the conformal invariance of the curvature energy. $A$ represents the contribution of the curvature energy without thermal fluctuations, and is a linear function of both $\kappa$ and $\bar{\kappa}$. $B$ represents the logarithmic corrections from the renormalization and is a linear function of temperature. Both $A$ and $B$ depend on the detailed geometrical structure of the sponge phase, and thus cannot be obtained from simple scaling arguments. It is important to note that the functional dependence of the free energy (135), which is based on the renormalization of the curvature elastic moduli due to *small-scale* membrane fluctuations, does not agree with the free energy (128) of the Gaussian random field model, which includes the topological entropy of a disordered bicontinuous phase.

## 5.3. *Triangulated-Surface Models for Membranes with Fluctuating Topology*

In order to use triangulated-surface models to study microemulsions, a new type of Monte Carlo step — in addition to the vertex moves and bond flips discussed above — has to be implemented. This topology-change step is shown schematically in Fig. 20. It consists of either inserting or cutting minimal necks between two distinct membrane segments.

Let the variable $s$ characterize the state of the system. The probability of state $s$ is $P(s) \sim \exp[-\mathcal{H}(s)/k_B T]$, where $\mathcal{H}$ is the curvature energy of the system in state $s$. Two states $s$ and $s'$ are linked by a transition probability $w(s \to s')$ of going from state $s$ to $s'$. The condition of "microscopic reversibility"

$$P(s)w(s \to s') = P(s')w(s' \to s) \tag{136}$$

is then sufficient to guarantee that the systems reaches the equilibrium state characterized by the Gibbs ensemble. For topology changes, Gompper and Kroll (1998) have used a Metropolis algorithm with

$$w(s \to s') = \begin{cases} a_{ss'} & \text{for } a_{s's}P(s') \geq a_{ss'}P(s) \\ a_{ss'}[a_{s's}P(s')]/[a_{ss'}P(s)] & \text{for } a_{s's}P(s') < a_{ss'}P(s) \end{cases} \tag{137}$$

The stochastic matrix $a_{ss'}$ for making a neck starting in state $s$ is $a_{ss'} = 2p/N_t(s)$; for destroying a neck when in state $s'$, it is $a_{s's} = 6(1-p)/N_t(s')$. Here, $p$ is the probability that we try to create a neck once we have chosen a surface triangle, and $(1-p)$ is the probability that we try to destroy a neck. In the first case, the factor 2 arises since when trying to create a neck, there are two surface triangles that would lead to the same result; in the second, the factor is 6 since there are six surface

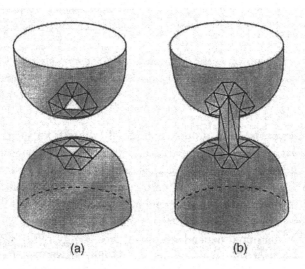

(a)                    (b)

Fig. 20. Monte Carlo step for changing the topology of a triangulated surface. (a) Two surface triangles are removed. (b) The two holes are connected by a prism of six triangles.

triangles in a neck, and the choice of any of the six will lead to the destruction
of the neck with equal probability. The choice (137) guarantees that Eq. (136) is
fulfilled. In the Metropolis algorithm, the move from $s$ to $s'$ should be accepted with
a probability (Allen and Tildesley 1992; Nicholson and Parsonage 1982)

$$\min \left( 1, \frac{a_{s's} P(s')}{a_{ss'} P(s)} \right). \tag{138}$$

In this algorithm, a surface triangle is chosen at random, and one attempts to
make a neck three times as often as one tries to destroy a neck. This means that
$p = 3/4$. When attempting to create a neck in the move $s \to s'$, the move is accepted
with the probability

$$\min \left( 1, \frac{N_t(s)}{N_t(s')} \exp[(\mathcal{H}(s) - \mathcal{H}(s'))/k_B T] \right), \tag{139}$$

where $N_t(s') = N_t(s) + 4$. Similarly, when attempting to destroy a neck, the move
is accepted with the probability (139), but where now $N_t(s') = N_t(s) - 4$.

In a simulation with a fixed number of vertices and variable topology, the number
of triangles and bonds must fluctuate as a consequence of Euler's theorem. This
implies that the number of attempted bond moves per Monte Carlo steps must now
adjusted in such a way that the probability for a bond-flip attempt is constant,
independent of the topology. This can easily be done by relating the number of
bond flips to the actual total number of bonds of each configuration.

### 5.4. *Simulation Results*

A typical configuration of a triangulated surface in a cubic box with parameters $\kappa$, $\bar{\kappa}$
and $\Psi$ chosen in the stability region of the microemulsion or sponge phase is shown in
Fig. 21. This configuration clearly illustrates the bicontinuous structure of balanced
microemulsions and sponge phases. The saddle-like geometry of the membrane can
also be easily seen. Finally, the figure shows that *locally*, the structure of the sponge
phase strongly resembles that of the cubic phases discussed above. A sponge phase
should therefore be considered as the molten state of the crystalline cubic phase.

A more quantitative comparison with the theoretical approaches of Secs. 5.2.1
and 5.2.2 can be made by determining the phase diagram of the randomly-
triangulated surface model, and by measuring the osmotic pressure, $p$, in the simulations as a function of the membrane volume fraction $\Psi$. From Eq. (135), we
obtain

$$p\delta^3/k_B T \equiv \frac{1}{k_B T}[\Psi \partial f/\partial \Psi - f] = [(2A + B) + 2B \ln \Psi]\Psi^3, \tag{140}$$

i.e. the same functional dependence as the free-energy density itself. This dependence has been confirmed by the simulation data (Gompper and Kroll 1998). The
simulations therefore provide strong evidence for the renormalization of the elastic
moduli of the curvature model and for the dependence (135) of the free energy on
the membrane volume fraction.

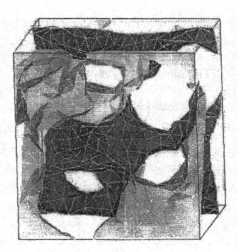

Fig. 21. A typical membrane configuration in a sponge phase for bending rigidity $\tau/k_B T = 3.0$ (where $\tau$ is the coefficient of the discretized bending energy (70)), and $\bar{\kappa}/k_B T = -0.7$. The two sides of the membrane are shaded differently in order to emphasize the bicontinuous structure of this phase. From Gompper and Kroll (1998).

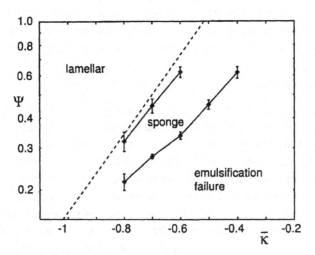

Fig. 22. The phase diagram as a function of membrane volume fraction $\Psi$ and saddle-splay modulus $\bar{\kappa}$, for bending rigidity $\tau/k_B T \simeq 3.0$. Note the logarithmic scale of the ordinate. The dashed line shows the theoretical prediction (134). From Gompper and Kroll (1998).

The phase diagram for fixed bending rigidity $\kappa$ is shown as a function of $\bar{\kappa}$ and $\Psi$ in Fig. 22. The simulation data are compared with the prediction (134) for the phase boundary. Since the slopes of the phase boundaries in this logarithmic plot are in good agreement, both the exponential dependence of the membrane volume fraction on the saddle splay modulus at the transition, as well as the value of the universal prefactor in Eq. (134) are confirmed.

## 5.5. *Comparison with Experiments*

Experimentally, phase diagrams and scattering intensities have been studied systematically for many different surfactant molecules. Qualitatively, the agreement with theory is very reasonable. For example, the scattering intensity in bulk contrast shows a peak at non-zero wave vector in the microemulsion phase, which moves out and decreases in height with increasing surfactant concentration. A quantitative comparison, however, is much more difficult. We discuss here three different classes of experiments where a such a quantitative comparison has been made.

The first type of experiments are scattering studies in bulk and film contrast. In *bulk* contrast, the scattering intensity $S_{ww}(\mathbf{q})$ is proportional to the spectral density $\nu(\mathbf{q})$ of the Gaussian random field model, compare Eq. (126), for wave vectors $q$ which are not much larger than the characteristic wave vector $k$ of the domain structure. The functional dependence of Eq. (126) describes the scattering data in this regime very well, as was first noted by Teubner and Strey (1987), who derived this result on the basis of a Ginzburg–Landau model. In *film* contrast, the scattering intensity $S_{ff}(\mathbf{q})$ is given in the limit of wave vector $q \to 0$ by (Porte *et al.* 1991)

$$S_{ff}(q \to 0) \sim \Psi \left( \frac{\partial p}{\partial \Psi} \right)^{-1}, \tag{141}$$

where $p$ is the osmotic pressure of Eq. (140). For the free energy (135), this implies

$$[\Psi S_{ff}(q \to 0)]^{-1} \sim \text{const} + \ln \Psi. \tag{142}$$

Such behavior has indeed been observed experimentally by Porte *et al.* (1991). However, this result has been questioned by Daicic *et al.* (1995b) and Daicic *et al.* (1995a). This has lead to a intensive debate, with arguments against and in favor (Roux *et al.* 1996; Porte *et al.* 1997) of the existence of a logarithmic renormalization of elastic moduli in the sponge phase.

In a second type of experiment, information about the average geometry of the surfactant film can be extracted from the scattering intensity in the regime $k < q \ll 1/\delta$. This information is contained in the corrections to the asymptotic $1/q^4$ law for smaller values of the wave vector (Teubner 1990). Experimentally, the average Gaussian curvature is found to be (Chen *et al.* 1996)

$$(V/S)^2 \langle K \rangle = 1.25 \pm 0.10, \tag{143}$$

which is in excellent agreement with the Gaussian random field result $(V/S)^2 \langle K \rangle = -\pi^2/8 = 1.23$, compare Sec. 5.2.1.

The third type of experiment concerns the phase behavior of mixtures of water and non-ionic surfactant as a function of temperature and surfactant concentration. On the basis of a model in which the bilayer is described by two parallel monolayer with spontaneous curvature $C_0^{(\text{mono})}$, it has been shown that $\bar{\kappa}^{(bi)} = 2\bar{\kappa}^{(\text{mono})} + 4\kappa^{(\text{mono})}C_0^{(\text{mono})}\delta$ where $\delta$ is the monolayer separation (Porte 1992). Since $C_0^{(\text{mono})} \sim (T - \bar{T})$ is observed experimentally (Strey 1994), the saddles play modulus $\bar{\kappa}$ should be a linear function of temperature in this case. Therefore, the concentrations at the phase boundaries of the lamellar and the sponge phase are expected from Eq. (134) to depend exponentially on temperature. A logarithmic plot of the phase diagram of $C_{12}E_5$ in water (Strey *et al.* 1990) is indeed consistent with this expectation.

The most detailed information about both the scattering intensities and phase behavior in these systems has been obtained very recently in ternary amphiphilic systems of water, oil and non-ionic surfactant $C_iE_j$, to which small traces of an amphiphilic block copolymer has been added (Jakobs *et al.* 1999; Endo *et al.* 2000; Endo *et al.* 2001; Gompper *et al.* 2001a; Gompper *et al.* 2001b). The results obtained in this system provide further, strong evidence for the existence of a logarithmic renormalization of $\kappa$ and $\bar{\kappa}$ in microemulsions and sponge phases.

## 6. Summary and Outlook

There has been enormous progress in the understanding of the behavior of fluctuating membranes in recent years. This is due to the fruitful interaction of experiment, theory and computer simulations.

Triangulated-surface models describe the behavior of membranes on length scales which are larger than the membrane thickness (Goetz *et al.* 1999). Surfaces with fixed connectivity describe polymerized membranes. When the triangulation becomes dynamic, the model describes fluid or crystalline membranes — depending on the tether length. Finally, with insertion and deletion of minimal necks, membranes of fluctuating topology can be modelled. Also, the vertices or triangles can have internal degrees of freedom, like a composition variable in the case of multi-component membranes. Triangulated surfaces have therefore proven to be very versatile, and should find application in studies of a wide range of problems in membrane physics and biology.

While the behavior of single-component membranes has been investigated in considerable detail, with many properties now well understood, several important properties still have to be clarified. For example, detailed field-theoretical calculations (David and Wiese 1996; Wiese and David 1997; David and Wiese 1998) predict a crumpled phase of self-avoiding polymerized membranes which has never been observed in simulations. Another example is the renormalization of the bending rigidity, where undulation modes for vesicles with considerably larger vertex number have to be simulated to unambiguously confirm the membrane softening on larger length scales.

Most importantly, however, triangulated surface models have reached a level of maturity where they can be used to study experimentally relevant questions. Examples are the spectrum of undulation modes of non-spherical vesicles, the budding of clathrin-coated pits, and the micropipette aspiration of red blood cells.

With increased computer power, these simulations can be extended to larger and more complex systems. This complexity involves not only internal degrees of freedom of the membrane, but also their interactions with integrated macromolecules and solvent mediated interactions with external objects and flows.

## Acknowledgments

DMK acknowledges support from the National Science Foundation under Grant No. DMR-0083219 and the donors of The Petroleum Research Fund, administered by the ACS.

## References

Abraham F F, *Phys. Rev. Lett.* **67**, 1669 (1991).

Abraham F F, in *Microscopic Simulations of Complex Hydrodynamic Phenomena*, eds. M Mareschal and B L Holian, pp. 361–401. Plenum Press, New York, 1992.

Abraham F F and Goulian M, *Europhys. Lett.* **19**, 293 (1992).

Abraham F F and Nelson D R, *Science* **249**, 393 (1990a).

Abraham F F and Nelson D R, *J. Phys. France* **51**, 2653 (1990b).

Abraham F F, Rudge W E and Plischke M, *Phys. Rev. Lett.* **62**, 1757 (1989).

Allen M P and Tildesley D J, *Computer Simulation of Liquids*. Clarendon Press, Oxford, 1992.

Ambjørn J, Durhuus B and Fröhlich J, *Nucl. Phys. B* **257**, 433 (1985).

Ambjørn J, Durhuus B and Jonsson T, *Nucl. Phys. B* **316**, 526 (1989).

Aronowitz J A and Lubensky T C, *Europhys. Lett.* **4**, 395 (1987).

Aronowitz J A and Lubensky T C, *Phys. Rev. Lett.* **60**, 2634 (1988).

Baig M, Espriu D and Travesset A, *Nucl. Phys. B* **426**, 575 (1994).

Baig M, Espriu D and Wheater J F, *Nucl. Phys. B* **314**, 609 (1989).

Baillie C F and Johnston D A, *Phys. Lett. B* **283**, 55 (1992).

Baillie C F, Johnston D A and Williams R D, *Nucl. Phys. B* **335**, 469 (1990).

Barsky S J and Plischke M, *Phys. Rev. E* **50**, 3911 (1994).

Baumgärtner A, *J. Phys. I France* **1**, 1549 (1991).

Baumgärtner A, *J. Chem. Phys.* **98**, 7496 (1993).

Baumgärtner A and Ho J S, *Phys. Rev. A* **41**, 5747 (1990).

Baumgärtner A and Renz W, *Europhys. Lett.* **17**, 381 (1992).

Berk N F, *Phys. Rev. Lett.* **58**, 2718 (1987).

Berk N F, *Phys. Rev. A* **44**, 5069 (1991).

Billoire A and David F, *Nucl. Phys. B* **275**, 617 (1986).

Billoire A, Gross D J and Marinari E, *Phys. Lett. B* **139**, 75 (1984).

Boal D H, *Biophys. J.* **67**, 521 (1994).

Boal D H, *Mechanics of the Cell*. Cambridge University Press, Cambridge, 2002.

Boal D H and Boey S K, *Biophys. J.* **69**, 372 (1995).

Boal D H, Levinson E, Liu D and Plischke M, *Phys. Rev. A* **40**, 3292 (1989).

Boal D H and Rao M, *Phys. Rev. A* **45**, R6947 (1992a).

Boal D H and Rao M, *Phys. Rev. A* **46**, 3037 (1992b).

Boal D H, Seifert U and Shillcock J C, *Phys. Rev. E* **48**, 4274 (1993).

Boal D H, Seifert U and Zilker A, *Phys. Rev. Lett.* **69**, 3405 (1992).

Boey S K, Boal D H and Discher D E, *Biophys. J.* **75**, 1573 (1998).

Boulatov D V, Kazakov V A, Kostov I K and Migdal A A, *Nucl. Phys. B* **275**, 641 (1986).

Bowick M J, Cacciuto A, Thorleifsson G and Travesset A, *Phys. Rev. Lett.* **87**, 148103 (2001a).

Bowick M J, Cacciuto A, Thorleifsson G and Travesset A, *Eur. Phys. J. E* **5**, 149 (2001b).

Bowick M J, Nelson D R and Travesset A, *Phys. Rev. B* **62**, 8738 (2001c).

Bowick M J and Travesset A, *Phys. Rep.* **344**, 255 (2001).

Cai W, Lubensky T C, Nelson P and Powers T, *J. Phys. II France* **4**, 931 (1994).

Camacho C J, Fisher M E and Singh R R P, *J. Chem. Phys.* **94**, 5693 (1991).

Canham P B, *J. Theor. Biol.* **26**, 61 (1970).

Cantat I and Misbah C, *Phys. Rev. Lett.* **83**, 880 (1999).

Cates M E, Roux D, Andelman D, Milner S T and Safran S A, *Europhys. Lett.* **5**, 733 (1988).

Cevc G, Blume G, Schätzlein A, Gebauer D and Paul A, *Adv. Drug Delivery Rev.* **18**, 349 (1996).

Chen S H, Lee D D, Kimishima K, Jinnai H and Hashimoto T, *Phys. Rev. E* **54**, 6526 (1996).

Daicic J, Olsson U, Wennerström H, Jerke G and Schurtenberger P, *Phys. Rev. E* **52**, 3266 (1995a).

Daicic J, Olsson U, Wennerström H, Jerke G and Schurtenberger P, *J. Phys. II France* **5**, 199 (1995b).

Dammann B, Fogeby H C, Ipsen J H and Jeppesen C, *J. Phys. I France* **4**, 1139 (1994).

David F, in *Statistical Mechanics of Membranes and Surfaces*, eds. D R Nelson, T Piran and S Weinberg, pp. 157–223. World Scientific, Singapore, 1989.

David F, Ginsparg P and Zinn-Justin J (eds.), *Fluctuating geometries in statistical mechanics and field theory*. Elsevier, Amsterdam, 1996.

David F, Guitter E and Peliti L, *J. Phys. France* **48**, 2059 (1987).

David F and Leibler S, *J. Phys. II France* **1**, 959 (1991).

David F and Wiese K J, *Phys. Rev. Lett.* **76**, 4564 (1996).

David F and Wiese K J, *Nucl. Phys. B* **535**, 555 (1998).

de Gennes P G, *J. Chem. Phys.* **60**, 5030 (1974).

Deem M W and Nelson D R, *Phys. Rev. E* **53**, 2551 (1996).

DiDonna B A, *Phys. Rev. E* **66**, 016601 (2002).

DiDonna B A and Witten T A, *Phys. Rev. Lett.* **87**, 206105 (2001).

DiDonna B A, Witten T A, Venkataramani S C and Kramer E M, *Phys. Rev. E* **65**, 016603 (2001).

Discher D E, Boal D H and Boey S K, *Biophys. J.* **75**, 1584 (1998).

Discher D E, Mohantes N and Evans E A, *Science* **266**, 1032 (1994).

Döbereiner H G, Evans E, Kraus M, Seifert U and Wortis M, *Phys. Rev. E* **55**, 4458 (1997).

Döbereiner H G, Gompper G, Haluska C, Kroll D M, Petrov P G and Riske K A, *Phys. Rev. Lett.* **91**, 048301 (2003).

Döbereiner H G, Selchow O and Lipowsky R, *Eur. Biophys. J.* **28**, 174 (1999).

Duplantier B, *Phys. Rev. Lett.* **58**, 2733 (1987).

Duwe H P, Käs J and Sackmann E, *J. Phys. France* **51**, 945 (1990).

Elgsaeter A, Stokke B T, Mikkelsen A and Branton D, *Science* **234**, 1217 (1986).

Endo H, Allgaier J, Gompper G, Jakobs B, Monkenbusch M, Richter D, Sottmann T and Strey R, *Phys. Rev. Lett.* **85**, 102 (2000).

Endo H, Mihailescu M, Monkenbusch M, Allgaier J, Gompper G, Richter D, Jakobs B, Sottmann T, Strey R and Grillo I, *J. Chem. Phys.* **115**, 580 (2001).

Espriu D, *Phys. Lett. B* **194**, 271 (1987).

Evans E and Rawicz W, *Phys. Rev. Lett.* **64**, 2094 (1990).

Evans E A, *Biophys. J.* **14**, 923 (1974).

Falcioni M, Bowick M J, Guitter E and Thorleifsson G, *Europhys. Lett.* **38**, 67 (1997).

Förster D, *Phys. Lett. A* **114**, 115 (1986).

Gefen Y, Aharony A and Alexander S, *Phys. Rev. Lett.* **50**, 77 (1983).

Gelbart W M, Ben-Shaul A and Roux D (eds.), *Micelles, Membranes, Microemulsions, and Monolayers*. Springer-Verlag, Berlin, 1995.

Glaser M A and Clark N A, *Adv. Chem. Phys.* **83**, 543 (1993).

Goetz R, Gompper G and Lipowsky R, *Phys. Rev. Lett.* **82**, 221 (1999).

Golubović L, *Phys. Rev. E* **50**, R2419 (1994).

Gompper G, Endo H, Mihailescu M, Allgaier J, Monkenbusch M, Richter D, Jakobs B, Sottmann T and Strey R, *Europhys. Lett.* **56**, 683 (2001a).

Gompper G and Goos J, *Phys. Rev. E* **50**, 1325 (1994).

Gompper G and Kroll D M, *J. Phys. I France* **1**, 1411 (1991a).

Gompper G and Kroll D M, *Europhys. Lett.* **15**, 783 (1991b).

Gompper G and Kroll D M, *Europhys. Lett.* **19**, 581 (1992a).

Gompper G and Kroll D M, *Phys. Rev. A* **46**, 7466 (1992b).

Gompper G and Kroll D M, *Phys. Rev. Lett.* **71**, 1111 (1993).

Gompper G and Kroll D M, *Phys. Rev. Lett.* **73**, 2139 (1994).

Gompper G and Kroll D M, *Phys. Rev. E* **52**, 4198 (1995a).

Gompper G and Kroll D M, *Phys. Rev. E* **51**, 514 (1995b).

Gompper G and Kroll D M, *J. Phys. I France* **6**, 1305 (1996).

Gompper G and Kroll D M, *Phys. Rev. Lett.* **78**, 2859 (1997a).

Gompper G and Kroll D M, *J. Phys. I France* **7**, 1369 (1997b).

Gompper G and Kroll D M, *J. Phys. Condens. Matter* **9**, 8795 (1997c).

Gompper G and Kroll D M, *Phys. Rev. Lett.* **81**, 2284 (1998).

Gompper G and Kroll D M, *Eur. Phys. J. E* **1**, 153 (2000).

Gompper G and Kroll D M, *Europhys. Lett.* **58**, 60 (2002).

Gompper G, Richter D and Strey R, *J. Phys. Condens. Matter* **13**, 9055 (2001b).

Gompper G and Schick M, in *Phase Transitions and Critical Phenomena*, edited by C Domb and J Lebowitz, vol. 16, pp. 1–176. Academic Press, London, 1994.

Goulian M, *J. Phys. II France* **1**, 1327 (1991).

Góźdź W T and Gompper G, *Phys. Rev. Lett.* **80**, 4213 (1998).

Góźdź W T and Gompper G, *Europhys. Lett.* **55**, 587 (2001).

Grest G S, *J. Phys. I France* **1**, 1695 (1991).

Grest G S and Murat M, *J. Phys. I France* **51**, 1415 (1990).

Grest G S and Murat M, in *Monte Carlo and Molecular Dynamics Simulations in Polymer Sciences*, ed. K Binder. Oxford University Press, New York, 1995.

Gross D J, *Phys. Lett. B* **139**, 187 (1984).

Guitter E and Kardar M, *Europhys. Lett.* **13**, 441 (1990).

Guitter E and Palmeri J, *Phys. Rev. A* **45**, 734 (1992).

Halperin B I and Nelson D R, *Phys. Rev. Lett.* **41**, 121 (1978).

Harden J L and MacKintosh F C, *Europhys. Lett.* **28**, 495 (1994).

Havlin S, Djordjevic Z V, Majid I, Stanley H E and Weis G H, *Phys. Rev. Lett.* **53**, 178 (1984).

Heinrich V, Sevšek F, Svetina S and Žekš B, *Phys. Rev. E* **55**, 1809 (1997).

Helfrich W, *Z. Naturforsch.* **28c**, 693 (1973).

Helfrich W, *Phys. Lett. A* **50**, 115 (1974).

Helfrich W, *J. Phys. France* **46**, 1263 (1985).

Helfrich W, *J. Phys. France* **47**, 321 (1986).

Helfrich W, in *Liquids at interfaces*, edited by J Charvolin, J F Joanny and J Zinn-Justin, Amsterdam. Les Houches, Session XLVIII, 1988, Elsevier, 1990.

Helfrich W, *Eur. Phys. J. B* **1**, 481 (1998).

Heuser J, *J. Cell Biol.* **108**, 401 (1989).

Ho J S and Baumgärtner A, *Phys. Rev. Lett.* **63**, 1324 (1989).

Ho J S and Baumgärtner A, *Europhys. Lett.* **12**, 295 (1990).

Hwa T, *Phys. Rev. A* **41**, 1751 (1990).

Ipsen J H and Jeppesen C, *J. Phys. I France* **5**, 1563 (1995).

Itzykson C, in *Proceedings of the GIFT Seminar, Jaca 85*, eds. J Abad, M Asorey and A Cruz, pp. 130–188. World Scientific, Singapore, 1986.

Jakobs B, Sottmann T, Strey R, Allgaier J, Willner L and Richter D, *Langmuir* **15**, 6707 (1999).

Jeppesen C and Ipsen J H, *Europhys. Lett.* **22**, 713 (1993).

Jülicher F, *Die Morphologie von Vesikeln*. Ph.D. thesis, Universität zu Köln (1994).

Jülicher F and Lipowsky R, *Phys. Rev. Lett.* **70**, 2964 (1993).

Jülicher F and Lipowsky R, *Phys. Rev. E* **53**, 2670 (1996).

Kantor Y, Kardar M and Nelson D R, *Phys. Rev. Lett.* **57**, 791 (1986).

Kantor Y, Kardar M and Nelson D R, *Phys. Rev. A* **35**, 3056 (1987).

Kantor Y and Kremer K, *Phys. Rev. E* **48**, 2490 (1993).

Kantor Y and Nelson D R, *Phys. Rev. Lett.* **58**, 2774 (1987).

Kardar M and Nelson D R, *Phys. Rev. Lett.* **58**, 1289 (1987).

Kazakov V A, Kostov I K and Migdal A A, 1985. *Phys. Lett. B* **157**, 295 (1985).

Kleinert H, *Phys. Lett. A* **114**, 263 (1986).

Kohyama T, Kroll D M and Gompper G, *Phys. Rev. E* **68**, 061905 (2003).

Komura S and Baumgärtner A, *J. Phys. France* **51**, 2395 (1990).

Kosterlitz J M and Thouless D J, *J. Phys. C* **6**, 1181 (1973).

Kramer E M and Witten T A, *Phys. Rev. Lett.* **78**, 1303 (1997).

Kraus M, Wintz W, Seifert U and Lipowsky R, *Phys. Rev. Lett.* **77**, 3685 (1996).

Kroll D M and Gompper G, *Science* **255**, 968 (1992a).

Kroll D M and Gompper G, *Phys. Rev. A* **46**, 3119 (1992b).

Kroll D M and Gompper G, *J. Phys. I France* **3**, 1131 (1993).

Kroll D M and Gompper G, *J. Chem. Phys.* **102**, 9109 (1995).

Kumar P B S, Gompper G and Lipowsky R, *Phys. Rev. Lett.* **86**, 3911 (2001).

Kumar P B S and Rao M, *Mol. Cryst. Liq. Cryst.* **288**, 105 (1996).

Kumar P B S and Rao M, *Phys. Rev. Lett.* **80**, 2489 (1998).

Le Doussal P, *J. Phys. A* **25**, L469 (1992).

Le Doussal P and Radzihovsky L, *Phys. Rev. Lett.* **69**, 1209 (1992).

Lenz P and Nelson D R, *Phys. Rev. Lett.* **87**, 125703 (2001).

Lenz P and Nelson D R, *Phys. Rev. E* **67**, 031502 (2003).

Lidmar J, Mirny L and Nelson D R, *Phys. Rev. E* **68**, 051910 (2003).

Lipowsky R, *Nature* **349**, 475 (1991).

Lipowsky R, *J. Phys. II France* **2**, 1825 (1992).

Lipowsky R, *Biophys. J.* **64**, 1133 (1993).

Lipowsky R and Giradet M, *Phys. Rev. Lett.* **65**, 2893 (1990).

Lipowsky R and Sackmann E (eds.), *Structure and dynamics of membranes — from cells to vesicles*, vol. 1 of *Handbook of biological physics*. Elsevier, Amsterdam, 1995.

Litster J D, *Phys. Lett. A* **53**, 193 (1975).

Lobkovsky A, *Phys. Rev. E* **53**, 3750 (1996).

Lobkovsky A, Gentges S, Li H, Morse D and Witten T A, *Science* **270**, 1482 (1995).

Lobkovsky A and Witten T A, *Phys. Rev. E* **55**, 1577 (1997).

Maggs A C, Leibler S, Fisher M E and Camacho C J, *Phys. Rev. A* **42**, 691 (1990).

Mashl R J and Bruinsma R F, *Biophys. J.* **74**, 2862 (1998).

Matan K, Williams R B, Witten T A and Nagel S R, *Phys. Rev. Lett.* **88**, 076101 (2002).

McDonald I R, *Mol. Phys.* **23**, 41 (1972).

Meleard P, Gerbeaud C, Pott T, Fernandezpuente L, Bivas I, Mitov M D, Dufourcq J and Bothorel P, *Biophys. J.* **72**, 2616 (1997).

Mermin N D, *Phys. Rev.* **176**, 250 (1968).

Miao L, Seifert U, Wortis M and Döbereiner H G, *Phys. Rev. E* **49**, 5389 (1994).

Milner S T and Safran S A, *Phys. Rev. A* **36**, 4371 (1987).

Morse D C, *Phys. Rev. E* **50**, R2423 (1994).

Morse D C, *Curr. Opin. Coll. Interface Sci.* **2**, 365 (1997).

Morse D C and Milner S T, *Phys. Rev. E* **52**, 5918 (1995).

Nelson D, Piran T and Weinberg S (eds.), *Statistical Mechanics of Membranes and Surfaces*. World Scientific, Singapore, 1989.

Nelson D R, in *Statistical Mechanics of Membranes and Surfaces*, eds. D Nelson, T Piran and S Weinberg, pp. 137–155. World Scientific, Singapore, 1989.

Nelson D R, In *Fluctuating Geometries in Statistical Mechanics and Field Theory*, eds. F David, P Ginsparg and J Zinn-Justin, pp. 423–477. North-Holland, Amsterdam, 1996.

Nelson D R and Halperin B I, *Phys. Rev. B* **19**, 2457 (1979).

Nelson D R and Peliti L, *J. Phys (Paris)* **48**, 1085 (1987).

Nicholson D and Parsonage N, *Computer Simulation and the Statistical Mechanics of Adsorption*. Academic Press, London, 1982.

Paczuski M, Kardar M and Nelson D R, *Phys. Rev. Lett.* **60**, 2638 (1988).

Park J M, *Phys. Rev. E* **54**, 5414 (1996).

Park J M and Lubensky T C, *J. Phys. I France* **6**, 493 (1996a).

Park J M and Lubensky T C, *Phys. Rev. E* **53**, 2665 (1996b).

Park J M and Lubensky T C, *Phys. Rev. E* **53**, 2648 (1996c).

Parrinello M and Rahman A, *Phys. Rev. Lett.* **45**, 1196 (1980).

Parrinello M and Rahman A, *J. Appl. Phys.* **52**, 7182 (1981).

Peliti L and Leibler S, *Phys. Rev. Lett.* **54**, 1690 (1985).

Petsche I B and Grest G S, *J. Phys. I France* **3**, 1741 (1993).

Pieruschka P and Marčelja S, *J. Phys. II France* **2**, 235 (1992).

Pieruschka P and Safran S A, *Europhys. Lett.* **22**, 625 (1993).

Pieruschka P and Safran S A, *Europhys. Lett.* **31**, 207 (1995).

Pinnow H A and Helfrich W, *Eur. Phys. J. E* **3**, 149 (2000).

Plischke M and Fourcade B, *Phys. Rev. A* **41**, 2056 (1991).

Porte G, *J. Phys.: Condens. Matter* **4**, 8649 (1992).

Porte G, Appell J and Marignan J, *Phys. Rev. E* **56**, 1276 (1997).

Porte G, Delsanti M, Billard I, Skouri M, Appell J, Marignan J and Debeauvais F, *J. Phys. II France* **1**, 1101 (1991).

Roux D, Cates M E, Olsson U, Ball R C, Nallet F and Bellocq A M, *Europhys. Lett.* **11**, 229 (1990).

Roux D, Coulon C and Cates M E, *J. Phys. Chem.* **96**, 4174 (1992).

Roux D, Nallet F, Coulon C and Cates M E, *J. Phys. II France* **6**, 91 (1996).

Sackmann E, Eggl P, Fahn C, Bader H, Ringsdorf H and Schollmeier M, *Ber. Bunsenges. Phys. Chem.* **89**, 1198 (1985).

Safran S A, *Statistical Thermodynamics of Surfaces, Interfaces, and Membranes*. Addison-Wesley, Reading, MA, 1994.

Safran S A, *Adv. Phys.* **48**, 395 (1999).

Safran S A, Roux D, Cates M E and Andelman D, *Phys. Rev. Lett.* **57**, 491 (1986).

Schmidt C F, Svoboda K, Lei N, Petsche I B, Berman L E, Safinya C R and Grest G S, *Science* **259**, 952 (1993).

Seifert U, *Z. Phys. B* **97**, 299 (1995).

Seifert U, *Adv. Phys.* **46**, 13 (1997).

Seifert U, *Phys. Rev. Lett.* **83**, 876 (1999).

Seifert U, Berndl K and Lipowsky R, *Phys. Rev. A* **44**, 1182 (1991).

Seung H S and Nelson D R, *Phys. Rev. A* **38**, 1005 (1988).

Shillcock J C and Boal D H, *Biophys. J.* **71**, 317 (1996).

Shillcock J C and Seifert U, *Biophys. J.* **74**, 1754 (1998).

Steck T L, in *Cell Shape: Determinants, Regulation, and Regulatory Role*, eds. W Stein
    and F Bonner, pp. 205–246. Academic Press, New York, 1989.

Strandburg K J, *Phys. Rev. B* **34**, 3536 (1986).

Strey R, *Colloid and Polymer Sci.* **272**, 1005 (1994).

Strey R, Jahn W, Skouri M, Porte G, Marignan J and Olsson U, in *Structure and
    Dynamics of Strongly Interacting Colloids and Supramolecular Aggregates in Solution*,
    eds. S H Chen, J S Huang and P Tartaglia, pp. 351–363. Kluwer, Dordrecht, 1992.

Strey R, Schomäcker R, Roux D, Nallet F and Olsson U, *J. Chem. Soc. Faraday Trans.*
    **86**, 2253 (1990).

Sukumaran S and Seifert U, *Phys. Rev. E* **64**, 011916 (2001).

Teubner M, *J. Chem. Phys.* **92**, 4501 (1990).

Teubner M, *Europhys. Lett.* **14**, 403 (1991).

Teubner M and Strey R, *J. Chem. Phys.* **87**, 3195 (1987).

van de Ven T G M, *Colloidal Hydrodynamics*. Academic Press, London, 1989.

Wheater J F, *J. Phys. A* **27**, 3323 (1994).

Wiese K J, in *Phase Transitions and Critical Phenomena*, eds. C Domb and J Lebowitz,
    vol. 19, pp. 253–480. Academic Press, London, 2001.

Wiese K J and David F M, *Nucl. Phys. B* **487**, 529 (1997).

Witten T A and Li H, *Europhys. Lett.* **23**, 51 (1993).

Yashonath S and Rao C N R, *Mol. Phys.* **54**, 245 (1985).

Young A P, *Phys. Rev. B* **19**, 1855 (1979).

Zhang Z, Davis H T and Kroll D M, *Phys. Rev. E* **48**, R651 (1993).

Zhang Z, Davis H T and Kroll D M, *Phys. Rev. E* **53**, 1422 (1996).

Zhang Z, Davis H T, Maier R S and Kroll D M, *Phys. Rev. B* **52**, 5404 (1995).

Zollweg J A and Chester G V, *Phys. Rev. B* **46**, 11186 (1992).

Printed in the United States
By Bookmasters